GEOTHERMAL, WIND AND SOLAR ENERGY APPLICATIONS IN AGRICULTURE AND AQUACULTURE

Sustainable Energy Developments

Series Editor

Jochen Bundschuh
University of Southern Queensland (USQ), Toowoomba, Australia
Royal Institute of Technology (KTH), Stockholm, Sweden

ISSN: 2164-0645

Volume 13

Geothermal, Wind and Solar Energy Applications in Agriculture and Aquaculture

Editors

Jochen Bundschuh

University of Southern Queensland, Toowoomba, Queensland, Australia
Royal Institute of Technology (KTH), Stockholm, Sweden

Guangnan Chen

University of Southern Queensland, Toowoomba, Queensland, Australia

D. Chandrasekharam

Indian Institute of Technology Hyderabad, Hyderabad, India

Janusz Piechocki

University of Warmia and Mazury, Olsztyn, Poland

CRC Press
Taylor & Francis Group
Boca Raton London New York Leiden

CRC Press is an imprint of the
Taylor & Francis Group, an **informa** business

A BALKEMA BOOK

Applied for

Published by: CRC Press/Balkema
                Schipholweg 107C, 2316 XC Leiden, The Netherlands
                e-mail: Pub.NL@taylorandfrancis.com
                www.crcpress.com – www.taylorandfrancis.com

First issued in paperback 2020

© 2017 Taylor & Francis Group, London, UK
CRC Press/Balkema is an imprint of the Taylor & Francis Group, an informa business

No claim to original U.S. Government works

ISBN 13: 978-0-367-57331-7 (pbk)
ISBN 13: 978-1-138-02970-5 (hbk)

Visit the Taylor & Francis Web site at
http://www.taylorandfrancis.com

and the CRC Press Web site at
http://www.crcpress.com

Typeset by MPS Limited, Chennai, India

Library of Congress Cataloging-in-Publication Data

About the book series

Renewable energy sources and sustainable policies, including the promotion of energy efficiency and energy conservation, offer substantial long-term benefits to industrialized, developing, and transitional countries. They provide access to clean and domestically available energy and lead to a decreased dependence on fossil fuel imports and a reduction in greenhouse gas emissions.

Replacing fossil fuels with renewable resources affords a solution to the increased scarcity and price of fossil fuels. Additionally, it helps to reduce anthropogenic emission of greenhouse gases and their impacts on climate. In the energy sector, fossil fuels can be replaced by renewable energy sources. In the chemistry sector, petroleum chemistry can be replaced by sustainable or green chemistry. In agriculture, sustainable methods can be used to enable soils to act as carbon dioxide sinks. In the construction sector, sustainable building practice and green construction can be used, replacing, for example, steel-enforced concrete by textile-reinforced concrete. Research and development and capital investments in all these sectors will not only contribute to climate protection but will also stimulate economic growth and create millions of new jobs.

This book series will serve as a multidisciplinary resource. It links the use of renewable energy and renewable raw materials, such as sustainably grown plants, with the needs of human society. The series addresses the rapidly growing worldwide interest in sustainable solutions. These solutions foster development and economic growth while providing a secure supply of energy. They make society less dependent on petroleum by substituting alternative compounds for fossil-fuel-based goods. All these contribute to minimize our impacts on climate. The series covers all fields of renewable energy sources and materials. It addresses possible applications not only from a technical point of view, but also from economic, financial, social, and political viewpoints. Legislative and regulatory aspects, key issues for implementing sustainable measures, are of particular interest.

This book series aims to become a state-of-the-art resource for a broad group of readers including a diversity of stakeholders and professionals. Readers will include members of governmental and non-governmental organizations, international funding agencies, universities, public energy institutions, the renewable industry sector, the green chemistry sector, organic farmers and farming industry, public health and other relevant institutions, and the broader public. It is designed to increase awareness and understanding of renewable energy sources and the use of sustainable materials. It also aims to accelerate their development and deployment worldwide, bringing their use into the mainstream over the next few decades while systematically replacing fossil and nuclear fuels.

The objective of this book series is to focus on practical solutions in the implementation of sustainable energy and climate protection projects. Not moving forward with these efforts could have serious social and economic impacts. This book series will help to consolidate international findings on sustainable solutions. It includes books authored and edited by world-renowned scientists and engineers and by leading authorities in economics and politics. It will provide a valuable reference work to help surmount our existing global challenges.

Jochen Bundschuh
(Series Editor)

Editorial board

Table of contents

List of contributors

Diogenes L. Antille   National Centre for Engineering in Agriculture, University of Southern Queensland, Toowoomba, QLD, Australia

Sarah Balistreri   University of Virginia, Charlottesville, Virginia, USA

Thomas Banhazi   National Centre for Engineering in Agriculture, University of Southern Queensland Toowoomba, QLD, Australia

Renata Brzozowska   Department of Water Protection Engineering Faculty of Environmental Sciences, University of Warmia and Mazury in Olsztyn, Poland

Jochen Bundschuh   Deputy Vice-Chancellor's Office (Research and Innovation) & Faculty of Health, Engineering and Sciences University of Southern Queensland Toowoomba, QLD, Australia & Royal Institute of Technology, Stockholm, Sweden

Francisco Javier Cabrera   Research Center on Solar Energy (CIESOL), University of Almería, Almería, Spain & Automatics, Robotics and Mecatronics Research Group, University of Almería, Agro Food International Excellence Campus (CeiA3), Almería, Spain

Adam Cenian   Physical Aspects of Ecoenergy Department; Institute of Fluid-Flow Machinery, Polish Academy of Science, Gdańsk, Poland

Guangnan Chen   Faculty of Health, Engineering and Sciences, & National Centre for Engineering in Agriculture, University of Southern Queensland, Toowoomba, QLD, Australia

Daniel Coaten   Faculty of Pharmaceutical Sciences, University of Iceland, Reykjavík, Iceland

Robert K. Dixon   Strategic Programs, US Department of Energy, Washington, DC, USA

Pilar Fernández-Ibáñez   Nanotechnology and Integrated BioEngineering Centre, School of Engineering, University of Ulster, Newtownabbey, Northern Ireland, United Kingdom

Noreddine Ghaffour   Water Desalination and Reuse Research Center, King Abdullah University of Science and Technology (KAUST), Thuwal, Saudi Arabia

Mattheus Goosen   Office of Research and Graduate Studies, Alfaisal University, Riyadh, Saudi Arabia

Ihsan Hamawand   Faculty of Health, Engineering and Sciences, University of Southern Queensland Toowoomba, QLD, Australia

Deborah Hines   United Nations World Food Programme, Bogota, Colombia

Maciej Klein	Physical Aspects of Ecoenergy Department; Institute of Fluid-Flow Machinery, Polish Academy of Science, Gdańsk, Poland
Anil Kumar	Energy Centre, Maulana Azad National Institute of Technology, Bhopal, India
Kamil Łapiński	Physical Aspects of Ecoenergy Department; Institute of Fluid-Flow Machinery, Polish Academy of Science, Gdańsk, Poland
Hacene Mahmoudi	Faculty of Science, Hassiba Benbouali University, Chlef, Algeria
Sixto Malato	CIEMAT-Plataforma Solar de Almería, Tabernas, Almería, Spain
Tek Maraseni	Institute for Agriculture and Environment, University of Southern Queensland Toowoomba, QLD, Australia
Yiheyis Maru	Commonwealth Scientific and Industrial Research Organisation (CSIRO), Alice Springs, NT, Australia
Supriya Mathew	Northern Institute, Charles Darwin University, Alice Springs, NT, Australia
Wojciech Miąskowski	Faculty of Technical Sciences, University of Warmia and Mazury in Olsztyn, Poland
Shahbaz Mushtaq	International Centre for Applied Climate Science, University of Southern Queensland, Toowoomba, QLD, Australia
Krzysztof Nalepa	Faculty of Technical Sciences, University of Warmia and Mazury in Olsztyn, Poland
Maciej Neugebauer	Department of Electric Engineering, Power Engineering, Electronics and Automatics, Faculty of Technical Sciences University of Warmia and Mazury in Olsztyn, Poland
Edoardo Pantanella	International Centre for Advanced Mediterranean Agronomic Studies (CIHEAM), Bari, Italy
István Patay	Department of Process Engineering, Szent István University, Budapest, Gödöllă, Hungary
Manuel Pérez-García	Research Center on Solar Energy (CIESOL), University of Almería, Almería, Spain & Automatics, Robotics and Mecatronics Research Group, University of Almería, Agro Food International Excellence Campus (CeiA3), Almería, Spain
Janusz Piechocki	Department of Electric Engineering, Power Engineering, Electronics and Automatics, Faculty of Technical Sciences University of Warmia and Mazury in Olsztyn, Poland
Paweł Pietkiewicz	Faculty of Technical Sciences, University of Warmia and Mazury in Olsztyn, Poland
María Inmaculada Polo-López	CIEMAT-Plataforma Solar de Almería, Tabernas, Almería, Spain
Om Prakarsh	Energy Centre, Maulana Azad National Institute of Technology, Bhopal, India
Digby Race	The Fenner School of Environment and Society, The Australian National University, Canberra, ACT, Australia

Kristin Vala Ragnarsdottir   Faculty of Earth Sciences and Institute of Sustainability Studies, University of Iceland, Reykjavík, Iceland

Kathryn Reardon-Smith   International Centre for Applied Climate Science, University of Southern Queensland, West Street, Toowoomba, QLD, Australia

Francisco Rodríguez-Díaz   Research Center on Solar Energy (CIESOL), University of Almería, Almería, Spain & Automatics, Robotics and Mecatronics Research Group, University of Almería, Agro Food International Excellence Campus (CeiA3), Almería, Spain

Alba Ruiz-Aguirre   Research Center on Solar Energy (CIESOL), University of Almería, Almería, Spain

Jorge Antonio Sánchez-Molina   Automatics, Robotics and Mecatronics Research Group, University of Almería, Agro Food International Excellence Campus (CeiA3), Almería, Spain

Erik Schmidt   National Centre for Engineering in Agriculture, University of Southern Queensland Toowoomba, QLD, Australia

Norbert Schrempf   Department of Process Engineering, Szent István University, Budapest, Gödöllő, Hungary

István Seres   Department of Process Engineering, Szent István University, Budapest, Gödöllő, Hungary

Atul Sharma   Non-Conventional Energy Laboratory, Rajiv Gandhi Institute of Petroleum Technology (RGIPT), Rae Bareli, India

Charlie Shultz   Greenhouse Management, Hydroponics and Aquaponics, Santa Fe Community College, Santa Fe, New Mexico, USA

Katarzyna Siuzdak   Physical Aspects of Ecoenergy Department; Institute of Fluid-Flow Machinery, Polish Academy of Science, Gdańsk, Poland

Bruno Spandonide   School of Computer Science, Engineering and Mathematics, Flinders University, SA, Australia & Ninti One Limited and the Cooperative Research Centre for Remote Economic Participation, Alice Springs, NT, Australia

Henk Stander   Faculty of AgriSciences, Department of Animal Sciences, Division of Aquaculture, University of Stellenbosch, Stellenbosch, Western Cape, South Africa

Ragnheidur Thorarinsdottir   Faculty of Civil and Environmental Engineering, University of Iceland, Reykjavík, Iceland

Barbara Tomaszewska   AGH University of Science and Technology; Mineral and Energy Economy Research Institute, Polish Academy of Sciences (PAN), Kraków, Poland

Ming Yang   Global Environment Facility, Washington, DC, USA

Guillermo Zaragoza   CIEMAT-Plataforma Solar de Almería, Tabernas, Almería, Spain & Research Center on Solar Energy (CIESOL), University of Almería, Almería, Spain

Foreword by Alessandro Flammini

Agriculture and global agrifood systems are at the same time major energy consumers and energy producers (through bioenergy) and they play therefore an important role in GHG emissions. Still, the relevance of energy input for agrifood systems is not fully understood and often underestimated, since those inputs are conventionally accounted for in a number of different sectors in national statistics (e.g. primary production, transport, food industry, chemical industry, energy generation, etc).

Today, the world's agrifood supply chains are being challenged. For several decades, the production, processing and distribution of food have been highly dependent on fossil fuel inputs (the exception being subsistence farmers who use only manual labor and perhaps animal power to produce food for their families that is then usually cooked on inefficient biomass cook-stoves). There has also been an ever growing demand for food as the world population grows, along with the increasing demand for higher protein diets.

In addition, the projected impacts of climate change show it is likely that in many regions, current food supply systems will be threatened, especially where more frequent floods and droughts are predicted. Diminishing fresh and clean water supplies in some countries will become a major threat to sustainable food production that is already constrained in some areas. Food supply systems need to become more secure, including through active management of water for irrigation, which require energy, whilst becoming more resilient to climate change impacts. The fertile soil available for crop and animal production is constrained and it is actually shrinking in some regions as the degraded land area increases. Further, land use change through deforestation for agricultural production is no longer acceptable. To meet the ever-growing food demand, the food productivity per hectare needs to be increased at the same time as energy, water, fertilizer and other inputs are reduced. As well as plant breeding to develop improved crop varieties, this can be achieved by more efficient production and processing systems and technologies that use energy more wisely and reduce waste of resources during each step of the agrifood chain.

Agrifood systems are therefore increasingly under pressure to achieve efficiency improvements and reduce their environmental footprint. Fostering the adoption of best available green technologies along agrifood supply chains is an essential step toward this objective.

Energy-smart food and sustainable agricultural production systems can be viable solutions for development and bring significant structural change in rural areas. Such systems could increase energy end-use efficiency and better manage demand to drive rural economic development along more climate-friendly pathways.

The open challenge for the agrifood sector is to decouple fossil fuel energy inputs from the increasing demands for food supply in the short term while ensuring food security. A central tenet is the promotion of technologies and practices that help make agricultural and food systems more efficient, while also making agriculture, forestry and fisheries more productive and sustainable. A change in agrifood technology inevitably has also an impact on the water, energy and food sectors as well as on the ability to mitigate and adapt to climate change. The introduction of clean technologies in the agrifood sector has therefore multifaceted implications, which require informed decision to manage possible economic, environmental and social trade-offs.

This book presents some of the above mentioned technologies needed to decouple the agrifood dependence from fossil fuels, and greatly contributes to a better understanding of their performance and potential to contribute to energy-smart food systems.

Alessandro Flammini
Natural Resources Officer
Investment Center Division, FAO
Rome, Italy
June 2017

Editor's preface

The increasing uses of energy resources are one of the major challenges facing agriculture, aquaculture and the agri-food chain in general. Continuously increasing scarcity of fossil fuels and rising high fuel cost together with the demands and needs for "green food" and significant reduction in the greenhouse gas emissions demand for an improvement of farming and food processing energy efficiency. Renewable energy use is essential for sustainable development in agriculture. Exploration of new alternative and renewable energy sources is also important.

This book is focused on using solar, wind and geothermal energy resources which could provide a large part of the energy needs in the agricultural sector and the agri-food chain but that are not used to a significant extent as they deserve. Only a diversified renewable energy approach will fulfill future needs of a modern sustainable agriculture, aquaculture and the agri-food chain.

This in mind, we provide a technological and scientific endeavor to assist the society and farming and related communities in different parts of the world and at different scales to improve their productivity and sustainability using solar, wind and geothermal energy resources. Through providing the latest technology and by deepening the understanding of possible cost-efficient techniques and application opportunities for different scales and situations, the book will showcase why the implementation of these technologies faces numerous technological, economic and policy barriers while giving suggestions and examples as to how these hurdles can be mitigated or overcome. Costs of novel technologies using both conventional and renewable energy sources are discussed and compared with those of other technologies for agricultural and aquacultural production, taking into account environmental and social costs that are caused by using fossil fuel based technologies. Also, applications of new and improved technologies powered by solar, wind and geothermal energy are discussed and promoted. Additionally, the book highlights which further R&D will need to be performed, and it hopefully will give readers appropriate stimuli. Such research is timely due to worldwide interest in sustainable agriculture and aquaculture.

The book has focused on critical and new areas of research and recent advances. This book has included chapters focusing on challenging issues in using solar, wind and geothermal energy directly as shaft power or heat or indirectly in the form of electricity in agriculture and its associated primary industries. It discusses both adapted classical and novel technologies and their applicability.

We hope that the book will be used as incentive for what and how modified or new technologies powered by solar, wind and geothermal energy can be applied as an effective solution in agricultural production to satisfy the continually increasing demand for food and fiber in an economically sustainable way, while contributing to global climate change mitigation.

We hope that this book will help all readers, in the professional, academics and non-specialists, as well as key institutions that are working in the fields of agriculture, aquaculture, and the agri-food chain. We further expect that it will be a useful source of information for leading decision and policy makers, municipalities and agriculture and energy sector representatives and administrators, business leaders, agricultural and power producers, from both industrialized and developing countries as well. It is expected that this book will become a standard and key source of information, used by educational institutions, and research and development establishments involved in the respective issues.

Jochen Bundschuh
Guangnan Chen
D. Chandrasekharam
Janusz Piechocki
(editors)
March 2017

About the editors

Jochen Bundschuh (1960, Germany), finished his PhD on numerical modeling of heat transport in aquifers in Tübingen in 1990. He is working in geothermics, subsurface and surface hydrology and integrated water resources management, and connected disciplines. From 1993 to 1999 he served as an expert for the German Agency of Technical Cooperation (GTZ, now GIZ) and as a long-term professor for the DAAD (German Academic Exchange Service) in Argentine. From 2001 to 2008 he worked within the framework of the German governmental cooperation (Integrated Expert Program of CIM; GTZ/BA) as adviser in mission to Costa Rica at the Instituto Costarricense de Electricidad (ICE). Here, he assisted the country in evaluation and development of its huge low-enthalpy geothermal resources for power generation. Since 2005, he is an affiliate professor of the Royal Institute of Technology, Stockholm, Sweden. In 2006, he was elected Vice-President of the International Society of Groundwater for Sustainable Development (ISGSD). From 2009–2011 he was visiting professor at the Department of Earth Sciences at the National Cheng Kung University, Tainan, Taiwan.

Since 2012, Dr. Bundschuh is a professor in hydrogeology at the University of Southern Queensland, Toowoomba, Australia where he leads the Platform for Water in the Nexus of Sustainable Development working in the wide field of water resources and low/middle enthalpy geothermal resources, water and wastewater treatment and sustainable and renewable energy resources. In November 2012, Prof. Bundschuh was appointed as president of the newly established Australian Chapter of the International Medical Geology Association (IMGA).

Dr. Bundschuh is author of the books "Low-Enthalpy Geothermal Resources for Power Generation" (2008) (CRC Press/Balkema, Taylor & Francis Group) and "Introduction to the Numerical Modeling of Groundwater and Geothermal Systems: Fundamentals of Mass, Energy and Solute Transport in Poroelastic Rocks". He is editor of 18 books and editor of the book series "Multiphysics Modeling", "Arsenic in the Environment", "Sustainable Energy Developments" and "Sustainable Water Developments" (all CRC Press/Balkema, Taylor & Francis Group). Since 2015, he is an editor in chief of the Elsevier journal "Groundwater for Sustainable Development".

Dr. Guangnan Chen graduated from the University of Sydney, Australia, with a PhD degree in 1994. Before joining the University of Southern Queensland as an academic in early 2002, he worked for two years as a post-doctoral fellow and more than five years as a Senior Research Consultant in a private consulting company based in New Zealand.

Dr. Chen has extensive experience in conducting both fundamental and applied research. His current research focuses on the sustainable agriculture and energy use. The researches aim to develop a common framework and tools to assess energy uses and greenhouse gas emissions in different agricultural sectors. These projects are funded by various government agencies and farmer organisations. Furthermore, Dr Chen has also conducted significant research to compare the life cycle energy consumption of alternative farming systems, including the impact of machinery operation, conservation farming practice, irrigation, and applications of new technologies and alternative and renewable energy.

Dr. Chen has so far published more than 100 papers in international journals and conferences, including one edited book on sustainable energy solutions in agriculture and eight invited book chapters on various topics. He serves as a Section Editor for the International Journal of Agricultural & Biological Engineering (IJABE), and was the Guest Editor of a special issue of the journal of Applied Energy in 2014. He is currently the Secretary of Board of Technical Section IV (Energy in Agriculture), CIGR (Commission Internationale du Génie Rural), which is one of the world's top professional bodies in agricultural and biosystems engineering.

Dornadula Chandrasekharam (Chandra: b1948, India), Chair Professor in the Department of Earth Sciences, Indian Institute of Technology Bombay (IITB) obtained his MSc in Applied Geology (1972) and PhD (1980) from IITB. He has been working in the fields of geothermal energy resources, volcanology, and groundwater pollution, for the past 30 years. Before joining IITB, he worked as a Senior Scientist at the Centre for Water Resources Development and Management, and Centre for Earth Science Studies, Kerala, India for 7 years.

He held several important positions during his academic and research career. He was a Third World Academy of Sciences (TWAS, Trieste, Italy); Visiting Professor to Sanaa University, Yemen Republic between 1996 and 2001; Senior Associate of Abdus Salam International Centre for Theoretical Physics, Trieste, Italy from 2002–2007; Adjunct Professor, China University of Geosciences, Wuhan from 2011–2012. Recently he has been appointed as a visiting Professor to King Saud University of Saudi Arabia. He received the International Centre for Theoretical Physics (ICTP, Trieste, Italy) Fellowship to conduct research at the Italian National Science Academy (CNR) in 1997.

Prof. Chandra extensively conducted research in low-enthalpy geothermal resources in India and is currently the Chairman of M/s GeoSyndicate Power Private Ltd., the only geothermal company in India. He is an elected board member of the International Geothermal Association, and has widely represented the country in several international geothermal conferences. He conducted short courses on low-enthalpy geothermal resources in Argentina, Costa Rica, Poland and China. He has supervised 18 PhD students and published 95 papers in international and 35 papers in national journals of repute and published 5 books in the field of groundwater pollution and geothermal energy resources. His two books on geothermal energy resources 1) "Geothermal Energy Resources for Developing Countries" Balkema Pub. (2002) and 2) "Low Enthalpy Geothermal Resources for Power Generation" Taylor and Francis, (2008) are widely read. Prof Chandra is currently on the Board of Director of 1) Oil and Natural Gas Corporation 2) Western Coal Fields Ltd., 3) India Rare Earths Ltd. and 4) Mangalore Refineries and Petrochemicals. He has been appointed as the Chairperson of the Geothermal Energy Resources and Management committee constituted by the Department of Sciences and Technology, Government of India.

Janusz Piechocki (1950, Olsztyn, Poland) was graduated from the Technical University in Gdansk (Poland) in 1973. First three years after graduated he has worked in industry. Since 1976 he has worked at the Agricultural and Technical University in Olsztyn (University of Warmia and Mazury in Olsztyn was established on the base this University in 1999). He was at St. Istvan University in Godollo (Hungary) in the years 1980–1981 according to PhD program. Janusz Piechocki finished his PhD in 1981. He is Professor Doctor since 1998.

He carries out research in the area of agricultural engineering, renewable energy sources, energy, electric engineering, assessment of the economic and energy effectiveness and sustainable energy development with reducing air pollutant emissions systems.

He is member of Agricultural Engineering Committee of Polish Academy of Science since 2007 and from 2011 he is Chairman of Energy Section inside this Committee. In the years 2003–2009 he was member of Scientific Committee in Institute of Agricultural Architecture, Mechanization and Electrification.

In 2000 year on CIGR Congress in Tsukuba (Japan) he was included as a member to Board Section IV (Energy in Agriculture) of CIGR (Commission International du Genie Rural – Committee of Agricultural Engineering and Biotechnology). In 2006 on CIGR Congress in Bonn (Germany) he was elected on the years 2006–2010 as a Vice-Chairman of the CIGR Section IV and from Congress of CIGR in Quebec (Canada) in 2010 he become Chairman of this Section until 2014. From Congress of CIGR in Bejing (China) in 2014 Prof. Dr. Janusz Piechocki is Honorary Vice-President of CIGR and Honorary Chairman of Section IV.

He is member of EurAgEng (European Committee of Agricultural Engineering), Polish BIOMAS Society, Board Member of Baltic Association of Mechanical Engineers in Kaliningrad (Russia), "Balttechmasz" Presidium at Technical University in Kaliningrad and member of Editorial Committee of "Journal of BAME" edited in Kaliningrad. He is member of Editorial Board "Industrial Power Engineering" in Moscow (Russia) – ISSN 0033-1155, IF = 0,464.

He has been working in international academic and technical co-operation programs in different fields of energy in agriculture and utilization of different renewable energy sources from agriculture and into agriculture. He was participated into educational and scientific works with students according to Erazmus Program in EU at Technical University in Graz (Austria) in 2004 and at University in Izmir (Turkey) in 2008 and 2011.

Prof. Dr. Janusz Piechocki was Vice-Rector of University of Warmia and Mazury in Olsztyn (Poland) in the years 1999–2008 and 2012–2016. He was Dean of Faculty of Technical Sciences at this University in years 2008–2012 and he is from 1998 Chairman of Electric, Energy, Electronics and Automation Engineering Department. He has published over 130 original research papers.

CHAPTER 1

Solar, wind and geothermal energy applications in agriculture: back to the future?

Jochen Bundschuh, Guangnan Chen, Barbara Tomaszewska, Noreddine Ghaffour, Shahbaz Mushtaq, Ihsan Hamawand, Kathryn Reardon-Smith, Tek Maraseni, Thomas Banhazi, Hacene Mahmoudi, Mattheus Goosen & Diogenes L. Antille

1.1 INTRODUCTION

The agri-food chain consumes about one third of the world's energy production with about 12% for crop production and nearly 80% for processing, distribution, retail, preparation and cooking (Fig. 1.1) (FAO, 2011a). The agri-food chain also accounts for 80–90% of total global freshwater use (Hoff, 2011) where 70% is for irrigation alone. Additionally, on a global scale, freshwater production consumes nearly 15% of the entire energy production (IEA, 2012). It can therefore be argued that making agriculture and the agri-food supply chain independent from fossil fuel use has huge potential to contribute to global food security and climate protection not only for the next decades, but also for the coming century. Provision of secure, accessible and environmentally sustainable supplies of water, energy and food must thus be a priority.

One of the major objectives of the world's scientists, farmers, decision-makers and industrialists is to overcome the present dependence on fossil fuels in the agri-food sector. This dependency increases the volatility of food prices and affects economic access to sustenance. For example, Figure 1.2 shows the close interrelationship between the crude oil price index and the cereals price index. An increasing energy demand for cultivation is particularly important in regions

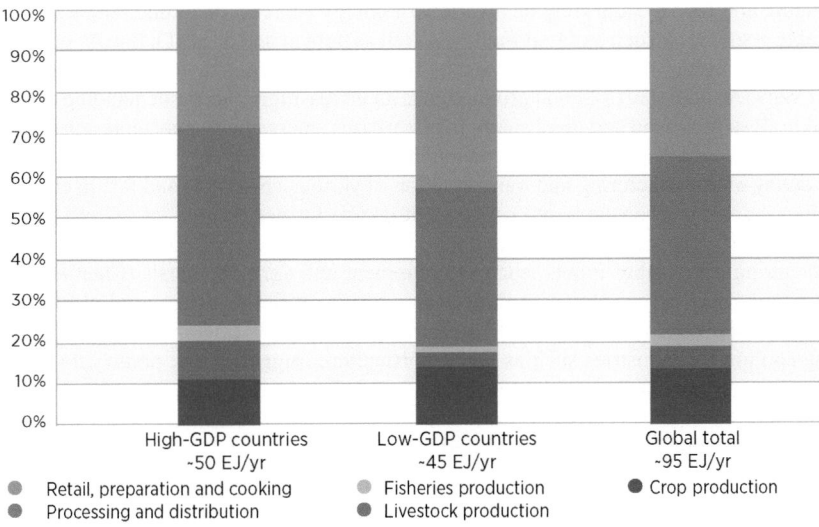

Figure 1.1. Direct and indirect energy inputs in the food sector (FAO, 2011a).

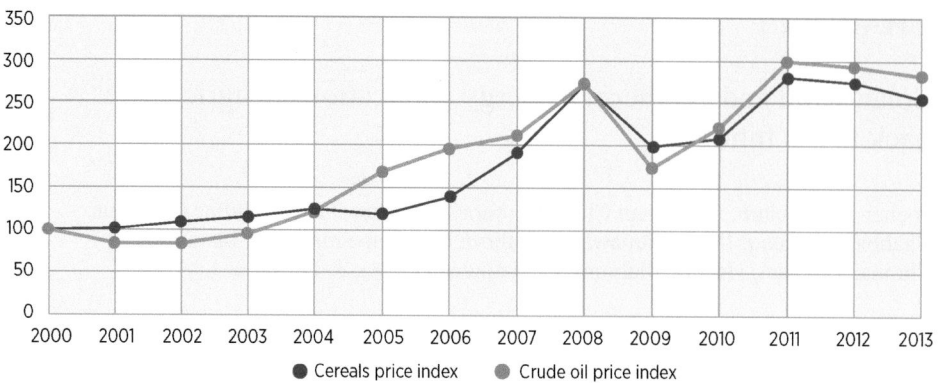

Figure 1.2.   Dependence of fossil fuel price index and the cereal price index for the period 2000–2013 (IRENA, 2015, based on FAO Food Price Index and BP Statistics Review of World Energy; base 2000 = 100).

with expanding irrigated agriculture using pumped water. This translates to a food-related risk to energy security.

The development and commercialization of renewable energy sources such as solar, wind and geothermal provides great potential to reduce costs in the agri-food sector. For instance, in addition to power generation, the main uses of geothermal waters are for space heating, district heating, spas balneology, aquaculture and greenhouse heating (Lund and Boyd, 2015). However, much work remains to be done to make better use of renewable energy in the agri-food sector.

The aim of this introductory chapter is to critically review recent developments in solar, wind and geothermal energy applications in agriculture and the agri-food sector such as processing, distribution, retail, preparation and cooking.

1.2   ENERGY DEMANDS IN AGRICULTURE

1.2.1   *Energy use in agriculture*

Agriculture and food systems rely on a variety of energy sources, including renewable and non-renewable resources – such as fossil fuels – as well as human and animal labor. At present, fossil fuels, in their various forms, supply most of the energy required by agriculture that feeds the world (Maraseni *et al.*, 2015). Food production is an increasingly energy-demanding sector and is needed in all stages of the agri-food chain. In many cases, energy costs may represent a significant proportion of the total agricultural production input cost, including the price of irrigation, as well as the outlay of manufacturing and transportation of various chemicals and fertilizers.

Energy is used both on-farm and off-farm. It can be further divided into direct energy used, i.e. the fuel and electricity consumed, and the indirect energy (embodied energy) involved in the manufacturing of all other inputs, such as equipment and agro-chemicals (Chen *et al.*, 2010). Direct energy may be consumed in three major forms on farms: (i) general electricity usage for lighting, appliances, irrigation; (ii) fuel use for machinery, tractors and vehicles; and (iii) heating/cooling for industries such as dairy, horticulture, piggeries and poultry. In field crops, irrigation energy may account for up to 85% of total direct energy use (Maraseni *et al.*, 2015). With the new technologies currently under development, Renewable Energy (RE) with its rapidly falling costs, is being increasingly employed in agriculture.

1.2.2   *The energy management process*

Energy audits are a crucial part of farm energy management (Chen and Baillie, 2009a). This type of audit refers to the systematic examination of a farm, facility or site, to determine whether, and to what extent, it has used energy efficiently. An energy review determines how efficiently it is being used. It also identifies energy cost-saving opportunities and highlights potential improvements in productivity and quality. An energy check may also assess potential savings through strategies such as fuel switching, tariff negotiation and demand-side management (e.g., by changing to alternative farming systems and farm layouts). An energy assessment may be undertaken as part of a broader plan to manage production inputs on-farm (Chen and Baillie, 2009a). The objectives of energy audits include:

- Conserve energy inputs.
- Reduce greenhouse gas emissions.
- Achieve operational and cost efficiencies with improved productivity and profitability.

Extensive research has been conducted on both energy use and conservation in agriculture. Recent results of energy efficiency programs have shown considerable variation in energy use between different crops as well as different farms of similar production systems (Chen *et al.*, 2015). Pellizzi *et al.* (1988) found that in Europe, the range of field energy consumption for wheat-like cereals varied from 2.5–4.3 GJ ha^{-1}. For cotton, a study by Chen and Baillie (2009b) showed that the direct energy inputs for cotton production in Australia ranged from 3.7–15.2 GJ ha^{-1}. Diesel energy inputs ranged from 95–365 L ha^{-1}, with most farms using between 120–180 L ha^{-1}. Dry land cotton was at the lower end of this range. The direct on-farm energy use of some nurseries may be up to 20,000 GJ ha^{-1} (Chen *et al.*, 2015).

1.3   THE WATER-ENERGY-FOOD-CLIMATE NEXUS

Agricultural productivity largely depends on the availability of water, energy and land resources. In the past, the agricultural sector enjoyed increasing productivity and was thus able to feed the growing world population. For example, in 2016, the world cereal production was 2526 Mt (million metric tons) compared with 2000 Mt in 2006 (FAO, 2016). However, the agriculture industry in the 21st century faces numerous challenges. It must feed a rapidly increasing population with a depleting availability of resources (FAO, 2013; Haddeland *et al.*, 2014). At the same time it needs to adapt to a changing climate and reduce greenhouse gas (GHG) emissions (Kulak *et al.*, 2013). Therefore, acclimatizing to climate transformation and the necessity to increase agricultural productivity with minimum use of these resources is a major requirement of the 21st century (Jackson *et al.*, 2010).

It can be argued that the irrigation industry is a significant contributor to the global economy. However, the industry is currently under pressure to cut water use as an adaptation to a reduction in water availability owing to climate variability and change, and competition from other sectors including demands for environmental water (Ward *et al.*, 2006). The conversion of less efficient flood irrigation systems to more efficient pressurized irrigation systems has been heralded as one way of increasing water use efficiency and creating water savings (Mushtaq *et al.*, 2013). However, pressurized irrigation systems may alter patterns of on-farm energy consumption and may increase cropping intensity. More intensive land use might involve more fuel, farm machinery and agrochemicals, and the production, packaging, transportation and application of these also requires significant energy resources, leading to an increase in GHG emissions (Maraseni and Cockfield, 2011a,b; 2012; Maraseni *et al.*, 2012a; 2012b).

This suggests that there is a potential conflict in terms of mitigation and adaptation policies, and thus warrants a comprehensive and robust investigation. This important policy area appears to be under-researched. Topak *et al.* (2005) and Baillie (2009) have compared the energy consumption of various irrigation systems, but these studies failed to provide a complete picture as they did not analyze soil carbon, GHG emissions associated with the use of primary farm inputs, or water consumption and GHG implications of more intensive cropping systems. Similarly, Maraseni *et al.* (2012a; 2012b) tried to assess GHGs and water saving implications of converting flood irrigation systems into pressurized irrigation systems through five case studies (three cotton, one lettuce and one lucerne) in southern Queensland, Australia. Yet, these studies considered only one crop and did not assess all crops in a rotation. Even within the same irrigation technology, the level of farm inputs (i.e. fuels, agrochemicals, machinery) and thus the energy consumption and GHG emissions due to production, consumption and use of those farm inputs, could vary significantly between crops in a rotation. A farmer may employ more agrochemicals in a first crop with the intention to use less in the following crop. Therefore, comprehensive studies of water and energy consumption, across full cropping rotations covering a wide geographical area, are necessary.

1.4   GREENHOUSE GAS EMISSIONS AND CARBON FOOTPRINT OF AGRICULTURE

1.4.1   *Sources of greenhouse gas emissions from agriculture*

Agriculture and food systems play an important role in climate change because of their significant energy use and also as a potential source for RE such as bio-ethanol and bio-diesel. Overall, the main sources of emissions in agricultural production are (Chen *et al.*, 2010):

- Emissions from energy used to power various machinery and processes which may include both the on-farm and post-farm activities.
- Emissions from energy used to produce agricultural inputs such as fertilizers and pesticides (i.e. pre-farm).
- Direct soil emissions or sequestration in soil (on-farm).
- Soil nitrous oxide (N_2O) emissions from the application of nitrogen fertilizer and manures (on-farm).
- Methane (CH_4) emissions from prolonged water-logging or from the digestion systems of livestock (on-farm).

With current technology, burning one liter of petrol or diesel would, on average, emit 2.3 and 2.7 kg CO_2, respectively. It is also noted that for the same power output, greenhouse gas emissions from electricity would vary considerably if different fuels were used to generate it. Furthermore, it is estimated that with the current manufacturing technology, the production of one kg of nitrogen fertilizer would require the energy input equivalent to 1.5–2 kg of fuel, while 1 kg of pesticides would require the energy input equivalent to up to 5 kg of fuel.

Finally, carbon emissions and sequestration in soil, as well as soil nitrous oxide (N_2O) emissions from the application of nitrogen fertilizer and manure, are difficult and expensive to measure accurately because of their spatio-temporal variability. Methane and N_2O have global warming potentials 25 and 298 times that of CO_2 respectively, over a 100-year time period (IPCC, 2007).

1.4.2   *Overview of global agricultural emissions*

There were approximately 50.1 GtCO_2e greenhouse gas (GHG) emissions from anthropogenic sources in 2010 (IPCC, 2014). Agriculture shares about 11% of these emissions (Tubiello *et al.*, 2015). This percentage relates to direct sources only. If emissions from indirect sources are considered, a further 3–6% of global emissions (Vermuelen *et al.*, 2012) are accounted for. Approximately 60% of all N_2O and 50% of all CH_4 emissions originate in the agricultural sector (Smith *et al.*, 2007).

Similarly, about 38% of direct agricultural emissions are attributed to N_2O from soils, 32% to CH_4 from ruminants, 12% to biomass burning, 11% to CH_4 from rice production and 7% to manure management (Bellarby *et al.*, 2008). There is evidence of an upsurge in GHG emissions with rising farm inputs, due to increased intensification of agriculture (Graham and Williams, 2003; Maraseni, 2007; Maraseni and Cockfield, 2011a; Mushtaq *et al.*, 2013). For example, during the period 1990–2005, global agriculture emissions increased by 14%, an average rate of 49 MtCO_2-e year^{-1} (USEPA, 2006). Therefore, meeting the proposed 2°C climate stabilization target is not possible without reducing agrarian discharges.

Agricultural emission from developing countries is much higher than that of developed nations. For example, in 2005, developing countries were responsible for 74% of global agricultural emissions, whereas developed states were accountable for only 26% (Smith *et al.*, 2007). The proportional contribution of agricultural emissions for developing countries has been increasing faster due to rapid increases in farm inputs to feed growing populations with limited farming areas (Tubiello *et al.*, 2015). Therefore, without developing countries shouldering responsibility, a reduction in global agricultural emissions is not possible.

FAO (2014) predicts that there will be over 9 billion people globally by 2050. In order to feed them, agricultural production should increase by 60% by 2050. About 80% of the yield rise will come from intensification (i.e. higher yields with more fertilizer, pesticide and water inputs, multiple cropping, shorter fallow periods and improved seed varieties) and another 20% will come from extensification (Johnson *et al.*, 2014). The ratio will be 70:30 in developing countries. Both these intensification and extensification processes will increase the share of agricultural emissions. However, research by Johnson *et al.* (2014) showed that the selective extensification of agriculture could save some 22 GtCO$_2$e emissions by 2050, compared with a business-as-usual approach. Consequently, a wise land-use planning and coordinated approach between government departments is necessary for reducing agricultural emissions.

1.4.3   *Life cycle assessment (LCA)*

To achieve sustainable development, the first step would be to understand and identify where the environmental impacts and damages actually occur so that targeted remedy actions can be taken. To determine the carbon footprint of agricultural products, it is necessary to identify and consider the full life cycle of the products, from growing to harvesting, as well as their waste management and finally disposal or recycling of the packaging (Klöpffer, 2012). Life cycle assessment (LCA) is a cradle to grave analysis. It examines a system performance starting with the extraction of raw materials from the earth followed by the operations, until the final disposal of the material as waste. This provides the knowledge of how much energy and raw materials are used at each stage of the product life and how much different types of waste are generated. LCA thus helps to analyze the process involved, and allows the consumer more information on the choices of the product. In particular, compared with other methods LCA analysis has the advantage of being able to quantify the magnitude of the potential environmental saving in each environmental category, and to avoid the pitfall of just shifting the environmental impacts from one category to another. An LCA project should comply with the international standards ISO 14040–14044 (Klöpffer, 2012).

Generally, an LCA analysis consists of the following four steps:

- Initiation (goal and scope definition, project planning).
- Inventory analysis (data collection, identifying all relevant input and output items).
- Impact assessment (analysis, quantification and evaluation of impacts on ecosystems, human beings and resources use).
- Interpretation and improvement assessment (sensitivity study, evaluation of options to reduce environmental loads).

Overall, LCA is often a computer-based data-intensive operation. It requires time, skill, a large amount of data and software to manipulate the data. Particularly, industry and region specific Life Cycle Inventory (LCI) is essential to undertake a proper LCA project, as differences in the input data as well as project methodologies can significantly affect results and lead to highly variable conclusions. Furthermore, international standardization is likewise necessary (Roy *et al.*, 2009). Decision support tools for best practices and most cost-effective designs and management must be developed. In addition to energy and greenhouse gas emissions, the quantification of the impacts of water and land footprints, and the effects of pesticides and biodiversity are also receiving increasing attention.

1.4.4   *Comparison of environmental impact of different foods*

LCA has now been widely used for process improvements and optimization, including in agriculture and in the food industry (Roy *et al.*, 2009; Chen *et al.*, 2010). For example, it has been estimated that to produce 1 kg of grain in Australia, overall 0.1 kg of fuel may be consumed and 0.3 kg of greenhouse gases emitted. However, the total emissions can be significantly influenced by the management and operation methods used and also the product types. Biswas *et al.* (2010) showed that the life cycle GHG emissions for 1 kg of wool is significantly higher than that for wheat and sheep meat. It was also identified that the on-farm stage contributed the most significant portion of total GHG emissions from the production of wheat, sheep meat and wool. It was further established that CH_4 emissions from enteric methane production and from the decomposition of manure accounted for a significant portion of the total emissions from sub-clover and mixed pasture production, while N_2O emissions from the soil under wheat production have been found to be the major source of GHG emissions. Use of RE may also have a significant benefit. With the advent of eco-labeling, LCA is being increasingly used as a reporting mechanism. It has been, however, argued by some, that carbon labeling may also have unintended consequences as a "green barrier" for international trade.

1.5   MERGING RENEWABLES WITH AGRICULTURE: THE SUSTAINABILITY APPROACH

A way to overcome the present dependence on fossil fuels is the integration of energy efficiency and particularly RE, the prices of which are – due to improved technologies, but especially due to increasing mass production – continuously decreasing into the agri-food chain sectors. Substituting fossil fuel energy resources with renewables becomes progressively important as agricultural intensity is always growing; i.e., increasing mechanization, expansion of irrigated land, fertilizer production and transportation require more and more energy (IRENA, 2015). Overcoming risks of future fossil fuel availability, fluctuating prices – often making them unavailable for the poor – and their contribution to global warming requires a decoupling of the agricultural sector from fossil fuels. This can be achieved by increasing energy efficiency and the introduction of Renewable Energy Sources (RES) providing mechanical energy, heat, and electricity as outlined in the FAO Energy-Smart Food Program (FAO, 2011a) (Fig. 1.3). Integration of RES, especially if they are available locally, into the food chain, i.e., integrated food-energy systems, can directly save the farmers' money, but most importantly, it can indirectly significantly contribute to creation of local development by creating jobs, poverty reduction, improved gender equity and food security and at the same time contribute to global food security and global climate protection. Renewable energy can be used either directly (on-site), or indirectly integrating RE into the existing conventional energy supply chain (Fig. 1.4).

1.5.1   *Solar photovoltaic energy applications*

Solar energy in the form of photovoltaic (PV)-produced electricity can supply and/or supplement many energy requirements both on the farm and the entire agri-food chain. All that is now powered by electricity (produced from fossil fuels) can be adapted to be powered by PV modules. These can be grid-connected PV systems, which allow the farmers to sell their not-used solar power to the electricity utility, or a stand-alone system which, if electricity is continuously used for a specific process, needs a storage system as solar is an intermittent available energy source. However, many electricity-driven processes can be performed during the day when solar power is available such as water pumping, and shearing of sheep. Depending on the distance to the grid and its voltage, PV systems could be much cheaper and of less maintenance needs in remote areas, than installing power lines for connecting to the grid, which further may include the need of installing step down transformers. Solar PV applications in agriculture are comprised of: (i) remote electricity supply,

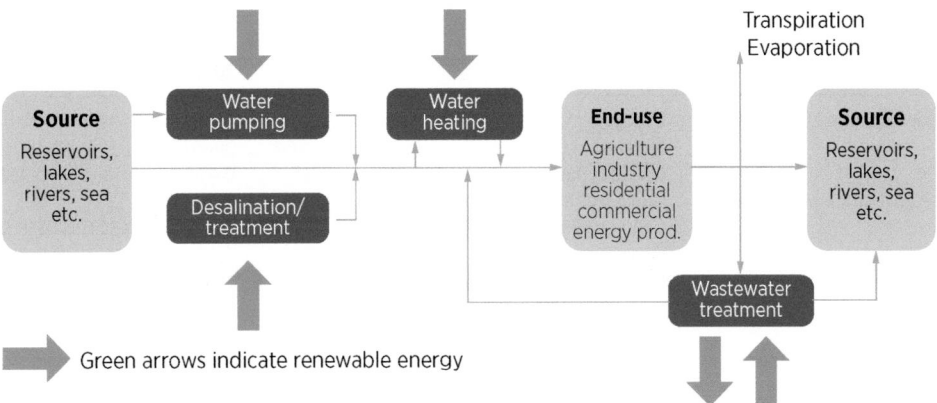

Figure 1.3.   Renewable energy applications of the water supply chain (IRENA, 2015).

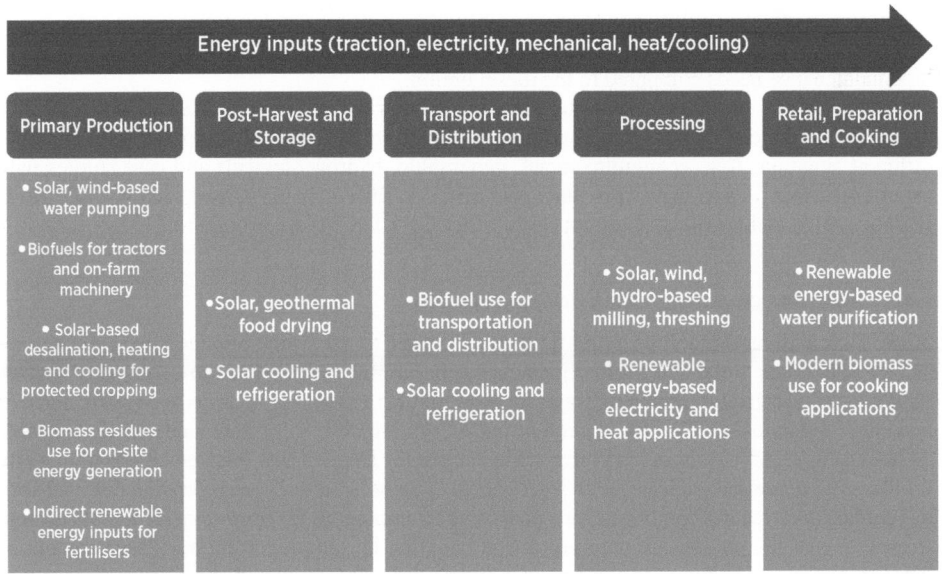

Figure 1.4.   Directly (on-site) or indirectly integrating of renewable energy into the existing conventional energy supply chain (IRENA, 2015, based on FAO, 2011b Practical Action 2012).

electric fencing and water pumping, and (ii) general applications, which are presently powered by electricity produced from fossil fuels.

1.5.1.1   *Water pumping*

In agriculture and the agri-food chain in general, PV can be employed for water pumping in different applications. Pumping from groundwater and surface water bodies, is a principal energy consumer and therefore a principal target for implementing RE technologies to substitute the electricity produced from the fossil fuels and diesel fuel commonly used for water pumping.

Even though solar water pumping is no novelty – it has existed since the early 1980s – it is not until recent decades that it has been accepted as an easy-to-install sustainable solution for different scales and for being environmentally friendly. Reasons for this were technical improvements, but especially the mass production of solar panels and their respective price drop (Varardi, 2014).

Table 1.1. Global use of geothermal heat in agriculture by application category (after Lund and Boyd, 2015).

Category		1995	2000	2005	2010	2015
Greenhouse heating	Capacity [MWt]	1085	1246	1404	1544	1830
	Utilization [TJ year^{-1}]	15742	17864	20661	23264	26662
	Capacity factor	0.46	0.45	0.47	0.48	0.462
Aquaculture pond heating	Capacity [MWt]	1097	605	616	653	695
	Utilization [TJ year^{-1}]	13493	11733	10976	11521	11958
	Capacity factor	0.39	0.61	0.57	0.56	0.546
Agriculture drying	Capacity [MWt]	67	74	157	125	161
	Utilization [TJ year^{-1}]	1124	1038	2013	1635	2030
	Capacity factor	0.53	0.44	0.41	0.41	0.400

Most solar water pumps have a power between 0.15–21 kW, with lifts of up to 350 m and flow rates of up to 130 m^3 h^{-1} (Varardi, 2014). PV water pumping systems are very reliable and generally the most cost-effective solution in locations where the electricity grid is non-existent or far away; in these areas, it has, when compared to diesel-powered pumps, the additional benefit of much less maintenance needs compared to the diesel option.

In Chapter 9 of this book, solar water pumping is described in more detail, and compared with diesel and grid-electricity-powered pumping in regard to its technical setup and suitability for different situations (e.g. economic, social, geographic, hydrogeological) and case studies with experiences from USA, New York state (NYSERDA, 2005), India (Casey, 2013; GIZ, 2013; IRENA, 2015) and Morocco (Lorentz GmbH, 2013; IFC, 2014) are presented.

1.5.2 *Solar and geothermal direct heat applications*

Since solar and geothermal heat can be generally used for the same applications, they will be discussed together. Solar and geothermal heat can be employed for: (i) heating/cooling of spaces, buildings, soil and water (including water for aquaculture), (ii) drying of crops and grains, and (iii) heating of greenhouses. The use of geothermal energy, which is a constant heat source, available 24 hours a day, 365 days a year, is often underestimated. The fact that this resource is practically limitless and renewable, together with – in contrast to the solar option – its independence of changing climate and weather conditions, presents clear advantages in favor of its extended utilization. Irrespective of the obvious climate protection benefits, the direct and indirect development of these resources for heating, agriculture and horticulture, should also respect other elements of the natural environment (Tomaszewska *et al.*, 2016). Table 1.1 shows global installed thermal capacities, thermal energy utilization and capacity factors for different applications in agriculture and the agri-food chain using geothermal resources and their change from 1995–2015. Integrated configurations for using geothermal heat for multiple use applications are presented separately in Section 1.5.5.

1.5.2.1 *Heating/cooling of spaces, buildings, soil and water*

Space heating (but also cooling using heat in mechanical or evaporative cooling cycles) is an important energy consumer in many applications, especially in pig and poultry production, which is mostly carried out in closed buildings and thus requires constant temperature and air quality conditions for the animals' well-being to optimize growth. Much energy is needed for replacing contaminated (moisture, gases, odor, dust, etc.) indoor air through ventilation and heating. A large part of this energy can be covered by RE using, e.g. solar or geothermal heat-driven air/space heaters/coolers/dehumidifier.

Heating and cooling of water (but also milk and agricultural products) for use in the agri-food chain is energy-intensive and an economic burden, especially in developing countries and areas

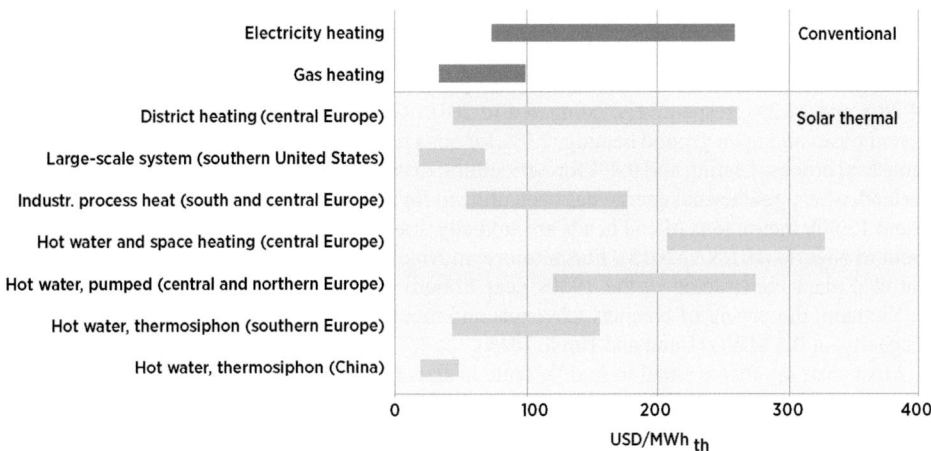

Figure 1.5.   Thermal solar water heating costs compared to conventional heating (IEA, 2014).

which are remote from electric power plants (as many agricultural areas are) which leads to further distribution losses in the electricity grid. Locally available RES, such as solar and geothermal, are economically and environmentally sustainable energy sources that are increasingly used to replace fossil fuels through electricity or – wherever possible as it has in many cases lower cost – directly to heat water. An example of high energy demands are commercial dairy farms where solar heat is used to heat water for cleaning equipment, warming and stimulating the cows' udders but also for cooling milk; these processes account for up to 40% of farms' total energy demand (IRENA, 2015). Considering all applications, the global capacity of solar water heating increased from 65 GWt (gigawatts-thermal) in 2000 to 326 GWt in 2013 (REN21, 2014). Considering further technological improvements and increasing mass production will result in a further increase in many areas, including different applications in the agri-food chain where water heating is required. As Figure 1.5 shows, water heating costs derived from different applications expand over large ranges and can economically compete with those using fossil fuels, either directly or through electricity for water heating. The same can be expected from applications within the agri-food chain industries.

1.5.2.2   *Drying of crops, fruits, grains and animal products*
There are many options for using heat from sun and geothermal sources for drying crops and grain, ranging from very simple and cheap to sophisticated and expensive. The sun has been employed for thousands of years for drying crops directly on the field or drying fruits on or in special devices after being harvested. Today's industrial-scale demands require generally fast and uniform drying to obtain high quality products suitable for the world market; however, for local or regional markets, in particular in developing countries, less sophisticated, cheaper drying designs are often more suitable.

Where geothermal sources are naturally available and accessible (at shallow depth, springs, etc.) geothermal drying is an excellent option. In contrast to solar, geothermal drying is also suitable in areas with little solar radiation such as colder climates. Fifteen countries report the use of geothermal energy for drying various grains, milk, vegetables and fruit crops. Examples include: seaweed (Iceland; Ragnarsson, 2015), onions (USA; Boyd *et al.*, 2015), wheat and other cereals (Serbia; Oudech and Djokic, 2015), fruit (El Salvador, Guatemala and Mexico; Montalvo and Gutierrez, 2015; Lund and Boyd, 2015; Gutiérrez-Negrin *et al.*, 2015), Lucerne or alfalfa (New Zealand; Carey *et al.*, 2015), coconut meat (Philippines; Fronda *et al.*, 2015), and timber

(Mexico, New Zealand and Romania; Gutiérrez-Negrin *et al.*, 2015; Carey *et al.*, 2015; Bendea *et al.*, 2015). The largest uses are in China, USA and Hungary (Lund and Boyd, 2015). By the end of 2014, a total of 161 MWt and 2030 TJ year^{-1} were being utilized, which is an increase of 28.8 and 24.2%, respectively, compared to 2010. Globally, 4.5% thermal energy is used for greenhouses and open ground heating, 2.0% for aquaculture pond and raceway heating, 1.8% for industrial process heating and 0.4% for agricultural drying (Lund and Boyd, 2015). An example is Iceland, where geothermal energy has been utilized for about 35 years for drying fish. Nowadays, about 15,000 metric tons of cod heads are annually dried using geothermal heat before exporting them to Nigeria (IRENA, 2015). Furthermore, in Nigeria pyrethrum drying is still being carried out at a plant constructed in the 1920s near Ebburu estimated at 1.0 MWt and 10 TJ year^{-1}. In Vietnam, the drying of bananas, coconuts and medicinal herbs occurs using 11.83 TJ year^{-1} (capacity: at 0.5 MWt) (Lund and Boyd, 2015).

Most solar dryers are small to middle scale in size. Large solar crop driers are in the minority as the example of the USA shows; the reason seems to be the high cost of the solar collector and since – in contrast to gas dryers – drying rates cannot be controlled (IRENA, 2015). However, it should be considered that these solar collectors could be used, in the non-harvesting period for other purposes, reducing the overall costs (IRENA, 2015). However, in hot arid and semi-arid climates, smaller sized dryers may not need a sophisticated collector (just using the glazed box of the system itself), and very low-cost dryers can be constructed using simple design and low-cost materials; such systems are very suitable for drying vegetables, fruits and animal products for home use (IRENA, 2015).

1.5.2.3   *Heating of greenhouses*

Solar and geothermal heat can be ideal and cost-effective energy sources for heating greenhouses. The heat can be employed to heat air and soil thus replacing or adapting heaters which are powered by fossil fuels or electricity produced from non-renewable conventional energy resources. The related cost-reduction of geothermal or solar-heated greenhouses is, in particular, high in remote areas without electricity access. Hence, the geothermal or solar option allows the installation and their economic operation in remote areas and production throughout the whole year.

The leading countries in the use of annual thermal energy being: Turkey, Russia, Hungary, China and The Netherlands (Lund and Boyd, 2015). The main crops grown in greenhouses are vegetables and flowers; however, tree seedlings (USA) and fruit such as bananas (Iceland) are also grown. Despite that the numbers with reported geothermal greenhouses decreased from 34 in 2010 to 31 in 2015, the worldwide use of geothermal energy utilized for green-houses and covered ground heating increased from 2010 to 2015 by 19% in installed capacity and 16% in annual thermal energy use. The installed capacity was 1830 MWt and energy use was 26,662 TJ year^{-1} (Lund and Boyd, 2015). In Turkey, greenhouse applications have reached 3 million m^2 (612 MWt) due to the great success. Tomatoes are mostly grown in these greenhouses with the major markets of Russia (60%), Europe 20%, around 10% elsewhere inter-nationally, and the remaining 10% sold domestically (Mertoglu *et al.*, 2015, Lund and Boyd, 2015).

Integrated operation and case studies of greenhouses using RE for geothermal and solar green-house heating/cooling, humidification, ventilation and water desalination for irrigation will be treated separately in Section 1.5.7.1. In Chapter 5 of this book, the energy usage patterns in the nursery industry in Queensland, Australia are examined. The opportunities of adopting renewable energy are also evaluated.

1.5.3   *Wind power applications*

Today, wind energy applications for sole agricultural applications are of reduced importance on a global scale, whereas in the past wind power was used for groundwater pumping. However, the mechanical sensitivity and respective maintenance needs of wind wheels today make diesel or – much better where possible – solar pumping the better option. In contrast, a combination of pumping using wind energy together with wind farming, i.e. selling energy to electricity companies or lease the land for installing their large-scale wind turbines is an increasing trend as outlined in Section 1.5.4.

1.5.4 *Multi-use of agricultural land for food and electric power production*

Agricultural areas can be used for multipurpose, i.e. food and energy production, where, for example, solar and wind energy are produced together with agricultural products on the same land; all in all being a winning combination by adding electricity generation – a long-term, stable, source of income – to the farmers' incomes from agricultural production. There are different options: (i) the farmers purchase wind turbines/solar panels, (ii) the farmers form wind/solar power cooperatives, and (iii) electricity utilities/developers install/operate the equipment and make payments to the farmers according to different criteria, such as generated electricity, impact on farming activities. According to an estimate by the US Department of Energy in the USA, wind energy alone could provide 80,000 new jobs for rural areas and US$ 1.2 billion in new income for farmers by 2020 (Cassaday, 2003). In areas with good wind conditions, electricity producers may pay US$ 2000–5000 per year for each wind turbine installed (USA example). Hence, such multi-use of agricultural land could be very attractive for farmers (Cassaday, 2003).

The concept of coproduction of electricity and food was first developed in Japan in 2004, where solar panels were installed in such a way that they did not significantly obstruct agricultural management, such as access of crops to sunlight, or movement of agricultural machinery (IRENA, 2015). In a similar way, wind turbines can be integrated into farming and grazing areas. Another interesting application is the installation of PV panels over irrigation canals to simultaneously produce electricity and reduce evaporative water losses. In an example from India, a 1 MW solar power plant was installed over a 750 m length of a canal system. In one year it produced 1.53 GWh of electricity while at the same time hindered the evaporation of about 3300 million of water, which can be used for irrigation (IRENA, 2015). It has been estimated that using the area of a 19,000 km canal network could save 20 billion liters of water per year (IRENA, 2015). Such multi-use of agricultural land for food and power production is becoming increasingly more common (IPCC, 2011). Multi-use could significantly increase the income for farmers as an example from Germany proves, where 11% of the installed RE capacity is in the ownership of farmers (2012).

1.5.5 *Agriculture within the cascade system of geothermal direct heat utilization*

When temperatures of geothermal sources are below 100°C, the warm water can be first run through a space heating installation and then cascaded to swimming pools, greenhouses and/or aquaculture pond heating, before being injected back into the aquifer. These kinds of projects maximize the use of the resource as well as improving the economics. An example of the use of geothermal energy in a cascade use is described by Bujakowski (2007). In 1993, The Mineral and Energy Economy Research Institute of the Polish Academy of Sciences (PAS MEERI) in Bańska Niżna (southern Poland, Podhale Basin) built and put into operation the first geothermal plant in Poland (Bujakowski, 2007; Bujakowski and Tomaszewska, 2012). During the 1995–98 period the installation has been developed and nowadays (2016) the cascaded heat supply includes five stages of heat distribution based upon a secondary circulation loop. These are: geothermal space heating system (85–65°C), timber drying building (60°C), parapet greenhouse (45°C), thermophilus fish farm (35°C), and finally, foil tunnels with ground heating (30°C) (Fig. 1.6). This setup enables studies on the multidirectional development of geothermal energy.

Gude (2016) suggested that in many cases the cascade system of geothermal energy use can also include desalination plants (see also Section 1.5.6). Figure 1.7 presents a conceptional example for an integrated configuration to produce power, followed by desalination using both thermal and membrane processes, then for applications in food processing, refrigeration plants and district heating or cooling systems, heating of building spaces, greenhouses and soil heating, industrial process heat and agricultural drying and fish farming. As shown in Figures 1.6 and 1.7, agricultural

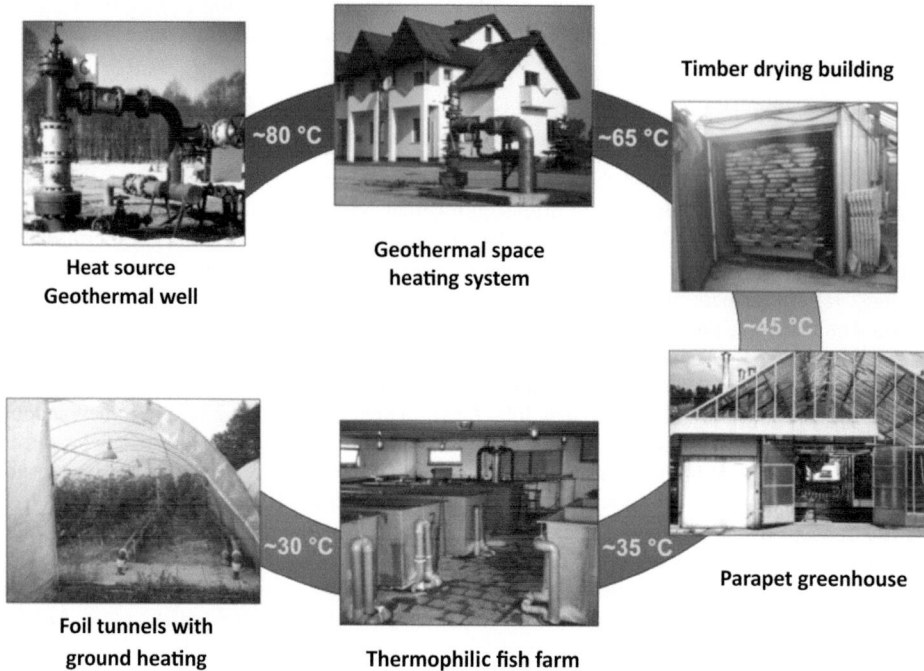

Figure 1.6.   The cascade system of geothermal energy utilization in PAS MEERI in Poland (based on Bujakowski, 2007).

and aquacultural uses require the lowest temperatures, with values from 25–90°C. Space heating requires temperatures in the range of 50–100°C, with 40°C useful in some marginal cases and ground-source heat pumps extending the range down to 5°C. Cooling and industrial processing normally require temperatures over 100°C (Gude, 2016; Lund, 2010).

1.5.6   *Renewables for water desalination and food security: decoupling freshwater production from fossil fuel supply*

As natural availability of freshwater is limited and unevenly distributed in space and time, energy-intensive desalination technologies are playing an increasing role in providing water and food security for future generations. At present, 70 million m^3 of freshwater are produced each day from the globally existing 16,000 desalination plants (IRENA, 2015). According to UN Water (2014), desalination annually consumes at least 75.2 TWh of electricity, which corresponds to about 0.4% of global electricity consumption (UN Water, 2014).

Conventional technologies for desalination of salt and brackish water are still limited since they have a high demand for energy, which is mostly provided by expensive fossil fuel, whereas less than 1% of the desalination capacity depends on renewables (IRENA, 2012). Energy costs represent as much as half of the production cost of desalination plants (Herndon, 2013). However, the upsurge in desalination capacity and the proportional energy demand rise make further use of fossil fuels increasingly economically and environmentally unsustainable. A massive shift from desalination powered by fossil fuel to RE-powered technologies will be essential to meet the growing demand for freshwater production by desalination. Decoupling freshwater production costs from the ever-increasing prices of fossil fuels is essential if freshwater is to be provided for agricultural purposes, such as irrigation where water is demanded in large quantities but at lowest possible cost.

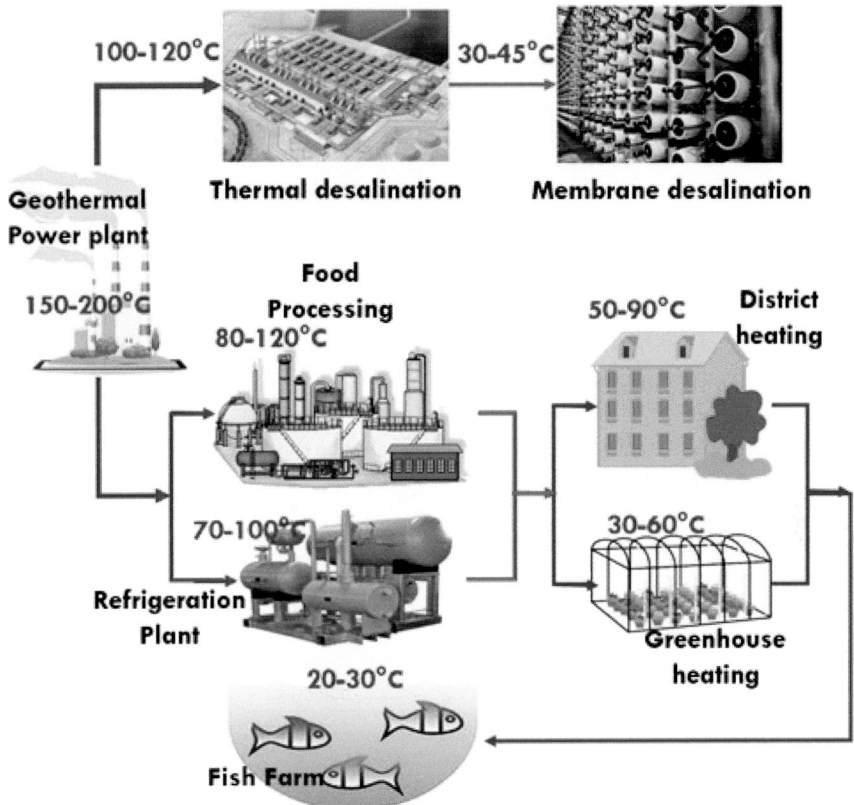

Figure 1.7. Integrated configurations for geothermal energy sources – polygeneration for multiple benefits (after Gude, 2016).

There exist different desalination technologies which can be powered by renewable energy sources (RES) (Ghaffour *et al.*, 2015; Goosen *et al.*, 2014). They comprise mature thermal technologies (multi-stage flash (MSF), multi-effect distillation (MED)), and mature membrane technologies (reverse osmosis (RO), electrodialysis (ED), and electrodialysis reversal (EDR)), as well as emerging technologies which are in their early stage and which require further research and development, such as membrane distillation (MD), vapor compression (VC), adsorption desalination (AD) (Ghaffour *et al.*, 2014) and others (Figs. 1.8 and 1.9).

Significant efforts for adapting conventional desalination technologies so that they can be powered by RES have been made in the last two decades; however, the upscaling to larger sized plants has been hampered by technological, economic and political/regulatory (e.g. subsidies for fossil fuels) challenges. In the last few years, several medium-scale RE-driven desalination plants have been installed worldwide. However, most of them are powered by electricity produced from solar PV and wind – commercial units which are available on the market (IRENA and IEA-ETSAP, 2012) – rather than solar or geothermal heat directly, which can provide more sustainable desalination processes. These electricity-powered desalination units are commonly deployed at the community level. However, they are increasingly being installed on a larger scale as the grid-connected desalination plants from Sydney and Perth demonstrate. Solar thermal desalination, which combines solar heat with desalination technologies such as MSF, MED, VC and MD, is obviously most suited for areas with high solar irradiation (e.g. northern Africa, Middle East, i.e. the MENA region, as well as parts of Australia: 2200–2400 kWh m^{-2} year and other world

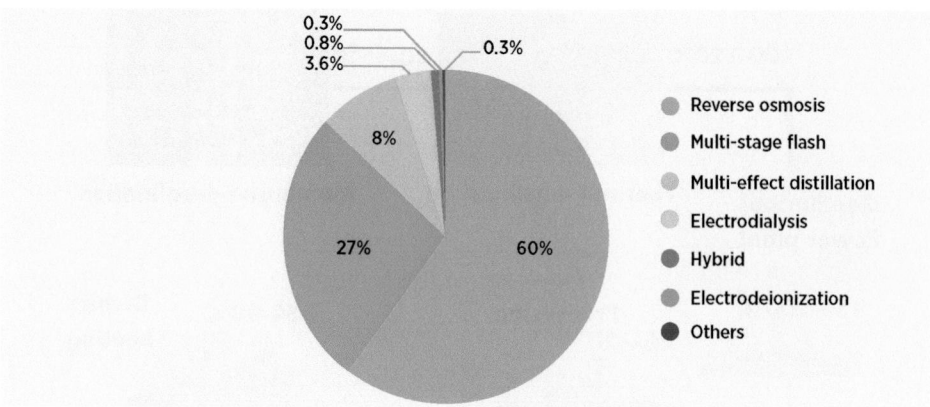

Figure 1.8.   Capacities for different technologies for desalination (Koschikowski, 2011).

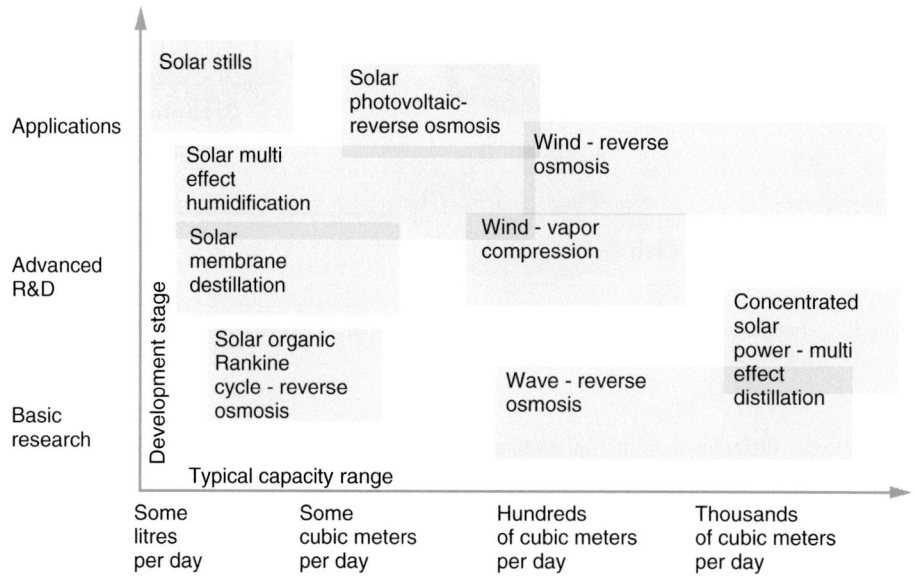

Figure 1.9.   Principal renewable energy powered desalination technologies and their development state (Papapetrou *et al.*, 2010).

regions) (Zaragoza *et al.*, 2014; Ghaffour *et al.*, 2014). Table 1.2 shows a list of relevant solar thermal plants implemented.

Energy requirements for desalination vary from process to process (Ghaffour *et al.*, 2013). Thermal desalination practices require both thermal and electrical energy for evaporation, process hydraulic flow and transport of the feed and product water. Pressure-driven membrane desalination processes necessitate electrical energy to supply the mechanical energy for membrane separation and pre-treatment, and pumping in and distribution out of the plant (Gude, 2016). Gude (2016) in his review, presented the specific energy consumption for thermal and membrane desalination processes in terms of kJ of energy required for producing one unit of freshwater in kilograms (Fig. 1.10; Table 1.3).

Table 1.2.   Summary of solar thermal desalination plants in the world.

Project	Capacity [$m^3 day^{-1}$]	Process
Margarita de Savoya, Italy [1]	50–60	MSF
Islands of Cape Verde [2]	300	Atlantis 'Autoflash'
Tunisia [3]	0.2	MSF
El Paso, Texas, USA [4]	19	MSF
University of Ancona, Italy [5]	30	MEB
Dead Sea, Jordan [6]	3000	MEB
Safat, Kuwait [6]	10	MSF
Takami Island, Japan [1]	16	ME-16 effects
Abu Dhabi, UAE [7]	120	ME-18 effects
Al-Ain, UAE [8]	500	ME-55 stages, MSF-75 stages
Arabian Gulf [9]	6000	MEB
Al Azhar University, Palestine [10]	0.2	MSF 4 stages
Almeria, Spain [11]	72	MED-TVC 14 effects
Berken, Germany [12]	10	MSF
Hzag, Tunisia [6]	0.1–0.35	Distillation
Gran Canaria, Spain [13]	10	MSF
La Desired Island, France [14]	40	ME-14 effects
Lampedusa Island, Italy [15]	0.3	MSF
Kuwait [1]	100	MSF
La Paz, Mexico [16]	10	MSF 10-stages

[1] Delyannis (1987); [2] Szacsvay *et al.* (1999); [3] Safi (1998); [4] Lu *et al.* (2000); [5] Caruso and Naviglio (1999); [6] El-Nashar (1985); [7] Hanafi (1991); [8] Abu-Jabal *et al.* (2001); [9] Borsani and Rebagliati (2005); [10] Zarza Moya (1991); [11] Kyritsis (1996); [12] Valverde Muela (1982); [13] Madani (1990); [14] Palma (1991); [15] Manjares and Galvan (1979); [16] Delgado-Torres and García-Rodríguez (2007).

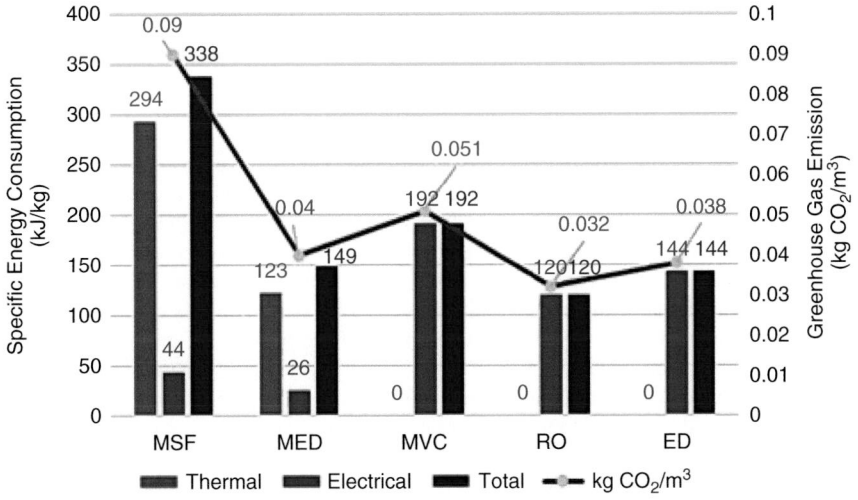

Figure 1.10.   Specific energy consumption for thermal and membrane desalination processes and greenhouse gas emissions for unit freshwater production (after Gude, 2016).

The MSF process has the highest specific energy consumption while the seawater reverse osmosis (SWRO) process has the lowest, followed by multi-effect distillation technology. For this reason, there is an increasing trend to replace thermal technologies with membrane methods, primarily reverse osmosis (RO) in cases where no cheap waste heat is available.

Table 1.3.   Typical capacity, energy demand and cost for some combinations of solar and desalination technologies (Ghaffour *et al.*, 2015 and references as specified).

	Typical capacity [m^3 day^{-1}]	Energy demand [kWh m^{-3}]	Water cost [US\$ m^{-3}]
Solar still	<0.1	Solar passive	1.3–6.5
Solar (MEH)	1–100	Ther: 100 Elec: 1.5	2.6–6.5
Solar MSF	1	Ther: 81–144[1]	1–5[1]
Solar tower MSF	1	T: 53.7[2]	–
Solar/CSP MED	>5000	Ther: 60–70; Elec: 1.5–2[3] T: 50–94[1]	2.3–2.8[3] (prospective cost) 2–9[1]
Solar tower MED	1	T: 42.4[2]	–
Solar tower VC	1	Elec: 55.5	–
PV-RO	<100	Elec: BW: 0.5–1.5; SW: 4–5[3]; BW-SW: 1.2–19[1]	BW: 6.5–9.1; SW: 11.7–15.6[3]; 3–27[1]
Solar tower RO	1	Elec: 41–45[2]	–
PV-EDR	<100	Elec: BW: 3–4[3] BW: 0.6–1[1]	10.4–11.7[3] 3–16[1]
Solar MD	0.15–10	Ther: 150–200[3]; 100–600[1]; 436[4]; 180–2200[5]	10.4–19.5[3] 13–18[1]
Solar AD	8	Elec: 1.38 T: 39.8[6]	0.7 (electrical cost only)[6]

Elec: BW: Brackish water; Electrical; SW: Seawater; T: Total; Ther: Thermal.
AD: adsorption desalination; CSP: concentrated solar power; ED: electrodialysis; EDR: electrodialysis reversal; MD: membrane distillation; MED: multi-effect distillation; MEH: multiple-effect humidification; MSF: multi-stage flash; PV: photovoltaic; RO: reverse osmosis; VC: vapor compression. [1]Ali *et al.* (2011); [2]Ahmad and Schmid (2002); [3]Paparetrou *et al.* (2010); [4]Kim *et al.* (2013); [5]Saffarini *et al.* (2012); [6]Ng et al. (2013).

The range of RE desalination technologies is large and each technology has particular characteristics which would need to be matched to a market analysis to enable investment decisions to be made. This level of detailed analysis is still missing as already mentioned earlier (Papapetrou *et al.*, 2010). On average, RE-based desalination is still expensive when compared to conventional desalination but numbers vary widely based on location and situation. Table 1.3 provides an overview of energy demand and water production costs for some possible combinations of solar and desalination technologies. A detailed assessment of the benefits of these technologies and their limitations are discussed in Tzen *et al.* (2012) using wind energy, Bundschuh *et al.* (2015) and Gude (2016) using geothermal heat, and by Ghaffour *et al.* (2015) for solar and geothermal.

Tzen (2012) provided a detailed review of usage of wind energy for freshwater production, the technology of which is mature and market available. The main problem in utilizing wind power in desalination applications is (i) the variable nature of the resource since storage wind energy as electricity is not sustainable at larger scale and (ii) the variable power input force to the desalination plant which may cause operational problems. Wind-powered desalination systems can be stand-alone or grid-connected. Most of the installed plants utilize the reverse osmosis desalination process. Their freshwater production capacity, electricity supply and year of installation are listed in Table 1.4. Other desalination technologies comprise vapor compression distillation and wind electrodialysis systems.

Figure 1.11 shows several possible combinations of solar and geothermal powered desalination, the individual suitability of which depends on specific site conditions (plant scale, feed water salinity, remoteness, electricity access or not, technical infrastructure, RES and its availability,

Table 1.4.   Wind reverse osmosis applications (modified from Tzen *et al.*, 2012).

Location	RO capacity [m^3 h^{-1}]	Electricity supply	Year of installation
Ile du Planier, France	0.5	4 kW W/T	1982
Island of Suderoog, Germany	0.25–0.37	6 kW W/T	1983
Island of Helgoland, Germany	40	1.2 MW W/T + diesel	1988
Fuerteventura, Spain	2.3	225 kW W/T + 160 kVA diesel, flywheel	1995
Pozo Izquierdo, Gran Canaria, Spain, SDAWES	8 × 1.0	2 × 230 kW W/T	1995
Therasia Island, Greece, APAS	0.2	15 kW W/T, 440 Ah batteries	1995/1996
Syros Island, Greece, JOULE	2.5–37.5	500 kW W/T, stand-alone + grid-connected	1998
Keratea, Greece, PAVET	0.13	900 W W/T, 4 kW$_p$ PV, batteries	2001/2002
Pozo Izquierdo, Gran Canaria, Spain, AEROGEDESA	0.80	15 kW W/T, 190 Ah batteries	2003/2004
Loughborough Univ., U.K.	0.5	2.5 kW, no battery	2001/2002
Milos Island, Greece, OPC Program	6 × 600	850 kW W/T, grid-connected	2007/2008
Heraklia Island, Greece	3.3	30 kW W/T, floating system, batteries	2007
Delft Univ., The Netherlands	0.2–0.4	Windmill, no battery	2007/2008
Perth (Kwinana Desalination Plant)	~140000 extended: ~250000	80 MW, W/T, grid-connected	2006
Sydney Desalination Plant	~250000 extended: ~500000	140 MW, W/T, grid-connected	2010

W/T: Wind turbines; PV: Photovoltaic.

potential and exploitation cost). Table 1.5 provides an overview of the fresh water productivity for different desalination technologies per square meter of solar collector area.

Ghaffour *et al.* (2015) provided a critical review of the status of RE-powered desalination technologies, highlighting integrated systems and potential applications together with current technological and economic limitations and challenges. Tzen (2012) and Ghaffour *et al.* (2015) concluded that the matching of the desalination process with a renewable energy source (RES) is technically feasible. However the problem lies in continuous versus non-continuous operation; desalination processes are best suited to continuous operation, whereas the principal renewable energy sources such as wind and solar are non-continuous. Thus, matching a desalination system with an RES source requires special design and operation which increases complexity and cost. At present, technological and economic constraints hinder large-scale applications. Despite that, renewables and desalinations as individual technologies are mature and are produced in mass.

Geothermal energy can be an attractive option if low-cost, low-enthalpy geothermal sources are available. These include geothermal resources at shallow depth, water coproduced from onshore and offshore hydrocarbon wells or from already existing deep wells, and residual heat from geothermal power plants (Bundschuh *et al.*, 2015). Geothermal energy is accessible day and night every day of the year and can thus serve as an add-on to energy sources which are only available intermittently (Goosen *et al.*, 2010). However, the application of geothermal energy in desalination is still a relatively unexplored technical concept (Davies and Orfi, 2014). Some experiments have been described: (i) electricity and freshwater production from geothermal brines using MSF unit (Awerbuch *et al.*, 1976); (ii) a case study of a low-enthalpy geothermal energy-driven MED unit on Milos Island in Greece (Karytsas, 1996); and (iii) one of the latest projects from USA, the VTE Geothermal Desalination Pilot, where geothermal steam will provide the

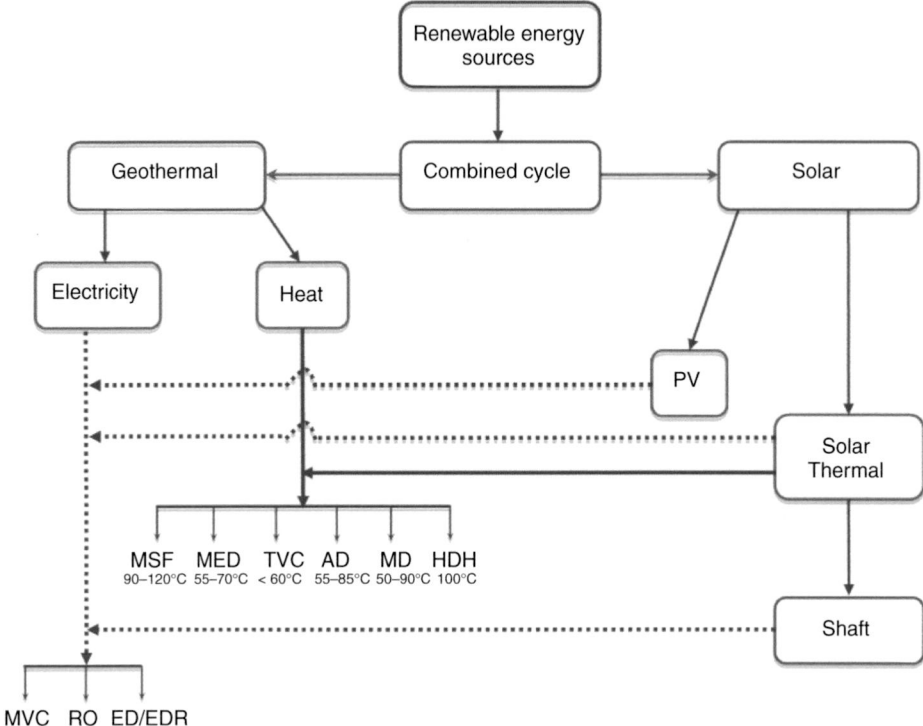

Figure 1.11.   Possible combinations of integrated systems: RES with conventional and innovative desalination processes (Ghaffour *et al.*, 2015). AD: Adsorption desalination; ED: Electrodialysis; EDR: Electrodialysis reversal; HDH: Humidification-dehumidification; MD: Membrane distillation; MED: Multi-effect distillation; MSF: Multi-stage flash; MVC: Mechanical vapor compression; PV: Photovoltaic; RO: Reverse osmosis; TVC: Thermal vapor compression.

Table 1.5.   Productivity of different desalination processes per square meter of solar collector area (Ghaffour *et al.*, 2015).

Desalination process	Water produced per solar collector area [L day^{-1}m^{-2}]
Simple solar still	4–5
H/D process-medium temperature solar thermal collector	12
MSF, MED with thermal storage – medium temperature solar thermal collector	40
SWRO-PV	200
VARI-RO DDE–Dish Sterling solar collector (only in concept stage)	1200

thermal energy source for the pilot VTE distillation (vertical tube evaporation) process, as the process applied to an MED plant design with low cost (Sephton Water Technology, 2012).

Taking into account a significant increase in global electricity production (Bertani, 2015), between 2015 and 2030, rapid expansion of geothermal electricity and heat production will take place, but be limited to areas where such resources are available. Gude (2016) suggested that high-enthalpy geothermal sources can be utilized in cogeneration schemes for simultaneous power and freshwater production using MSF and MED technologies. Low temperature desalination processes can be coupled with low-enthalpy geothermal sources. Low temperature desalination processes have lower specific energy requirements and a higher thermodynamic efficiency.

One should notice that geothermal systems not only provide a valuable RES, but can also be considered as the source and solution for freshwater production, including irrigation water being the main freshwater consumer. The desalination of geothermal waters used for energy purposes could be seen as one of the methods for securing high quality water for various purposes. In countries with warm climates it is mainly used for the irrigation of agricultural crops (Kabay *et al.*, 2004a, 2004b, 2009; 2013; Koseoglu *et al.*, 2010; Tomaszewska *et al.*, 2016). Given the increasing deficit of fresh water worldwide, the possibilities for desalinating and treating geothermal waters for drinking and household purposes should be considered (Gallup 2007; Tomaszewska and Bodzek, 2013a, 2013b; Tomaszewska *et al.*, 2014; 2016; Gude, 2016). Alternative solutions such as using cooled water directly for drinking or household purposes are advantageous ones in certain cases as confirmed by the activities of Geotermia Mazowiecka S.A. (Mszczonów, central Poland). Water with a low mineral content (ca. $0.5\,g\,L^{-1}$) and with an intake temperature of 42°C has been extracted since 2000 from the Mszczonów IG-1 well – from a Lower Cretaceous horizon composed of sandstone interbedded with mudstone and claystone. These are high quality $Cl\text{-}HCO_3\text{-}Na\text{-}Ca$ waters that are fed to the municipal water supply network as drinking water following cooling and simple treatment. The extraction of these waters in an open system with a maximum capacity of $60\,m^3\,h^{-1}$ (without re-injecting cooled water into the formation) has significantly improved the economic performance of the project and the utilization of cooled water as drinking water has additionally enhanced the management of ordinary water resources (Tomaszewska and Szczepański, 2014). Applications for agriculture can be similarly developed.

Freshwater production by RE-powered desalination is a technological-sound option at a small- or medium-scale and economically viable for water supply in remote areas. However, upscaling to a large size is still hindered, due to the intermittent availability of wind and solar energy, a disadvantage which geothermal does not have. This also suggests the implementation of a combined-cycle solar and geothermal powered desalination process without the need for energy storage. Also, the development and improvement of innovative desalination technologies which do not need continuous operation such as AD and MD, and which consequently are more suitable for RE use can be utilized to overcome this limitation.

Ongoing research and development of concentrated solar power (CSP) based desalination is also promising. CSP with thermal energy storage shows a large potential for powering large-scale desalination plants and simultaneously producing electricity for other purposes (Ghaffour *et al.*, 2015; Goosen *et al.*, 2014; IRENA, 2015). According to Trieb *et al.* (2011), CSP-based desalination could produce a major part of the freshwater in the MENA region amounting to about 16% of its total water production in 2030 and 22% in 2050.

1.5.7 *Geothermal and solar greenhouse heating/cooling, ventilation, humidification, desalination*

1.5.7.1 *Solar and geothermal based greenhouse development*
Several recent reviews have appeared on the state-of-the-art solar and geothermal technology in agricultural greenhouse development (Mahmoudi *et al.*, 2010, Hassanien *et al.*, 2016; Harjunowibowo *et al.*, 2016; Lund and Boyd, 2016). The integration of thin film solar PV panels into the roof area of glass greenhouses was one of the advances reported by Hassanien *et al.* (2016). The authors noted that while there was some loss in the availability of solar radiation inside the greenhouse it was more than offset by electrical power generation (Fig. 1.12). The conversion of agricultural land into PV plants can cause friction between farmers and energy producers. By combining PV panels and crops on the same area of land the increasing competition for land between food and energy production can be alleviated.

Harjunowibowo *et al.* (2016) presented the latest technological developments used in greenhouses to control the microclimate by focusing on passive techniques. For example, heat can be

Figure 1.12.   Thin film PV solar glass greenhouse (adapted from Hassanien *et al.*, 2016).

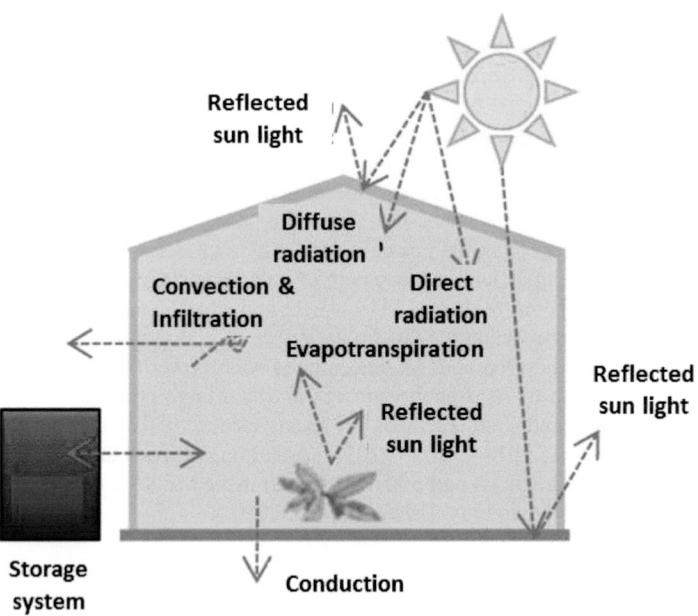

Figure 1.13.   Closed greenhouse thermal flow with heat storage system (adapted from Harjunowibowo *et al.*, 2016; Vadiee and Martin 2013).

taken from the greenhouse during the day and deposited in a thermal storage system. This heat is then used at night in accordance with the required heat in the greenhouse (Fig. 1.13). The storage system can use, for instance, water, rocks, phase-change material and soil water collectors. Harjunowibowo *et al.* (2016) found that heat storage systems containing phase-change materials (PCM) could provide both heating and cooling for closed greenhouses. For greenhouses in northern climates, the authors claimed a reduction in energy demands by 80% with a potential payback of six years. In a related study by Bouadila *et al.* (2014) the excess heat in the greenhouse was stored in a packed bed through the daytime period and extracted at night. Additionally, Çakır and

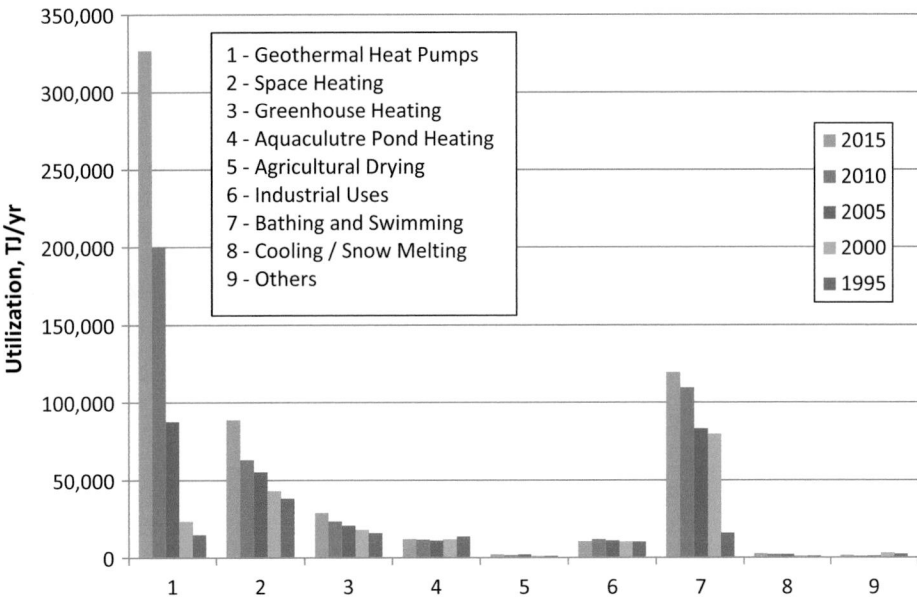

Figure 1.14.   Comparison of worldwide direct-use geothermal energy in TJ year^{-1} from 1995, 2000, 2005, 2010 and 2015 (Lund and Boyd, 2016).

Şahin (2015) assessing solar greenhouses in cold climates, evaluated the optimum type according to sizing, position and location. They concluded that one of the chief operational parameters for solar energy acquisition rates of greenhouses is the roof shape. Elliptic greenhouses were preferred (at least for cold climates).

Global consumption of geothermal energy for greenhouses and covered ground heating has increased by 28% in installed capacity and 25% in annual energy use over the past decade (Lund and Boyd, 2016). As already mentioned, the leading countries in annual energy use being: Turkey, Russia, Hungary, China and The Netherlands. The authors noted that most countries do not distinguish between covered greenhouses versus uncovered ground heating, and only a few reported the actual area heated. The main crops grown in greenhouses were vegetables and flowers. However, tree seedlings (USA) and fruit such as bananas (Iceland) were also grown. Not surprisingly, developed countries experience competition from developing countries due to lower labor costs for the latter. Over a 20-year period, from 1995–2015, greenhouse heating using geothermal energy has seen about a 50% increase (Fig. 1.14).

Improved greenhouse heating by coupling of geothermal heat pumps with solar collectors has been described (Awani *et al.*, 2015; Ghosal and Tiwari, 2004; Kondili and Kaldellis, 2006; Ozgener, 2010). In many cases modeling was employed to try and optimize the performance. Ozgener (2010), for example, reported on the use of solar-assisted geothermal heat pump and small wind turbine systems for heating agricultural and residential buildings. The main objectives of their study were to analyze thermal loads of geothermally and passively heated solar greenhouses and to investigate wind energy utilization in greenhouse heating which was modeled as a hybrid solar-assisted geothermal heat pump, and a small wind turbine system which was separately installed. The main conclusion of the investigation was that a modeled passive solar pre-heating technique, combined with a geothermal heat pump system and a small wind turbine system could be economically superior to conventional space heating/cooling systems used in agricultural and residential building heating applications if these buildings are installed in a region which has a good wind resource. Additionally, Chinese *et al.* (2005) developed technical and economic optimization models in order to exploit a renewable energy source represented by

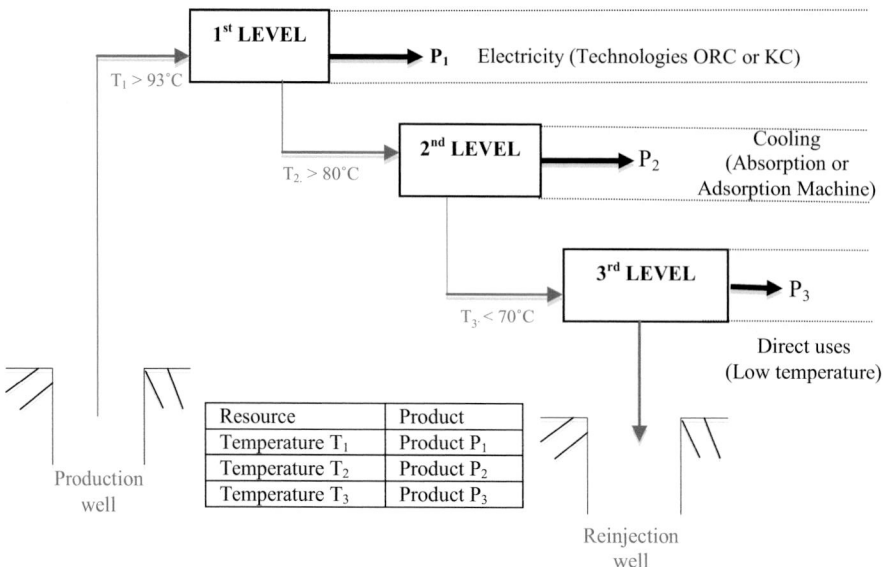

Figure 1.15.   Conceptual diagram of the cascade utilization of geothermal energy (Adapted from Rubio-
Maya *et al.*, 2015).

waste heat coming from the condenser of a waste-to-energy plant built to convert wood scraps from
a chair manufacturing industrial district in north-eastern Italy. The authors argued that coupling
a greenhouse with a waste-to-energy plant could represent an important step towards sustainable
development of such industrial structures, since it encourages both business diversification and
full exploitation of internal resources.

The concept and application of cascade utilization of low and medium enthalpy geothermal
resources have been described previously in Bujakowski (2007). It can be argued that this is a
crucial idea that essentially couples different technologies and applications in order to utilize
all the available geothermal heat by integrating different technologies for electricity generation,
distribution and use of thermal energy, drying and dehydration processes, recreational uses, and
any other direct use of geothermal heat (Fig. 1.15). For greenhouse systems geothermal cascade
technology can combine electricity production for running pumps and providing lighting, along
with direct heating of the greenhouse. Rubio-Maya *et al.* (2015) presented a comprehensive
review of different regions around the world employing geothermal resources of medium and low
enthalpy in a cascade manner. It is possible that the fluid temperature at the outlet of the generation
process has sufficient thermodynamic quality and can be useful at a sequential process in a second
or third level of temperature. Such processes or direct uses are, for instance: heating systems, hot
water supply, intermediate drying processes of food or wood, and other direct uses of geothermal
heat for aquaculture and greenhouses.

Figure 1.15 is presented in order to describe better the concept of cascade utilization of geothermal resource. This particular system is a three-level cascade with electricity production and some
thermal applications. In this example, the geothermal resource of medium enthalpy is utilized
in the first level of the cascade for the production of electricity. Afterwards, the geothermal
resource which is derived from this process feeds the second level of the cascade for freezing or
cooling purposes using thermally activated technologies, such as absorption or adsorption effect
machines. After this second use, the fluid can be employed further for additional purposes with
lower temperature requirements to form the third level of the cascade. Such application might
be dehydration processes and greenhouses. Rubio-Maya *et al.* (2015) concluded that the main
benefits are an increased profitability of the facility, maximized use of geothermal resources of

Table 1.6.   Profitability analysis data of various applications of renewable energies in greenhouses in Crete-Greece (Vourdoubas, 2015).

Type of renewable energy	Energy generated	Operating period [years]	Initial investment [€]	Payback period [years]	NPV [€]
Solar PV	Electricity	20	13062	16.98	1622
				12.30*	6030*
Solid biomass	Heat	15	14864	2.95	53880
Direct heating with geothermal fluid	Heat	20	12500	1.25	129268
Geothermal heat pumps	Heat (and cooling)	20	122130	–	−19116

Note. *In the case of 30% higher than current electricity prices.

Table 1.7.   Environmental benefits due to renewable energy use in greenhouses (Vourdoubas, 2015).

Type of renewable energy used	Energy used initially	Energy generated	Initial CO_2 emissions [tons year^{-1} $\times 10^3$ m^2]	CO_2 emissions due to renewables [tons year^{-1} $\times 10^{-3}$ m^2]	Reduction of CO_2 emissions [tons year^{-1} $\times 10^{-3}$ m^2]
Solar PV	Grid electricity	Electricity	13.85	0	13.85
Solid biomass	Fuel oil	Heat	85.60	0	85.60
Direct heating with geothermal fluid	Fuel oil	Heat	85.60	0	85.60
Geothermal heat pumps	Fuel oil	Heat (and cooling)	85.60	62.17	23.43

medium and low-enthalpy, local development of communities and cities, as well as social and environmental benefits.

Lastly, economic and environmental assessments of solar-geothermal greenhouse systems are critical in their effective development and commercialization (Vourdoubas, 2015; Russo *et al.*, 2014; Ozgener and Hepbasli, 2006). In a case study of Crete-Greece by Vourdoubas (2015), cost analysis of the use of solid biomass and geothermal energy for direct heating and cooling greenhouses showed that these investments are very profitable and attractive (Tables 1.6 and 1.7). However, the author reported that the use of geothermal heat pumps for heating and cooling them was not cost-effective. Furthermore, the use of solar photovoltaic cells for power generation was also not cost-effective, particularly when electricity generation in greenhouses was subsidized by the government. On the positive side the decrease of CO_2 emissions due to the use of renewables in the greenhouses was considered as an additional benefit.

Profitability analysis data of various applications of renewable energies in greenhouses in Crete-Greece by Vourdoubas (2015) are presented in Table 1.6. Direct heating with geothermal fluid was found to be the most profitable. The net present value (NPV) was the highest at € 129,268. In finance, the NPV or net present worth (NPW) is a measurement of the profitability of an undertaking that is calculated by subtracting the present value (PV) of cash outflows (including initial cost) from the present values of cash inflows over a period of time. Similarly, the environmental benefits, due to the use of renewable energy sources in agricultural greenhouses are presented in Table 1.7. Use of the renewable energy systems for heat and power generation in greenhouses will result in a reduction of greenhouses gases emitted due to energy use in them. The reductions were estimated by Vourdoubas (2015) as the difference of the emissions due to fossil fuels use minus the emissions due to renewable energy use. In the case of using PV cells, CO_2 emissions were zero when using direct heating with geothermal fluids and heating with solid biomass. In the case of employing a geothermal heat pump the emissions were estimated from the consumption of grid electricity for the operation of the heat pump.

1.5.7.2   *Closed seawater greenhouses for meeting water, energy and food security*

In contrast to the natural environment where growing conditions may not be optimum for a given crop, glasshouses – when appropriately managed – provide optimal growth conditions and all-year-round operation since parameters such as air temperature, relative humidity, carbon dioxide concentration, soil temperature and soil moisture can be controlled (Cooper and Fuller, 1983). In particular in arid areas, closed seawater greenhouses which use only 10% of the water that open farming requires for obtaining the same yield (Masudi, 2014) provide excellent opportunities to supply food to the domestic market and to produce high-priced export products throughout the year (IRENA, 2015).

If the required solar/geothermal sources are available, it is a feasible solution for providing food, and freshwater to a large part of global inhabitants given the fact that over 70% of the population lives within a distance of less than 70 km (El-Dessouky and Ettouney, 2002), or 80% within 100 km (Ghaffour, 2009) from the seashore.

Due to high freshwater and energy demands, solar or geothermal heat can be employed for heating/cooling/humidification and for freshwater production through desalination of saline or brackish water, as well as producing PV energy for other purposes. A promising example for such a completely solar power operated greenhouse is the Sundrop System, which harnesses solar thermal energy to desalinate seawater to produce fresh water for irrigation (Saumweber, 2013). According to Saumweber (2013), the upfront costs of solar-based greenhouses are lower than the present-value annual cost of fossil fuels for traditional greenhouses in the same location (Saumweber, 2013). In Port Augusta (Australia), a pilot greenhouse was upgraded to a 20-hectare facility with a capacity of producing daily 10,000 L desalinated water, and more than 15,000 t of vegetables per year (WWF and CEEW, 2014).

Using hydroponics that utilize nutrient-laden water, no fertile soil is required and the facilities can also be installed in areas otherwise not suitable for cultivation (e.g. deserts, arctic regions). Since the nutrients can be well-dosed, and subsequently the nutrients not consumed by the crops can be reused, this eliminates the negative impacts on surface water resources such as eutrophication and contamination of land and water resources, which are disadvantages of conventional agriculture. Additionally, both, aeroponic and hydroponic systems (i) require much less space and water than conventional farming methods, and (ii) do not require pesticides to make them a true renewable and sustainable farming system. Finally, by applying vertical farming (Growing Power, 2011) and providing artificial light, space demands can be further reduced thus making them ideal horticulture systems for cities or small islands. The electricity demand of aeroponic and hydroponic cultivations can be provided from renewable energy systems.

1.5.8   *The present market of renewable energy technologies*

Integrating RES in agriculture and the wider agri-food chain is a key goal to assuring food security and protecting the earth's climate. Use of RES for freshwater production is essential since crop irrigation consumes about 70% of all produced freshwater.

Renewable energy technologies, suitable for direct agricultural purposes and within the wider area of the agri-food chain, are mature technologies. Mass production has resulted in massive price drops and it can be expected that prices will further decrease as the technologies are further improved. Solar energy can provide electricity, heat and shaft energy; wind can provide electricity and shaft energy, and geothermal can provide electricity and heat energy (Fig. 1.16). For all of these RES, there exist mature technologies and mature markets.

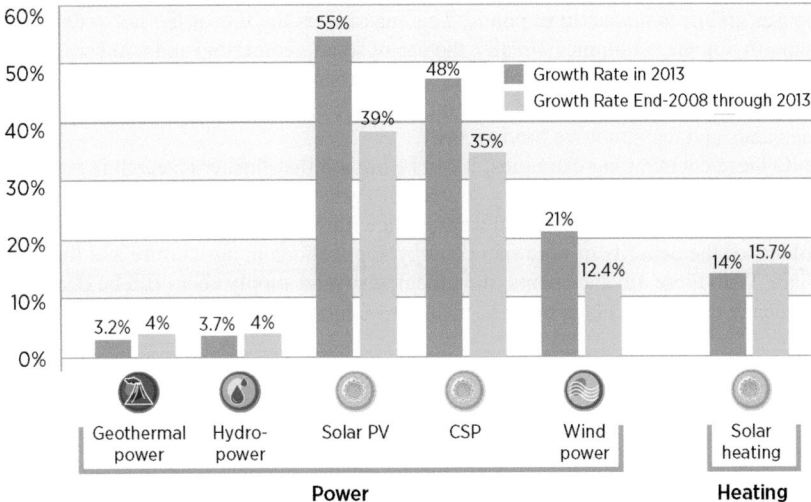

Figure 1.16. Average annual growth of installed geothermal, hydro, wind and solar (thermal and photo-voltaic) capacity of the end-users of the power and heating sectors (modified from REN21, 2014).

1.6 CONCLUSIONS

Globally, the agri-food chain in the world today consumes about one-third of the total world energy production, with about 12% for crop production and nearly 80% for post-harvest related activities. Overall, agriculture and agri-food industry is often highly energy-intensive, particularly in developed countries, and reliant on fossil fuels that are employed (e.g. for running farm equipment, generation of electricity for powering fans and pumps, and for refrigeration of food supplies). Currently, conventional fossil fuel sources, however, are being depleted at a progressively rapid rate not only due to global population growth, but also owing to the growing energy demands in the agri-food production chain in general. The highest share of on-farm energy for instance, is required for irrigation by groundwater pumping. As a solution to this problem, the use of locally available renewable energy sources, together with energy-efficient technologies, has become increasingly attractive for minimizing the impacts of rising energy costs on agri-food profitability and competitiveness as well as contributing to climate protection. Renewable energy includes solar, wind and geothermal sources. These renewable energy sources can be utilized through a number of well-proven established and new technologies, depending on the site-specific conditions and needs. Such technologies can be either integrated into existing agricultural operations or designed in the planning phase of innovative projects.

On a global scale a number of successful examples already exist where solar, wind and geothermal energy sources have been used either directly or indirectly for electricity generation in agriculture. Their economic viability and competitiveness against fossil fuels has been demonstrated. The financial benefits are highest in cases where costly electricity generation by diesel or petrol generators is replaced by renewable energy sources. Furthermore, solar or geothermal direct heat applications can be employed for heating/cooling of spaces, buildings, soil and water; drying of crops and grains; and heating/cooling during food processing; and ventilation and humidification of greenhouses. In addition, multi-use of agricultural land for food and electric power production also constitutes an attractive way for farmers as it could significantly increase their income. Another promising option is a cascade system, where solar-heated water or geothermal water is utilized stepwise for applications requiring temperature gradients. For example, from a space heating system installation the residual water is cascaded to swimming pools, to greenhouses and/or to aquaculture ponds. This maximizes the use of the hot water resource as well as improving the economics. Finally, the use of solar, geothermal and wind energy for water desalination has also received considerable attention since it decouples freshwater production from the fossil fuel supply and therefore is an attractive alternative to guarantee future irrigation water demands and thus improve food security.

Despite these encouraging examples, it can be argued that further research is required on the life cycle impacts of renewable energy sources of agricultural products. Life cycle assessment (LCA) will help to identify suitable pathways, policy frameworks and convince farmers and other stakeholders of the benefits of renewable energy applications in agriculture and the entire agri-food chain. With these advancements, the global agri-food supply chain can be decoupled from its dependency on fossil fuels in order to meet future food demands.

REFERENCES

Abu-Jabal, M.S., Kamiya, I. & Narasaki, Y. (2001) Proving test for a solar-powered desalination system in Gaza-Palestine. *Desalination*, 137, 1–6.

Ahmad, G.E. & Schmid, J. (2002) Feasibility study of brackish water desalination in the Egyptian deserts and rural regions using PV systems. *Energy Conversion and Management*, 43, 2641–2649.

Ali, M.T., Fath, H.E.S. & Armstrong, P.R. (2011) A comprehensive techno-economical review of indirect solar desalination. *Renewable and Sustainable Energy Reviews*, 15, 4187–4199.

Awani, S., Chargui, R., Kooli, S., Farhat, A. & Guizani, A. (2015) Performance of the coupling of the flat plate collector and a heat pump system associated with a vertical heat exchanger for heating of the two types of greenhouses system. *Energy Conversion and Management*, 103, 266–275.

Awerbuch, L., Lindemuth, T.E., May, S.C. & Rogers, A.N. (1976) Geothermal energy recovery process. *Desalination*, 19(1), 325–336.

Baillie, C. (2009) Energy and carbon accounting case study on Keytah. Project report for the Cotton Research and Development Corporation (CRDC). University of Southern Queensland, Toowoomba, QLD, Australia.

Bellarby, J., Foereid, B., Hastings, A. & Smith, P. (2008) Cool farming: climate impacts of agriculture and mitigation potential. Greenpeace International, Amsterdam, The Netherlands.

Bendea, C., Antal, C. & Rosca, M. (2015) Geothermal energy in Romania: country update 2010–2014. *Proceedings World Geothermal Congress 2015, 19–25 April 2015, Melbourne, Australia*. Paper 1013.

Bertani, R. (2015) Geothermal power generation in the world 2010-2014 update report. *Proceedings World Geothermal Congress 2015, 19–25 April 2015, Melbourne, Australia*. Paper 1001.

Biswas, W.K., Graham, J., Kelly, K. & John, M.B. (2010) Global warming contributions from wheat, sheep meat and wool production in Victoria, Australia – a life cycle assessment. *Journal of Cleaner Production*, 18, 1386–1392.

Borsani, R. & Rebagliati, S. (2005) Fundamentals and costing of MSF desalination plants and comparison with other technologies. *Desalination*, 182, 29–37.

Bouadila, S., Lazaar, M., Skouri, S., Kooli, S. & Farhat, A. (2014) Assessment of the greenhouse climate with a new packed-bed solar air heater at night, in Tunisia. *Renewable and Sustainable Energy Reviews*, 35, 31–41.

Boyd, T.L., Sifford, A. & Lund, J.W. (2015) The United States of America country update 2015. *Proceedings World Geothermal Congress 2015, 19–25 April 2015, Melbourne, Australia*. Paper 1009.

Bujakowski, W. (2007) Energia geotermalna [Geothermal energy]. In: Sapinska-Śliwa, A. (ed.) *Odnawialne źródła energii w Małopolsce* [*Renewable Energy Sources in Małopolska*]. Wyd. Stowarzyszenie Gmin Polska Sieć „Energie Cités" [The Association of Municipalities Polish Network „Energy Cities"]. Kraków, Poland.

Bujakowski, W. & Tomaszewska, B. (2012) Geothermal energy used in Podhale region. *Technologia Wody*, 5(19), 30–36.

Bundschuh, J., Ghaffour, N., Mahmoudi, H., Goosen, M., Mushtaq, S. & Hoinkis, J. (2015) Low-cost low-enthalpy geothermal heat for freshwater production: innovative applications using thermal desalination processes. *Renewable and Sustainable Energy Reviews*, 43, 196–206.

Çakır, U. & Şahin, E. (2015) Using solar greenhouses in cold climates and evaluating optimum type according to sizing, position and location: a case study. *Computers and Electronics in Agriculture*, 117, 245–257.

Carey, B., Dunstall, M., McClintock, S., White, B., Bignall, G., Luketina, K., Robson, B., Zarrouk, S. & Seward, A. (2015) 2010–2015 New Zealand Country Update. *Proceedings World Geothermal Congress 2015, 19–25 April 2015, Melbourne, Australia.* Paper 1052.

Caruso, G. & Naviglio, A. (1999) A desalination plant using solar heat as s heat supply, not affecting the environment with chemicals. *Desalination*, 122, 225–234.

Casey, A. (2013) Reforming energy subsidies could curb India's water stress. Worldwatch Institute, Washington, DC. Available from: http://www.worldwatch.org/reforming-energy-subsidies-could-curb-india's-water-stress-0 [accessed October 2016].

Cassaday, A. (2003) USPIRG renewables.report. U.S. PIRG Education Fund. Available from: https://www.hks.harvard.edu/hepg/Papers/Cassady_USPIRG.renewables.report_4-03.pdf [accessed October 2016].

Chen, G. & Baillie, C. (2009a) Agricultural applications: energy uses and audits. In: Capehart, B. (ed.) *Encyclopedia of Energy Engineering and Technology*, 1(1), 1–5. Taylor & Francis Books, London, UK.

Chen, G. & Baillie, C. (2009b) Development of a framework and tool to assess on-farm energy uses of cotton production. *Energy Conversion and Management*, 50(5), 1256–1263.

Chen, G., Maraseni, T.N. & Yang, Z. (2010) Life-cycle energy and carbon footprint assessments: agricultural and food products. In: Capehart, B. (ed.) *Encyclopedia of Energy Engineering and Technology*, 1(1), 1–5. Taylor & Francis Books, London, UK.

Chen, G., Maraseni, T.N., Banhazi, T. & Bundschuh, J. (2015) Benchmarking energy use on farm. Publication No. 15/059, Rural Industries Research and Development Corporation, Canberra, ACT, Australia.

Chinese, D., Meneghetti, A. & Nardin, G. (2005) Waste-to-energy based greenhouse heating: exploring viability conditions through optimization models. *Renewable Energy*, 30(10), 1573–1586.

Cooper, P.I. & Fuller, R.J. (1983) A transient model of the interaction between crop, environment and greenhouse structure for predicting crop yield and energy consumption. *Journal of Agricultural Research*, 28, 401–417.

Davies, P.A. & Orfi, J. (2014) Self-powered desalination of geothermal saline groundwater: technical feasibility. *Water*, 6(11), 3409–3432.

Delgado-Torres, A.M. & García-Rodríguez, L. (2007) Status of solar thermal-driven reverse osmosis desalination. *Desalination*, 216, 242–251.

Delyannis, E.E. (1987) Status of solar-assisted desalination: a review. *Desalination*, 67, 3–19.

El-Dessouky, H.T. & Ettouney, H.M. (2002) *Fundamentals of Salt Water Desalination*. Elsevier Science, Amsterdam, The Netherlands and New York, NY.

El-Nashar, A.M. (1985) Abu Dhabi solar distillation plant. *Desalination*, 52, 217–234.

FAO (2011a) Energy-smart food for people and climate. Issue paper, Food and Agriculture Organization of the United Nations, Rome, Italy. Available from: http://www.fao.org/docrep/014/i2454e/i2454e00.pdf [accessed October 2016].

FAO (2011b) The state of the world's land and water resources for food and agriculture (SOLAW) – managing systems at risk. Food and Agriculture Organization of the United Nations, Rome, Italy and Earthscan, London, UK.

FAO (2013) Global forest resources assessment: forests future. Food and Agriculture Organization of the United Nations, Rome, Italy. Available from: http://www.fao.org/forestry/fra/85504/en/ [accessed October 2016].

FAO (2014) Walking the nexus talk: assessing the water-energy-food nexus in the context of the Sustainable Energy for All Initiative. Food and Agriculture Organization of the United Nations, Rome, Italy. Available from: http://www.fao.org/3/a-i3959e.pdf [accessed October 2016].

FAO (2016) World food situation. FAO cereal supply and demand brief. New season production prospects improve, stocks to remain high. Food and Agriculture Organization of the United Nations, Rome, Italy. Available from: http://www.fao.org/worldfoodsituation/csdb/en/ [accessed October 2016].

Fronda, A.D., Marasigan, M.C. & Lazaro, V.S. (2015) Geothermal development in the Philippines: the country update. *Proceedings World Geothermal Congress 2015, 19–25 April 2015, Melbourne, Australia.* Paper 1053.

Gallup, D.L. (2007) Treatment of geothermal waters for production of industrial, agricultural or drinking water. *Geothermics*, 36, 473–483.

Ghaffour, N. (2009) The challenge of capacity building strategies and perspectives for desalination for sustainable water use in MENA. *Desalination & Water Treatment*, 5, 48–53.

Ghaffour, N., Missimer, T.M. & Amy, G.L. (2013) Technical review and evaluation of the economics of water desalination: current and future challenges for better water supply sustainability. *Desalination*, 309, 197–207.

Ghaffour, N., Lattemann, S., Missimer, T.M., Ng, K.C., Sinha, S. & Amy, G. (2014) Renewable energy-driven innovative energy-efficient desalination technologies. *Applied Energy*, 136, 1155–1165.

Ghaffour, N., Bundschuh, J., Mahmoudi, H. & Goosen, M.F.A. (2015) Renewable energy-driven desalination technologies: a comprehensive review on challenges and potential applications of integrated systems. *Desalination*, 356, 94–114.

Ghosal, M.K. & Tiwari, G.N. (2004) Mathematical modeling for greenhouse heating by using thermal curtain and geothermal energy. *Solar Energy*, 76(5), 603–613.

GIZ (2013) Solar water pumping for irrigation: potential and barriers in Bihar, India. Deutsche Gesellschaft für Internationale Zusammenarbeit (GIZ) GmbH, Indo-German Energy Programme (IGEN), New Delhi, India, In cooperation with the Ministry of New and Renewable Energy (MNRE), New Delhi, India. Available from: http://www.igen-re.in/files/giz__2013__factsheet_solar_water_pumping_for_irrigation_in_bihar.pdf [accessed October 2016].

Goosen, M., Mahmoundi, H. & Ghaffour, N. (2010) Water desalination using geothermal energy. *Energies*, 3, 1423–1442.

Goosen, M.F., Mahmoudi, H. & Ghaffour, N. (2014) Today's and future challenges in applications of renewable energy technologies for desalination. *Critical Reviews in Environmental Science and Technology*, 44(9), 929–999.

Graham, P.W. & Williams, D.J. (2003) Optimal technological choices in meeting Australian energy policy goals. *Energy Economics*, 25(6), 691–712.

Growing Power (2011) Growing power vertical farm. Available from: http://www.tkwa.com/growing-power-vertical-farm/ [accessed August 2016].

Gude, V.G. (2016) Geothermal source potential for water desalination – current status and future perspective. *Renewable and Sustainable Energy Reviews*, 57, 1038–1065.

Gutiérrez-Negrin, L., Maya-González, R. & Quijano-León, J.L. (2015) Present situation and perspectives of geothermal in Mexico. *Proceedings World Geothermal Congress 2015, 19–25 April 2015, Melbourne, Australia*. Paper 1002.

Haddeland, I., Heinke, J., Biemans, H., Eisner, S., Flörke, M., Hanasaki, N., Konzmann, M., Ludwig, F., Masaki, Y., Schewe, J., Stacke, T., Tessler, Z.D., Wada, Y. & Wisser, D. (2014) Global water resources affected by human interventions and climate change. *PNAS*, 111(9), 3251–3256.

Hanafi, A. (1991) Design and performance of solar MSF desalination system. *Desalination*, 82, 165–174.

Harjunowibowo, D., Erdem, C.U.C.E., Omer, S. & Riffat, S.B. (2016) Recent passive technologies of greenhouse systems: a review. *Proceedings of the fifteenth International Conference on Sustainable Energy Technologies*. pp. 19–22.

Hassanien, R.H.E., Li, M. & Lin, W.D. (2016) Advanced applications of solar energy in agricultural greenhouses. *Renewable and Sustainable Energy Reviews*, 54, 989–1001.

Herdon, A. (2013) Energy makes up half of desalination plant costs. Bloomberg. Available from: http://www.bloomberg.com/news/articles/2013-05-01/energy-makes-up-half-of-desalination-plant-costs-study [accessed December 2016].

Hoff, H. (2011) Understanding the nexus. Stockholm Environment Institute (SEI), Stockholm, Sweden. Available from: http://www.water-energy- food.org/documents/understanding_the_nexus.pdf [accessed August 2016].

IEA (2012) World Energy Outlook 2012. International Energy Agency, Paris, France. Available from: http://www.worldenergyoutlook.org/publications/weo-2012/ [accessed August 2016].

IEA (2014) Heating without global warming. International Energy Agency, Paris, France. Available from: http://www.iea.org/publications/freepublications/publication/FeaturedInsight_HeatingWithoutGlobalWarming_FINAL.pdf [accessed August 2016].

IFC (2014) Scaling up opportunities for solar-powered irrigation pumps. World Water Week, Stockholm, Sweden. International Finance Corporation. Available from: http://programme.worldwaterweek.org/sites/default/files/colback_stockholm_presentation.pdf [accessed August 2016].

IPCC (2007) Climate change 2007: the physical science basis. Contribution of Working Group I to the Fourth Assessment Report of the Intergovernmental Panel on Climate Change. Cambridge University Press, Cambridge, UK and New York, NY.

IPCC (2011) Renewable energy sources and climate change mitigation: summary for policy makers and technical summary. Intergovernmental Panel on Climate Change. Cambridge University Press, Cambridge, UK and New York, NY. Available from: www.ipcc.ch/pdf/special-reports/srren/SRREN_FD_SPM_final.pdf [accessed October 2016].

IPCC (2014) Technical summary of climate change 2014: mitigation of climate change, Working Group III Contribution to the IPCC Fifth Assessment Report (AR5). Intergovernmental Panel on Climate Change, Cambridge University Press, Cambridge, UK and New York, NY.

IRENA (2012) Key findings and recommendations. *International Off-grid Renewable Energy Conference 2012, 1–2 November 2012, Accra, Ghana*. International Renewable Energy Agency, Masdar City, UAE. Available from: http://www.irena.org/DocumentDownloads/Publications/IOREC_Key%20Findings%20 and%20Recommendations.pdf [accessed August 2016].

IRENA (2015) Renewable energy in the water, energy & food nexus. International Renewable Energy Agency, Masdar City, UAE.

IRENA and IEA-ETSAP (2012) Water desalination using renewable energy: technology brief. Available from: www.irena.org/DocumentDownloads/Publications/IRENA-ETSAP%20Tech%20Brief%20I12% 20Water-Desalination.pdf [accessed October 2016].

Jackson, T.M., Khan, S. & Hafeez, M. (2010) A comparative analysis of water application and energy consumption at the irrigated field level. *Agricultural Water Management*, 97, 1477–1485.

Johnson, J.A., Runge, C.F., Senauer, B., Forley, J. & Polasky, S. (2014) Global agriculture and carbon trade-offs. *PNAS*, 111(34), 12,342–12,347.

Kabay, N., Yılmaz, I., Yamac, S., Samatya, S., Yuksel, M., Yuksel, U., Arda, M., Sağlam, M., Iwanaga, T. & Hirowatari, K. (2004a) Removal and recovery of boron from geothermal wastewater by selective ion exchange resins. I. Laboratory tests. *Reactive and Functional Polymers*, 60, 163–170.

Kabay, N., Yılmaz, I., Yamac, S., Samatya, S., Yuksel, M., Yuksel, U., Arda, M., Sağlam, M., Iwanaga, T. & Hirowatari, K. (2004b) Removal and recovery of boron from geothermal wastewater by selective ion-exchange resins. II. Field tests. *Desalination*, 167, 427–438.

Kabay, N., Yilmaz-Ipek, I., Soroko, I., Makowski, M., Kirmizisakal, O., Yag, S., Bryjak, M. & Yuksel, M. (2009) Removal of boron from Balcova geothermal water by ion exchange-microfiltration hybrid process. *Desalination*, 241, 167–173.

Kabay, N., Köseoğlu, P., Yavuz, E., Yüksel, U. & Yüksel, M. (2013) An innovative integrated system for boron removal from geothermal water using RO process and ion exchange-ultrafiltration hybrid method. *Desalination*, 316, 1–7.

Karytsas, C. (1996) Low-enthalpy geothermal energy-driven sea-water desalination plant on Milos island – a case study. *Proceedings of Mediterranean Conference on Renewable Energy Sources for Water Production, 10–12 June 1996, Santorini, Greece*. pp. 128–131.

Kim, Y.D., Thu, K., Ghaffour, N. & Ng, K.C. (2013) Performance investigation of solar-assisted hollow fiber DCMD desalination system. *Journal of Membrane Science*, 427, 345–364.

Klöpffer, W. (2012) The critical review of life cycle assessment studies according to ISO 14040, and 14044. *International Journal of Life Cycle Assessment*, 17(9), 1087–1093.

Kondili, E. & Kaldellis, J.K. (2006) Optimal design of geothermal-solar greenhouses for the minimisation of fossil fuel consumption. *Applied Thermal Engineering*, 26(8), 905–915.

Koschikowski, J. (2011) Water desalination: when and where will it make sense. Presentation at the *2011 Annual Meeting of the American Association for the Advancement of Science, Fraunhofer ISE (Institute for Solar Energy Systems), Freiburg, Germany*.

Koseoglu, H., Harman, B.I., Yigit, N.O., Guler, E., Kabay, N. & Kitis, M. (2010) The effects of operating conditions on boron removal from geothermal waters by membrane processes. *Desalination*, 258, 72–78.

Kulak, M., Graves, A. & Chatterton, J. (2013) Reducing greenhouse gas emissions with urban agriculture: a life cycle assessment perspective. *Landscape and Urban Planning*, 111, 68–78.

Kyritsis, S. (1996) *Proceedings of the Mediterranean Conference on Renewable Energy Sources for Water Production, European Commission, EURORED Network, CRES, EDS, 10–12 June 1996, Santorini, Greece*. pp. 265–270.

Lorentz GmbH (2013) Solar water pumping for irrigation in Oujda, Morocco. Bernt Lorentz GmbH & Co. KG, Henstedt-Ulzburg, Germany. Available from: http://www.lorentz.de/pdf/lorentz_casestudy_ irrigation_morocco_en-en.pdf [accessed August 2016].

Lu, H., Walton, J.C. & Swift, A.H.P. (2010) Zero discharge desalination.. *International Desalination and Water Reuse Quarterly*, 3, 35–43.

Lund, J.W. (2010) Direct utilization of geothermal energy. *Energies*, 3(8), 1443–1471.

Lund, J.W. & Boyd, T.L. (2015) Direct utilization of geothermal energy 2015: worldwide review. *Proceedings World Geothermal Congress 2015, 19–25 April 2015, Melbourne, Australia*. Paper 1000.

Lund, J.W. & Boyd, T.L. (2016) Direct utilization of geothermal energy 2015 worldwide review. *Geothermics*, 60, 66–93.

Madani, A.A. (1990) Economics of desalination systems. *Desalination*, 78, 187–200.

Mahmoudi, H., Spahis, N., Goosen, M.F., Ghaffour, N., Drouiche, N. & Ouagued, A. (2010) Application of geothermal energy for heating and fresh water production in a brackish water greenhouse desalination unit: a case study from Algeria. *Renewable and Sustainable Energy Reviews*, 14(1), 512–517.

Manjares, R. & Galvan, M. (1979) Solarmultistage flash evaporation (SMSF) as a solar energy application on desalination processes, description of one demonstration project. *Desalination*, 31, 545–554.

Maraseni, T.N. (2007) *Re-Evaluating Land Use Choices to incorporate Carbon Values: a Case Study in the South Burnett Region of Queensland*. PhD Thesis, University of Southern Queensland, Toowoomba, QLD, Australia.

Maraseni, T.N. & Cockfield, G. (2011a) Crops, cows or timber? Including carbon values in land use choices. *Agriculture Ecosystems & Environment*, 140, 280–288.

Maraseni, T.N. & Cockfield, G. (2011b) Does the adoption of zero tillage reduce greenhouse gas emissions? An assessment for the grains industry in Australia. *Agricultural Systems*, 104, 451–458.

Maraseni, T.N. & Cockfield, G. (2012) Including the costs of water and greenhouse gas emissions in a reassessment of the profitability of irrigation. *Agricultural Water Management*, 103, 25–32.

Maraseni, T.N., Mushtaq, S. & Reardon-Smith, K. (2012a) Integrated analysis for a carbon- and water-constrained future: an assessment of drip irrigation in a lettuce production system in eastern Australia. *Journal of Environmental Management*, 111, 220–226.

Maraseni, T.N., Mushtaq, S. & Reardon-Smith, K. (2012b) Climate change, water security and the need for integrated policy development: the case of on-farm infrastructure investment in the Australian irrigation sector. *Environmental Research Letters*, 7, 1–12.

Maraseni, T.N., Chen, G., Banhazi, T., Bundschuh, J. & Yusaf, T. (2015) An assessment of direct on-farm energy use for high value grain crops grown under different farming practices in Australia. *Energies*, 8, 13,033–13,046.

Masudi, F. (2014) Greenhouses key to water and food security in UAE, expert says. Gulf News, http:// gulfnews.com/news/uae/environment/greenhouses-key-to-water-and-food-security-in-uae-expert-says-1. 1349562 [accessed October 2016].

Mertoglu, O., Simsek, S. & Basarir, N. (2015) Geothermal country update report of Turkey (2010–2015). *Proceedings World Geothermal Congress 2015, 19–25 April 2015, Melbourne, Australia*. Paper No 1046.

Montalvo, F. & Gutierrez, H. (2015) El Salvador country update. *Proceedings World Geothermal Congress 2015, 19–25 April 2015, Melbourne, Australia*. Paper 108.

Mushtaq, S., Maraseni, TN. & Reardon-Smith, K. (2013) Climate change and water security: estimating the greenhouse gas costs of achieving water security through investments in modern irrigation technology. *Agricultural Systems*, 117, 78–89.

Ng, K.C., Thu, K., Kim, Y.D., Chakraborty, A. & Amy, G. (2013) Adsorption desalination: an emerging low-cost thermal desalination method. *Desalination*, 308, 161–179.

NYSERDA (2005) Guide to solar-powered water pumping systems in New York State. New York State Energy Research and Development Authority (NYSERDA), Albany, NY. Available from: http://water.epa. gov/infrastructure/sustain/upload/2005_1_27_publications_solarpumpingguide.pdf [accessed August 2016].

Oudech, S. & Djokic, I. (2015) Geothermal energy use, country update for Serbia. *Proceedings World Geothermal Congress 2015, 19–25 April 2015, Melbourne, Australia*. Paper 1041.

Ozgener, O. (2010) Use of solar assisted geothermal heat pump and small wind turbine systems for heating agricultural and residential buildings. *Energy*, 35(1), 262–268.

Ozgener, O. & Hepbasli, A. (2006) An economical analysis on a solar greenhouse integrated solar assisted geothermal heat pump system. *Journal of Energy Resources Technology*, 128(1), 28–34.

Palma, F. (1991) Seminar on new technologies for the use of renewable energies in water desalination. 26–28 September, Athens, Greece. Commission of the European Communities, DG XVII for Energy, Centre for Renewable Energy Sources (CRES).

Papapetrou, M., Wieghaus, M. & Biercamp, Ch. (eds.) (2010) Roadmap for development of desalination powered by renewable energy, promotion of renewable energy for water production through desalination. Fraunhofer Verlag, Stuttgart, Germany. Available from: http://www.prodes-project.org/ fileadmin/Files/ProDes_Road_map_on_line_version.pdf [accessed August 2016].

Pellizzi, G., Cavalchini, A.G. & Lazzari, M. (1988) *Energy Savings in Agricultural Machinery and Mechanization*. Elsevier Science Publishing Co. New York, NY.

Ragnarsson, A. (2015) Geothermal development in Iceland 2010–2014. *Proceedings World Geothermal Congress 2015, 19–25 April 2015, Melbourne, Australia*. Paper 1077.

REN21 (2013) Renewables 2013: Global status report. Renewable Energy Policy Network for 21st Century, Paris, France. Available from: www.ren21.net/Portals/0/documents/Resources/GSR/2013/GSR2013_lowres.pdf [accessed August 2016].

REN21 (2014) Renewables 2014: Global status report. Renewable Energy Policy Network for 21st Century, Paris, France. Available from: www.ren21.net/Portals/0/documents/Resources/GSR/2014/GSR2014_full%20report_low%20res.pdf [accessed October 2016].

Roy, P., Nei, R., Orikasa, T., Xu, Q., Okadome, H., Nakamura, N. & Shiina, T. (2009) A review of life cycle assessment (LCA) on some food products. *Journal of Food Engineering*, 90, 1–10.

Rubio-Maya, C., Díaz, V.A., Martínez, E.P. & Belman-Flores, J.M. (2015) Cascade utilization of low and medium enthalpy geothermal resources – a review. *Renewable and Sustainable Energy Reviews*, 52, 689–716.

Russo, G., Anifantis, A.S., Verdiani, G. & Mugnozza, G.S. (2014) Environmental analysis of geothermal heat pump and LPG greenhouse heating systems. *Biosystems Engineering*, 127, 11–23.

Saffarini, R.B., Summers, E.K., Arafat, H.A. & Lienhard, J.H. (2012) Technical evaluation of stand-alone solar powered membrane distillation systems. *Desalination*, 286, 332–341.

Safi, M.J. (1998) Performance of a flash desalination unit intended to be coupled to a solar pond. *Renewable Energy*, 14, 339–343.

Saumweber, P. (2013) Farming sustainability Part III: Sundrop farms – growing with seawater and sunlight. Available from: http://www.futuredirections.org.au/publications/food-and-water-crises/1337-sundrop-farmsgrowing-with-seawater-and-sunlight-2.html [accessed October 2016].

Sephton Water Technology (2012) VTE geothermal desalination pilot/demonstration project. Project summary. Kensington, CA. Available from: http://sephtonwatertech.com/DocumentsPDF/VTE_Geothermal_Desalination_Project_Summary_2012_02_05.pdf [accessed August 2016].

Smith, P., Martino, D., Cai, Z., Gwary, D., Janzen, H., Kumar, P., Mccarl, B., Ogle, S., Mara, F., Rice, C., Scholes, B., Sirotenko, O., Howden, M., Mcallister, T., Pan, G., Romanenkov, V., Schneider, U. & Towprayoon, S. (2007) Policy and technological constraints to implementation of greenhouse gas mitigation options in agriculture. *Agriculture Ecosystems & Environment*, 118, 6–28.

Szacsvay, T., Hofer-Noser, P. & Posnansky, M. (1999) Technical and economic aspects of small-scale solar-pond-powered seawater desalination systems. *Desalination*, 122, 185–193.

Tomaszewska, B. & Bodzek, M. (2013a) Desalination of geothermal waters using a hybrid UF-RO process. I. Boron removal in pilot-scale tests. *Desalination*, 319, 99–106.

Tomaszewska, B. & Bodzek, M. (2013b) The removal of radionuclides during desalination of geothermal waters containing boron using the BWRO system. *Desalination*, 309, 284–290.

Tomaszewska, B. & Szczepański, A. (2014) Possibilities for the efficient utilisation of spent geothermal waters. *Environmental Science and Pollution Research*, 21, 11,409–11,417.

Tomaszewska, B., Pająk, L. & Bodzek, M. (2014) Application of a hybrid UF-RO process to geothermal water desalination. Concentrate disposal and costs analysis. *Archives of Environmental Protection*, 40(3), 137–151.

Tomaszewska, B., Rajca, M., Kmiecik, E., Bodzek, M., Bujakowski, W., Wator, K. & Tyszer, M. (2016) The influence of selected factors on the effectiveness of pre-treatment of geothermal water during the nanofiltration process. *Desalination*, 406, 74–82.

Topak, R., Suheri, M. & Calisir, S. (2005) Investigation of the energy efficiency for raising crops under sprinkler irrigation in a semi-arid area. *Applied Engineering in Agriculture*, 21(5), 761–767.

Trieb, F., Moser, M. & Fichter, T. (2011) MENA Regional Water Outlook – desalination using renewable energy. Overview of DLR work within the MENA Regional Water Outlook study. Workshop, 22–23 February 2011, Muscat. German Aerospace Center (DLR), Köln, Germany. Available from: http://elib.dlr.de/72591/1/Workshop_Oman_Final_DLR.pdf [accessed October 2016].

Tubiello, F.N., Cóndor-Golec, R.D., Salvatore, M., Piersante, A., Federici, S., Ferrara, A., Rossi, S., Flammini, A., Cardenas, C., Biancalani, R., Jacobs, H., Prasula, P. & Prosperi, P. (2015) Estimating greenhouse gas emissions in agriculture: a manual to address data requirements for developing countries. Food and Agriculture Organization of the United Nations, Rome, Italy.

Tzen, E. (2012) Wind energy powered technologies for freshwater production: fundamentals and case studies. In: Bundschuh, J. & Hoinkis, J. (eds.) *Renewable Energy Applications for Freshwater Production*. CRC Press, Boca Raton and IWA, London, UK. pp. 161–180.

UN Water (2014) Statistics detail. Available from: http://www.unwater.org/statistics/statistics-detail/pt/c/211827/ [accessed October 2016].

USEPA (2006) Global anthropogenic non-CO_2 greenhouse gas emissions: 1990–2020. EPA 430-R-06-003, United States Environmental Protection Agency, Washington, DC.

Vadiee, A. & Martin, V. (2013) Energy analysis and thermoeconomic assessment of the closed greenhouse – the largest commercial solar building. *Applied Energy*, 102, 1256–1266.

Valverde Muela, V. (1982) Planta desaladora con energıa solar de Arinaga (Las Palmas de Gran Canaria) [The solar desalination plant of Arinaga (Las Palmas, Gran Canaria)]. Departamento de Investigacion y Nuevas Fuentes, Centro de Estudios de la Energıa, Las Palmas de Gran Canaria, Spain.

Varadi, P.F. (2014) *Sun above the Horizon: Meteoric Rise of the Solar Industry*. Pan Stanford Series on Renewable Energy, Pan Stanford, Singapore.

Vermuelen, S.J., Campbell, B.M. & Ingram, J.S.I. (2012) Climate change and food systems. *Annual Review of Environment and Resources*, 37, 195–222.

Vourdoubas, J. (2015) Economic and environmental assessment of the use of renewable energies in greenhouses: a case study in Crete-Greece. *Journal of Agricultural Science*, 7(10), 48–57.

Ward, F.A., Booker, J.F. & Michelsen, A.M. (2006) Integrated economic, hydrologic, and institutional analysis of policy responses to mitigate drought impacts in the Rio Grande Basin. *Journal of Water Resources and Planning and Management* – ASCE, 132, 488–502.

WWF and CEEW (2014) Renewables beyond electricity: solar air conditioning and desalination in India. Available from: http://awsassets.wwfindia.org/downloads/solar_air_conditioning_desalination_in_india.pdf [accessed October 2016].

Zaragoza, G., Ruiz-Aguirre, A. & Guillén-Burrieza, E. (2014) Efficiency in the use of solar thermal energy of small membrane desalination systems for decentralized water production. *Applied Energy*, 130, 491–499.

Zarza Moya, E. (1991) Solar thermal desalination project: first phase and results and second phase description. Secretaría General Técnica del CIEMAT, Madrid, Spain.

CHAPTER 2

Agriculture sector modernization and renewable energy development: perspectives from developing countries

Robert K. Dixon & Ming Yang

2.1 INTRODUCTION

In making decisions about energy supplies and sources, many resource managers lack the means to make environmentally sustainable choices that provide for their energy needs, while simultaneously protecting public health and the environment. For example, in 2012 approximately 1.4 billion people (about 17% of the world's population) lived without access to electricity (GEA, 2012; IEA, 2016), and 2.8 billion people relied on traditional fuels such as wood, charcoal, and animal and crop waste to cook and heat their homes (The Economist, 2013). This reliance on traditional forms of energy limits social and economic development, slows down poverty elimination, affects human health, and aggravates global deforestation. In many instances, locally available renewable energy is a viable alternative for traditional fuels and rural electrification and helps the poor to meet their needs for important domestic energy services such as heat, light, and refrigeration.

Renewable energy sources are valuable because they increase the independence from and resilience to the volatility of the fossil fuel markets, help protect the environment, bring in added benefits of stimulating local employment and technological development, and contribute to economic growth by reducing the movement of rural population to cities. In particular, small-scale renewable energy technologies – such as mini hydro power, biomass digesters, improved cooking stoves, solar photovoltaic (PV), and mini windmills for power generation – offer rural households access to cleaner, and in some cases, the most cost-effective energy resources. Renewable energy investment projects and programs that involve governments, businesses, non-profit organizations, commercial and development banks, and community partners have expanded access to these technologies and their services over the past few decades (Dixon et al., 2011).

The variability of renewable energy technology advantages is largely a function of the type of power generation that it displaces. Siler-Evans et al. (2013) evaluated how existing power generation facilities are distributed across the United States and analyzed the impacts of replacing those facilities powered from wind or solar sources. They found uneven but significant social and environmental benefits resulting from the adoption of these technologies. They also discussed how the production tax credits for wind energy generation was valued, how its effectiveness could be improved by regional differentiation of the policy, and how the large-scale adoption of wind or solar energy might affect their site-specific analysis.

2.1.1 Challenges to renewable energy development

Robust, reliable and cost-competitive technologies are essential, but other ingredients are necessary to enhance development. Rural electrification and agricultural modernization also requires support from policy improvement, capacity-building and institutional development. There are instances where energy suppliers defer expanding access to disadvantaged areas because such expansion is not yet widely regarded as financially viable. In most developing countries, the public sector can be thwarted by its inability to implement or finance rural electrification projects, and is always under pressure to satisfy other urgent public financing needs. In many cases, national planners may hesitate to promote off-grid renewable energy projects because renewable energy technologies are generally more expensive than fossil fuel energy technologies and the systems must be sourced from overseas suppliers. And, for a variety of social and political reasons, government agencies give higher priority to expanding access for urban, rather than rural areas.

As developing countries expand their economies and reduce poverty, they face major climate change and energy challenges, including the need to:

- Improve living conditions and enhance economic opportunities for the 1.4 billion people without access to electricity (GEA, 2012).
- Increase energy security for all nations, regions, and communities, while at the same time reducing greenhouse Gas (GHG) emissions. The International Energy Agency (IEA, 2012) estimates that energy use accounts for approximately 65% of the world's GHG emissions.
- Reduce the adverse environmental effects and risks associated with current energy systems and increase prosperity.
- Supply clean fuel to about three billion people worldwide who rely on solid fuels for cooking. This reduces indoor air pollution resulting from cooking with biomass and coal that has caused almost two million deaths annually through pneumonia and chronic lung disease (UNDP, 2009).

Despite global growth in renewable energy generation, particularly in developed countries, renewable energy resources are virtually untapped in the least developed countries (LDCs). The primary barrier to the widespread adoption of renewable energy in LDCs is its high initial construction cost, particularly for installing equipment. As such, grants, subsidies, and concessional loans are necessary for these LDCs to overcome the financial barriers to investing in renewable energy technologies. Significant expansion of renewable power production will require the mobilization of considerably more investment in renewable energy, of which at least 75% should be directed to developing countries (IEA, 2009). In addition to capital mobilization, other elements, for example, strengthening institutional capacities, promoting enabling environments, developing policy frameworks, and improving demand for renewable energy technologies could help unlock other barriers, mitigate steep transaction costs, and promote underdeveloped markets.

2.1.2 *Opportunities for renewable energy in developing countries and countries with economies in transition*

Renewable energy opportunities are growing around the world. According to the IEA (2012), the share of non-hydroelectric renewable energy in power generation is expected to increase from 3% in 2009 to 15% in 2035. This growth will be underpinned by annual subsidies to renewables that will rise almost five times to US$ 180 billion by 2035. Developing countries will take the lead in renewable power generation. Figure 2.1 shows the IEA's projected marginal power generation from renewable energy between 2011 and 2035.

Even though the subsidy cost per unit of energy produced is expected to decline, most renewable energy sources will need continued support in order to compete in energy markets. While remaining costly in the beginning until the emergence of mass production and cost reduction, the renewable energy development is expected to bring long-term benefits of both increased energy security and independence, and greater climate and environmental protection. Market launching and penetration that leads to mass production and significant cost reduction will be the key factors for renewable energy development in the next decades, which will make renewable technologies cost-competitive or cheaper than fossil energy technologies.

Accommodating more energy from renewable sources, sometimes in remote locations, will require additional investment in energy storage and transmission networks, which amounts to 10–25% of the total investment. The contribution of hydropower, a conventional renewable energy source, to global power generation remains at around 15%, with China, India, Brazil and other developing countries accounting for more than half of the 680 gigawatts (GW) of new capacity (IEA, 2012).

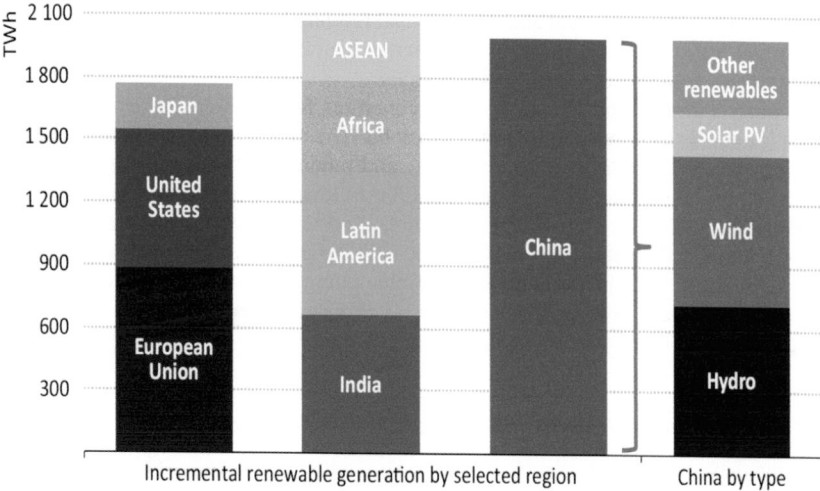

Figure 2.1. Incremental electricity generation from renewables in selected regions, 2011–2035. Source: IEA (2013).

Energy demand and supply patterns can be altered with the development of renewable energy on a large scale. This is a major challenge that demands comprehensive and sustainable solutions. Renewable energy technologies are vital to alleviating poverty, expanding rural development, and improving environmental quality. The productive use of renewable energy and implementation of energy-efficient technologies will, in turn, reduce costs as well as GHG emissions in rural areas, help raise incomes and improve health, and provide cheap energy in order to, for example, pump water for irrigation, process crops, power cottage industries, and light homes, schools, and hospitals – all services of first importance to remote rural areas.

Advanced technologies for new renewable power generation integrated with smart grids have been tested and proven in the field on many occasions. Experiences have shown that, if renewable energy technologies are optimized and applied widely around the world, substantial economic and environmental benefits will be generated.

Renewable energy and energy-efficient technologies can also play crucial roles in the creation of employment and economic growth. These technologies are more labor-intensive than conventional technologies for the same energy output – employing a mixture of local and decentralized workers. For an investment in renewable energy and energy-efficient technologies of US$ 1 million over 10 years (Pachauri, 2009), both wind energy and solar PV generate 5.7 person-years of employment each, while the fossil energy industry only generates 4.0 person-years. Although the data are not very different, the rationale is that renewable energy and energy-efficiency jobs are created in local communities and rural areas while fossil energy jobs are usually created in urban areas or in mines far away from their original homes. Renewable and energy-efficiency jobs will help keep the rural population in the local communities.

2.2   THE ROLE OF THE GLOBAL ENVIRONMENT FACILITY

To overcome the technological, market, institutional and regulatory barriers, and to promote wider adoption of renewable energy and energy-efficient technologies around the world, the international community created the Global Environment Facility (GEF) in 1991. The GEF unites 183 countries in partnership with national governments, international institutions, civil society organizations, and the private sector to address global environmental issues and to support national sustainable development initiatives. The GEF operates as a financial mechanism for international cooperation aiming to provide additional grant and concessional funding to meet the agreed incremental costs of measures to achieve agreed global environmental benefits in seven focal areas. Clean energy technologies and policies are a primary GEF focus.

Developing countries and countries with economies in transition are keen to access GEF resources to address technical, policy, and market barriers and to boost sustainable energy development for economic, environmental, security, and public health reasons. In agricultural and renewable energy areas, the GEF has provided the following types of services to participating countries: (i) grants to overcome project incremental costs; (ii) support for policy development, capacity-building, and institutional development; and (iii) assistance in accomplishing technology transfer. As a result, GEF investments in renewable energy and energy-efficiency technologies in rural areas have had great impact, not only on climate change mitigation, but also on rural electrification, agriculture development, and poverty eradication.

2.2.1   *The GEF's renewable energy and energy-efficiency strategies*

In managing these investments, the GEF pursues strategies to boost renewable energy demonstration and deployment. These include: (i) supporting the development of functional markets, and (ii) providing finance for critical opportunities.

Support for the development of functional markets includes efforts to enable appropriate policies and regulatory frameworks, more effective standards and certification, greater levels of awareness and know-how, and greater capacities of institutions, businesses, and workers. In supporting these activities, the GEF views national policies as a critical element in creating favorable conditions for renewable energy development. For example, many GEF renewable energy projects have contributed directly to the development of such policies, for example, by drafting or revising national strategies, or by developing roadmaps and national action plans for renewable energy development.

The GEF has also been successful in developing standards, and testing and identifying renewable energy technologies (Dixon *et al.*, 2011). This has proven to be an important success factor in sustaining market development as more effective standards and testing can significantly improve quality, reliability, and consumer acceptance. In parallel, most GEF renewable energy projects have supported awareness-raising activities such as distribution of promotional material and production of audiovisual tools that help build community trust in renewable energy technologies. The GEF also helps recipient countries build technical and institutional capacity by organizing workshops and training opportunities for government officials, engineers and technicians, and business staff.

The availability of affordable finance remains a key barrier to renewable energy investments, especially in developing countries and rural areas. GEF projects focus on understanding the nature of financial barriers so that effective barrier removal can be targeted – whether at financial intermediaries (e.g., banks, development finance institutions, and microlenders), equipment suppliers, dealers, energy service companies, end-users, or a combination of the above.

One of the GEF's most effective strategies is to test innovative financing approaches in order to increase access to local funding. Such approaches differ according to the status of the local financial sector, the type of financial barriers to be overcome, and the business model employed. For example, in the case of distributed small-scale power generation, sales-based business models may require financing for suppliers and dealers. However, the main need is microfinance for consumers. Since 1991, the GEF has sought to:

- Provide grants and contingent financing for project preparation and investment. The GEF offers contingent loans and grants to cover front project development capital costs. A contingent loan has an interest rate and payment schedule similar to a traditional loan, but the loan can be forgiven if certain conditions are met.
- Mitigate technology-specific project risks. For example, the highest risk during geothermal plant development occurs when the first well is drilled, even if there has been

Table 2.1.   GEF/IFC China Utility-Based Energy Efficiency Finance Program.

- The GEF/IFC/World Bank has won the best investment project award for its China Utility-Based Energy Efficiency Finance Program (CHUEE), which has generated loans worth US$ 783 million through its partner banks and cut more than 18 million tons of GHG emissions a year.
- In 2006, the GEF invested US$ 16.9 million as a seed capital to de-risk energy efficiency and renewable energy project investments in this project. The CHUEE program was also endorsed by China's Ministry of Finance, the Ministry of Employment and Economy of Finland, and the Norwegian Agency for Development Cooperation.
- Local commercial banks participating in the GEF/IFC's CHUEE program have provided loans to 178 energy efficiency and renewable energy projects. The latest phase of the program, targeted at small and medium enterprises, is expected to help finance energy efficiency and renewable energy projects of around 175 companies in China. In addition to providing risk-sharing facilities, the program offers expert advice to ensure that lending to climate-friendly projects is profitable for businesses.
- The Environment, Ethical and Social Global Investment Awards, organized by Investment Week, a British weekly magazine, were awarded on November 26, 2013 in London, to give recognition to companies that display high environmental, ethical and corporate governance standards, and engage in social-impact investing. GEF's CHUEE program, stood out in a hotly contested category to win the coveted award.
- "The best investment project award recognizes the importance of finance in achieving sustainable projects on a global scale," said Deborah Benn, chair of the awards' judging panel. "Congratulations go to (GEF/) IFC's CHUEE program, which the judges felt provided an excellent sustainable-energy-financing model to help mitigate climate change in one of the fastest-growing economies in the world."

Source: IFC (2014).

successful surface-based geophysical exploration. The GEF projects in Africa, the Caribbean, and Eastern Europe are developing risk mitigation facilities to insure investors against the geological and technical risks during development of such projects. Table 2.1 shows a GEF example in promoting energy efficiency in China.

- Initiate microfinance schemes. Extending financing to private consumers – such as households and small enterprises – for the purchase of renewable energy equipment is often considered a low priority by Financial Institutions (FIs), especially in the developing world. The GEF has supported existing FIs or new microfinance institutions to provide lending to such recipients for the purchase of, for example, solar home energy systems in Bangladesh and Uganda.

2.2.2   *The GEF's renewable energy portfolio*

From October 1991 to June 2012, the renewable energy portion of the GEF's portfolio amounted to approximately US$ 1.2 billion (approximately 30% of the total portfolio), with an average of approximately US$ 4.9 million per project. These investments included 186 stand-alone renewable energy projects, as well as 69 other projects that had renewable energy components. The projects were undertaken in 160 countries, including those in developing countries and countries with economies in transition. Figure 2.2 is a map showing the location of GEF renewable energy projects around the world. For these projects GEF funding has been supplemented with approximately US$ 9.6 billion of co-financing.

Figure 2.3 shows how GEF funding for renewable energy projects has changed over time. As shown, funding for the renewable energy portfolio increased from the pilot phase (1991–1994) through GEF-1 and GEF-2 to GEF-3 replenishment periods. The percentage of the GEF climate change portfolio for renewable energy projects peaked in GEF-2 at over 50%, but has fallen steadily since. This is due to a variety of factors, including the expansion of other climate change portfolios, especially energy efficiency; the high amount of funding directed to renewable energy projects – such as concentrating solar power (CSP) projects – under GEF-3; and the decision not to pursue the strategic objective of promoting off-grid renewable energy technologies in GEF-4 (2006–2010).

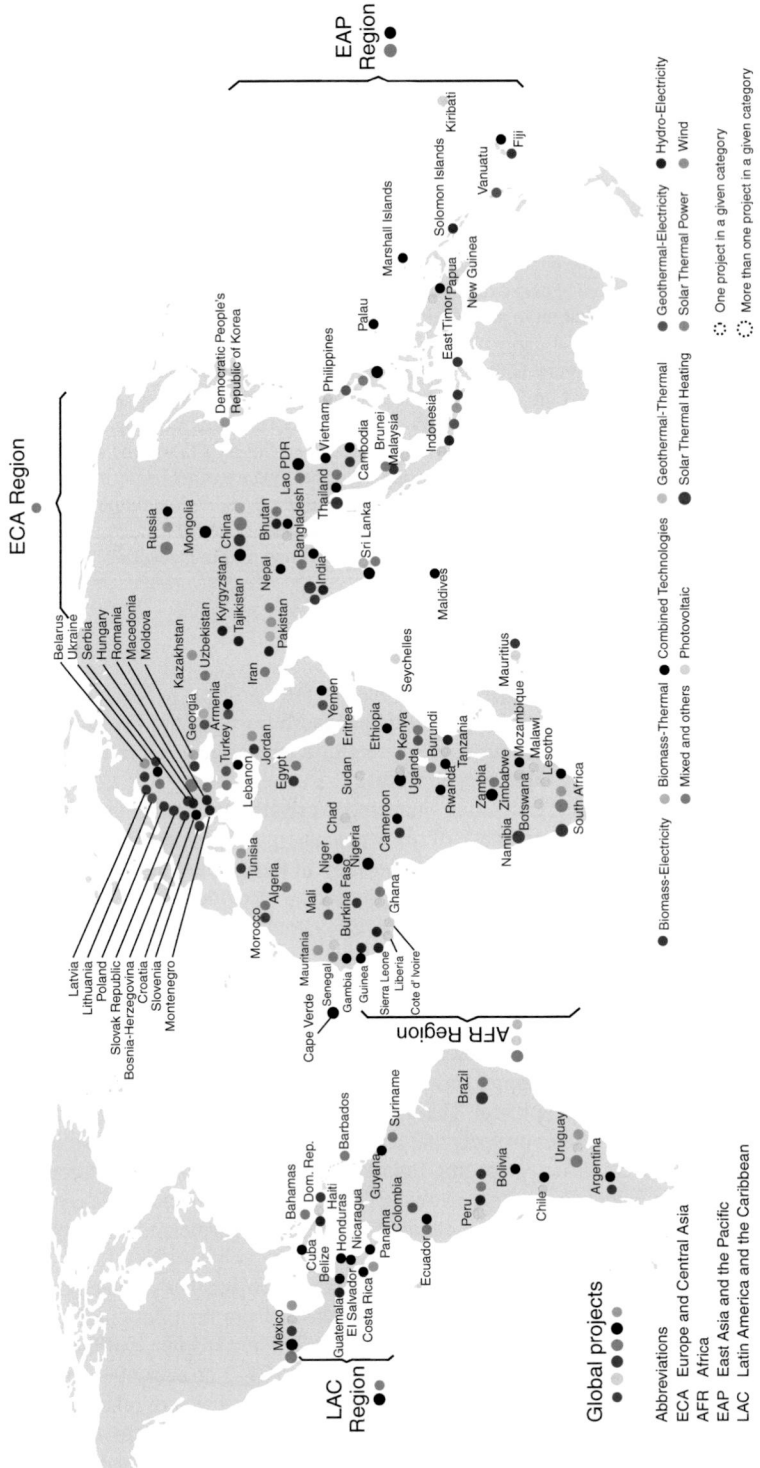

Figure 2.2.   GEF renewable energy projects around the world.

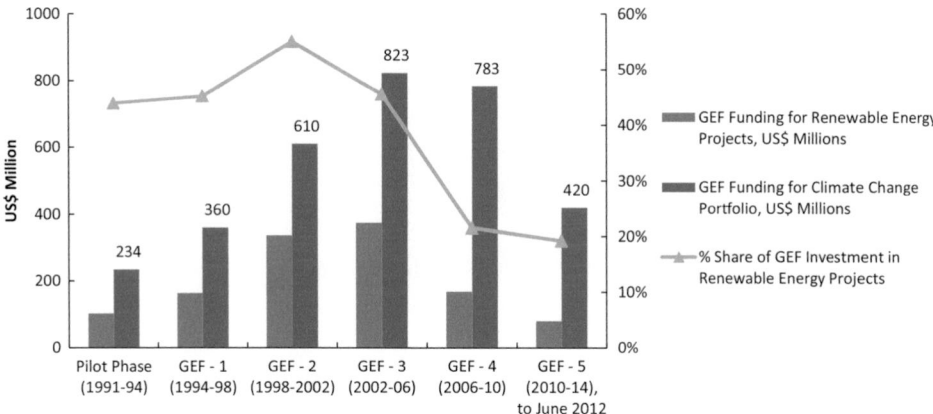

Figure 2.3.   GEF renewable energy investment by GEF phases.

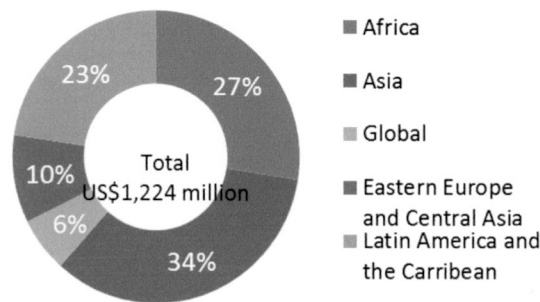

Figure 2.4.   Regional distribution of the GEF renewable energy portfolio.

Figure 2.4 shows the regional distribution of the GEF's renewable energy portfolio. As shown, most of the renewable energy investments take place in Asia, Africa, Latin America, and Eastern Europe and Central Asia. A relatively smaller share of the GEF's renewable energy portfolio has been invested in Eastern Europe and the Caribbean.

Figure 2.5 presents the shares of GEF investments in renewable energy technologies. As shown, GEF funding has been invested in a variety of different technologies. The variety is broad because it is the GEF's role to address barriers to investment, and to catalyze and transform energy markets generally, not to single out individual renewable energy technologies for support. In fact, GEF renewable energy investments have been targeted towards those technologies that are best suited and most cost-effective as determined by the participating countries, based on local renewable energy sources, available energy-efficient technologies, capacities, financial market, climate, energy market, and policy conditions.

Figure 2.6 shows the amount of renewable energy-generating capacity in megawatts [MW] that has been installed around the world by the type of renewable energy technology. From October 1991 to July 2012, through direct investments alone, GEF projects have contributed to the installation of approximately 6.2 GW of renewable energy-generating capacity. This capacity consists of solar energy, wind power, geothermal energy, small hydropower, and biomass for either heat or power generation. Some projects involved the installation of more than one renewable energy technology, and therefore are classified as combined technologies as shown in Figure 2.6.

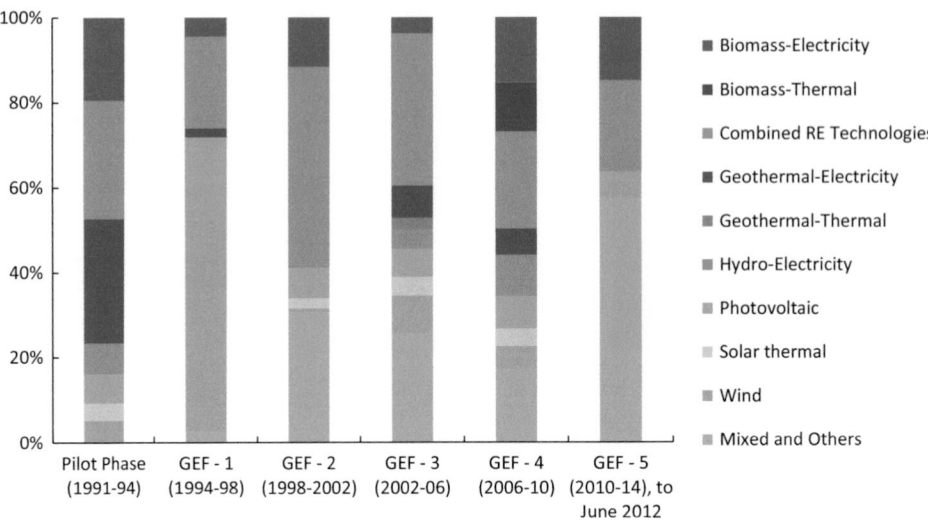

Figure 2.5.   Percentages of GEF investments in renewable energy technologies.

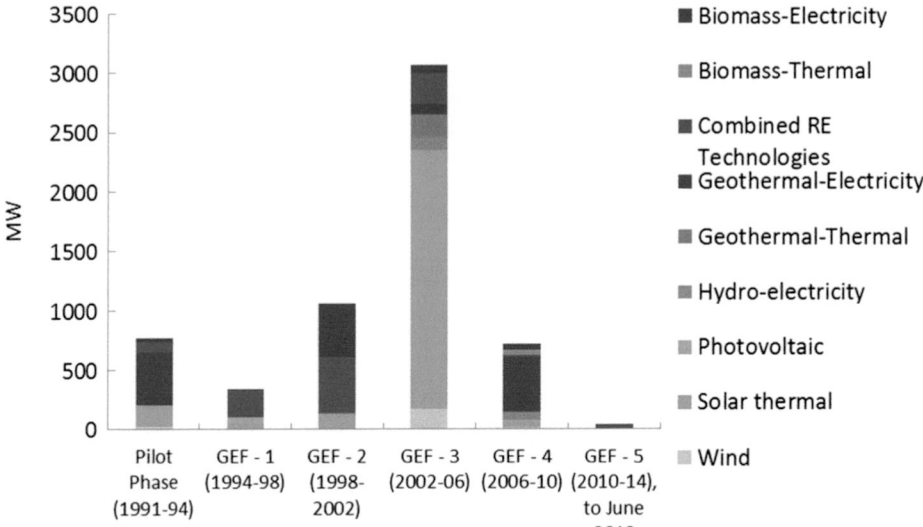

*These MWs (includes both thermal and electric energy) correspond to 186 standalone RE projects (as of June 2012).

Figure 2.6.   Installed renewable generating capacity by type of renewable energy technology.

2.2.3   *The GEF's renewable energy portfolio and the modernization of agriculture*

The GEF renewable energy portfolio has had an impact on rural development and the modernization of agriculture in developing countries and countries with economies in transition. This is discussed in the following sections in a series of case studies for the following technologies:

- Biomass energy.
- Combined renewable energy technologies.
- Geothermal energy.
- Small hydropower.

- Off-grid solar PV.
- On-grid solar PV.
- Solar thermal heating.
- Solar thermal power.
- Wind power.

2.3 BIOMASS ENERGY

Bioenergy projects are of interest to the GEF because biomass represents an energy source with zero net carbon emissions if produced sustainably. The GEF-funded biomass projects include power production (combustion, gasification, co-generation, and waste-to-energy) from forestry and agricultural wastes, including sugarcane bagasse and waste, husks, palm oil residues, wood chips, sawmill waste, municipal waste, and production of biofuels. Many of these projects focus on technology demonstrations, but they also include activities that seek to address enabling policies, availability of finance, business infrastructure development, awareness-raising, capacity development, and technology transfer.

In more than 40 countries, the GEF has funded 62 projects with 722 MW electric and 212 MW thermal with US$ 275 million that leveraged US$ 2.1 billion of co-financing. Technology itself is frequently no longer the barrier and can be obtained on a commercial basis. Rather, the challenge is demonstration of the commercial and institutional frameworks under which the technologies can be profitably deployed and replicated.

2.3.1 *Case study: Thailand – biomass co-generation*

From 2001–09, the GEF provided US$ 6.8 million and mobilized US$ 102 million for the project "Removal of Barriers to Biomass Power Generation and Cogeneration" in Thailand. The objectives were to (i) build capacity to provide information and services for potential biomass power project investors; (ii) improve the regulatory framework to provide financial incentives for biomass co-generation and power projects; (iii) increase access to commercial financing for biomass co-generation and power projects; and (iv) facilitate the implementation of two initial biomass power pilot plants by supporting commercial guarantees to reduce the technical risks associated with deployment of this new technology.

The outcomes of the project included two pilot biomass power plants with a total electric capacity of 32 MW, which serve as valuable demonstration plants for rural and agricultural communities. Significant project impacts included influencing government policies, such as adding feed-in tariffs to make biomass power generation more commercially viable. A Biomass One-Stop Clearing House was created and responded well to biomass investors and public interest. The project facilitated follow-on investments including the installation of 398 MW of electricity capacity that generates over 358 GWh of electricity annually from biomass power plants and avoids the emission of 194,722 metric tons of carbon dioxide equivalent (CO_2-eq) per year (GEF PMIS, 2014).

2.3.2 *Case study: India – biomass gasification*

Between the years 2001 and 2010, the GEF provided a US$ 4.0 million grant for the project of "Biomass Energy for Rural India". The Indian government put forward US$ 4.6 million as co-financing for the project. The objectives of the project were to (i) demonstrate the technical feasibility and financial viability of bioenergy technologies, including biomass gasification for power generation, on a significant scale; (ii) build capacity and develop mechanisms for project implementation, management, and monitoring; (iii) develop financial, institutional, and market strategies to overcome barriers to large-scale replication of the bioenergy package for decentralized applications; and (iv) disseminate bioenergy technology and relevant information on a large scale in 24 villages in Karnataka's Tumkur district.

The outcomes of the project included significant forest growth in the form of energy plantations (2965 acres), forest regeneration (2100 acres), and tree-based farming (about 2471 acres) by villagers. The wood was used to generate electricity in locally manufactured gasifiers. The power generated was sold to the regional electrical distribution company to supply the local population. The project also resulted in 171 families replacing fuel wood with biogas, reducing GHG emissions by 256 metric tons annually (GEF PMIS, 2014).

2.3.3   *Case study: Latvia – biomass combustion*

In the early 2000s, the GEF/UNDP developed a project entitled "Economic and Cost-Effective Use of Wood Waste for Municipal Heating Systems". The GEF invested US$ 0.75 million and mobilized US$ 2.73 million as co-financing for the project. The objectives of the project were to (i) promote the use of wood waste by removing barriers to the replacement of imported heavy fuel oil (mazut) with local sustainably-produced wood waste in municipal heating systems; (ii) promote the development and implementation of an economical commercially-run municipal heating system, providing generation, transmission, and distribution in the municipality of Ludza; and (iii) help remove or reduce technical, legislative, institutional, organizational, economic, information-related, and financial barriers pertinent to the replication of a pilot project in the municipality.

The outcomes of the project included 11,200 metric tons of CO_2-eq emissions avoided annually in Ludza since 2005, equating to about 80% of the emissions from using heating oil. The project and the financial scheme developed through it have encouraged more than 12 other municipalities to use forest wastes in their district heating networks, resulting in over 100,000 metric tons of CO_2-eq avoided annually.

2.4   COMBINED RENEWABLE ENERGY TECHNOLOGIES

In more than 50 countries, the GEF has financed 70 projects that involve more than one type of renewable energy technologies. The investments in these combined renewable technology projects are often more complicated than investments in single-technology projects because there could be needs to provide integrated systems for taking advantage of opportunities to optimize outputs or other potential synergies. Total GEF investments in these combined renewable energy technologies amounted to more than US$ 444 million and these investments mobilized US$ 3.1 billion of co-financing by the end of 2012 (GEF PMIS, 2014).

2.4.1   *Case study: India – combined renewable energy technologies*

In 2008, the GEF financed US$ 0.95 million and leveraged US$ 1.1 million of co-financing for "Energy Conservation in Small Sector Tea Processing Units in South India". In South India, tea is grown on large estates as well as small farms. Large tea estates have captive tea factories for processing while small farms sell tea leaves for processing to 'bought-leaf' factories. In total, there were approximately 266 tea factories in South India in 2010. All of these tea factories relied heavily on biomass to meet their thermal energy requirements for tea drying. Tea processing, based on fossil fuels and inefficient technologies, is energy intensive with energy costs contributing around 30–40% of the total production cost, second only to labor costs in tea-making.

The objective of the project was to remove barriers and to develop replicable strategies for energy conservation and energy efficiency interventions – including renewable energy – in the tea-processing industry in South India (UNDP, 2012). The outcomes include the following:

- Each of the 266 tea factories in South India has been made aware of energy conservation activities through awareness programs targeted under the project. The project developed a highly informative and innovative tool – Energy Score Card – for the factories to assess their performance against baseline energy use and renewable energy development and to communicate their progress with other interested groups.

- In total the project introduced 17 energy efficiency and seven renewable energy interventions to reduce energy use (fuels, electricity) in various sections of tea processing.
- The project target of introducing energy conservation measures to 30 tea factories was also achieved, as by June 2011, energy conservation measures with project support were demonstrated in 90 units. Thanks to the increased awareness, another 24 factories have adopted energy reforms without project support, bringing to the total 114 factories adopting energy conservation and energy efficiency (EC/EE) measures.
- Total private investments in EC/EE measures by these 114 factories aggregate to US$ 3.0 million with US$ 2.54 million derived from the subsidy of the Tea Board, an administrative commission of the tea industry in South India.
- The estimated cumulative direct CO_2-eq savings are 277,255 tons of CO_2-eq vs. target of 30 factories with cumulative potential savings of 56,925 tons direct CO_2-eq.
- The project is successful in strengthening the value chain of suppliers that are energy-efficiency and renewable energy driven. By customizing equipment to the capacity needs of tea sector, suppliers have shifted from being merely "salesmen of equipment" to value-adding energy service providers (GEF PMIS, 2014).

2.5   GEOTHERMAL ENERGY

The GEF has supported 11 projects to help 31 countries exploit their geothermal energy potential in the past 21 years. The projects were financed with US$ 90.9 million of GEF resources, which mobilized US$ 1.70 billion of co-financing. The energy extracted was more than 1059 MW of electricity plus 117 MW of thermal energy.

This experience has shown that, in addition to barriers of access to the grid faced by renewable energy generators, an additional – and especially difficult – barrier is the cost of confirming the presence and location of exploitable geothermal resources. Traditionally, each site is confirmed exploitable by drilling – at a cost of up to several million dollars. To deal with this barrier, the GEF has established several contingent-funding mechanisms to reimburse the costs of drilling non-productive wells. The mechanisms are similar to that used in the CHUEE project in China (see Table 2.1).

Another way of reducing the risk of drilling non-productive wells may be found by using joint geophysical imaging for geothermal reservoir assessment, as was done in a GEF project in East Africa. Advanced geophysical imaging techniques have been used to locate commercially exploitable geothermal power. Results to date indicate that wells found using this approach – when combined with directional drilling – can achieve 4–6 MW per well as opposed to the previous 2 MW per well.

2.5.1   *Case study: The Philippines – geothermal power*

From 1995–2000, the GEF and the World Bank supplied US$ 30 million and US$ 1.3 billion jointly to finance "Republic of the Philippines Leyte-Luzon Geothermal Project." The objectives were to (i) meet the rapidly increasing demand for power in Luzon using indigenous and environmentally superior geothermal energy; (ii) strengthen the energy sector by implementing institutional, planning, and financial policies recommended by the Energy Sector Plan of the government of the Philippines; (iii) support large ongoing private sector participation in power generation – and facilitate this by extending the national grid; (iv) strengthen the National Power Corporation's (NPC) capabilities in environmental and social-impact analysis; (v) introduce enhanced co-financing operations in the Philippines; and (vi) ensure the financial viability of NPC and the Philippines National Oil Corporation by undertaking a long-term investment program.

The outcomes of the project included new geothermal capacity of 385 MW and 59 producer and injector wells drilled. The combined system surpassed the required annual energy output specified under the agreement with NPC, with the power plants operating within the plant factor commitment in the build-operate transfer contract. Further, the project significantly mitigated GHG emissions, as an alternative coal-fired plant would have meant incremental CO_2-eq of about 2.2 million tons per year (GEF PMIS, 2014).

2.6   SMALL HYDROPOWER

Small hydropower is a mature technology, but it has not been deployed on a large scale anywhere around the world. The GEF has supported the technology in 20 countries and has identified numerous barriers to its adoption, including a lack of information about the technology and underlying resources, unfavorable institutional frameworks, regulatory obstacles, and absence or inadequacy of financing.

In general, small hydropower projects are gaining interest in rural and agricultural areas as pilot demonstration projects have progressed in some places to full-scale policy options for rural villages. Hydroelectric resources often require joint community management, participation, leadership, teamwork, and coordination. Take the project in Sri Lanka as an example; small hydropower installations were built, owned, and operated by the communities through electric cooperatives that were specifically formed.

From October 1991 to July 2012, small hydropower was supported by the GEF through 61 small hydro projects in more than 50 countries, with US$ 230.5 million of GEF funding and US$ 1.84 billion of co-financing. Among other outcomes, these projects led to investments in 414 MW of installed capacity, most of which led to decentralized electrification in rural areas and farming communities (GEF PMIS, 2014).

In West Africa, for example, several countries implemented projects to develop a market environment for improved access to small hydropower-based energy services. The essential elements of the market-based approach involved developing a critical mass of skilled and knowledgeable technicians, increasing awareness of appropriate technologies and best practices, and generating access to innovative financial mechanisms. In fact, the West Africa projects aimed to establish two to three pilot demonstration sites in isolated off-grid communities and implement them using a learning-by-doing approach in order to build local capacity.

2.6.1   *Case study: Indonesia – small hydropower*

In 2007, the GEF invested US$ 2.1 million and mobilized US$ 18.5 million from the UNDP and other partners in co-finance for the "Integrated Micro-Hydro Development and Application Program, Part I." The objectives were to remove key market, policy, technical, and financial barriers to micro-hydro development and utilization, and to reduce GHG emissions from fossil fuel-based power generation.

The project had four main outcomes: (i) enhanced private sector interest and involvement in capacity-building within the micro-hydro business community; (ii) increased number of community-based micro-hydro projects as a result of institutional capacity-building; (iii) improved availability of local knowledge of the technology and its applications; and (iv) increased implementation of micro-hydro projects for electricity and production. From 2007–10, this project created approximately 40 community-based micro-hydro sub-projects for productive use and mitigated 60,800 tons of CO_2-eq per annum (GEF PMIS, 2014).

2.7   OFF-GRID SOLAR PHOTOVOLTAIC

Between October 1991 and July 2012, the GEF helped deploy solar energy technologies to many people in rural areas and farming communities that lacked access to electricity. Since connecting rural and remote areas to the power grid were expensive, off-grid solar was regarded as a highly beneficial solution to rural needs for electricity. In response to these needs, the GEF-funded over 20 stand-alone and around 45 mixed (i.e., including additional GEF program areas, such as energy efficiency and transport) technology transfer projects in 68 countries that now provide expanded access to electricity through the use of solar home systems and off-grid PV electricity.

The GEF has supported these projects by investing US$ 379 million and mobilizing approximately US$ 2.9 billion of co-financing. These projects have led to the installation of approximately 125 MW of peak power capacity. GEF projects have accelerated the growth of local PV industries in several countries, improving the quality of production and reducing costs, thereby expanding the market for solar home systems and other off-grid PV applications.

2.7.1 *Case study: India – off-grid photovoltaic*

Between 1993–2002, the GEF provided US$ 26 million and the World Bank released a loan of US$ 424 million to "Alternate Energy" in India to promote off-grid PV applications. The objectives of the project were to (i) promote commercialization of renewable energy technologies by strengthening the Indian Renewable Energy Development Agency's (IREDA) capacity to promote and finance entrepreneurial investments in alternate energy; (ii) create marketing and financing mechanisms for the sale and delivery of alternate energy systems based on cost-recovery principles; (iii) strengthen the institutional framework for encouraging private sector investments in nonconventional power generation; and (iv) promote environmentally sound investments to reduce the energy sector's dependence on fossil fuels.

The outcomes of the project included 2.1 MW of peak off-grid PV power generation capacity. Solar PV products from the project ranged from 5 W solar lanterns, 900 W PV irrigation pumps, 500–2500 W solar power packs, and 25 kW village power schemes to a 200 kW grid-tied system. The IREDA financed an additional 4 MW of PV irrigation pumps with assistance from the Ministry of Nonconventional Energy Sources. Evidence of positive development impacts from PV use among poorer consumers are emerging, including a five-fold income increase among farmers using PV pumps, a 50% increase in net income among some traders using solar instead of kerosene lighting; income increases of 15–30% in some rural households because home industry output ratcheted up, and longer study hours with better lighting conditions for children (GEF PMIS, 2014).

2.8 ON-GRID SOLAR PHOTOVOLTAIC

The GEF has supported the market transfer and installation of grid-connected PV systems in 31 projects. Approximately 52 MW of peak electric capacity has been installed, mostly in combination with small wind and hydro, and often to support mini-grids. The GEF invested US$ 172.7 million, which mobilized co-financing of approximately US$ 1.9 billion.

2.8.1 *Case study: Philippines – on-grid photovoltaic*

From 2003–04, the GEF and the IFC financed US$ 4 million and US$ 1.8 million, respectively, to "CEPALCO Distributed Generation PV Power Plant." The objectives of the project were to (i) act as a demonstration plant for grid-connected applications of PV power plants in the developing world; and (ii) demonstrate the principle of conjunctive PV-hydro peak power generation (GEF PMIS, 2014).

The outcomes include the construction of a 1 MW (6500 solar panels on 2 hectares of land) PV power plant that was integrated into the 80 MW distribution network of CEPALCO, a private utility on the Philippine island of Mindanao. The PV system operates in conjunction with a 7 MW hydroelectric plant with dynamic load control, enabling the joint PV-hydro resource to reduce distribution-level and system-level demand, effectively providing reliable electric services for consumers in that area. The PV plant helped postpone the need for additional substation installations in the distribution system for up to three years, reducing the need for CEPALCO to purchase additional thermal-plant-based power and lowering its GHG emissions by 1200 tons per year.

Significantly, the plant provided the first full-scale demonstration of the environmental and economic benefits of the joint development of PV-hydro resources and represented the first example of grid-connected PV in developing countries. This project also marked a milestone towards solving storage issues faced by solar and wind energy projects – the energy production varied depending on the availability of the wind and sun. This arrangement established useful models for using local water resources as a cost-effective way to store solar energy. (Further information is available at: http://www.cepalco.com.ph/solar.php).

2.9   SOLAR THERMAL HEATING

The GEF has supported 14 national and multinational solar thermal projects in 30 countries with financing of US\$ 186.1 million. GEF investments have mobilized US\$ 393.1 million of co-financing and have led to the installation of approximately 2.5 GW of thermal energy capacity.

Although solar water heating installations are normally trouble-free, in rural areas cost-effectiveness can be challenging when inexperienced companies and technicians are involved, and the quality of fittings, solar collectors, and installation techniques fall short of best practice. Inexpensive but low quality materials, poor workmanship, and shoddy installation practices are known to have resulted in non-functional units and abandoned systems. The GEF's experience has shown that well-trained technicians and quality assurance practices are critical to the successful dissemination of this technology.

2.9.1   *Case study: Tunisia – solar water heating*

From 1994–2004, with co-financing of US\$ 16.9 million from the World Bank and the Belgian government, the GEF invested US\$ 4 million to "Solar Water Heating in Tunisia." The objective of the project was to help Tunisia substitute solar energy for fossil fuels in public and commercial institutions and to demonstrate solar water heating's potential to reduce global warming.

The outcomes of the project included: (i) contributing 35% of the cost of investment before taxes in solar water heaters (including installation); and (ii) assisting incentivized users to invest in solar water heaters rather than in conventional water heating systems. During project implementation, solar water heater installations tripled as approximately 80,000 m^2 (56 MW) of solar water heater panels were installed, of which 51,060 m^2 (35 MW) were installed by the project. Greenhouse gas emission reductions that can be attributed to the GEF project amount to about 25,000 tons of CO_2-eq annually (GEF PMIS, 2014). Quality control and system maintenance ensured efficient and effective ongoing operation.

2.10   SOLAR THERMAL POWER

Most of the GEF's investment in solar thermal power has been for CSP systems. The GEF has supported five countries in harnessing the potential of solar thermal power. Collectively, the GEF has invested approximately US\$ 152.5 million, while mobilizing approximately US\$ 1.3 billion in co-financing. The projects are leading to the installation of 75 MW of solar thermal electric-generating capacity. The GEF, in partnership with the World Bank, developed a portfolio of three CSP demonstration plants in Mexico, Morocco, and Egypt. The projects built solar fields, typically approximately 30 MW, as part of hybrid gas-turbine plants. Successful hybridization of gas turbines and solar thermal power plants enables such projects to dispatch power when needed, making them more financially attractive and valuable for grid support (Yang *et al.*, 2013).

However, these hybrid power plants require sophisticated engineering and are often slow to develop and bring to completion. One lesson from these experiences is that it is difficult for developing countries to adopt technologies that are not fully commercialized, and failure to achieve market viability in developed countries damages the technology's credibility elsewhere. In the case of the CSP plants, construction costs increased as the projects progressed. Host countries were burdened with both additional costs and the risk that the projects might not produce the rated power on a firm basis. In fact, in two cases, the additional costs exceeded the GEF's investment. In these cases the countries had to provide significant cash subsidies to enable the plants to move forward.

2.10.1   *Case study: Egypt – solar thermal power*

From 2007–11, the GEF and the World Bank financed US$ 49.8 million and US$ 277 million, respectively, to "Solar Thermal Hybrid Project" in Egypt. The objectives of the project were to (i) provide modern infrastructure by efficient private suppliers and operators; (ii) increase energy-generating capacity from renewable resources that can reduce local and regional pollution; (iii) increase capacity to develop large-scale innovative renewable energy projects; (iv) position Egypt as a source of expertise and equipment in future solar thermal power projects internationally; and (v) develop renewable energy supplies with private investments (GEF PMIS, 2014).

The outcomes of the project included: (i) demonstration of viability of hybrid solar thermal power generation in Egypt; (ii) accelerated market penetration of large-scale back-up power generation technologies; and (iii) reduction of GHG emissions. The incremental physical benefits of the project included the following: (i) increased renewable electricity production by 35.1 GWh year^{-1}, and (ii) reduced GHG emissions by 15,410 tons of CO_2-eq (World Bank, 2012).

2.10.2   *Case study: Morocco – concentrating solar power*

In 1999, the GEF financed US$ 44 million and mobilized US$ 70.4 million from the African Development Bank for a CSP plant in Morocco. The objectives of the project were to (i) identify key issues in financing CSP plants in Morocco; (ii) demonstrate the GEF's investment in CSP plants in Morocco and its impacts on climate change mitigation; and (iii) present GEF's projects in Morocco by way of demonstration (Yang *et al.*, 2013).

At the time the project was initiated, Morocco relied heavily on energy imports with 95% of its energy needs derived from fossil fuel sources, and 44% of the population living in rural areas and farming communities with little or no access to electricity. In addition, GHG emissions from fossil fuel combustion in Morocco was growing at an average rate of approximately 6.3% annually. In November 2009, the government of Morocco initiated the Moroccan Solar Plan aiming at achieving a 42% renewable power target by 2020, which included 2 GW of solar power capacity, including CSP.

The outcome of the project included: (i) improved air quality and/or reduction in fuel costs due to solar power capacity expansion to rural households and agricultural production; (ii) providing momentum for the development of the global solar market, and revealing the barriers in earlier stages of CSP projects; (iii) building capacity for the country for CSP development; (iv) transferring CSP technology to the country; (v) developing a policy framework for CSP promotion in the country; and (vi) building up a financing mechanism in the country.

2.11   WIND POWER

Wind power is the fastest-growing renewable energy resource in the world. By the end of 2011, worldwide capacity reached 197 GW with 3.6 GW added in 2010 alone. Wind power showed a growth rate of 23.6% in 2010, the lowest growth since 2004 and the second lowest growth of the past decade. All wind turbines installed by the end of 2010 worldwide can generate

430 Terawatt-hours (TWh) per annum, more than the total electricity demand of the United Kingdom, the sixth largest economy of the world, and equaling 2.5% of global electricity consumption. The wind sector in 2010 had revenue of over US$ 50 billion and employed 670,000 persons worldwide (WWEA – World Wind Energy Association, 2011).

As of July 2012, the wind power industry faces a large number of technical, economic, financial, institutional, market, and other barriers. To overcome these barriers, many countries have employed policy instruments, including capital subsidies, tax incentives, tradable energy certificates, feed-in tariffs, grid access guarantees, and mandatory standards.

The GEF has financed 54 wind power projects in more than 34 countries. GEF funds and co-financing in these projects are US$ 352 million and US$ 2.53 billion, respectively. These investments have led to the installation of almost 972 MW of wind power capacity (GEF PMIS, 2014). Experience shows that resource availability and familiarity with the technologies are key considerations. However, the most significant barriers to successful growth of the wind market are the ones that interfere with wind power getting access to the grid, including the need for new transmission and distribution facilities, and the siting, permitting, and cost-recovery issues that are involved with new construction projects.

Worldwide experience demonstrates several successful regulatory solutions to this problem of access, including the creation of renewable portfolio standards and guaranteed renewable feed-in tariffs. The GEF has helped countries understand and adopt appropriate policies and regulations.

2.11.1   *Case study: China – wind power*

During the period 2000–11, the GEF invested US$ 40.2 million and the World Bank invested US$ 188.6 million for "Renewable Energy Scale Up Program (CRESP), Phase 1" in China. The objective was to scale up China's renewable energy market by developing the mandated market share policy of China and engaging other interventions and commercial renewable energy suppliers in the market.

The outcomes of the project included: (i) a successful policy dialog between the World Bank and the Chinese government, and (ii) the introduction of technologies, policies, and practices drawn from experiences from around the world to assist the Chinese government in developing "best in class" renewable energy policy frameworks. As a result, China passed a renewable energy law to require mandatory purchase of renewable energy and to allow the incremental costs to be shared nationwide. The policy dialog was complemented by three World Bank scaling-up projects: (i) a World Bank investment (US$ 173 million) in 200 MW wind farms; (ii) a 25 MW biomass power plant; and (iii) small hydro projects (GEF PMIS, 2014).

These projects were among the first set of large-scale wind and biomass power plants in China. The World Bank's carbon financing program played a key role in improving the financial viability of the Inner Mongolia wind farm, which was not financially viable at electricity rates of US$ 0.06 kWh^{-1} that had been set by the government. In addition, the CRESP provided cost-shared research and development (R&D) to domestic wind manufacturers supporting joint design with international design institutes to transfer international wind turbine technologies to China. The domestic wind manufacturing industry was boosted by the "market-pull" approach guaranteed by more favorable electric rate policies; and the government's requirement of 70% local content, as well as the "technology-push" approach were a result of the cost-shared R&D activities (World Bank, 2010).

2.11.2   *Case study: Mexico – wind power*

In Mexico, the GEF has financed three projects with a total of US$ 35.24 million for wind power development with the UNDP, the World Bank, and the Inter-American Development Bank from 2004–15. The total of mobilized co-financing has reached US$ 288.2 million.

The objectives of these projects were to (i) create an enabling environment for the development and use of distributed wind power that is connected to the grid by removing the existing regulatory and technical barriers; (ii) assist Mexico in stimulating and accelerating the commercialization of renewable energy applications and markets – particularly at the grid-connected level; and (iii) assist Mexico in developing initial experience in commercially-based grid-connected renewable energy applications by supporting the construction of an approximately 101 MW wind farm financed by independent power producers.

The outcomes of the projects included: (i) development of an action plan for removing barriers to the full-scale implementation of wind power in Mexico; (ii) creation of a Regional Wind Technology Centre (Centro Regional de Tecnología Eólica), which offered support to interested wind turbine manufacturers and provides training to local technicians; (iii) development of the 83.5 MW La Venta II wind power project in Oaxaca that became operational in January 2007; (iv) effective regulatory and institutional frameworks to enable distributed wind power development; (v) project development capacities among national stakeholders, including improved access to finance to allow widespread replication of distributed wind power; (vi) a successful independent power producer tender resulting in construction and operation of a 101 MW wind farm; and (vii) institutional capacity to issue subsequent tenders for additional wind farms and other renewable energy resource development at a higher reference price and/or lower incentive support level (World Bank, 2006).

2.12   SUMMARY AND CONCLUSIONS

Providing sustainable energy to poor subsistence farmers in rural area has been a global challenge for many decades. In 2012, approximately 2.8 billion people – 34% of the world population – relied on traditional fuels such as wood, charcoal and animal and crop waste to cook and heat their homes; and three-quarters of the above population congregated in just 20 countries in Asia and Africa. Lack of information, scarcity of capital investment, insufficiency of policy and institutional supports, shortage of capacity, and inadequacy of technologies are major barriers to the sustainable energy supplies needed to serve the needs of these people.

Since its inception in 1998, the GEF has provided assistance in rural economic development and electrification and poverty reduction from investments in renewable energy projects in the context of mitigating climate change. From October 1991 to June 2012, the GEF invested approximately US\$ 1.2 billion, with an average of US\$ 4.9 million per project, for 186 stand-alone renewable energy projects, as well as in 69 mixed projects with renewable energy components, in 160 developing countries and countries with economies in transition.

The GEF's experience from these projects has shown that renewable energy can be a catalyst for changes in rural areas including possibilities for spillover effects that benefit rural communities and farm-based economies. Key lessons learned from these experiences include:

- Demonstrating the cost and performance of new technologies is important. It is also necessary to consider policies and practices for creating markets conducive to renewable energy. As a result, the GEF has learned to include technical assistance for policies and regulatory frameworks, training and other efforts to build technical and institutional capacities, and innovative approaches to establishing financing mechanisms for investment in the deployment and diffusion of renewable energy technologies.
- Project investment is a necessary first step; catalyzing follow-on investments is also an important goal for success to be fully realized. The GEF has learned that investments need to include effective technology transfer components for renewable energy technologies. The GEF has invested in the transfer of commercially-proven renewable energy technologies in rural areas and has emphasized market demonstration and commercialization of the most promising technologies, tools, and techniques. The GEF has focused its efforts on promoting successful demonstrations of cost-effective applications with the aim of removing further barriers to commercialization and cost reduction over time.
- To achieve successful implementation in rural areas and remote farming communities, promotion of access to modern energy services is paramount. Given the acute demand for energy

access and services in rural areas in developing countries, GEF support has focused on bringing modern energy services to these areas by emphasizing decentralized production of electricity and thermal energy using local resources, businesses, and workers that suit local needs. As a result, GEF investments have produced a track record of success, particularly in areas where needs are the greatest, including Sub-Saharan Africa, South Asia, and Small Island Developing States.

ACKNOWLEDGMENTS

Acknowledgments are due to Dr. Zhihong Zhang, Ms. Hang Yin, and Mr. Karan Chouksey of the World Bank Group, and Mr. Marcel Alers of the UNDP for their contribution to the book chapter. The authors wish to thank Mr. Richard Scheer for his English editing of this chapter.

REFERENCES

Dixon, R.K., Scheer, R.M. & Williams, G.T. (2011) Sustainable energy investments: contributions of the Global Environment Facility. *Mitigation and Adaptation Strategies for Global Change* 16, 83–102.

GEA (2012) Global energy assessment: toward a sustainable future. International Institute for Applied Systems Analysis, Laxenburg, Austria.

GEF PMIS – Global Environment Facility Project Management Information System, updated in May, 2014. Available from: https://www.gefpmis.org [accessed October 2016].

IEA (2009) Special early excerpt of the World Energy Outlook. OECD, International Energy Agency, Paris, France.

IEA (2012) World Energy Outlook. OECD, International Energy Agency, Paris, France.

IEA (2013) World Energy Outlook. OECD, International Energy Agency, Paris, France. Available from: http://www.worldenergyoutlook.org/publications/weo-2013/ [accessed October 2016].

IEA (2016) World Energy Outlook OECD, International Energy Agency, Paris, France.

IFC (2014) China Utility-Based Energy Efficiency Finance Program (CHUEE). International Finance Corporation. Available from: http://www.ifc.org/wps/wcm/connect/regprojects_ext_content/ifc_external_ corporate_site/home_chuee/news/ifc+wins+investment+award+for+mitigating+climate+change+in+ china [accessed on October 2016].

Pachauri, R.K. (2009) Climate change, energy and the green economy. Presentation at *Global Renewable Energy Forum*, Mexico. IPCC.

Siler-Evans, K., Azevedo, I.L., Morgan, M.G. & Apt, J. (2013) Regional variations in the health, environmental, and climate benefits of wind and solar generation. *PNAS*, 110, 11,768–11,773.

The Economist (2013) Access to energy. May 29th 2013, 13:07 by Economist.com. Available from: http://www.economist.com/blogs/graphicdetail/2013/05/focus-6?fsrc=rss [accessed October 2016].

UNDP (2009) The energy access situation in developing countries: a review focusing on the least developed countries and sub-Saharan Africa. November 2009. United Nations Development Programme, New York, NY. Available from: www.undp.org/energyandenvironment [accessed on October 2016].

UNDP (2012) Energy conservation in small sector tea processing units in South India. Terminal Evaluation (TE) review report, 7 November 2012, United Nations Development Programme: UNDP/GEF Project 00057404, New York, NY.

World Bank (2006) Project document on a proposed grant from the Global Environment Trust Fund in the amount of US$ 25 million to the United Mexican States for a large-scale renewable energy development project. June 1, 2006, Washington, DC.

World Bank (2010) Development and climate change: a strategic framework for the World Bank Group. Interim progress report April 20, 2010. Report No. 54190, The World Bank Group, Washington, DC.

World Bank (2012) On a grant in the amount of US$ 49.8 million to the Arab Republic of Egypt for the Kureimat Solar Thermal Hybrid Project. Implementation completion and results report (TF-91289) April 30, 2012, Report No: Icr2173, Washington, DC.

WWEA (2011) World wind energy report. World Wind Energy Association, Bonn, Germany. Available from: http://www.wwindea.org/home/images/stories [accessed on October 2016].

Yang, M., Dixon, R.K., Wu, Y. & Diarra M. (2013) Mitigating GHGs with solar power in Africa. World Energy Congress, Daegu, S. Korea.

CHAPTER 3

Linking food and nutrition security, urban and peri-urban agriculture, and sustainable energy use: experiences from South America

Deborah Hines & Sarah Balistreri

3.1 INTRODUCTION

3.1.1 *The global food and nutrition security context*

Despite the fact that our world produces enough to feed the entire global population, this food and the technologies used to produce it do not always reach those who need it most. While global hunger and malnutrition rates have been steadily decreasing since 1990, an estimated 795 million people continue to suffer from chronic hunger (Food and Agricultural Organization of the United Nations [FAO] *et al.*, 2015), and 98% of the world's undernourished population resides in developing countries (FAO *et al.*, 2015). Approximately two billion people around the world are anemic, a condition that stems from a number of nutrition and social determinants (World Health Organization [WHO], 2014).

Greater demand for energy and food, higher agriculture production costs, and the effects of climate change all impede progress towards a food and nutrition secure future. The number of people at risk of food and nutrition insecurity is projected to increase by 10–20% by 2050 as a direct result of climate change on food systems (Met Office and United Nations World Food Programme [WFP], 2012). This increase could lead to a 21% rise in the number of malnourished children according to the International Food Policy Research Institute's (IFPRI) IMPACT projections (IFPRI, 2012).

Approximately 52% of the world's population currently live in urban areas, and it is projected that by 2050, 67% of the global population will live in cities (United Nations Department of Economic and Social Affairs [UN DESA], 2012). Rural-urban migration patterns are contributing to a growing class of poor and vulnerable people living in urban and peri-urban areas. Urbanization tends to go hand in hand with food insecurity and malnutrition, as the urban poor live in overpopulated areas with high levels of unemployment, inadequate infrastructure, and unhealthy living conditions. Newcomers to urban areas often have insufficient access to quality foods, especially upon arrival. As they adapt to urban environments, they tend to eat fewer basic staples and begin to consume more fruits, livestock products, and cereals, as well as fast food, all of which require less preparation. Thus, rapid urbanization and migration from rural to urban areas have a significant impact on food consumption patterns, and consequently on food and nutrition security. If these energy, food and nutrition, climatic, and population trends continue, 60% more food will be needed to feed the world in 2050 (Alexandratos and Bruinsma, 2012).

Globally, the urban poor spend 60–80% of their household expenditure on food and are therefore disproportionately affected by rising food prices and shortages (FAO, 2014c). A decrease in purchasing power for urban families translates more directly into poorer food consumption and nutritional status than it does for rural populations, as urban dwellers often have fewer informal safety nets and are significantly more dependent on purchased food.

Latin America and the Caribbean (LAC) is the most urbanized region in the world, with an average urbanization rate of 80% (UN Habitat, 2012). The number of cities in Latin America has increased sixfold in the past half-century primarily due to migration to cities from rural areas. Since the 1990s, the percentage of people living in cities relative to the rural population has continued to increase, but at a slower pace than in previous decades. Even if the current rate of urbanization continues to decelerate, it is estimated that by 2050, 90% of the LAC region's population will reside in urban areas (UN Habitat, 2012). About 124 million urban residents, or one in four city dwellers, live in poverty, and the average per capita income of the richest 20% of the population is nearly 20 times the per capita income of the poorest 20% (UN Habitat, 2012).

Although the number of undernourished people in LAC has been steadily declining since 1990, most countries continue to confront high levels of inequality and unequal access to quality and nutritious foods. Thus, the prevalence of malnutrition, including micronutrient deficiencies and over-nutrition, are serious and there are growing concerns that this will pose substantial challenges to public health systems throughout the region. For example, 65.6% of adult Ecuadorian women are either overweight or obese, and young children are increasingly affected by these forms of malnutrition (Encuesta Nacional de Salud y Nutrición [ENSANUT-ECU], 2014).

Global energy prices have increased substantially in the last decade, and it is anticipated that they will continue to rise, although at a slower rate. Higher energy prices increase farmers' costs, as the rural agriculture sector has become energy-intensive at all stages of the production, marketing, and consumption chain. Specifically, energy prices affect the cost of inputs and water. Rural farmers use diesel fuel for tillage, planting, transportation, and harvesting, in addition to commercial fertilizers and traditional irrigation systems that consume a significant amount of energy. However, higher energy prices, combined with the transfer of appropriate technologies, could provide incentives for farmers to shift to less energy-intensive crops and agricultural practices such as conservation tillage, lower water pressure irrigation systems, alternative fertilizing methods, and Urban and Peri-Urban Agriculture (UPA). Thus, the impact of higher energy prices could be more gradual if aggressive investments are made in renewable energies and alternative fuels.

UPA offers an innovative solution to address food and nutrition insecurity while also reducing the carbon footprint of agriculture. This chapter highlights successful UPA practices in LAC that have (i) improved food and nutrition security, and specifically increased dietary diversity; (ii) introduced methods and systems that rely on sustainable production techniques; (iii) decreased dependence on traditional energy sources by promoting conservation agriculture techniques; (iv) reduced energy costs related to supplying food long distances from rural areas; and (v) promoted renewable energy sources with the aim of stimulating crop production in difficult environments. While food and nutrition security has many dimensions, this chapter focuses on the linkages between food production and availability, food consumption and the effects on nutritional status, and increased purchasing power and food access for poor urban and peri-urban households.

3.2   MULTIPLE BENEFITS FROM URBAN AND PERI-URBAN AGRICULTURE ACTIONS

Urban and peri-urban agriculture has been defined as,

> *an industry that produces, processes and markets food and fuel, largely in response to the daily demand of consumers within a town, city or metropolis, on land and water dispersed throughout the urban and peri-urban area, applying intensive production methods, using and reusing natural resources and urban wastes, to yield a diversity of crops and livestock* (FAO, 2001).

It is not a new system, as many urban communities have long-standing traditions of farming and raising livestock in urban areas. UPA occurs both within city limits and in peripheral urban areas, and farmers utilize many different types of spaces and agricultural techniques to cultivate crops and produce food. Most urban farmers specialize in short cycle, high value, and low-input crops, the raising of small animals, or the intercropping of food crops with trees in peri-urban agro-forestry systems (Resource Centers on Urban Agriculture and Food Security [RUAF], 2009).

Estimates suggest that urban and peri-urban agriculture comprises around 15% of the world's food production (Baker, 2008). The urban poor are significantly more likely to participate in some form of agricultural activity than their wealthier counterparts, and in many countries, over half of all urban households in the poorest expenditure quintile engage in some form of agriculture to meet their food needs (FAO, 2010).

The case studies presented in this chapter provide sound evidence that UPA activities can be used as a tool to reduce poverty and hunger in urban areas. UPA activities can decrease malnutrition by giving under-resourced urban residents access to fresh and nutritious foods at reasonable prices, while offering alternative livelihoods to urban residents working in the informal sector, which is driven mainly by young people and women. As the UPA experiences of Colombia and Ecuador show, UPA has proven to be a particularly relevant solution to food and nutrition security challenges in Latin America and the Caribbean, given the region's high rates of urbanization and urban poverty, the increasing burden of over-nutrition on public health systems, and its overreliance on energy-intensive agricultural systems.

3.2.1 *UPA and food and nutrition benefits*

UPA can generate numerous food and nutrition security benefits for poor urban households, in addition to creating additional employment and sources of income for under-resourced urban families. Urban households that practice some sort of agriculture activity tend to consume greater quantities of food, in some cases as much as 30% more, and often have a more diversified diet. These households are more likely to eat vegetables, fruits, and meat products, which provide a higher number of calories and energy (FAO, 2010a).

From a social, food, and nutrition security perspective, urban farming benefits poor urban households by:

- Improving family food consumption by diversifying diets and increasing the quantity of nutritious food consumed.
- Generating savings on food expenditure, which can in turn be spent on livelihood and educational needs.
- Providing a source of income through the sale of surpluses or specialized products, especially for people working in the informal sector.
- Promoting gender empowerment by increasing and diversifying women's incomes and giving women a more dynamic role within the household.
- Expanding local economies through the production, packaging, and marketing of food products.

Importantly, foods grown in close proximity to markets often have a higher nutritional value than those cultivated far from points of sale. Non-local foods spend significantly more time in transit and therefore are more likely to lose nutrients before reaching consumers. In local markets, produce is typically sold within 24 hours of being harvested when it is most fresh, ripe, and attractive to consumers (Harvard School of Public Health, 2012). Farmers that cultivate food for local markets tend to favor nutrition, diversity, and taste over durability for shipping purposes, which can translate into increased dietary diversity for consumers.

3.2.2 *UPA and women's empowerment*

Approximately 65% of urban farmers are women (van Veenhuizen, 2006). If women farmers were given access to the appropriate resources, on-farm yields could increase by 20–30%, leading to a 12–17% reduction in the number of hungry people in the world (Consultative Group on International Agricultural Research [CGIAR], 2014). UPA can provide women with a source of income when they are faced with limited employment prospects, thus increasing the economic and food security of their families and themselves. Their position as income generators allows them to have a stronger role in the use of household finances, which often increases their leverage in decision-making within households. This often leads to improvements in the well-being of families.

Figure 3.1.   Smallholder farmers provide fresh foods to WFP program participants in Carchi, Ecuador. WFP (2013) *Local production and consumption empower rural women.*

Findings from a WFP evaluation in Carchi, Ecuador showed that the income of one farmers' association, the majority of whose members were women, increased by 79% solely from selling their fresh products to a WFP-assisted school feeding program (WFP, 2013). The study also showed that these women were more likely to spend their incomes on food and their children (see Fig. 3.1). Extensive research has confirmed that when mothers have control of the household budget, a child's chances of survival increase by 20% (FAO, 2014a).

Small-scale urban farming is especially appealing to women because it can be practiced as part-time work and complements other domestic activities such as food preparation and child care. For example, mothers residing on the outskirts of Mexico City often traveled long distances to work as domestic servants. Their children were left alone, and many eventually joined street gangs. By engaging in urban farming, these mothers were able to cultivate food for their families, generate income in close proximity to their homes, and spend more time with their children (FAO, 2008).

3.2.3   *UPA and environmental benefits*

In addition to playing an important role in improving the food and nutrition security of poor urban families and the empowerment of women, UPA also has the potential to reduce the agricultural sector's reliance on traditional energy sources and their related environmental impacts when compared to mechanized rural farming systems. Furthermore, UPA can mitigate some of the negative impacts of climate change and contribute to the advancement of more sustainable food systems and cities. Some of the environmental benefits of urban and peri-urban agriculture include:

- Reducing the carbon footprint of agriculture production systems by eliminating the need for packaging, storing, and transporting food.
- Making efficient use of organic household waste for fertilizer and recyclable storage containers such as plastic bottles and boxes for planting spaces and garden beds.
- Decreasing the amount of food lost or wasted due to inadequate storage, packaging, and transportation facilities.
- Using more efficient irrigation technologies that permit crops to grow in dry or marginal areas.
- Increasing resilience to intense rain events by expanding the amount of pervious surface area in cities, which partially reduces the need for excess demand on municipal water systems and decreases the quantity of polluted water that enters neighboring water bodies.
- Off-setting carbon pollution by sequestering carbon dioxide.

3.2.3.1   *UPA and carbon footprint mitigation*

As agriculture is the largest contributor of non-CO_2 greenhouses gases in the atmosphere, innovative UPA solutions can begin to help redress the environmental distortions in food production systems. Food systems produce 19–29% of anthropogenic greenhouse gas emissions, or a total of 9800–16,900 million tons of carbon dioxide each year, through the energy used in fertilizer manufacturing, food production, processing, transport, and distribution, household food management, and waste disposal (Vermeulen *et al.*, 2012). The production phase of the food production system is the greatest contributor of emissions, generating 7300–12,700 million tons of CO_2 each year. This translates into about 80–86% of total food systems emissions and 14–24% of total global emissions (Vermeulen *et al.*, 2012). Food transportation alone produces 5% of total global carbon dioxide emissions (Harvard School of Public Health, 2012).

Roughly one-third – or 1.3 billion tons – of all the food produced for human consumption is wasted, contributing approximately 500 kilograms of carbon dioxide to the atmosphere per person per year (Gustavsson *et al.*, 2011). In high-income countries, food waste occurs mainly at the consumer level, while developing countries generate high food losses at the post-harvest and processing stages due to a lack of modern transportation and food storage infrastructure, and financial, managerial, and technical limitations in difficult climatic conditions (Gustavsson *et al.*, 2011; Venkat, 2011). It is estimated that approximately 40% of food waste in low-income countries occurs during the transportation, storage, and processing stages of the food supply chain (Gustavsson *et al.*, 2011). UPA practices tend to have a significantly smaller carbon footprint than large-scale agriculture due to their proximity to urban marketplaces, which generates energy savings in food packaging, transportation, storage, and distribution and decreases greenhouse gas emissions by reducing dependence on the use of commercial fertilizers and pesticides.

Urban farmers generally make productive use of under-utilized resources, such as unoccupied land, treated wastewater, and inorganic and organic household waste, often using farming practices that respect local knowledge and traditions. Such practices in urban and peri-urban environments can lead to in an increase in productivity (Moreno-Peñaranda, 2011). For example, the use of compost produced by organic household waste in urban farms and gardens is an effective alternative to commercial man-made fertilizers and can significantly reduce the amount of nitrous oxide released into the atmosphere. Homemade organic compost also reduces methane emissions from landfills and lessens the amount of energy used in the production of commercially-manufactured fertilizers.

Many of the urban agriculture projects sponsored by city municipalities promote organic farming as a safer, healthier, and more sustainable alternative to conventional farming practices. It is estimated that organic agriculture systems produce 48–66% less CO_2 than conventional agricultural systems, and they are also able to sequester carbon and act as carbon sinks if appropriate techniques are employed (El-Hage *et al.*, 2002). The natural carbon sequestration process carried out by plants, combined with specialized agricultural techniques, can increase carbon removal from the atmosphere and prevent the release of CO_2 at harvest (Rowe, 2010). For example, green roofs made of uncut grass measuring 2000 square meters have the capability to remove up to 4000 kilograms of particulate matter. Green roofs measuring one square meter are able to offset the annual particular matter emissions of one car (Mayer, 1999). In addition, increasing the level of organic matter in the soil by using organic compost and utilizing low- or no-tillage practices allows soil to act as a sink for atmospheric CO_2.

3.2.3.2   *UPA and sustainable water systems*

Agriculture accounts for 70% of the world's water withdrawal, and both the absolute scarcity of renewable water resources and the relative scarcity of reliable water services are serious concerns for the 21st century (AQUASTAT and FAO, 2014). The reuse of wastewater in UPA farms that meets adequate sanitation standards and the harvesting of rainwater for the irrigation of urban gardens and farms can optimize water resources and ensure the availability of year-round water supply (FAO, 2008). These practices also allow fresh water to be utilized for higher value purposes and reduce the amount of greenhouse gas emissions that are generated from the treatment of wastewater. Domestic wastewater is another viable and sustainable alternative to commercially-made fertilizer and, when treated appropriately, has the potential to supply all of the nitrogen and a significant portion of the phosphorous and potassium that are normally needed in agricultural production (Westcot, 1997).

Approximately 7–8% of the world's energy is used to lift groundwater and pump it through pipes to treat both groundwater and wastewater (UN World Water Assessment Programme, 2012). By lowering the quantity of water that must be treated, cities use less tax money in water treatment and generate significant savings in energy usage. Water is allowed to return to the underlying water table and undergoes a natural filtration process through the soil.

Water released during intense rain events poses serious environmental problems for many urban areas, as water treatment centers are often unable to handle extremely high volumes of water in short time periods. With their vast expanses of impervious surfaces, municipalities construct complex sewage systems to prevent city streets from flooding during storms and periods of high rainfall, which are expected to increase in intensity due to the effects of global climate change (Trenberth, 2011). Urban farms and gardens can increase pervious surface areas in cities and reduce the amount of water that passes through municipal treatment centers. They make efficient use of rainwater, and their absorption of rain decreases the amount of polluted water that flows into bodies of water located within and around urban areas, protecting local aquatic ecosystems and water resources (Gittleman, 2009). The absorption of rainfall also mitigates the risk of landslides that could occur during periods of high rainfall (Dubbeling, 2011).

Several different sustainable irrigation and water management techniques are available to urban farmers. Bio-intensive gardens, which are typically 60 cm deep, 1.2 m wide, and 8 m long, need less water than traditional gardens, as their shape allows for optimal water circulation (WFP and Ministry of Agriculture, Livestock, Aquaculture and Fisheries [MAGAP], 2012). Drip irrigation systems are also an important water management technique, as they deliver water slowly to the plant root zone and reduce water losses from runoff and evaporation. Drip irrigation is typically more than 90% efficient in ensuring that the water is received by the plant, compared to other irrigation techniques such as sprinklers, which are only 65–75% efficient (University of Rhode Island, 2006).

Hydroponics, a cultivation technique in which the roots of the plant are placed into a liquid nutrient solution rather than in soil, also has important benefits for farmers and the environment. Hydroponic systems are considerably less labor-intensive than other farming practices, as there is no need to weed, use herbicides, or till the soil. In closed hydroponic systems, surplus liquid is recovered, replenished, and recycled back through the system. UPA farms and gardens that use hydroponic production techniques require less water than traditional irrigation systems (Hayden, 2000).

3.3   URBAN AGRICULTURE: EXAMPLES FROM SOUTH AMERICA

In April 2000, local government representatives from nine different Latin American countries convened in the city of Quito, Ecuador to discuss urban agriculture in the 21st century. The conference produced the landmark Quito Declaration, a document that encouraged Latin American cities to embrace urban agriculture as a tool for alleviating urban poverty, improving the food security of urban residents, and preventing further environmental degradation in cities. UPA's value in social and economic development was acknowledged, and participants urged cities to include urban agriculture in their agendas and to develop tax incentives and policies that facilitate the practice of urban agriculture in Latin American cities (FAO, 2014b).

Throughout the region, many urban agriculture programs are increasingly promoting the use of technologies and agricultural practices that produce higher quality crops in greater quantities, while making efficient use of natural and energy resources and agricultural inputs. The following section presents examples of good practices in UPA from two South American countries: Ecuador and Colombia.

3.3.1   *UPA in Ecuador*

Urban and peri-urban agriculture projects introduced in Ecuadorian municipalities, though not without their challenges, have shown great promise with respect to improving the food and

Figure 3.2. WFP works with Ecuador's Ministry of Agriculture to set up gardens in schools with Colombian refugee children. WFP (2014) *Niños y huertos en Carchi, Ecuador.*

nutrition security of poor urban residents, reducing energy costs, and creating more environmentally sustainable and resilient cities. This section discusses two successful UPA programs that have been implemented in the cities of Quito and Tulcán. These programs can be viewed as models for other projects throughout the region and the developing world.

3.3.1.1 *Participative urban agriculture in Quito, Ecuador*

In 2000, the United Nations Urban Management Programme and the municipality of Quito formed an alliance to implement an urban agriculture project in the neighborhood of El Panecillo in central Quito. The program was considered a success, and after consulting with several different actors, including universities and NGOs, the municipality of Quito launched its citywide urban agriculture project known as *Agricultura Urbana Participativa* (Participative Urban Agriculture) or AGRUPAR. In 2005, the municipality's corporation for economic development, CONQUITO, assumed full responsibility for the project, and the municipality of Quito currently allocates US$ 250,000 of its annual budget to AGRUPAR. This budget covers the costs of training sessions, technical advice, and project logistics, and partially covers the cost of inputs, equipment, seeds, and animals such as poultry, bees, and guinea pigs.

AGRUPAR gardens are present in all eight of Quito's administrative zones. In 2015, the project reported over 2700 active urban gardens in Quito, including community, school and family gardens adding up to 29 hectares of cultivated area (CONQUITO, 2016). CONQUITO estimates that the project has directly benefited more than 61,000 people and indirectly benefited 400,900 people (CONQUITO, 2016). Through partnerships with community-based organizations, the project works with the elderly, single mothers, abandoned children, migrants, refugees, the disabled, and the ill (see Fig. 3.2). AGRUPAR urban gardens have been set up in schools, prisons, government-sponsored day care centers, and institutions for the disabled, the homeless, and the elderly.

Figure 3.3.   AGRUPAR gardens provide nutritious food to vulnerable families throughout Quito, Ecuador.
CONQUITO (2014) Huertos urbanos en Quito, Ecuador.

The project contributes to greater social cohesion within neighborhoods, as these gardens now occupy a special position within communities, directly connecting food producers with consumers. AGRUPAR also includes a microcredit, microenterprise management, and a marketing component. Project technicians seek to show participants that urban farming is feasible and can be set up with relatively few expenses and resources. The majority of participants have low levels of educational attainment, as most have only completed primary school (FAO, 2014b).

Approximately 19,180 people have received training in organic urban agriculture from AGRUPAR technicians since the project's inception, and an estimated 500 metric tons of food have been produced by project participants (CONQUITO, 2016). According to the results of a 2010 survey of 480 project participants, urban farmers in Quito produce about 70 different crop species, which include potatoes, maize, Swiss chard, broccoli, cabbage, tomatoes, carrots, lemons, passion fruit, blackberries, aromatic plants, and spices. Participants also cultivate ancestral crops such as quinoa and amaranth (see Fig. 3.3). An estimated 47% of products are sold, while the rest is consumed by the household. AGRUPAR urban farmers earn an average of US$ 55 per month by selling their surpluses and save at least US$ 72 per month on food expenses by consuming the products they cultivate in their gardens. Approximately 97% of interviewees that participated in the 2010 survey reported that household consumption of vegetables increased after joining the project (Carvajal, 2010).

AGRUPAR gardens follow Ecuador's national standards for organic gardening and train participants in environmentally sustainable and organic agriculture techniques, promoting production systems that enhance soil health, biodiversity, and natural biological cycles. AGUPAR discourages the use of genetically modified organisms (GMOs) and pesticides and pests, and weeds are managed using natural control methods. Participants learn how to maintain soil health through the use of organic compost, green manure, crop rotation, and soil protection via physical and live barriers.

Technicians encourage the use of recyclable materials such as plastic soda bottles, which can be transformed into small garden beds or watering cans. Boxes and tires are used in dense urban areas so that gardens can be developed on balconies, terraces, and patios (see Fig. 3.4). In addition,

Figure 3.4.   AGRUPAR gardens utilize materials such as plastic bottles and tires as receptacles to cultivate crops. This practice allows urban farmers to take advantage of available materials and space. CONQUITO (2014) *Huertos urbanos en Quito, Ecuador.*

participants are taught how to compost their organic food waste to use as fertilizer in their gardens. Urban farming families produce on average 12.5 kitchen organic materials per week that they then use as organic compost. It is estimated that 1820 metric tons of organic waste is diverted from landfills and used in AGRUPAR gardens each year (FAO, 2014b). Project participants are also trained to harvest rainwater, which can in turn be used to irrigate their gardens. As of 2013, drip irrigation systems had been installed in 70 AGRUPAR gardens.

Challenges
While urban farming in Quito receives financial support from the municipal government, the city lacks an established legal framework that explicitly supports urban agriculture. Quito's Development Plan for 2012–2022 envisions an environmentally sustainable and socially equitable city, yet it does not recognize urban farmers as legitimate stakeholders in Quito's economic development. The use of municipal lands for food production faces challenges due to existing municipal laws that require urban farmers to form a legal association before they can be granted a lease (FAO, 2014b). Stronger legal and political support from the municipality is critical to ensuring the future success of urban agriculture in Quito.

Best practices
AGRUPAR has proven itself to be a successful and sustainable project that increases the food security, dietary diversity, and income of its participants, while also contributing to the sustainable management of Quito's urban landscape. One successful feature of the project is AGRUPAR's implementation model, which stipulates that project participants pay a small fee to receive technical training and related resources. In most cases, participants contribute 50% of the resources necessary to cultivate a garden, while CONQUITO provides the other half. The participation fees are symbolic in value, as AGRUPAR will not recover its investment with such small fees. However, they serve to remind participants that they are making a joint investment in the project (Rodríguez Dueñas, 2014).

Another effective practice of AGRUPAR is the establishment of farmers' markets, known as *bioferias*, which are open exclusively to project participants. These markets have been very successful, and there are currently 17 *bioferias* operating in several different neighborhoods in Quito (CONQUITO, 2016). Before urban farmers are permitted to take part in the markets, they must participate in the program for a minimum of six months so that AGRUPAR technicians can certify that their gardens follow the regulations for organic gardening and that participants have incorporated the appropriate production techniques. It is also important that participants not only learn how to effectively farm and garden, but that they improve their dietary diversity and food consumption levels during the first six months of participation (Rodríguez Dueñas, 2014). In addition to these 17 *bioferias*, WFP and AGUPAR are currently collaborating to open a store where urban farmers can sell their products to WFP's program participants, most of whom are Colombian refugee families.

The creation of farmers' associations and microfinance institutions has been a successful component of the program and has permitted urban farmers to establish small businesses and increase their income. Three farmers' associations now have better market opportunities to sell their products, and a total of ten microfinance associations have been created to provide credit services to participants (Carvajal, 2010). Thanks to AGRUPAR's training sessions and technical assistance, participants have opened more than 200 small businesses in Quito (CONQUITO, 2016). Adding value to agricultural products by using food-processing techniques is also an effective feature of AGRUPAR's training, as this practice provides full- and part-time employment to half of the project's participants.

3.3.1.2   *WFP urban agriculture initiatives in Ecuador*

In February 2011, over 1000 people assembled at a conference in New Delhi, India organized by IFPRI to discuss the potential of agriculture to contribute to improved nutrition and health. Delegates called for comprehensive approaches to food and nutrition security that integrate supply-side, demand-side, and market measures, aligning the goals of the agriculture, health, and nutrition sectors. In addition to calling for a renewed investment in smallholder production and productivity, the aim was to strengthen national and regional strategies to promote the participation of farmers, particularly smallholders, in markets.

WFP initiated the Purchase for Progress (P4P) program to enhance the socioeconomic development impact of its procurement activities, while also providing food assistance in a timely and cost-effective manner. As a pilot global project, P4P emphasized the importance of learning from experience based on three pillars: (i) WFP's demand platform; (ii) supply-side partnerships and capacity-building activities; and (iii) learning and sharing best practices. Through P4P strategies, WFP uses its demand for staple commodities to enhance the capacity of smallholders to access formal markets, such as the WFP market, government programs, and the food-processing sector. The ultimate goal of P4P is to identify proven implementation models which national governments can adopt and bring to scale. Importantly, P4P aims to increase household incomes through country-specific approaches that stimulate markets and improve food quality by promoting the development of smallholder farmers' associations and creating marketing opportunities for their farmers.

In the period 2008–14, WFP has contracted more than 72,000 metric tons of food commodities from smallholder farmers in Central America, with an investment of US$ 40 million in rural economies (WFP, 2014a). Changes are apparent for the more than 36,000 smallholder farmers in the region who have participated in the program due to their increased market opportunities and incomes (WFP, 2014b). As well, governments have benefited from the program by learning from the initiative's experiences, which show how to better link farmers to markets through actions such as removing obstacles and cultural barriers which restrict women's access to markets. Urban and peri-urban farmers have also experienced cultural benefits, as many have migrated to cities from rural areas and have previously practiced agriculture. UPA allows them to reconnect with their farming traditions and can also help to strengthen family relationships.

WFP Ecuador adapted the P4P concept model to focus on the supply of fresh products, including fruits, vegetables, and proteins. Strengthening smallholder farmers' associations was a first step, combined with WFP's demand platform and the organization's ability to help stimulate other local markets in partnership with decentralized governments. The program involves Colombian refugees and vulnerable Ecuadorian populations in four provinces of Ecuador. WFP's support gives smallholder farmers association's stable access to secure markets by linking their production to WFP's procurement activities and other local markets, including primary schools. Producers often diversify their crop production in response to WFP's demand and receive better prices for their products. According to a 2013 project evaluation, WFP's support allowed smallholder associations to increase their incomes by an average of US$ 25 per month (WFP, 2013). Through local governments and other partners, smallholders receive training on organic farming techniques, marketing strategies and requirements for the legalization of their associations.

Figure 3.5.   WFP program participants can purchase nutritious foods produced by smallholder farmers at the Shushufindi Vegetable Market in Sucumbíos, Ecuador. WFP (2013) *La verdulería Shushufindi en Sucumbíos, Ecuador.*

WFP has supported local governments in establishing points of sale suited to a particular location or context. Participants in WFP's protracted relief and recovery program receive an electronic voucher which enables them to purchase food produced by smallholder farmers' associations (see Fig. 3.5). A conditioned voucher transfer was found to be a very efficient assistance mechanism when the objective is to improve dietary diversity and food consumption within the family, according to the results of a randomized experimental study carried out by IFPRI and WFP in the provinces of Carchi and Sucumbíos (Hidrobo *et al.*, 2012). As a complement to WFP's local purchase model that works to provide fresh foods to beneficiaries, the program also supports peri-urban and rural small-scale farmers' associations and family garden initiatives with the aim of increasing the dietary diversity and food and nutrition security of vulnerable families. In 2013, approximately 80% of participants saw a substantial improvement in their food consumption patterns. Upon entering the program, the average household consumed products from only four food groups; after receiving vouchers restricted to the purchase of nutritious foods for a period of at least six months, families consumed on average seven of the ten recommended recognized food groups (WFP, 2013).

In addition to improving the food and nutrition security of vulnerable households, WFP's family garden activities have also generated wider social benefits. For example, WFP worked with the Office of the Vice-president's Joaquín Gallegos Lara on an urban gardening initiative for people with disabilities living in the city of Tulcán. Vulnerable families were assisted in setting up family gardens to generate income, diversify family diets, and create opportunities for families to spend time together (Arteaga, 2014). Gardens also provided disabled children the opportunity to take responsibility for the growing of the plants, which improved their self-esteem.

All WFP Ecuador's UPA initiatives emphasize environmentally sound production systems that require little to no energy or chemical inputs. To transfer these practices to program participants, WFP and Ecuador's Ministry of Agriculture, Livestock, Aquaculture and Fisheries (MAGAP) developed an instructive organic farming manual for project participants detailing several different sustainable farming techniques and methodologies for use in rural, peri-urban, and urban settings (see Fig. 3.6) (WFP and MAGAP, 2012).

El huerto familiar
orgánico y nutritivo

Manual técnico para mejorar la alimentación de las familias vulnerables

Ministerio de
Agricultura, Ganadería,
Acuacultura y Pesca

WFP
Programa
Mundial de
Alimentos
wfp.org/es

Figure 3.6.   WFP and Ecuador's Ministry of Agriculture collaborated to create an organic gardening manual
(WFP and MAGAP, 2012).

The manual is used in training where practices such as rotating crops to reduce the incidence of plant diseases and pest infestations are promoted (see Fig. 3.7). Participants learn about organic soil disinfection techniques, such as pouring boiling water on the garden plot before planting to kill pests present in the soil and applying vegetable ash to the surface of the soil to prevent the growth of weeds. Participants also see firsthand how to arrange plants in a hexagonal pattern or to place them at the appropriate distance from other plants to better manage pests and plant diseases. They also learn which plant species attract certain harmful insects and therefore keep pests away from crops. For example, dill, parsley, mint, rue, coriander, borage, and nettles keep harmful insects away from the garden.

In several urban and peri-urban initiatives, WFP and its local partners assist participants in setting up bio-intensive gardens, which are optimal in size and resource use for small urban farmers. In bio-intensive gardens, plants can be placed close together using intensive and companion planting techniques. Crops will cover most of the garden after a few weeks, thus making it harder for weeds to grow. Participants also learn to practice hilling, or arranging small hills of dirt around plants, which is another organic technique that can be used for weed control and prevention (WFP and MAGAP, 2012). Such practices reduce the need for chemical fertilizers and insecticides and therefore decrease energy input requirements.

Challenges
Although the organic gardening training manual has been a very useful resource in WFP's urban, peri-urban, and rural gardening initiatives throughout Ecuador, finding committed partners

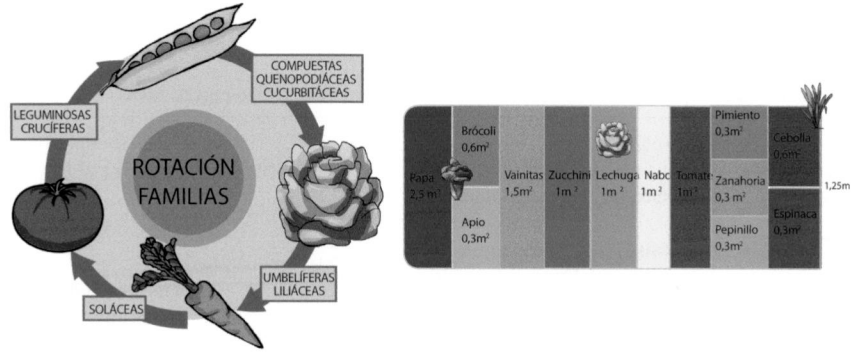

Figure 3.7.   The manual serves as a resource for WFP program participants and trains them in organic gardening techniques using easily understandable instructions and diagrams (WFP and MAGAP, 2012).

willing to provide continued technical assistance and training to participants limits the scaling up such activities. Providing technical assistance to both smallholder farmers' associations and vulnerable families is vital to improving productivity and ensuring the environmental sustainability of farms and gardens.

In addition, while women's empowerment is an important objective in all WFP activities in Ecuador, it was initially a significant challenge to purchase locally-grown foods from female-majority smallholder farmers' associations (see Fig. 3.8). To promote women's participation in these associations and support their income-earning and decision-making roles within their families, WFP stipulated that foods should be purchased from associations that have at least 50% women in their membership and from women producers. While WFP has reached this target in all of its programs, WFP and the provincial government continue to work primarily with female-majority associations in the province of Carchi.

Best practice
The comprehensive assistance package of training, food assistance contingent on participation in training, and technical support for the implementation of gardens, has been a successful practice that effectively contributes to the food security and dietary diversity of program participants. Success can be attributed to the strong partnerships between local governments and WFP in which these partners work jointly at all stages of the project: from its design, through the implementation phase to the evaluation of project activities.

In addition, the project has sought innovative solutions, for example, establishing both fixed and mobile sales points to deliver foods produced by smallholder farmers' associations to program participants. In the province of Carchi, farmers' associations bring their products to Carchi Productivo, a store that was established specifically to link smallholder associations directly with WFP and local government programs. Both stationary and mobile marketplaces have allowed WFP and local governments to stimulate local economies by strengthening smallholder farmers' associations, advocating for the production of nutritious foods produced with environmentally sustainable techniques. Because the sale of products was secure, farmers were more inclined to accept new technologies and move towards organic farming practices. Finally, advocacy efforts on the demand-side complemented production activities, promoting the consumption of nutritious, healthy, environmentally-friendly and locally-grown and consumed products.

3.3.2   *Innovative UPA practices in Colombia*

Rapid urbanization puts great pressure on the supply, cost, and quality of food in Colombia, where urban dwellers spend 30% more on food than rural populations. However, the consumption of

Table 3.1.   Sustainable organic gardening techniques.

Garden management	Techniques	Energy savings benefits
Soil management	• Crop rotation • Planting crops at appropriate distances (interspersing them with other plants or planting them in a hexagonal pattern) • Application of organic compost	• Emissions savings from decreased dependence on chemical fertilizers
Pest control	• Crop rotation • Diversifying plant species within garden • Application of boiling water to garden bed before planting • Maintaining plants in optimum health • Planting pest-repelling plants • Removing infected plants • Cultivating local plant species • Spraying pest-repelling solutions made from dish soap, tobacco or cigarette butts, or pest-repelling plants • Placing containers of fermented liquid (such as beer) in the garden to attract slugs • Placing ash, salt, sand or lime around plants • Placing containers of liquid with potato peels in garden to attract centipedes • Using copper-based fungicides to kill fungus growth	• Emissions savings from decreased dependence on chemical pesticides
Weed control	• Bio-intensive garden bed • Application of boiling water to garden bed before planting • Application of vegetable ash to garden bed before planting • Manual weeding • Hilling	• Emissions savings from decreased dependence on chemical herbicides
Sustainable water management	• Bio-intensive garden bed • Hilling • Application of organic compost • Drip irrigation	• Savings in water inputs • Energy savings from decreased dependence on fuel inputs for irrigation • Emissions savings from decreased dependence on

fruits, vegetables, and animal protein only meets about 35% of the recommended daily minimum requirements (Corporation Colombian de Investigation Agropecuria [Corpoica], 2010). In major urban centers such as Bogotá, Medellín, Cali and cities on the Caribbean coast, the urban poor invest almost 60% of their income in food, while the average intake of fruits and vegetables is only about 30 grams per day. This amount is significantly less than WHO's recommendation of a minimum of 56 grams per day (Corpoica, 2010). The quality of food available also is a concern, especially for poor urban dwellers who cannot afford organic products. A 2011 study conducted by the Universidad Nacional de Colombia found alarming levels of heavy metal contamination in fresh vegetables, dairy products, fruits and tubers, as well as evidence of fecal waste, coliforms, and traces of radioactive particles (Corpoica, 2010).

Figure 3.8. Women from smallholder farmers' associations provide food to WFP's food assistance programs WFP (2014). *Rural women feed Ecuador's cities.*

In response to concerns surrounding food access and quality in urban areas, over ten institutions came together in 2010 to create the Interagency Support Group for Urban and Peri-Urban Agriculture (GIAUP) under the leadership of the Colombian Cooperation of Agricultural Research (Corpoica), which forms part of Colombia's Ministry of Agriculture.

GIAUP projects work with single mothers, people displaced by the armed conflict, the unemployed, and small-scale urban and peri-urban farmers. It is estimated that there are about 10,000 stable UPA production units in Colombia, 3500 of which are located in Bogotá. These UPA farms and gardens benefit more than 120,000 people (Corpoica, 2010). Corpoica develops and disseminates technologies for use in urban and peri-urban farming systems. As many of the technologies were originally developed for rural areas, they were adapted to urban and peri-urban setting in order to effectively contribute to the supply of fresh, healthy foods to urban areas.

Another interesting example of UPA in Colombia is WFP and FAO's food security project in the southwestern coastal department of Nariño. The project supports the Women's Producer Association of Ricaurte, which was formed by approximately 27 female-headed families who have been affected by the armed conflict. Many of the association's members live in the densely populated city center of Ricaurte and use their backyards for agricultural production. The organizational process has taken a significant amount of time, in particular for the planning of joint production and marketing activities. However, the women have come to value the benefits of working together and have planted a variety of crops, including tubers, maize, cassava, banana, sugar cane, china potato, and beans for household consumption and sale at local markets.

3.3.2.1 *UPA powered by renewable energy in La Guajira, Colombia*

The department of La Guajira in northern Colombia is home to a significant portion of the country's indigenous population. The Fundación Cerrejón Guajira Indígena (FCGI) works to improve the quality of life for the Guajira's Wayuu indigenous communities, Colombia's largest indigenous population, through the implementation of sustainable, innovative, and participative programs.

Figure 3.9.   The difficulty of growing crops La Guajira's dry desert climate has been exacerbated by a protracted drought (Hines, 2017).

FCGI has promoted UPA activities with the specific objective of decreasing malnutrition rates among La Guajira's indigenous populations. From 2014 until early 2016, the department suffered from a protracted drought due to El Niño that left many children severely undernourished and child mortality at an unacceptably high rate (see Fig. 3.9).

Since 2008, 25 urban gardens and five peri-urban gardens have been implemented with FCGI's support, with WFP assistance for a number of them. In addition to garden activities, FCGI conducts studies regarding the recovery and promotion of native plant species, especially those with high nutritional values, and uses for traditional medicine and agriculture diversification. The organization has also constructed an experimental farm, which is used to test the viability of different crops in La Guajira's climate and acts as an educational center promoting local communities' agricultural knowledge and techniques. FCGI implemented a goat-raising program in ten different communities as well as on its experimental farm, given that goats are an important part of the diet and culture of the Wayuu people (Vergara and Tiller Ipuana, 2014).

Challenges
Given the department's dry climate and high propensity towards drought, the availability of clean water for human consumption and for agricultural purposes is a constant concern for residents. In addition, many families are not familiar with planting crops, as they have traditionally been part of herding and nomadic communities. Therefore, it is very important to focus UPA efforts on improving planting techniques, as well as food consumption patterns.

Best practices
FCGI's gardening initiatives promote the use of alternative agricultural production methods such as drip irrigation and irrigation with porous clay pots, which serve the dual purpose of storing and distributing water, thus increasing the availability of moisture in the soil. Through these practices, they have managed to maintain family agricultural production in times of drought.

To further strengthen communities' resilience against drought and promote sustainable energy use, the United States Agency for International Development's Colombia Clean Energy Program

Figure 3.10.   FCGI's wells powered by sustainable energy sources will provide water sources for agricultural use and human consumption to the Wayuu people (Hines, 2016).

and FCGI formed an official partnership to install water management systems powered by renewable energy in indigenous Wayuu communities throughout La Guajira. After conducting an extensive review of existing water systems and technologies, it was decided to construct solar-powered water systems in 11 deep community wells, two community artisanal wells, and three community reservoirs, including one on FCGI's experimental farm.

Photovoltaic systems, bike-powered pumps linked to eight artisanal wells, and 23 manual pumps connected to community artisanal wells, were installed on the experimental farm (see Fig. 3.10). An existing solar pump was further adapted to fit community needs. These renewable management systems served as demonstration pumps for participant communities. The program directly benefits 1500 people from Wayuu communities throughout La Guajira by (i) giving them access to clean and renewable technologies, (ii) optimizing the efficiency and sustainability of their water management systems, and (iii) preventing the pollution of valuable water resources (Tiller Ipuana, 2014). Efficient water systems help the Wayuu diversify their livelihoods and build more resilient production systems. As variable climatic conditions have devastated many traditional livelihoods, these technologies offer opportunities to improve the food and nutrition security of the Wayuu people.

WFP has promoted an integrated model to enhance resilience to climate change and to natural disasters with a cultural and gender-sensitive approach. Innovation is the cornerstone of this initiative, which includes documentation and dissemination of traditional and ancestral knowledge to reduce climate risks. It also supports livelihood diversification and prevents malnutrition through actions that increase community awareness and knowledge of climate change risks as well as food security and nutrition. Innovative community and ecosystem-based adaptation measures help reduce recurrent climate vulnerabilities and food insecurity.

Protection of ecosystems and improvement in their service provision strengthen community capacities to implement resource management and protection plans. The integrated model enhances community capacities, and increases family income through training provision to indigenous communities and women. Good practices inform public policies design and implementation. This intervention not only prevents loss of life, but also reports economic, environment and social benefits for local indigenous communities.

3.4   SHAPING THE FOOD AND NUTRITION SECURITY AGENDA

Mounting pressure on urban and peri-urban resources due to the unprecedented levels of migration from rural areas to urban centers necessitates the introduction of policies and programs that address both human and environmental development concerns. UPA initiatives throughout Latin America provide a wealth of experience regarding successful practices and techniques that can serve as models in other regions of the world. While these actions have yielded promising results in terms of increasing the food and nutrition security of the urban poor, reducing energy consumption, and contributing to the environmental sustainability of cities, UPA programs and initiatives still face several challenges. Practicing agriculture in urban and peri-urban environments is quite simply more complex and faces a different set of challenges from rural agriculture.

The absence of national and municipal legislation regarding the use of urban land for agriculture may affect the future of UPA if it is not addressed at both national and local levels. Formulating and revising government policies and legal frameworks to prioritize urban agriculture and proactively enable the poor to participate in their own food production are fundamental to UPA's success. The future of UPA will depend on the ability of municipal governments and urban farmers to minimize environmental and health risks and to rely on farming techniques that make responsible and efficient use of scarce natural resources. Of particular concern is the potential toxicity of urban soils and the additional pressure urban agriculture puts on water systems that are often already strained in city environments. Raising animals in urban areas can also pose health risks to urban residents. To overcome such challenges, low-income urban dwellers must receive support from city municipalities and others actors to access the land, water, credit, farming tools, and technical training required to effectively practice UPA. Conducting inventories of available public lands that can be utilized for food production should be a policy priority if vulnerable urban populations are to secure access to the land and the required natural resource inputs needed to practice sustainable UPA. Urban planning and zoning should also incorporate UPA considerations into municipal laws and policies.

The case studies presented in this chapter highlight that training urban farmers in organic agriculture methodologies and renewable energy farming techniques is critical to ensuring the sustainability of UPA interventions. Organic pest and weed control, along with sustainable water management techniques and practices such as drip irrigation, hydroponics, and rainwater harvesting have been successfully implemented in these projects. Using organic household waste and recycling plastic bottles as seedbeds and watering cans shows that UPA has great potential to reduce waste. Another important success factor in UPA is the development of marketing strategies and the creation of marketplaces for producers to sell their products directly and quickly to consumers. WFP's P4P model demonstrates that supporting smallholder farmers, and female farmers in particular, can generate benefits at the household, community, and regional levels in terms of improving food and nutrition security and creating sustainable livelihoods. Such practices can be replicated in UPA initiatives within the region and throughout the world.

Integrated approaches that incorporate sustainable agricultural, energy, land, and water management practices with poverty alleviation goals are more likely to succeed and have a positive and sustainable impact in urban and peri-urban areas. This integration requires incorporating urban agriculture into national and local development agendas. Such a focus is clearly articulated in the Sustainable Development Goals (SDGs) which recognize that access to quality food deserves the same amount of attention as food production (United Nations Sustainable Development Knowledge Platform, 2014b). Goal 2 seeks to end hunger, achieve food security and improved nutrition, and promote sustainable agriculture, while Goal 12 aims to ensure sustainable consumption and production patterns.

WFP and FAO have led a review of Goal 2, aligning the goal's objectives with those of the Zero Hunger Challenge (ZHC). The ZHC is composed of a broad coalition of NGOs, governments, businesses, and individuals that have come together to work towards ending chronic hunger and malnutrition. The relevant aligned goals going forward include:

2.1 Ending hunger by 2030 and ensuring year-round access for all people to sufficient quantities of safe and nutritious foods;

2.2 Ending all forms of malnutrition by 2030, including achieving by 2025 the internationally agreed targets on stunting and wasting in children under five years of age;

2.3 Doubling the agricultural productivity and the incomes of small-scale food producers by 2030, particularly that of women, indigenous peoples, family farmers, pastoralists, and fishers;

2.4 Ensuring sustainable food production systems and implementing resilient agricultural practices that increase productivity and production, help maintain ecosystems, strengthen capacities for adaptation to climate change, extreme weather, drought, flooding and other disasters, and progressively improve land and soil quality by 2030; and,

12.3 Halving per capita global food waste at the retail and consumer level by 2030, along with reducing food losses along production and supply chains, including post-harvest losses (United Nations Sustainable Development Knowledge Platform, 2014a).

Urban and peri-urban agriculture can make significant contributions towards the achievement of these goals. UPA can sustainably improve the food and nutrition security of urban dwellers and provide an important safety net for the most vulnerable populations. It also can improve the resilience of cities against the negative effects of climate change and reduce the carbon footprint of food-related energy consumption. Specifically, UPA can make more efficient use of organic urban waste, strengthen capacities to handle intense rain events, improve water management systems, and contribute to a healthier urban living environment by creating green spaces, capturing carbon dioxide, and increasing urban biodiversity.

As the most urbanized region in the world, Latin America has the potential to make important progress in meeting the goals outlined in the ZHC and the SDGs. The region's cities have served as a laboratory for many agricultural, energy, and environmental innovations and can continue to lead these types of initiatives in the future. The stark reality of increasing rates of urban poverty coupled with the negative environmental impacts and the energy-intensive practices of conventional large-scale agriculture necessitates a shift with respect to where and how our world's food is produced. National and local governments, donors, and development actors must continue to support UPA as an opportunity for addressing human and environmental development challenges in areas where the majority of the world's population lives.

ACKNOWLEDGMENTS

The authors would like to thank Alexandra Dueñas Rodríguez of CONQUITO's AGRUPAR project, Ramasio Tiller Ipuana of the Fundación Cerrejón Guajira Indígena, and Andrés Mosquera of WFP Colombia for their contributions this chapter.

The authors would also like to acknowledge the support and valuable comments received from the following WFP colleagues: Jorge Arteaga, Raphael Chuinard, Andrés Garzón, Inés López, Gabriel Martinez, Wilson Salazar, and Lilian Velasquez.

REFERENCES

Alexandratos, N. & Bruinsma, J. (2012) World agriculture towards 2030/2050: the 2012 revision. ESA working paper No. 12-03, June 2012, Food and Agricultural Organization of the United Nations (FAO), Agricultural Development Economics Division. Available from: http://www.fao.org/docrep/016/ap106e/ap106e.pdf [accessed August 2014].

AQUASTAT and FAO (2014) Water uses. Food and Agriculture Organization of the United Nations (FAO), Rome, Italy. Available from: http://www.fao.org/nr/water/aquastat/water_use/index.stm [accessed December 2016].

Baker, J. (2008) Impacts of financial, food and fuel crisis on the urban poor. Directions in Urban Development, World Bank Urban Development Unit, Washington, DC. Available from: https://openknowledge.world bank.org/bitstream/handle/10986/10263/475250BRI0GLB01ections020Box334118B.pdf?sequence=1 [accessed December 2016].

Carvajal, E. (2010) Informe de la evaluación final: Proyecto: producción y comercialización de productos orgánicos [Final evaluation report: Project: production and commercialization of organic products]. CONQUITO and FOMIN (Fondo Multilateral de Inversions), Quito, Ecuador.

CGIAR (2014) Big facts and infographics on climate change, agriculture and food security. Consultative Group on International Agriculture Research, Consortium of International Agricultural Research Centers: Research Program on Climate Change, Agriculture and Food Security, Montpellier, France. Available from: http://ccafs.cgiar.org/bigfacts2014/# [accessed December 2016].

CONQUITO (2014) Guápulo tendrá su bioferia [Guápulo has its bio-fair]. Available from: http://www.conquito.org.ec/guapulo-tendra-su-bioferia/ [accessed April 2017].

CONQUITO (2016) Quito entre las ocho ciudades en el mundo que forman parte del programa de Sistemas Alimentarios Ciudad – Región [Quito, among the 8 cities in the world participating of the program City-Region Food Systems]. Available from: http://www.conquito.org.ec/tag/agrupar/ [accessed December 2016].

Corpoica (2010) Nace grupo Interinstitucional Colombiano de Apoyo a la Agricultura Urbana y Periurbana, liderado por CORPOICA [Interagency Support Group for Urban and Peri-Urban Agriculture led by CORPOICA is born]. Corporación Colombiana de Investigación Agropecuria [Colombian Corporation of Agricultural Research]. Available from: http://www.agronet.gov.co/Noticias/Paginas/Noticia326.aspx [accessed March 2017].

Dubbeling, M. (2011) Urban agriculture, climate change and food security: responses in northern and southern cities. Resource Centres on Urban Agriculture and Food Security and START. Available from: http://resilient-cities.iclei.org/fileadmin/sites/resilient-cities/files/Resilient_Cities_2011/Presentations/A4_Dubbeling.pdf [accessed December 2016].

El-Hage Scialabba, N. & Hattam, C. (2002) Organic agriculture, environment and food security. Food and Agricultural Organization of the United Nations, Environment and Natural Resources Service Sustainable Development Department. Rome, Italy. Available from: http://www.fao.org/docrep/005/y4137e/y4137e02b.htm [accessed December 2016].

ENSANUT-ECU (2014) Encuesta nacional de salud y nutrición – resumen ejecutivo [National health and nutrition survey – executive summary], Volume 1. Quito, Ecuador. Available from: http://instituciones.msp.gob.ec/images/Documentos/varios/ENSANUT.pdf [accessed December 2016].

FAO (1999) Issues in urban agriculture. Food and Agricultural Organization of the United Nations, Agriculture and Consumer Protection Department, Rome, Italy. Available from: http://www.fao.org/unfao/bodies/COAG/COAG15/X0076e.htm#P135_15324 [accessed April 2017].

FAO (2001) Urban and peri-urban agriculture: a briefing guide for the successful implementation of urban and peri-urban agriculture in developing countries and countries of transition. Food and Agricultural Organization of the United Nations, Special Programme for Food Security, Rome, Italy. Available from: http://www.fao.org/docs/eims/upload/215253/briefing_guide.pdf [accessed December 2016].

FAO (2008) Urban agriculture for sustainable poverty alleviation and food security. Food and Agricultural Organization of the United Nations, Rome, Italy. Available from: http://www.fao.org/fileadmin/templates/FCIT/PDF/UPA_-WBpaper-Final_October_2008.pdf [accessed December 2016].

FAO (2010a) Fighting poverty and hunger: what role for urban agriculture? Economic and Social Perspectives Policy Brief 10, Food and Agricultural Organization of the United Nations, Rome, Italy. Available from: http://www.fao.org/docrep/012/al377e/al377e00.pdf [accessed December 2016].

FAO (2014a) Gender: food security. Food and Agricultural Organization of the United Nations, Rome, Italy. Available from: http://www.fao.org/gender/gender-home/gender-programme/gender-food/en/ [accessed December 2016].

FAO (2014b) Urban and peri-urban agriculture in Latin America and the Caribbean: Quito. Food and Agricultural Organization of the United Nations, Rome, Italy. Available from: http://www.fao.org/ag/agp/greenercities/en/GGCLAC/quito.html [accessed December 2016].

FAO (2014c) Urban and peri-urban horticulture: greener cities. Food and Agricultural Organization of the United Nations, Rome, Italy. Available from: http://www.fao.org/ag/agp/greenercities/en/whyuph/foodsecurity.html [accessed December 2016].

FAO, IFAD and WFP (2015) The State of Food Insecurity in the World 2015. Meeting the 2015 international hunger targets: taking stock of uneven progress. Food and Agricultural Organization of the United Nations, Rome, Italy.

Gittleman, M. (2009) The role of urban agriculture in environmental and social sustainability: case study of Boston. Undergraduate Thesis, Tufts University, Medford, MA. Available from: http://dl.tufts.edu/catalog/tufts:UA005.008.048.00001 [accessed December 2016].

Gustavsson, J., Cederberg, C., Sonesson, U., van Otterdijk, R. & Meybeck, A. (2011) Global food losses and food waste. Swedish Institute for Food and Biotechnology and Food and Agricultural Organization of the United Nations. Rome, Italy. Available from: http://www.fao.org/fileadmin/user_upload/suistainability/pdf/Global_Food_Losses_and_Food_Waste.pdf [accessed December 2016].

Harvard School of Public Health (2012) Local and urban agriculture. Harvard School of Public Health, Center for Health and the Global Environment, Boston, MA. Available from: http://www.chgeharvard.org/topic/local-and-urban-agriculture [accessed December 2016].

Hayden, A.L. (2000) Overview of hydroponics. University of Arizona, College of Agricultural and Life Sciences, Tuscon, AZ. Available from: http://ag.arizona.edu/hydroponictomatoes/overview.htm [accessed December 2016].

Hidrobo, M., Hoddinott, J., Peterman, A., Margolies, A. & Moreira, V. (2012) Cash, food, or vouchers? Evidence from a randomized experiment in northern Ecuador. Discussion Paper 01234, International Food Policy Research Institute (IFPRI). Available from: http://www.ifpri.org/sites/default/files/publications/ifpridp01234.pdf [accessed December 2016].

Hines, D. (Photographer). (2016) FGCI wells in La Guajira [photograph].

Hines, D. (Photographer). (2017) Untitled illustration of a drought-affected area in La Guajira [photograph].

IFPRI (2011) Leveraging agriculture for improving nutrition and health: highlights from an international conference. Washington, DC. Available from: http://www.fao.org/fileadmin/user_upload/wa_workshop/PPT/PPT-Levergaing_agriculture_for_improving_nutrition.pdf [accessed December 2016].

IFPRI (2012) Sustainable food security under land, water and energy stresses. International Food Policy Research Institute. Available from: http://www.ifpri.org/publication/leveraging-agriculture-improving-nutrition-and-health [accessed December 2016].

Mayer, H. (1999) Air pollution in cities. *Atmospheric Environment*, 33, 4029–4037.

Met Office and United Nations World Food Programme [WFP] (2012) Climate impacts on food security and nutrition: a review of existing knowledge. Met Office and World Food Programme Office for Climate Change, Environment and Disaster Risk Reduction. Available from: http://www.metoffice.gov.uk/media/pdf/k/5/Climate_impacts_on_food_security_and_nutrition.pdf [accessed December 2016].

Moreno-Peñaranda, R. (2011) Japan's urban agriculture: cultivating sustainability and well-being. United Nations University, Tokyo, Japan. Available from: http://unu.edu/publications/articles/japan-s-urban-agriculture-what-does-the-future-hold.html [accessed December 2016].

Rowe, D.B. (2010) Green roofs as a means of pollution abatement. *Environmental Pollution*, 159(8–9), 2100–2110.

RUAF (2009) Cities, food and agriculture: challenges and the way forward. Resource Centres on Urban Agriculture and Food Security. Available from: http://www.ruaf.org/sites/default/files/Working%20paper%203%20%20Cities%20Food%20and%20Agriculture.pdf [accessed December 2016].

Tiller Ipuana, R. (2014) Proyecto instalación de tecnología limpias para la comunidad Wayuu [Clean technology installation project for the Wayuu community]. Fundación Cerrejón Guajira Indígena, Bogotá, Colombia.

Trenberth, K.E. (2011) Changes in precipitation with climate change. *Climate Research*, 47, 123–138.

Truitt Nakata, G.A. (2014) The next global breadbasket: how Latin America and the Caribbean can feed the world: a call to action for addressing challenges and developing solutions. Global Harvest Initiative and Inter-American Development Bank, Washington, DC, 13 May 2014.

UN DESA (2012) World urbanization prospects, the 2011 revision. United Nations Department of Social and Economic Affairs, New York, NY. Available from: http://www.un.org/en/development/desa/population/publications/pdf/urbanization/WUP2011_Report.pdf [accessed April 2017].

UN Habitat (2012) State of Latin American and Caribbean Cities 2012: towards a new urban transition. United Nations Habitat, Nairobi, Kenya. Available from: http://www.citiesalliance.org/sites/citiesalliance.org/files/SOLAC-ProjectOutput.pdf [accessed April 2017].

United Nations Sustainable Development Knowledge Platform (2014a) Outcome document. Open Working Group on Sustainable Development Goals. Available from: http://sustainabledevelopment.un.org/focussdgs.html [accessed December 2016].

United Nations Sustainable Development Knowledge Platform (2014b) Sustainable Development Goals. Available from: http://sustainabledevelopment.un.org/?menu=1300 [accessed December 2016].

United Nations World Water Assessment Programme (2012) Managing water under uncertainty and risk. The United Nations Water Development Report 4. Available from: http://unesdoc.unesco.org/images/0021/002154/215492e.pdf [accessed December 2016].

University of Rhode Island (2006) Sustainable landscaping: drip irrigation for the home garden. Available from: http://web.uri.edu/safewater/protecting-water-quality-at-home/sustainable-landscaping/drip-irrigation/ [accessed April 2017].

van Veenhuizen, R. (ed.) 2006 Cities farming for the future: urban agriculture for green and productive cities. International Institute of Rural Reconstruction, ETC Urban Agriculture and the RUAF Foundation, Philippines. Available from: http://www.ruaf.org/publications/cities-farming-future-urban-agriculture-green-and-productive-cities [accessed December 2016].

Venkat, K. (2011) The climate change and economic impacts of food waste in the United States. *International Journal on Food System Dynamics*, 2(4), 431–446.

Vermeulen, S.J., Campbell, B.M. & Ingram, J.S. (2012) Climate change and food systems. *Annual Review of Environment and Resources*, 37, 195–222.

Westcot, D.W. (1997) Quality control of wastewater for irrigated crop production. Water Reports 10, Food and Agricultural Organization of the United Nations, Rome, Italy. Available from: http://www.fao.org/docrep/w5367e/w5367e00.HTM [accessed December 2016].

WFP (2013) Evaluación de la situación de seguridad alimentaria de la población refugiada en el Ecuador – PRRO 200275 [Food security evaluation of the refugee population in Ecuador – PRRO 200275]. United Nations World Food Programme, Monitoring and Evaluation Unit, Rome, Italy.

WFP (2014a) United Nations World Food Programme Food Procurement Tracking System Database.

WFP (2014b) United Nations World Food Program Purchase for Progress 2nd Quarterly Report.

WFP & MAGAP (2012) El huerto familiar orgánico y nutritivo: manual técnico para mejorar la alimentación de las familias vulnerable [Organic and nutritious family gardens: manual to improve the food consumption of vulnerable families]. United Nations World Food Programme, Rome, Italy and MAGAP – Ministry of Agriculture, Aquaculture, Livestock and Fisheries, Quito, Ecuador.

WHO (2014) Micronutrient Deficiencies. World Health Organization, Geneva, Switzerland. Available from: http://www.who.int/nutrition/topics/ida/en/ [accessed December 2016].

CHAPTER 4

Renewable energy use for aquaponics development on a global scale towards sustainable food production

Ragnheidur Thorarinsdottir, Daniel Coaten, Edoardo Pantanella,
Charlie Shultz, Henk Stander & Kristin Vala Ragnarsdottir

4.1   INTRODUCTION

Aquaponics – the combination of recirculating aquaculture systems (RAS) and hydroponics for plant cultivation – has received increased interest from researchers and entrepreneurs globally over the last decade. The driving force for this sustainable design of combined production is the need for new approaches to future food production. The aquaponics methodology mimics natural circular processes of water, nutrients and energy and has been known for a long time in various countries in South America and Asia. The world is facing problems due to the scarcity of these natural resources together with an increasing population. The world's population is now 7.4 billion and is expected to reach 9.6 billion around 2050 whereof more than two-thirds is projected to be living in urban areas (Jahan, 2015). Conventional methods in aquaculture and horticulture are challenged by its environmental impact, resource depletion and fluctuating production costs. Therefore, it is crucial to implement new production technologies which not only help to secure food levels, but which also have a minimal impact on, and help to maintain a healthy environment. Sustainable agriculture has been defined as a process that does not deplete any non-renewable resources that are essential to agriculture in order to sustain the agricultural practices (Lehman *et al.*, 1993). This can be achieved by implementing production systems based on natural ecosystem's closed nutrient cycles (Francis *et al.*, 2003). This is the case in aquaponics where nutrient rich wastewater from aquaculture is used to grow plants, limiting the need for mineral fertilizers and optimizing water use (Turcios and Papenbrock, 2014). An important issue related to this development is direct use of geothermal-, hydro-, wind- and solar renewable energy.

Aquaponics has been developed by a few pioneering researchers in the US (Mageau *et al.*, 2015) and the US Virgin Islands (UVI) (Rakocy *et al.*, 1992, 2005, 2007). Simple family scale units have received increased popularity, not least in the US and Australia, where backyard private aquaponics have been quite popular in temperate areas (Lennard and Leonard, 2005). Aquaponic systems originally developed in these regions due to limited land and freshwater resources. Beginning in 1998, Dr. James Rakocy and his colleagues began teaching the first intensive aquaponic training program. Since that time, many aquaponic systems have been spawned across the US, North America and globally. In the US, both backyard and commercial aquaponic systems can be found in every state, including in many educational institutions. To support this growing field, the Aquaponics Association was created in 2011. As of November, 2016, the association is stronger than ever. The goals of the association are: to promote the benefits of aquaponic growing; to educate the general consumer and food safety officials about the inherent safety of aquaponically grown food; and to dispel the myths and rumors about aquaponicaly grown food.

Commercial facilities now exist not only outdoors in warm climates such as Puerto Rico, Hawaii and Florida, but also in more temperate climates such as Illinois, Indiana, Wisconsin and Minnesota, sited in climate controlled greenhouses or inside buildings. Indoor production facilities using vertical rack systems and artificial lighting have continued to increase in popularity. Urban Organics in St. Paul, Minnesota, for example, has been operating in a retrofitted

Figure 4.1.   Green River Greenhouses, Indiana, USA (Photo: Shultz).

brewery building for three years (News, 2015). Currently, they are expanding to a 7500 m^2 indoor production facility. Other commercial farms like Farmed Here in Chicago, IL and the Farming Fish in Oregon currently certify their crops as USDA Certified Organic. During 2015–2016, the US National Organics Program formed a task force to develop a report on how water-based crop production aligns with the current organic rules. Resistance from soil-based farmers led to this call of action. The report was finalized in July, 2016 (USDA, 2016).

Figures 4.1–4.3 show examples of aquaponics in Indiana, Texas and California.

To distinguish organic hydroponics (including aquaponics) from traditional hydroponics, the term "bioponics" was defined as a water-based crop production system that relies on biological activity for mineralization of organic nutrients into available nutrients for plants. This microbial identification and characterization in aquaponic systems has become a topic of interest by many US researchers, including NASA and Purdue University, as well as many independent researchers.

US universities and colleges are slowly beginning to offer courses specific to aquaponics. At Santa Fe Community College, students can earn a certificate or an associate's degree in aquaponics. Currently, the school is building a 1200 m^2 greenhouse that will be used for aquaponics, hydroponics and aquaculture. The greenhouse will incorporate a micro-grid for energy production, utilizing a variety of alternative fuel sources, allowing the production facility to operate independently from the city's electrical grid. Kentucky State University will soon release the first accredited online aquaponics course, available for undergraduate and graduate students in the US and abroad. Other schools with aquaponic curriculum and/or research opportunities include University of Wisconsin-Stevens Point, Cabrillo College, Mesa Community College, Auburn University, Temple University Ambler Campus, University of Arizona, Iowa State, University of Minnesota, Loyola University, Berea College, Chicago State University, University of California Davis, and many others. In addition, thousands of aquaponic educational systems can be found across the US in elementary, middle and high schools.

In Asia, aquaponics is being widely considered as a production mainly to address food security, water scarcity, salinity and land access issues. The energy efficiency and labor intensity vary to a

Figure 4.2.   Sustainable Harvesters, Texas, USA (Photo: Shultz).

Figure 4.3.   Ouroborous Farms, California, USA (Photo: Shultz).

large degree and is dependent on the production methods. However, labor costs do not have a high incidence in the farming due to the lower wages in the region. In South East Asia, aquaponics is gaining interest among countries as a strategy to address food insecurity and to guarantee production in land scarcity conditions. Interesting examples are in Indonesia, where national programs are supporting the development of aquaponics at village and community level with

Figure 4.4.   BAK Pemeliharaan ikan (aquaponics with tilapia). Subirrigation of media beds on canals (Photo: Pantanella).

low-tech systems that mainly emulate traditional farming (e.g. floating pots on ponds, small closed systems with lined ponds feeding potted plants or NFT pipes). Such systems are farming portion-size fish, which guarantee fast turnovers (growing periods of 3–4 months). In the case of tilapia, this solution is easier to manage since it does not require male-only fish. In Thailand, the high concerns on the massive use of pesticides is pulling up aquaponics at small-scale level to allow access to the benefits of chemical-free vegetables.

Figures 4.4–4.7 show examples of aquaponic systems in Indonesia.

In many Pacific Islands, aquaponics is being considered for the production of fresh vegetables, to reduce the pressure on overexploited fishery resources by also integrating fishermen's income and to promote the consumption of healthy food as a program to reduce the incidence of non-communicable diseases (NCD), such as diabetes. In Samoa, there are currently three registered aquaponic farms, while in the Marshall Islands and Fiji, local NGOs or small-scale enterprises are running systems either for communities or commercial level.

Beside applied and academic research in many Asian institutions addressing production issues, there is the need to prove the economic sustainability of fish being produced in commercial aquaponics, given the fierce competition with inexpensive pond culture. Therefore, the economic success is seen in high-value fishes or in the nursing of fish with very short turnover, as the Asian Institute of Technology recently demonstrated with trials on tilapia fry.

In the Middle East, aquaponics is regarded as a strategic solution to saving water, bringing back salinated farms into production and guaranteeing food security/self-sufficiency, which are part of the national priorities in the whole region. Beyond the pioneering farm, built at the Zayed Higher Agricultural Centre for Development and Rehabilitation, in Abu Dhabi (consisting of six UVI-size systems), there are many public or private ventures in Oman, Bahrain and Kuwait that are commercially operating and offer high quality products.

Figure 4.5.   Wolkaponic. Aquaponic with catfish. Pots are irrigated by gravity (Photo: Pantanella).

The interest in aquaponics has also started in Africa (Dediu *et al.*, 2012), but it is still in its infancy. Production stats on aquaponic operations are non-existent due to the small number of systems. The systems that are known to exist are generally small-scale backyard do-it-yourself systems and those developed by NGO's to feed several families in a community. Companies promoting aquaponics are very active in Africa, so there is an expectation that these systems

Figure 4.6.   Hydroponik Sayuran, a small NFT in Indonesia (Photo: Pantanella).

will become more prominent over time. African countries where aquaponics is known to exist include South Africa, Lesotho, Botswana, Malawi, Kenya, Zambia, Ghana, Zimbabwe, Namibia, Uganda and Rwanda. Many of the initiatives are those by groups concerned about alleviating local poverty and nutritional deficiencies. As with most development, primary human needs related to health and survival will be the dominant driving force with social benefits following.

Figure 4.7.   Bumina tanks made with bamboo and liners. Lateral pots with plants are directly irrigated with surface flow system (Photo: Pantanella).

A lot of interest in aquaponics was shown recently in South Africa. A couple of commercial aquaponic operations are running now all over the country and a local Aquaponics Association was established earlier this year (http://www.aquaponicssa.org). The biggest commercial system is located on a farm outside of Grabouw, in the Western Cape Province. This system was designed and constructed by a private company called "Full Circle Integrated Agriculture". It includes a 2500 m^2 area, all under greenhouse tunnels, with deep-water culture grow beds, and an additional 0.5 ha open area with passion-fruit production in aquaponics media based grow beds. The main crops in the deep-water culture grow beds are different high value herb varieties and four different lettuce groups. Figure 4.8 shows an aquaponics system in Grabouw in South Africa.

Aquaponics development in Europe has been driven by small-scale hobby systems and research units along with a few commercial aquaponics companies which exist with limited production capacity (Goddek *et al.*, 2015, Thorarinsdottir *et al.*, 2015). The interest in aquaponics has been increasing in Europe, with the research focus moving toward implementing larger scale aquaponics production units built on the recent development in RAS and modern hydroponics. At present, the largest aquaponics facility in Europe is the 6000 m^2 NER BREEN farm in Hondarribia, northern Spain, see Figure 4.9 (Thorarinsdottir *et al.*, 2015). Also, urban aquaponics has received increased interest (Hall *et al.*, 2014). Earlier this year, Urban Farmers built a 1500 m^2 rooftop farm in The Hague, Netherlands, see Figure 4.10. Other commercial aquaponics facilities in Europe are still on a few hundred square meter pilot scale, but are planning expansion, e.g. in Denmark, France, Germany, Iceland, Norway, Slovenia, Spain, Sweden, Switzerland and UK (Thorarinsdottir *et al.*, 2015). Figure 4.11 shows an aquaponics system at the Laugarmyri farm in Skagafjordur, Iceland. In the UK The Bristol Fishproject social enterprise has been built by an enthusiastic entrepreneur in collaboration with her local community (A.-M. Archer, BristolFish, pers. Comm., 2016).

Figure 4.8.   Full Circle Integrated Agriculture (Pty) Ltd, Aquaponics System, Grabouw, South Africa
(Photo: Stander).

This year, the Association of Commercial Aquaponics Companies (ACAC) was established, based on the work carried out in the European funded project EcoPonics
(http://www.aquaponics.is/ecoponics) (2013–2016) and the COST Action FA1305 EU Aquaponics Hub (http://www.euaquaponicshub.com). ACAC's main aim is to provide a network of SMEs
whose primary focus is the commercialization of aquaponics in Europe. The group will col-
laborate and share information relating to current best practice in aquaponics production (e.g.
regarding regulations, licenses and related issues in each country, and to promote standardization
and certification in aquaponics).

More recently, the development of specialism within this field has begun to grow. Some
aquaponics systems have now been merged with existing knowledge of mariculture to evolve the
culturing of purely seawater or saline tolerate species (Joesting et al., 2016), known as salt water
aquaponics or integrated mariculture (Wilson, 2005). Others have integrated crayfish, worms
or insects in the production circle (Thorarinsdottir et al., 2015), or have redirected the focus of
their target crops from more traditional species, cultivated purely for use as food, to more diverse
organisms such as algae (algaeponics). In this way, algae produced can then be extracted to afford

Figure 4.9. NER BREEN Aquaponics in Hondarribia, Basque country, Spain (Photo: Fernando Sustaeta).

Figure 4.10. UrbanFarmers roof-top aquaponics in The Hague, Netherlands (Photo: Thorarinsdottir).

higher value products such as biofuels or raw materials for the cosmetic-, nutraceutical- and pharmaceutical industries (Bux and Chisti, 2016); the extra revenue therefore helping to offset the running costs of the system.

This chapter gives an overview of the present aquaponics techniques available today, focusing on the direct use of renewable resources and replication potential in different climates and

Figure 4.11.   Aquaponics integrated in the greenhouse farm Laugarmyri in Iceland (Photo: Thorarinsdottir).

economic environments. The main bottlenecks of each production method are presented, and suggestions of research areas for future development connected to UN Sustainable Development Goals 2015–2030 are given.

4.2   AQUAPONICS TECHNOLOGY

4.2.1   *Simple recirculating units*

A simple schematic aquaponics system is shown in Figure 4.12. It consists of a fish tank, a sedimentation unit, a bio-filter and a hydroponics system. The water is moved from the fish tank, through the filtering units, and on to the hydroponics area, before then returning back to the fish tank. The design, size, and complexity can vary greatly. In most cases, a special sump tank collects the water from the filtering units. The water can run in a simple circle or may be pumped directly from the sump tank to the three main units: (i) the fish tanks, (ii) the hydroponics, and (iii) the bio-filter. In all cases, the same water is delivered to all parts of the system, and so it may be difficult to maintain optimum conditions for both the fish and plant production, as most often there will be a large variation between these growing environments (e.g. optimum temperature and pH intervals for fish versus plant).

Removal of solids (i.e. excess fish feed and feces) is essential for maintaining good water quality. This can be facilitated via simple sedimentation tanks, sand filters, radial flow filters, bead filters, or a drum filter. The choice between passive sedimentation and mechanical filtration depends on the degree of intensification (e.g. fish stocking density and feeding rates) of the farm (Lekang, 2013; Timmons and Ebeling, 2010).

The bio-filter is necessary for the nitrification processes, as it converts ammonia excreted by fish to nitrate that can be taken up by plants, thereby helping to maintain a healthy system (Timmons and Ebeling, 2010). The bio-filter can be a simple tank with aerated porous media, or a more complex structure. The main issue is to have a suitable large surface area, and maintain stable and good conditions for the nitrifying micro-organisms to grow.

Depending on the density of the system, the use of algae as bio-filters (or bio-filter promoters) could be an option, and have been commonly employed by open-water aquaculture systems for

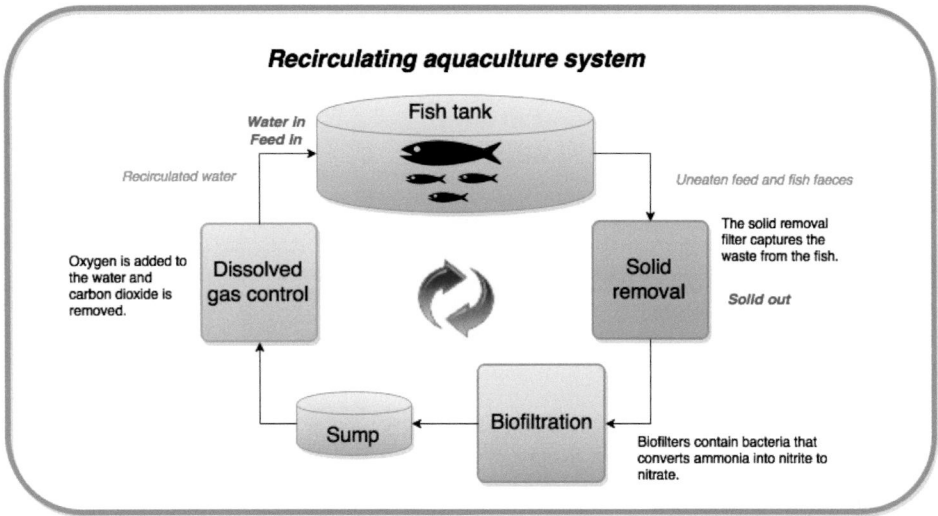

Figure 4.12. Schematic overview of a simple aquaponics system (Thorarinsdottir *et al.*, 2015).

many years with good results (Neori *et al.*, 2004). Either freshwater or seawater (micro or macro) species are selected, depending upon their compatibility with tank cultivation, the specific setup used, and also in relation to the other species cultivated within the same system (Amosu *et al.*, 2016; Lin *et al.*, 2016; Rabiei *et al.*, 2016). As such, inorganic compounds present within the wastewater are absorbed by the algae. This not only helps to clean and buffer the recirculated water, thereby protecting the other species from exposure to potentially fatal levels of these compounds (Amosu *et al.*, 2016), but at the same time it promotes their own healthy growth, which in turn reduces the likelihood of colonization of opportunistic, undesirable algae strains.

Many macroalgae species are seen to harbor a variety of microbiota, which are not only essential to the health of their hosts, but in themselves, may also act as promotors of the biofiltration process through their contributions to the acceleration of nutrient cycling (Tapia *et al.*, 2016). Additionally, studies have shown some algae varieties to be effective in the absorption and bioremediation of many types of pollutants including heavy metals (Singh *et al.*, 2016), and even dioxins and Bisphenol A from aquatic environments (Subashchandrabose *et al.*, 2013), however, the onward use of the obtained algae biomass would be questionable.

The hydroponics unit can be based on nutrient film technique (NFT), deep-water culture (DWC) or media-based grow beds (Lennard and Leonard, 2006). The choice will depend on the plant species and the size of the system.

4.2.2 *Modern RAS and hydroponics*

Larger scale commercial aquaponic production units have been built on recent developments in RAS technology and modern hydroponics. The main principles are the same as described for the simple recirculating units. However, the emphasis is on keeping optimum farming conditions in both the fish and plant production units, and thereby eliminating the trade-offs. This method is described as *decoupled aquaponics*, see Figure 4.13 (Thorarinsdottir *et al.*, 2015), and like the traditional aquaponics design, is based on using nutrient rich wastewater from the fish for growing plants. However, in decoupled systems, the same water is not then recirculating again from the plants to the fish, except for possible condensed water from the plant evaporation (Kloas *et al.*, 2015). By this method, optimum growing conditions can be maintained in both the aquaculture and hydroponics. Modern RAS and hydroponics in a decoupled system have higher capital costs and higher energy intensity, but require less manpower than the simple systems described above.

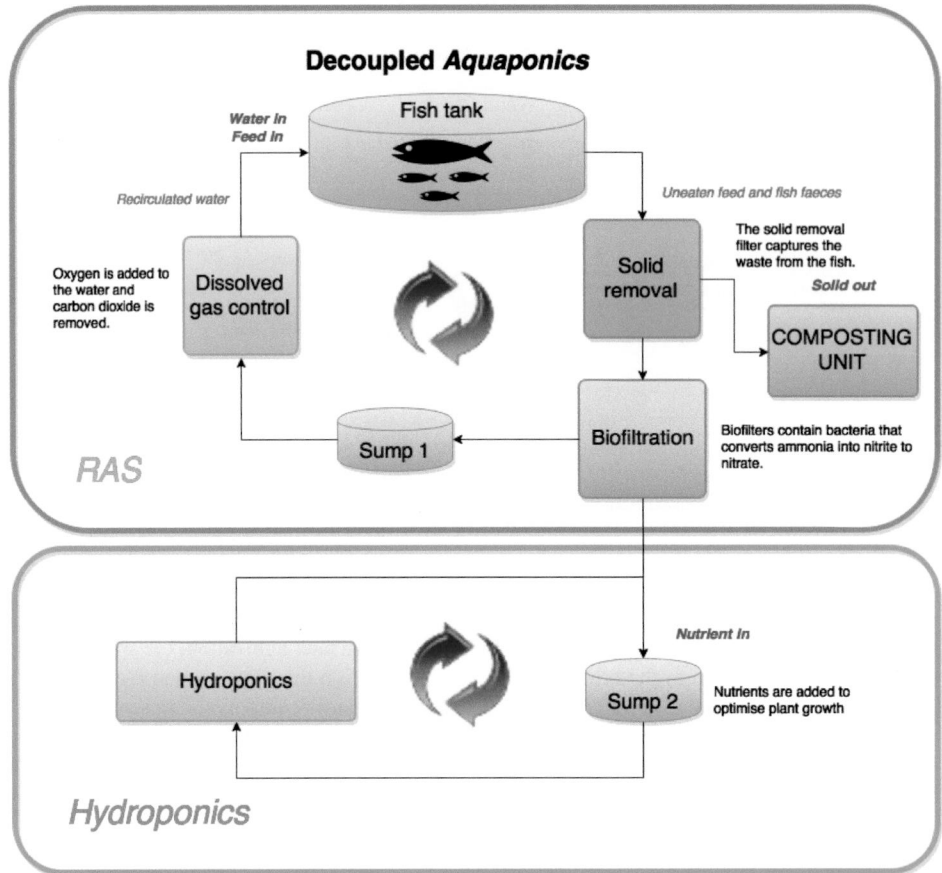

Figure 4.13.   Modern RAS connected to hydroponics in a decoupled system (Thorarinsdottir *et al.*, 2015).

Decoupled modern systems include more automatization and control, and can maintain higher densities and productivity to make the business financially viable.

4.2.3   *Resource intensity*

With the symbiotic interaction between the production units, the need for input of nutrients as well as the output of waste is minimized. Aquaponics recirculate all the water, nutrients, and output waste. Examples of such waste are in the form sludge, and fish and plant waste, which can be used as fertilizers, or in the cultivation of added-value by-products such as crayfish, worms or insects. Modern RAS alone have a high degree of water reuse (i.e., 95–99%) (Al-Hafedh *et al.*, 2003; Dalsgaard *et al.*, 2013). However, in aquaponics, nutrients are used for valuable plant production instead of being removed as waste (Martins *et al.*, 2010). This is important, not least concerning the emerging problem of lack of availability of phosphorous, one of the important macro-nutrients in agriculture (Ragnarsdottir *et al.*, 2011; Sverdrup and Ragnarsdottir, 2011, 2014).

Simple, small scale aquaponics can be built with cheap materials and passive filtration, thereby minimizing the energy use (Sommerville *et al.*, 2014). Larger scale commercial production facilities, with a higher density of fish and plants, need more effective filtration and lighting. As such, they are more energy demanding, and require more expensive design and materials (Thorarinsdottir *et al.*, 2015).

4.2.4 *Integrated multi-trophic approach*

The inclusion of other species from varying trophic levels within an enclosed, recirculating system, not only allows for a more complete use of by-products produced between the cultured species present, thereby greatly reducing waste streams, but may also affords other advantages over monoculture techniques as well (Cubillo *et al.*, 2016). In this respect, a symbiosis can often be observed between co-cultured species, helping to balance biological and chemical processes by mimicking that observed in natural ecosystems, leading to the promotion of healthy growth of both plants and fish and increased survival rates (Barrington *et al.*, 2009).

Using algae species within aquaponics systems can greatly help to balance nutrient levels within a given system. Which, in turn, reduces running costs, and in some cases can increase the overall net production value of a setup, thereby making its operation more economically viable (Packer *et al.*, 2016). Additionally, algae may produce secondary metabolites and/or harbor essential microorganisms which can further help to increase growth rates and even boost immunity of the other species present (Pardee *et al.*, 2004; Petit and Wiegertjes, 2016).

Over the past two decades, researchers have shown that ecosystems with more species are more efficient at removing nutrients from soil and water than are ecosystems with fewer species. Multi-species ecosystems are therefore more resilient. Increasing the biodiversity in aquaponic systems may therefore not only increase the efficiency of the system, but may also aid in the preservation of biodiversity – which is important because we are in the era of decreasing biodiversity (Cardinale, 2011; Naem *et al.*, 1994).

4.3 AQUAPONICS AS A SUSTAINABLE FOOD PRODUCTION METHOD

4.3.1 *Types of products*

4.3.1.1 *Fish*
Several types of fish have been successfully farmed in aquaponic systems under different conditions (see Table 4.1).

4.3.1.2 *Terrestrial plants*
Initially, the main focus of aquaponic development was in the production of high-quality, high-value, specialty fish species for food. Conversely, the plant species grown were seen more as a by-product; their role being more important regarding biofiltration of the fish wastewater within the closed system. However, more recently, the full potential of aquaponics systems has been realized, and as such, a large variety of fruit, vegetables and herbs have now been tested for their suitability to aquaponics (see Table 4.2).

As can be seen in the case of saltwater fish species, the following halophytic plants have also been recently studied with regard to their suitability for potential aquaponics application: sea purslane (*Sesuvium portulacastrum*) (Rutger, 2014) and saltwort (*Batis maritima*) (Gomes, 2012), black needlerush (*Juncus roemerianus*) (Joesting *et al.*, 2016), salicornia (*Salicornia europea*) (Pantanella *et al.*, 2013), seabeet (*Beta vulgaris* subsp. *maritima*) (Rutger, 2014) and sweet basil (*Ocimum basilicum*) (Rutger, 2014).

4.3.1.3 *Other species*
Originally, fish and land plants were the main focus of aquaponics production (hence the name "aquaponics" being derived from the words *aqua*culture and hydro*ponics*) (Tyson *et al.*, 2011). However, through research and a better understanding of the significance of operating more integrated multitrophic systems, tests have been carried out with cultivation of a variety of other species alongside these staple goods. This includes freshwater crustaceans such as the red claw crayfish (*Cherax quadricarinatus*) (Diver, 2006), redworms (*Eisenia fetida*) (Rakocy *et al.*, 2006), and freshwater prawns (*Macrobrachium rosenbergii*) (Uddin *et al.*, 2007).

Table 4.1.   Examples of edible and ornamental fish species cultivated using aquaponic systems.

Edible species

Fresh water	Marine	Ornamental species
Barramundi (*Lates calcarifer*)	Barramundi (being able to	Angelfish (*Pterophyllum* sp.)
Bluegill (*Lepomis macrochirus*)	survive in both fresh water	Goldfish (*Carassius auratus*)
Carp (*Ctenopharyngodon*	and saltwater conditions)	Guppy (*Poecilia reticulata*)
idella × *Aristichthys nobilis*)	(Auaponics, 2014)	Hatchet fish (*Gasteropelecus* sp.)
Channel Catfish (*Ictalurus*	Milkfish (*Chanos chanos*)	Koi (*Cyprinus carpio*
punctatus)	(Gomes, 2012)	*haematopterus*)
Crappie (*Pomoxis* sp.)	Red drum (*Sciaenops*	Molly (*Poecilia sphenops*)
Golden Perch (*Macquaria*	*ocellatus*) (Rutger, 2014)	Swordtail fish (*Xiphophorus* sp.)
ambigua)		Tetra (species of the
Largemouth Bass (*Micropterus*		Alestiidae, Characidae and
salmoides)		Lebiasinidae families)
Murray cod (*Maccullochella peelii*		Zebra fish (*Danio rerio*)
peelii)		(Selock, 2003; Weller, 2005).
Nile Tilapia (*Oreochromis niloticus*)		
Pacu (species of the Serrasalminae		
subfamily)		
Rainbow trout (*Oncorhynchus mykiss*)		
Silver Perch (*Bidyanus bidyanus*)		
Striped Bass (*Morone*		
chrysops × *Morone saxatilis*)		
Hybrid Tilapia (*Oreochromis urolepis*		
hornorum × *Oreochromis mossambicus*)		
Mozambique Tilapia (*Oreochromis*		
mossambicus)		
Redbreast Tilapia (*Tilapia rendalii*)		
African Sharptooth Catfish (*Clarias*		
gariepinus)		
Jade Perch (*Scortum barcoo*)		
White Sturgeon (*Acipenser*		
transmontanus)		
Pangas Catfish (*Pangasius pangasius*)		
Yellow Perch (*Perca flavescens*)		
(Adler *et al.*, 2000; Selock, 2003; Weller,		
2005).		

In regard to saline aquaponics, the production of salt tolerant plants or even algae are increasing in popularity. This is in contrast with fresh water aquaponics systems, which algae often encounter potentially harmful contamination by opportunistic microalgae species. Species such salicornia (*Salicornia europea*), salsola (*Salsola soda*), seabeet (*Beta maritima*) (Pantanella and Bhujel, 2015; Pantanella and Rakocy, 2012; Pantanella *et al.*, 2011), and Ogo (*Gracilaria parvisipora*) (Gomes, 2012), have all been utilized with remarkably good results.

Other halotolerant species, which have been successfully introduced in order to create multitrophic polyculture systems, are marine shrimp (*Litopenaeus vanname*) (Kuhn *et al.*, 2007), abalone (*Haliotis discus hannai*), sea urchin (*Paracentrotus lividus*) (Wikfors and Ohno, 2001), and sea cucumber (*Parastichopus californicus*) (Ahlgren, 1998).

4.3.2   *Control and safety*

The benefits of employing bi- and multitrophic systems are well documented. However, as mentioned previously, the reality of maintaining such systems are far from easy. Environmental

Table 4.2.   Examples of edible and ornamental plant species cultivated using aquaponic systems.

Commonly cultivated species		Other species (Nelson and Pade Inc, n.d.)
Any sized system	Commercial systems only	
Arugula (*Eruca sativa*)	Bean (*Phaseolus vulgaris*)	Beet (*Beta vulgaris*)
Basil (*Ocimum basilicum*)	Bell pepper (*Capsicum annuum*)	Cannabis (*Cannabis* sp.)
Chives (*Allium schoenoprasum*)	Broccoli (*Brassica oleracea*)	Carrot (*Daucus carota*
Kale (cultivar of *Brassica oleracea*)	Cabbage (cultivar of *Brassica oleracea*)	subsp. *sativus*)
Lettuce (*Lactuca sativa*)	Cauliflower (cultivar of	Micro greens
Pak choi (*Brassica rapa*)	*Brassica oleracea*)	Onion (*Allium cepa*)
Spearmint (*Mentha spicata*)	Cucumber (*Cucumis sativus*)	Radish (*Raphanus sativus*)
Spinach (*Spinacia oleracea*)	Melon (species of the	Sweet corn (*Zea mays* convar.
Swiss chard (*Beta vulgaris* subsp. *vulgaris*)	Cucurbitaceae family)	*Saccharata* var. *rugosa*)
Watercress (*Nasturtium officinale*)	Pea (*Pisum sativum*)	Dwarf trees, such as:
	Squash (*Cucurbita* sp.)	Bananas (*Musa* sp.)
Ornamental species, such as:	Tomato (*Solanum lycopersicum*)	Bonsai (various species)
Ivy (*Hedera* sp.)	(Diver, 2006; Weller, 2005).	Lemon (*Citrus × limon*)
Philodendron (*Philodendron* sp.)		Lime (*Citrus aurantifolia*)
(Diver, 2006; Rakocy *et al.*, 1992; Weller, 2005).		Orange (*Citrus × sinensis*)
		Pomegranate (*Punica granatum*)
		Edible flowers such as:
		Orchids (*Dendrobium* sp.)
		Violets (*Viola* sp.)

variables such as nutrient balance, illumination, pH, temperature and liquid gas concentrations (i.e. O_2 and CO_2) all need to be carefully monitored on a regular basis to maintain the optimal health of all the species present (Shafeena, 2016).

Another major consideration is that of contamination. This could be in the form of toxins leaching from the materials used to construct the recirculating system (e.g. aluminum or plastics), or may be introduced via externally sourced materials (such as in tap water or feed). More often, due to the nature of these systems, they are biologically based. Such contaminants include filamentous algae, bacteria and viruses; they not only can be detrimental to the species present within an aquaponics system, but if not quickly identified and adequately addressed, could also have potentially serious health implications for the people consuming these produce (Chalmers, 2004; Fox *et al.*, 2012).

4.3.3   *Efficient use of resources*

4.3.3.1   *Renewable energy sources*
Direct use of sunlight, geothermal heat, and waste heat can all be used in food production. However, these are currently only used to a limited extent, when compared to the huge potential these renewable energy sources offer.

Solar energy is one of the most important sources of renewable energy. The passive use of solar energy is employed actively in greenhouse production. Furthermore, solar energy can be effectively captured and distributed using photovoltaic systems. Even so, the potential solar energy is highly dependent on the geography. For example, countries far from the equator are less able to depend on solar energy, especially during the dark winters, or even on cloudy summer days. Whole year production in these locations therefore depend on artificial lighting to maintain growth when the sun light is not sufficient.

Geothermal areas are found widely, although they are mostly known to be close to tectonic plate boundaries that have associated volcanic activity (Palmason, 2005). Countries with large geothermal sources include California, Mexico, Hawaii, Israel, Jordania, Iceland, Italy, France, Germany,

Austria, Turkey, Russia, China, Japan, Thailand, New Zealand, Central America, Western coast of South America and Africa (e.g. Kenya, Ethiopia). The direct use of geothermal energy is sustainable and cost effective, and includes heating buildings, greenhouses and aquaculture water, as well as offering several useful industrial processes (see below for more details). Geothermal energy is also often used for electricity production. The potential of direct use of geothermal energy is large in countries that have volcanic activity, and many countries are just in the starting phase of learning to use this abundant and environmentally friendly energy source. Taking Iceland as an example, with abundant geothermal resources, the main emphasis thus far has been on house heating and electricity generation. Only a limited percentage of the energy is used for food production; horticulture, aquaculture, and processing. However, the use of geothermal resources for food production and processing is expanding.

Waste heat from combined heat and power plants, or other industries, are used in several places for heating up or cooling down greenhouses and aquaculture water. Some large greenhouse companies, as for example Thanet Earth in the UK (http://www.thanetearth.com), even use their own combined heat and power plant to produce electricity, heat and CO_2 for the production. The aquaponics company Breen (http://www.nerbreen.com), in Spain, has used solar energy and waste heat for its production since 2010.

Bioenergy can be used for combined heat and power production. In aquaponics, organic waste from the farmed products could be looked at as a potential source for bioenergy. One example might be the production of methane as is done on some animal farms (Mageau *et al.*, 2015). Although the organic waste would probably be better used as fertilizer, or as an input source for other high-value, innovative production.

Wind power is the fastest-growing energy source in the world, with a huge potential to be directly linked to the agricultural sector (UCSUSA, n.d.). Farmers can receive an important extra income by including windmills on their land at the same time as using it for food production. A wind turbine could be easily incorporated into an aquaponics system to gain the benefits of the extra energy harvested in this way.

As the name "hydropower" suggests, it is the process of obtaining power (either kinetic or electrical) from the movement of water. As most aquaponic designs involve the movement of water to and from the vegetative growth and fish run areas, it has therefore been suggested that energy from this recirculating internal flow of water could be harvested and used to offset electricity (or at least a portion of it) needed to power the unit. Experiments have employed the use of bell-siphons and mini-hydroturbines (Rama, 2014). However, so far, only minimal amounts of power have been gained.

Possibly a better solution could be achieved by locating an aquaponics system close by to an external source of constantly running water (possibly even using an open system as opposed to a closed loop aquatic system), or even near to a hydroelectric dam. In this way, the extra energy which could be harvested would be far greater than that gained via the aquaponics setup alone.

Ocean energy derives power through a combination of wave and tidal movements, which are then converted into electricity. Saline aquaponic systems located close to (or just off) the shore may be able to benefit from energy derived from this source.

The use of renewable energy sources may also provide a further potential use for aquaponics via their direct application. Such actions as preservation, food processing and extraction/production of high-value materials obtained from cultivated species could be easily facilitated by the heat and/or mechanical force afforded by these resources (Riva, 1992). For example, geothermal steam/water could provide an alternative source of heat, vacuum (via the use of an ejector) and extraction (as superheated water), which in turn could be used in the process of vacuum packing, (vacuum) drying, freeze drying, distillation/fractionation, and vacuum filtration (Palmason, 2005). In a similar way, wind/ocean energy, and hydropower could all provide the ability to crush and grind materials, such as was done in the early 19th century via the use of water and windmills.

The recirculation in aquaponics systems helps to reduce water use/waste. This is an extremely important value in today's age, as potable water is becoming an increasingly scarce and precious resource globally (Bigas, 2012).

4.3.3.2 *Energy efficiency and waste streams*

When integrating RAS and hydroponics, energy should be saved through the use of proper design and technology. Simple aquaponics setups with a natural flow of water through the system require a limited amount of energy. As the aquaponics design becomes more complex with more mechanical filtration such as drum filters and oxygenation, the energy requirements go up. At the same time, production capacity increases and the need for manpower decreases. Depending on the economic and environmental parameters, manpower cost, access to sunlight, and other energy sources, different aquaponics designs will need to be considered.

In greenhouses, energy is needed to control the temperature, and in northern areas, additional lighting is required. In recent years, plant production in industrial buildings has developed with species growing on shelves with LED lights using the space efficiently. Urban food production also lowers the need for transportation and its associated environmental and economic costs.

Aquaponics design is, in principle, based on optimum use of all resources the minimization of waste (e.g. CO_2 from the aquaculture part of the system can be used by the plants, thereby lowering the need for additional CO_2). This relates to the cradle-to-cradle design of life cycle analysis (LCA), and presents eco-effectiveness which moves beyond zero emissions and produce services, resulting in products which take into account social, economic and environmental benefits (Braungart *et al.*, 2007; Kumar and Putnam, 2008; McDonough and Braungart, 2002). Creative systems which include not only fish and plants, but also additional innovative cultures (e.g. vermiculture, insect production, and/or crayfish), have been constructed (Thorarinsdottir *et al.*, 2015). Also, as has been mentioned previously, the inclusion of algae bio-filters give rise to the conversion of unassimilated and/or excreted phosphorus and nitrogen waste into higher value biomass such as food, feed, and biochemicals. This generates simultaneous provision of a more stable and habitable environment for the other organisms in the integrated system, as well as potentially generating a higher net worth of the system as a whole (Shpigel and Neori, 2007).

4.3.3.3 *Savings through well-designed integration*

It is important to remember that all three of the key aquaponics components; bacteria, fish, and plants, have a peak performance temperature at which their metabolism occurs optimally. The bacteria function best at around 27°C, whereas different fish and plant species each have their preferred temperature. For example, tilapia should be kept at around 28°C, which would require the co-cultivation of warm climate plants, such as cucumbers, herbs, tomato, peppers etc. On the other hand, when farming under cooler conditions, such as with trout as the fish species, then typically peas, lettuce, and watercress would be grown, in order to maintain temperatures of around 18°C. However, under these cooler conditions, bacteria activity is far slower, therefore larger colonies are required for the conversion of toxic ammonia into plant food. The result is that a higher grow bed to fish ratio is necessary when cultivating under cool water conditions.

In semi-tropical climatic areas, keeping the water cool in summer is best achieved by shading the fish (they do not require sunlight), and opening the doors and vents of the greenhouse tunnel. Under extreme conditions, it is possible to use geothermal cooling or evaporative cooling to prevent the temperature of the air, or water, respectively, from exceeding the desired levels.

Urban aquaponics, in general, has received increased interest (Kledal, 2012; Graber *et al.*, 2014). In the UK especially, several young people have started their own companies using old industrial buildings (A.-M. Archer, BristolFish, pers. Comm., 2016). Of note is Archer's company, which is a social enterprise, exemplifying the rising development from commercial- to social enterprises (Clark, 2009). The energy use in the production may be higher compared to growing in greenhouses, however, the energy use for transport is less, and also packaging costs may be decreased.

Roof gardens are becoming more popular, even in colder climates. A showcase can be found on top of the newest building at the University of Greenwich UK (B. Kotzen Kotzen, University of Greenwich, pers. comm., 2016). The gardens have been a success with all year round growing of herbs and vegetables, and the University also has a small greenhouse with an aquaponics pilot unit.

A good example of effective use of space (where it is in very short supply), is the Spirulina rooftop production project in Bangkok (Brooks, 2013). Due to high temperatures and constant sunlight, both the yield and grow time for this species is very high. The company "Energia" first tested its growing system on a skyscraper. However, owing to its huge success, it is now preparing to expand its operation to include hotels, businesses, and other building rooftops in the capital.

Effective use of algae within these integrated systems can help them function more efficiently by lowering running costs, and in some cases, may even generate revenue (Griffiths *et al.*, 2016). In this respect, reductions in water consumption and energy requirements can be seen (Soto, 2009), and overall food production and profits may be increased via the introduction of lucrative macroalgaevores in order to form multitrophic systems (Bansemer *et al.*, 2016). Additionally, as tighter environmental regulations come into force, the bio-filtration action of seaweeds may facilitate higher compliance with these new standards, thus helping to avoid the potentially costly polluter pays principle (Neori *et al.*, 2004).

4.4   CONNECTION TO UN SUSTAINABLE DEVELOPMENT GOALS 2015–2030

4.4.1   *UN Sustainable Development Goals*

Global food security is characterized by four major distinctions, namely food availability, physical and economic access to food, food utilization, and food stability (Alsanius, 2014).

Both food quality and quantity are paramount to the public health. Therefore, developing integrated food production systems, as seen in the example of aquaponics, is especially interesting as they are able to simultaneously produce high yields of excellent quality food substances, materials and bioactive compounds. Additionally, these types of systems may offer solutions to global challenges including climate change, resource sustainability, urbanization, and population growth (Alsanius, 2014). The continued degradation of soils globally due to chemical industrial agriculture threatens food security of the future (Bindraban *et al.*, 2012). Similarly, it is now estimated that over 80% of fish stocks are overfished. Therefore, there is need to shift food production from being global and unsustainable to being local and sustainable, and to integrate protein (e.g. fish) and plant systems with efficient cycling of nutrients and water as aquaponics can provide.

The United Nations Sustainable Development Goals (UN SDGs) were signed by all member nations of the UN in 2015. The 17 goals present mark a significant new direction of global development as they are not only aimed at improving the living standards of people in the global South, but the SDGs are for all UN member nations and focus on living within the planetary limits, and achieving wellbeing for all people as well as nature. The recent uptake of aquaponics in urban and rural environments will help to achieve many of the UN SDGs as outlined below.

4.4.1.1   *Goal 2 End hunger, achieve food security and improved nutrition and promote sustainable agriculture*
Goal 2 has targets towards ending hunger and malnutrition, increasing the incomes of small-scale food producers, and sustaining food production. Aquaponic food production systems can aid in achieving all of these targets.

4.4.1.2   *Goal 3 Ensure healthy lives and promote well-being for all at all ages*
Aquaponics can support the achievement of the targets in Goal 3 by 2030, by helping to substantially reduce the number of deaths and illnesses from hazardous chemicals and air, water and soil pollution and contamination. This is because aquaponics does not use any hazardous chemicals in the production system.

4.4.1.3   *Goal 6 Ensure access to water and sanitation for all*
Goal 6 has targets relating to access to safe drinking water, improvement of water quality by reducing pollution, increasing water-use efficiency and integrating water resources management. Aquaponics can help achieve all of these targets.

4.4.1.4   *Goal 7 Ensure access to affordable, reliable, sustainable and modern energy for all*
Aquaponics can aid in achieving the target of doubling the global rate of improvement in energy efficiency.

4.4.1.5   *Goal 8 Promote inclusive and sustainable economic growth,*
              employment and decent work for all
Goal 8 has goals relating to job creation, decoupling economic growth from environmental degradation, achieve employment for all women and men with equal pay, and decrease youth not in employment. Aquaponics provide an opportunity to grow food in larger or smaller scale and even in cities providing new opportunities for job creation.

4.4.1.6   *Goal 9 Build resilient infrastructure, promote sustainable industrialization*
              and foster innovation
Aquaponics can aid in achieving a target pertaining to innovation.

4.4.1.7   *Goal 11 Make cities inclusive, safe, resilient and sustainable*
Goal 11 has targets pertaining to reducing the adverse per capita environmental impact of cities and resource efficiency and aquaponics can help achieving these.

4.4.1.8   *Goal 12 Ensure sustainable consumption and production patterns*
Goal 12 has targets for sustainable consumption and production, sustainable management and efficient use of natural resources, halving the per capita global food waste, reducing waste generation, increasing sustainability information into companies reporting cycle, promoting public procurement practices that are sustainable, and awareness for sustainable development patterns of consumption and production. Aquaponics contributes to all of these targets.

4.4.1.9   *Goal 13 Take urgent action to combat climate change and its impacts*
Goal 13 has targets relating to mitigation and adaptation to climate change. Aquaponics helps achieve these targets.

4.4.1.10   *Goal 14 Conserve and sustainably use the oceans, seas and marine resources*
Goal 14 has a target of reducing marine pollution, ending overfishing. Aquaponics can aid in achieving these targets.

4.4.1.11   *Goal 15 Sustainably manage forests, combat desertification, halt and reverse land*
                degradation, halt biodiversity loss
Aquaponics can aid in achieving the targets of sustainable use of inland freshwater ecosystems and their services, halt loss of biodiversity, and integrate ecosystem and biodiversity values into national and local planning.

4.5   CONCLUSIONS AND FUTURE WORK

4.5.1   *Conclusions*

Aquaponics is promising in regard to its potential contribution to both global and urban sustainable food production, whilst at the same time could help to reduce the pressure on ecosystems and the natural environment in general. The methodology is non-linear, based on circular economy principles, with zero waste and no pollution. Waste streams from one production unit are recirculated to become valuable inputs to other production areas. Thus, the methodology supports the development of sustainable food production and supports many of the important UN Sustainable Development Goals put forward for 2015–2030.

The development of aquaponics is ongoing all over the world. The design and scale of the customized systems are advancing in relation to available resources, climate, and economical and environmental parameters in the different countries. At the same time, globally, the interest is raising within research groups and industries as well as policy developers.

4.5.2   *Future work*

The future work of developing aquaponics into a profitable food production method includes several factors, a few are listed below:

- Improved integration of RAS and hydroponics: Aquaponics aims to optimize the use of resources to produce zero-waste (Nichols and Savidov, 2011). There is a strong need for better knowledge about the integrated production method under different conditions to maximize the capacity and keeping a good balance in the systems. This includes optimum ratio between fish and plants and securing good growth parameters.
- Improved safety protocols: Due to the multitrophic nature of aquaponics setups, it is essential that more research and monitoring is done to ensure best practice consumer safety protocols are in place for the products produced (Love *et al.*, 2015; Pantanella *et al.*, 2015).
- Certified organic production: One area in which aquaponics has great potential is that of the production of certified organic goods. Consumers are in general willing to pay more for products which are free from pesticides and herbicides (Milicic *et al.*, 2016). The whole aquaponic system is based on a holistic thinking in terms of recycling optimizing the use of resources with no unhealthy chemicals added. Aquaponics can be certified organic in the US, however, the regulatory framework in Europe for organic farming requires the plants to be grown in soil and RAS is not allowed. There is a need for a review of the organic standards and securing common understanding worldwide about the definitions used so that the aquaponic method is universally accepted as a form of organic production.
- Novel byproducts: Several added value products and services can be combined to the aquaponics development to establish a viable business. This includes educational and experience tourism, but also further innovation through added value byproducts such as crayfish, insect or fertilizer production (Thorarinsdottir *et al.*, 2015).
- More research into optimal use of algae in aquaponics systems and exploration of the potential of algaeponics: The advantages of the addition of algae species to salt water aquaponics units is slowly becoming better recognized, especially with regard to their action as natural biofilters. It would therefore be of much benefit for further research to be undertaken surrounding the optimum parameters, effective species and efficient use of these organisms within the aquaponics context. Furthermore, the study of high value compounds derived from selected algae species could lead to the progression of systems specifically designed to accommodate the requirements of these organisms as primary products.

ACKNOWLEDGEMENTS

This work was funded in part by the EASME funded project EcoPonics and the Icelandic Centre for Research (Rannis) funded Aquaponics.is project.

REFERENCES

Adler, P.R., Harper, J.K., Wade, E.M., Takeda, F. & Summerfelt, S.T. (2000) Economic analysis of an aquaponic system for the integrated production of rainbow trout and plants. *International Journal of Recirculating Aquaculture*, 1, 15–34.

Ahlgren, M.O. (1998) Consumption and assimilation of salmon net pen fouling debris by the red sea cucumber *Parastichopus californicus*: implications for polyculture. *Journal of the World Aquaculture Society*, 29(2), 133–139.

Al-Hafedh, Y.S., Alam, A. & Alam, M.A. (2003) Performance of plastic biofilter media with different configuration in a water recirculation system for the culture of Nile tilapia (*Oreochromis niloticus*). *Aquacultural Engineering*, 29, 139–154.

Alsanius, B.W. (2014) Sustainable systems for integrated fish and vegetable production – new perspectives on aquaponics. In: Glynn, C. & Planting, M. (eds) The SLU Global Food Security Research and Capacity Development Program 2012–2014. SLU-Global Report 2014:6, Swedish University of Agricultural Sciences, Uppsala, Sweden. pp. 60–65.

Amosu, A.O., Robertson-Andersson, D.V., Kean, E., Maneveldt, G.W. & Cyster, L. (2016) Biofiltering and uptake of dissolved nutrients by *Ulva armoricana* (Chlorophyta) in a land-based aquaculture system. *International Journal of Agriculture and Biology*, 18, 298–304.

Aquaponics (2014) Aquaponic seaweed: the new superplant? Available from: http://aquaponicsinindia.com/aquaponics/seaweed-the-new-superplant/ (accessed February 2017).

Bansemer, M.S., Qin, J.G., Harris, J.O., Duong, D.N., Hoang, T.H., Howarth, G.S. & Stone, D.A. (2016) Growth and feed utilisation of greenlip abalone (*Haliotis laevigata*) fed nutrient enriched macroalgae. *Aquaculture*, 452, 62–68.

Barrington, K., Chopin, T. & Robinson, S. (2009) Integrated multi-trophic aquaculture (IMTA) in marine temperate waters. In: Soto, D. (ed) Integrated mariculture: a global review. FAO Fisheries and Aquaculture, Technical Paper, 529. pp. 7–46.

Bigas, H. (ed) (2012) The global water crisis: addressing an urgent security issue. Papers for the InterAction Council, 2011–2012, UNU-INWEH, Hamilton, Canada. Available from: http://inweh.unu.edu/wp-content/uploads/2013/05/WaterSecurity_The-Global-Water-Crisis.pdf (accessed February 2017).

Bindraban P.S., Van der Velde, M., Ye, L., Van den Berg, M., Materechera, S., Kiba, D.I., Tamene, L., Ragnarsdottir, K.V., Jongschaap, R., Hoogmoed, M., Van Beek, C. & Van Lynden, G. (2012) Assessing the impact of soil degradation on food production. *Current Opinion in Environmental Sustainability*, 4, 478–488.

Braungart, M., McDonough, W. & Bollinger, A. (2007) Cradle-to-cradle design: creating healthy emissions – strategy for eco-effective product and system design. *Journal of Cleaner Production*, 15(13–14), 1337–1348.

Brooks, R. (2013) Urban rooftop farms produce edible, nutritious algae. PSFK LLC, New York, NY. Available from: http://www.psfk.com/2013/09/urban-rooftop-algae-farm.html (accessed February 2017).

Bux, F. & Chisti, Y. (eds) (2016) *Algae Biotechnology: Products and Processes*. Springer International, Switzerland.

Cardinale, B.J. (2011) Biodiversity improves water quality through niche partitioning. *Nature*, 472(7341), 86–89.

Chalmers, G.A. (2004) Aquaponics and food safety. Available from: http://www.fastonline.org/images/manuals/Aquaculture/Aquaponic_Information/Aquaponics_and_Food_Safety.pdf (accessed February 2017).

Clark, M. (2009) *The Social Entrepreneur Revolution: Doing Good by Making Money, Making Money by Doing Good.* Marshall Cavendish, Singapore.

Cubillo, A.M., Ferreira, J.G., Robinson, S.M., Pearce, C.M., Corner, R.A. & Johansen, J. (2016) Role of deposit feeders in integrated multi-trophic aquaculture: a model analysis. *Aquaculture*, 453, 54–66.

Dalsgaard, J., Lund, I., Thorarinsdottir, R., Drengstig, A., Arvonen, K. & Pedersen, P.B. (2013) Farming different species in RAS in nordic countries: current status and future perspectives. *Aquacultural Engineering*, 53, 2–13.

Dediu, L., Cristea, V. & Xiaoshuang, Z. (2012) Waste production and valorization in an integrated aquaponic system with bester and lettuce. *African Journal of Biotechnology*, 11(9), 2349–2358.

Diver, S. (2006) Aquaponics – integration of hydroponics with aquaculture. Publication No. IP163, Slot 54, Version 090606, ATTRA, National Sustainable Agriculture Information Service, Fayetteville, AZ.

Fox, B.K., Tamaru, C.S., Hollyer, J., Castro, L.F., Fonseca, J.M., Jay-Russell, M. & Low, T. (2012) A preliminary study of microbial water quality related to food safety in recirculating aquaponic fish and vegetable production systems. College of Tropical Agriculture and Human Resources, University of Hawaii at Manoa. Honolulu, HI. Available from: http://agrilife.org/fisheries/files/2013/10/A-Preliminary-Study-of-Microbial-Water-Quality-Related-to-Food-Safety-in-Recirculating-Aquaponic-Fish-and-Vegetable-Production-Systems.pdf (accessed February 2017).

Francis, C., Lieblein, G., Gliessman, S., Breland, T.A., Creamer, N., Harwood, R., Salomonsson, L., Helenius, J., Rickerl, D., Salvador, R. & Wiedenhoeft, M. (2003) Agroecology: the ecology of food systems. *Journal of Sustainable Agriculture*, 22, 99–118.

Goddek, S., Delaide, B., Mankasingh, U., Ragnarsdottir, K.V., Jijakli, H., & Thorarinsdottir, R.I. (2015) Challenges of sustainable and commercial aquaponics. *Sustainability*, 7(4), 4199–4224.

Gomes, A. (2012) Can commercial aquaponics be profitable (in Hawaii). Japan Aquaponics. Available from: http://www.japan-aquaponics.com/economics-of-aquaponics.html (accessed February 2017).

Graber, A., Durno, M., Gaus, R., Mathis, A. & Junge, R. (2014) The first commercial rooftop aquaponic farm in Switzerland. Oral presentation. *2014 International Conference on Vertical Farming and Urban Agriculture (VFUA 2014), 9–10 September 2014, Nottingham, UK.*

Griffiths, M., Harrison, S.T., Smit, M. & Maharajh, D. (2016) Major commercial products from micro- and macroalgae. In: Bux, F. & Chistim, Y. (eds) *Algae Biotechnology*. Springer International Publishing, Switzerland. pp. 269–300.

Hall, S.G., Constant, D., Philippe, D., Starring, H., Beyer, D., Patel, A., Castillo, T. & Malone, R. (2014) Optimizing biofiltration in a synergistic vertical aquaponics system to improve urban sustainability. *ASABE and CSBE/SCGAB Annual International Meeting, 13–16 July 2014, Montreal, Quebec, Canada.*

Jahan, S. (ed) (2015). Human development report 2015. United Nations Development Programme (UNDP), New York, NY. Available from: http://hdr.undp.org/en/media/HDR_2013_EN_complete.pdf (accessed February 2016).

Joesting, H.M., Blaylock, R., Biber, P. & Ray, A. (2016) The use of marine aquaculture solid waste for nursery production of the salt marsh plants *Spartina alterniflora* and *Juncus roemerianus*. *Aquaculture Reports*, 3, 108–114.

Kledal, P.R. (2012) Green cities and future food supplies: Aquaponics as part of urban design and environmental city planning. *AQUA2012, Global Aquaculture securing our future, 1–5 September 2012, Prague, Czech Republic.* European Aquaculture Society.

Kloas, W., Groβ, R., Baganz, D., Graupner, J., Monsees, H., Schmidt, U., Staaks, G., Suhl, J., Tschirner, M., Wittstock, B., Wuertz, S., Zikova, A. & Rennert, B. (2015) A new concept for aquaponic systems to improve sustainability, increase productivity, and reduce environmental impacts. *Aquaculture Environment Interactions*, 7, 179–1992.

Kuhn, D.D., Boardman, G.D., Craig, S.R., Flick, G.J. & McLean, E. (2007) Evaluation of tilapia effluent with ion supplementation for marine shrimp production in a recirculating aquaculture system. *Journal of the World Aquaculture Society*, 38(1), 74–84.

Kumar, S. & Putnam, V. (2008) Cradle-to-cradle: reverse logistics strategies and opportunities across three sectors. *International Journal of Production Economics*, 115(2), 305–315.

Lehman, H., Clark, E.A. & Weise, S.F. (1993) Clarifying the definition of Sustainable agriculture. *Journal of Agricultural and Environmental Ethics*, 6, 127–143.

Lekang, O.-I. (ed) (2013). *Aquaculture Engineering*. John Wiley & Sons, Oxford, UK.

Lennard, W.A. & Leonard, B.V. (2005) A comparison of reciprocating flow versus constant flow in an integrated, gravel bed, aquaponic test system. *Aquaculture International*, 12, 539–553.

Lennard, W.A. & Leonard, B.V. (2006) A comparison of three different hydroponic subsystems (gravel bed, floating and nutrient film technique) in an aquaponic test system. *Aquaculture International*, 14, 539–550.

Lin, J., Ju, B., Yao, Y., Lin, X., Xing, R., Teng, L. & Jiang, A. (2016) Microbial community in a multi-trophic aquaculture system of *Apostichopus japonicus*, *Styela clava* and microalgae. *Aquaculture International*, 4, 1119–1140.

Love, D.C., Fry, J.P., Li, X., Hill, E.S., Genello, L., Semmens, K. & Thompson, R.E. (2015) Commercial aquaponics production and profitability: findings from an international survey. *Aquaculture*, 435, 67–74.

Mageau, M.T., Radtka, B., Fazendin, J. & Ledin, T. (2015) The aquaponics solution. *Solutions*, 6(3), 51–59.

Martins, C.I.M., Eding, E.H., Verdegem, M.C.J., Heinsbroek, L.T.N., Schneider, O., Blancheton, J.P., D'Orbcastel, E.R. & Verreth, J.A.J. (2010) New developments in recirculating aquaculture systems in Europe: a perspective on environmental sustainability. *Aquacultural Engineering*, 43, 83–93.

McDonough, W. & Braungart, M. (2002) *Cradle to Cradle: Remaking the Way we make Things*. North Point Press, New York.

Milicic, V., Thorarinsdottir, R., Skar, S.L.G. & Hancic, M.T. (2016) Consumer acceptance of aquaponics products. *Aquaculture Europe 2016, 20-23 September 2016, Edinburgh, Scotland.*

Naem, S., Thompson, L.J., Lawler, S.P., Lawton, J.H. & Woodfin, R.M. (1994) Declining biodiversity can alter the performance of ecosystem. *Nature*, 368, 734–737.

Nelson and Pade Inc. (n.d.) Recommended plants and fish in aquaponics. Nelson and Pade, Inc., Montello, WI. Available from: http://aquaponics.com/recommended-plants-and-fish-in-aquaponics/ (accessed February 2017).

Neori, A., Chopin, T., Troell, M., Buschmann, A.H., Kraemer, G.P., Halling, C., Shpigel., M. & Yarish, C. (2004) Integrated aquaculture: rationale, evolution and state of the art emphasizing seaweed biofiltration in modern mariculture. *Aquaculture*, 231(1), 361–391.

News (2015) Pentair partners with urban organics to advance aquaponics. *Pump Industry Analyst*, 9, 12–13.

Nichols, M.A. & Savidov, N.A. (2011) Aquaponics: a nutrient and water efficient production system. *II International Symposium on Soilless Culture and Hydroponics, 15–19 May 2011, Pueble, Mexico*. pp. 129–132.

Packer, M.A., Harris, G.C. & Adams, S.L. (2016) Food and feed applications of algae. In: Bux, F. & Chisti, Y. (eds) *Algae Biotechnology*. Springer International Publishing, Switzerland. pp. 217–247.

Palmason, G. (2005) *Jardhitabok – Edli og nyting audlindar* (in Icelandic), Orkustofnun, Iceland.

Pantanella, E. & Bhujel, C.R. (2015) Saline aquaponics-potential player in food, energy production. *Global Aquaculture Advocate*, January/February 2015, 42–43.

Pantanella, E. & Rakocy, J.E. (2012) Production and quality of salicornia grown under different salinity levels and aquaponic sub-systems. AQUA2012, *Global Aquaculture securing our future, 1–5 September 2012, Prague, Czech Republic*. European Aquaculture Society.

Pantanella, E., Bergonzoli, S., Fabrizi, F., Cardarelli, M. & Colla, G. (2011) Saline aquaponics, new challenges for marine aquaculture. *Aquaculture Europe 2011, 18–21 October 2011, Rhodes, Greece*.

Pantanella, E., Bhujel, R.C. & Rakocy, J.E. (2013) Saline aquaponics: bringing fish and profits out of the sea. WAS, Asia Pacific Aquaculture 2013, 10-13 December 2013, Ho Chi Minh City, Vietnam. Available from: https://www.was.org/meetings/mobile/MG_Paper.aspx?i=30510 (accessed February 2017).

Pantanella, E., Cardarelli, M., Colla, G. & Di Mattia, E. (2015) Aquaponics and food safety: effects of UV sterilization on total coliforms and lettuce production. *Acta Horticulturae* (ISHS), 1062, 71–76.

Pardee, K.I., Ellis, P., Bouthillier, M., Towers, G.H. & French, C.J. (2004) Plant virus inhibitors from marine algae. *Canadian Journal of Botany*, 82(3), 304–309.

Petit, J. & Wiegertjes, G.F. (2016) Long-lived effects of administering β-glucans: indications for trained immunity in fish. *Developmental & Comparative Immunology*, 64, 93–102.

Rabiei, R., Phang, S.M., Lim, P.E., Salleh, A., Sohrabipour, J., Ajdari, D. & Zarshenas, G.A. (2016) Productivity, biochemical composition and biofiltering performance of agarophytic seaweed, *Gelidium elegans* (red algae) grown in shrimp hatchery effluents in Malaysia. *Iranian Journal of Fisheries Sciences*, 15(1), 53–74.

Ragnarsdottir, K.V., Sverdrup, H.U. & Koca, D. (2011) Challenging the planetary boundaries. I. Basic principles of an integrated model for phosphorous supply dynamics and global population size. *Applied Geochemistry*, 26, S303–S306.

Rakocy, J.E., Losordo, T.M. & Masser, M.P. (1992) Recirculating aquaculture tank production systems: integrating fish and plant culture. SRAC Publication No. 454, Southern Region Aquaculture Center, Mississippi State University, Stoneville, MS.

Rakocy, J.E., Bailey, D.S., Shultz, R.C. & Danaher, J.J. (2005) Preliminary evaluation of organic waste from two aquaculture systems as a source of inorganic nutrients for hydroponics. *Acta Horticulturae* (ISHS), 742, 201–207.

Rakocy, J.E., Masser, M.P. & Losordo, T.M. (2006). Recirculating aquaculture tank production systems: aquaponics—integrating fish and plant culture (November, 2006 Revision). SRAC publication No. 454, Southern Region Aquaculture Center, Mississippi State University, Stoneville, MS. pp. 1–16.

Rakocy, J.E., Bailey, D.S., Shultz, R.C. & Danaher, J.J. (2007) Fish and vegetable production in a commercial aquaponic system: 25 years of research at the University of the Virgin Islands. *Proceedings of the 2007 National Canadian Aquaculture Conference, 23–26 September 2007, Edmonton, Alberta, Canada*.

Rama, Y. (2014) Hydro-powered aquaponics farming system. 5th International GreenTech & Eco Products Exhibition and Conferences Malaysia (IGEM2014), 16–19 October 2014, Kuala Lumpur, Malaysia. Available from: http://umpir.ump.edu.my/7552/ (accessed February 2017).

Riva, G. (1992) Utilization of renewable energy sources and energy-saving echnologies by small-scale milk plants and collection centres. FAO Animal Production and Health Paper 93. Food and Agriculture Organization of the United Nations, Rome, Italy. Available from: http://www.fao.org/docrep/004/t0515e/T0515E00.HTM (accessed February 2017).

Rutger, H. (2014) Aquaponic fish & veggies worth their salt. MOTE – Marine Laboratory & Aquarium, Sarasota, FL. Available from: https://mote.org/news/article/aquaponic-fish-veggies-worth-their-salt (accessed February 2017).

Selock, D. (2003) An introduction to aquaponics: the symbiotic culture of fish and plants. Rural enterprise and alternative agricultural development initiative, Report No. 20. Southern Illinois University, IL.

Shafeena, T. (2016) Smart aquaponics system: challenges and opportunities. *International Journal of Advance Research in Computer Science and Management Studies*, 3(2), 52–55.

Shpigel, M. & Neori, A. (2007) Microalgae, macroalgae, and bivalves as biofilters in land-based mariculture in Israel. In: Bert, T.M. (ed) *Ecological and Genetic Implications of Aquaculture Activities*. Springer, The Netherlands. pp. 433–446.

Singh, V., Tiwari, A. & Das, M. (2016) Phyco-remediation of industrial waste-water and flue gases with algal-diesel engenderment from micro-algae: a review. *Fuel*, 173, 90–97.

Sommerville, C., Cohen, M., Pantanella, E., Stankus, A. & Lovatelli, A. (2014) Small-scale aquaponic food production. Integrated fish and plant farming. FAO Fisheries and Aquaculture Technical Paper No. 589, Food and Agricultural Organisation of the United Nations, Rome, Italy.

Soto, D. (ed) (2009) Integrated mariculture a global review. Fisheries and Aquaculture Technical Paper 529, Food and Agricultural Organisation of the United Nations, Rome, Italy.

Subashchandrabose, S.R., Ramakrishnan, B., Megharaj, M., Venkateswarlu, K. & Naidu, R. (2013) Mixotrophic cyanobacteria and microalgae as distinctive biological agents for organic pollutant degradation. *Environment International*, 51, 59–72.

Sverdrup, H.U. & Ragnarsdottir, K.V. (2011) Challenging the planetary boundaries. II. Assessing the sustainable global population and phosphate supply, using a systems dynamics assessment model. *Applied Geochemistry*, 26, S307–S310.

Sverdrup, H.U. & Ragnarsdottir, K.V. (2014) Natural resources in a planetary perspective. *Geochemical Perspectives*, 3(2), 129–341.

Tapia, J.E., González, B., Goulitquer, S., Potin, P. & Correa, J.A. (2016) Microbiota influences morphology and reproduction of the brown alga *Ectocarpus* sp. *Frontiers in Microbiology*, 7, 197.

Thorarinsdottir, R.I., Kledal, P.R., Skar, S.L.G., Sustaeta, F., Ragnarsdottir, K.V., Mankasingh, U., Pantanella, E., van de Ven, R. & Shultz, C. (2015) Aquaponics guidelines. University of Iceland, Reykjavík, Iceland.

Timmons, M.B. & Ebeling, J.M. (2010) *Recirculating aquaculture*. Northeastern Regional Aquaculture Center (NRAC), Publication no. 01-002, College Park, MD.

Turcios, A.E. & Papenbrock, J. (2014) Sustainable treatment of aquaculture effluents – what can we learn from the past for the future? *Sustainability*, 6, 836–856.

Tyson, R.V., Treadwell, D.D. & Simonne, E.H. (2011) Opportunities and challenges to sustainability in aquaponics. *HortTechnology*, 21(1), 6–13.

UCSUSA (n.d.) Farming the Wind: Wind Power and Agriculture. Union of Concerned Scientists. Cambridge, MA. Available from: http://www.ucsusa.org/clean_energy/smart-energy-solutions/increase-renewables/farming-the-wind-wind-power.html#.WAeyjyShvFM (accessed February 2017).

Uddin, M.S., Farzana, A., Fatema, M.K., Azim, M.E., Wahab, M.A. & Verdegem, M.C.J. (2007) Technical evaluation of tilapia (*Oreochromis niloticus*) monoculture and tilapia-prawn (*Macrobrachium rosenbergii*) polyculture in earthen ponds with or without substrates for periphyton development. *Aquaculture*, 269(1), 232–240.

USDA (2016) Hydroponic and aquaponic task force report. National Organic Standards Board (NOSB), United Stated Department of Agriculture, July 21, 2016. Available from: https://www.ams.usda.gov/sites/default/files/media/2016%20Hydroponic%20Task%20Force%20Report.PDF (accessed February 2017).

Weller, T. (2005) *The Best of the Growing Edge International, 2000-2005: Select Cream-of-the-crop Articles for Soilless Growers*. The Growing Edge International. New Moon Publishing Inc, 3:360.

Wilson, G. (2005) Seaweed is the common denominator in exciting saltwater aquaponics. *Aquaponics Journal*, 36, 12–16.

Wikfors, G.H. & Ohno, M. (2001) Impact of algal research in aquaculture. *Journal of Phycology*, 37(6), 968–974.

CHAPTER 5

Renewable energy use and potential in remote central Australia

Yiheyis Maru, Supriya Mathew, Digby Race & Bruno Spandonide

5.1 INTRODUCTION

High electricity prices coupled with a high level of energy use by Australian households is likely to exert more pressure on low-income households, raising concerns about whether Australia's high standard of living can be sustained, particularly for disadvantaged indigenous people living in remote Australia. Australia's average residential electricity prices have recently been higher than many other developed countries such as Japan, United States of America, Canada and others in the European Union (EC, 2014; Mount, 2012). In addition, Australia's energy consumption per capita is much higher than most other countries, including more than ten times that of some rapidly developing countries, such as India (EIA (US Energy Information Agency), 2011).

Remote Australia covers more than 86% of the surface area of the continent but is only sparsely populated by 2.3% of the Australian population. Remote Australia is quite different from other parts of Australia, with both rich opportunities and considerable challenges (Commonwealth, 2014a, 2014b). Stafford Smith (2008) discussed a number of linked factors such as climate variability, scarce resources, sparse population, distant markets and decision-making centers, social variability and cultural differences as characteristics of the remote regions of Australia. A quarter of remote Australia residents are indigenous people with a majority of them living in more than 1200 discrete settlements. Remote Australia residents experience high energy and transport costs, as well as limited access to many goods and services (ABS, 2013).

This dispersed settlement pattern – distant from major energy grids – presents significant challenges for energy provision. A majority of remote settlements, mining, tourism and pastoral operations rely on off-grid diesel generators which are expensive to run and depend on government subsidy (Byrnes, 2014), which in recent years has come under increasing political pressure (Altman, 2014). The cost of ensuring a reliable supply of energy is a major concern often raised by governments about the viability of small discrete settlements.

For inland remote Australia, there is a high level of certainty about the projected increases of temperature, frequency and intensity of heatwaves, as well as a likelihood of increasing droughts, floods and other extreme weather events under changing climate (Green and Minchin, 2014; Watterson, 2015). These changes are in turn likely to increase energy demand and interruptions to the supply of diesel to remote locations, adding to the cost and vulnerability of energy in remote settlements (Maru *et al.*, 2012).

In this chapter we discuss a range of opportunities and challenges, and suggest development pathways, for renewable energy (RE) in remote Australia. A shift to RE would contribute to mitigation of greenhouse gases, adaptation to climate change and support a transformation in livelihood options for remote Australian residents. This study is based on a review of literature on the current state of the energy supply mix, and the potential opportunities and structural challenges for the provision of solar energy in remote Australia. We also draw on two case study communities in inland Australia to further explore the opportunities and challenges for increasing the use of RE.

5.2   RENEWABLE ENERGY SUPPLY: STATE AND TRENDS
IN REMOTE AUSTRALIA

5.2.1   *Current state of energy use in remote Australia*

Around 93% of all energy consumed in Australia is provided by two major electricity markets – the national electricity market (NEM) and the south-west interconnected system (SWIS), both of which utilize power generated from coal and natural gas (AECOM, 2014). There are other multiple but small independent interconnected systems including the North West Interconnected System (NWIS) in Western Australia, and Darwin to Katherine System (DKS) and Alice Springs to Katherine (ASK) in the Northern Territory, that are operated by local network providers. However, the energy supply in these systems is still considered 'off-grid' by definition as they do not feature sophisticated market structures (AECOM, 2013).

Approximately 75% of Australia's electricity is generated from coal-fired generation (AECOM, 2013). In 2014, a little more than 13% of Australian electricity was generated from RE sources. Solar energy constituted only 2% of the total energy generation (Clean Energy Council, 2015).

The off-grid areas (see gray shaded regions in Fig. 5.1) correspond closely to remote regions of Australia. Although the off-grid area is home to only 2% of Australia's population, over 6%

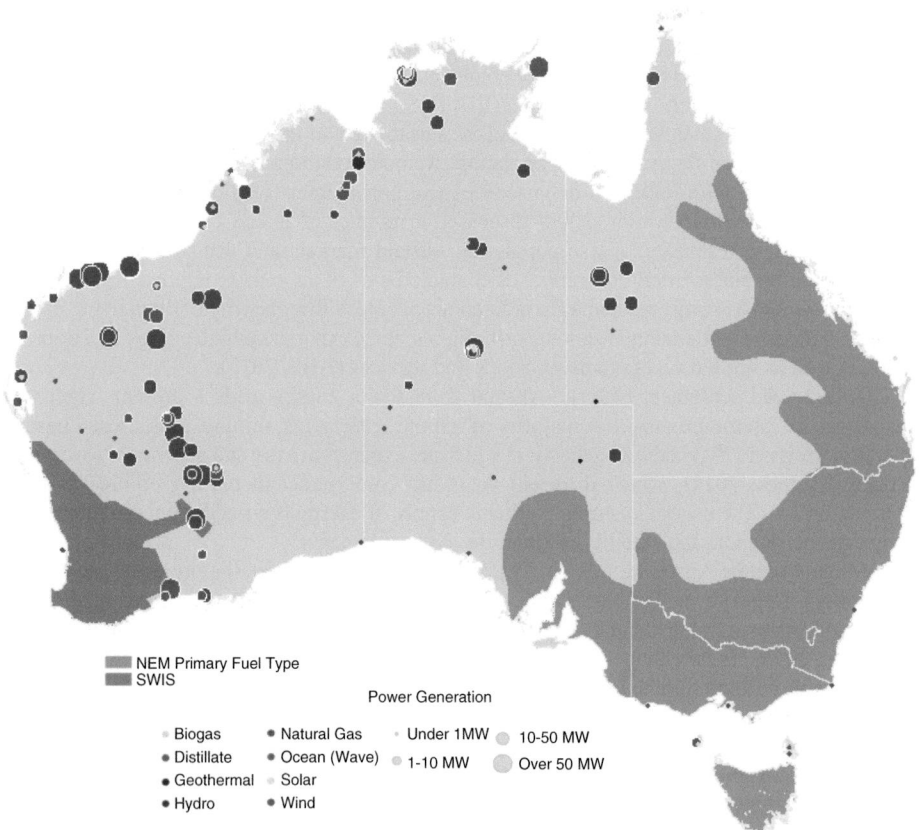

Figure 5.1.   Distribution of power sources in Australia including off-grid and fringe-of-grid (note: remote areas that are not connected to NEM or SWIS: orange and red shaded).
(Source: AECOM, 2014, page ii, Data based on Geoscience Australia 2006 and 2012 power generation database).

of the country's total electricity is consumed in off-grid areas, most of it (around 98%) generated from natural gas and diesel fuel. These factors make it Australia's most expensive electricity due to the underlying high gas and diesel prices in remote areas. Only 1% of electricity is generated from renewable sources (AECOM, 2014).

The populations on the fringe of the grid regions (i.e. that are part of the remote regions of Australia), can still be long distances from the electricity generators and hence are prone to loss of supply during transmission and distribution. However, fringe-of-grid populations can be more likely to face issues of reliability of supply, with 'brown-outs' (partial or temporary reduction in system voltage often experienced as a dimming of lighting) much more common than 'blackouts' (complete interruption of power in a given service area).

5.2.2   *Potential and opportunities for solar energy supply*

Several factors provide remote Australia with significant potential to scale up the supply of energy from solar generation. These factors pertain mainly to the location and the increasing energy demand of remote indigenous communities. Other factors include the declining cost, improving efficiency and penetration capacity of solar energy technology as well as the growing confidence and experience of a network of government, non-government and private sector RE providers (AECOM, 2013; Bahadori and Nwaoha, 2013; Frischknecht, 2014; Lowe and Lloyd, 2001; Maru *et al.*, 2012; Pittock, 2011; Shafiullah *et al.*, 2012). Solar energy penetration refers to the amount of solar energy that can be used as a proportion of the total energy of a mixed energy source supply system. The cost transitioning from low to medium and high penetration of solar energy supply is however not linear. Increasing solar penetration requires large power storage capacity, which is costly. This leads to sharp increases in total cost supply with high solar energy penetration. Drawing from a review of the literature, Table 5.1 is a summary of potential opportunities for RE supply, especially solar energy.

Table 5.1.   Drivers of potential opportunities for transitioning to RE supply in remote Australia.

Factors	Detail description
Location	Most of remote Australia has abundant solar radiation and most of it is off-grid, thus "low hanging fruit" in terms of trying new RE sources.
Community energy needs	Increasing energy demand from communities and industry sectors operating in the regions. Awareness and demand for clean energy sources increasing. Potential for low-cost energy and reduction of community dependence on government subsidy of energy costs.
Cost of unit of RE	Declining price of unit of solar energy as technology improves and initial capital cost reduces. Comparable with cost of producing energy from diesel.
Cost to government	Potential to substantially reduce high cost of subsidized and imported diesel energy supply to remote communities. Contribution to greenhouse gas (GHG) emission reduction.
Experience & capacity	Growing experience and skills of a network of actors in installing, operating and maintenance of variety of scales of solar plants.
Penetration	ARENA* is supporting a trial of high solar energy penetration plants and testing new storage technologies.
Storage & integration	Solar energy to play a significant role in the future of integrated RE supply to domestic and transport facilities.

* = Australian Renewable Energy Agency (ARENA, see www.arena.gov.au).

5.2.2.1 *Location*

Many off-grid remote communities rely on diesel generators where fuel needs to be transported over long distances. Long distance transport of fuel is dangerous and at risk of disruption due to extreme weather events and poor road conditions. This situation often requires many remote sites to store fuel during wet seasons when many roads are closed. Also, the fuel costs are particularly high in remote Australia. The monopolistic nature of transport services (public transport, fuel companies), implies that prices are less likely to reflect demand levels: the cost elasticity of transport is particularly low and public transport fare strategies tend to be less price sensitive than in urban environments. Thus, solar energy could become a more reliable and economically viable option in the vast area of inland Australia. It is the same area which has abundant potential for solar power generation – it is among the regions in the world with the highest solar radiation per unit area. The annual solar radiation reaching Australia (~58 million peta joules, is approximately 10,000 times Australia's annual energy consumption (see Geosceince Australia n.d.). Solar energy resource concentration is higher in areas where there is often low access to a central power grid (north-west and central Australia). High solar radiation levels, proximity to local loads, and a high cost for alternative sources of electricity are all decisive factors in determining locations that are suitable for solar thermal power plants (Wyld Group and MMA, 2008). Based on these characteristics, a high potential for thermal power exists within the NEM (e.g. Port Augusta in South Australia, north-west Victoria, central and north-west New South Wales), within the SWIS (Kalbarri in Western Australia), within the Darwin-Katherine interconnected system (Northern Territory), and within the Alice Springs-Tennant Creek interconnected systems (Northern Territory).

Along the fringe of the established grid areas, the use of renewables can increase the reliability of supply or defer costly investments to upgrade the existing power networks. For example, Queensland (in contrast to Northern Territory and Western Australia) has a long and narrow grid with more fringe-of-grid areas (see Fig. 5.1) that have reliability issues, so these areas would benefit from additional supply from renewables (AECOM, 2014). Further afield, energy markets in China, India and Indonesia might have a significant influence over future technological developments of large-scale solar electricity generation and subsequent projects in remote Australia (Pittock, 2011).

5.2.2.2 *Community needs*

Demand for energy supply in off-grid remote areas of Australia is increasing. In recent decades, this has been largely driven by an expansion of the mining industry, but also industrialization of other rural sectors and increasing energy needs of communities. With growing mining towns and over 1000 small remote communities and outstations across Australia, the community segment represents a relatively untapped RE market (AECOM, 2014).

Transport is particularly important for fulfilling livelihood aspirations and maintaining cultural practices and obligations in very remote communities, and also an essential component of social and business enterprises. Remote Australians travel two to three times more than the national average (35,000 to 45,000 km per year). This high level of transport and the reliance on fossil fuel imply that emission levels per capita in remote Australia are among the highest in the world.

In the future, synergetic battery systems for electric vehicles and the solar energy generation facilities could lead to a seamless integration between household and transport energy, resulting in substantial energy cost reduction, which currently forms a significant proportion of expenses for remote households and businesses.

One of the main benefits of an increasing use of solar energy in a remote context is the resulting increased resilience and autonomy of remote communities (Pittock, 2011). This element is rarely factored into the calculation of benefits of solar energy facilities but has been demonstrated through the successful implementation of the Bushlight program that has installed standalone solar or solar-diesel hybrid electricity systems in over 130 indigenous communities across northern Australia. In addition to providing efficient energy and a user-friendly demand management interface, it has also substantially increased the reliability of supply of energy to

Table 5.2. Comparison of conventional and renewable energy sources for remote agriculture (Kaushik, 2011; Lowe and Lloyd, 2001).

Kerosene/diesel	Solar
Transport of fuel over long distances and challenging roads (especially during times of extreme weather)	Transportation of fuel not required; Remote Australian farms have a lot of open spaces required by solar photovoltaic (PV) systems
Noise and pollution (air pollution and spills) associated with the use of these fuels can affect livestock	Pollution is minimal due to the use of clean energy
Fuel costs can be dependent on the remoteness of the location	No fuel costs and hence low operating costs
Maintenance to generators and availability of spare parts/technicians	Low maintenance, though capital costs may be high; Energy supply is dependent on the weather; Innovation in adequate storage of energy can make energy supply more reliable

many remote communities as it can continue to generate power when diesel supply has been lost for a variety of reasons including floods, or insufficient diesel storage on site. This is not an unusual situation in some very remote communities – an A\$ 1.3 million project was recently completed to extend the fuel storage capacity at Maningrida (NT) to mitigate supply risks (AECOM, 2014).

5.2.2.3 *Potential applications of renewable energy in agriculture in remote communities*
There are several important industries in remote Australia, including irrigated and pastoral agriculture, mining, services (education, health, housing) and tourism. Various sources of RE, including solar and wind, offer the potential to provide a sustainable solution to the energy requirements for remote communities and industries. RE can assist in achieving sustainable agriculture which relies on balanced soil nutrient and water cycles, balanced energy flows, beneficial soil organisms, and natural pest controls, as much as possible (Kaushik, 2011). A number of challenges exist with the use of non-RE sources, such as kerosene and diesel, for agriculture in remote communities. A brief comparison of renewable and non-renewable energy sources is provided in Table 5.2.

RE has a number of applications in the agriculture sector (Kaushik, 2011). A few of the applications of solar and wind energy installations include:

- Electricity generated through solar photovoltaic (PV) cells that can be used for water pumping and crop irrigation, powering of electric fans in closed buildings, lighting of buildings, product refrigeration, product processing, and electric fencing.
- Thermal energy generated by solar plants may be used for crop drying and water heating.
- Wind turbines could help in water pumping, electricity generation, and grinding grains and legumes.

The use of reliable and cost-effective RE in inland Australia can assist in the development of a sustainable arid environment-based agricultural industry. Bushfood harvesting (both customary harvest by Aboriginal communities and commercial harvest) is one such agricultural industry that could be further expanded in central Australia (Mathew *et al.*, 2016).

5.2.2.4 *Cost of solar technology*
Most remote Australian communities supply their electricity from expensive diesel-fueled power systems. Depending on the remoteness and the size of the community, diesel prices vary from

approximately A\$ 1.10 to A\$ 1.70 L^{-1} after rebates. This corresponds to a fuel cost between A\$ 300 and A\$ 450 MWh^{-1} generation (AECOM, 2014, p. 71).

The capital costs of solar energy technology have been declining to an extent that it already makes economic sense for small installations of solar technology to meet new off-grid energy demands and complement existing aging diesel operations in many remote settlements and operations (Frearson, 2014). In some areas, such as the remote Pilbara region of Western Australia, solar PV is either at, or approaching, parity with diesel for off-grid energy supply. The levelized cost of electricity (LCOE – total lifetime costs in dollars per unit of energy [A\$ MWh^{-1}] that enables comparison between technologies) of solar PV as compared with diesel in Pilbara can be expected to be approximately A\$ 230 MWh^{-1} for projects with a low penetration of RE (AECOM, 2014). In addition to being able to reduce energy costs, solar energy technology has seen significant improvements over the last decade in efficiency, durability and ease of maintenance as well as reliability of supply through enhanced and new energy storage capacities.

Economic and trend analysis has also indicated that the price of RE technology, particularly solar and wind, will continue to fall rapidly in price over the next decade and will be among the lowest cost of all types of electricity generation within 10 to 20 years (AECOM, 2013). Also, the cost of solar energy is expected to fall as economies of scale reduce supply chain costs and experience with construction, installation and maintenance increases.

Solar technologies with adequate storage systems could work in conjunction with fossil fuel powered plants to provide the base load electricity generation. There is also an opportunity to fully commercialize solar technology in the future. Solar thermal systems convert solar radiation into thermal energy (heat) which could be used for space heating or generating electricity, using the steam and turbines. Concentrating solar thermal plants can result in large-scale electricity production. The development of cost-effective storage technologies and low-cost solar technologies may enable a much higher uptake of solar thermal power in the future and this may be achieved with the help of funding from agencies such as ARENA, who currently play an important role in delivering cost-effective solar technology and addressing the barriers to the uptake of solar technology (Byrnes and Brown, 2015).

5.2.2.5 *Increasing solar energy supply penetration*

The cost of energy supply increases with higher penetration of RE into the system. However, ARENA's Regional Australia Renewables program is providing financial support to enable RE systems to remain competitive, even at higher penetration levels. This plays an important role in delivering an enabling environment for cost-effective solar technology and addressing the barriers to the uptake of solar technology (Clean Energy Council, 2015).

The AECOM 2014 report on Australia's off-grid energy forecasts that hybridizing of renewables will happen at low penetration in Western Australia, the Northern Territory and Queensland. This short-medium term forecast also indicates that around 150–200 MW (~A\$ 450–A\$ 600 million in capital value) of energy will be produced. In the longer term, as research progresses, cost-effectiveness and advanced technologies together should result in around 1 GW (~A\$ 2 billion in capital value) of off-grid RE production.

Opportunities for high penetration will be more evident as pilot cases in various power intensive industries are trialed. The Alice Springs airport's solar project, which meets around 50% of the airport's energy requirements, is a good example of how off-grid airports could function in remote Australia. Also, a 6.7 MW solar farm project was recently announced by ARENA and Rio Tinto at the Weipa bauxite mine on Cape York to supply energy to the mine, processing facilities and township – estimated to displace about 20% of current diesel use (ARENA, 2015a). A 10.6 MW solar power installation at Degrussa Cooper-Gold Mine is expected to be operational in 2016. This off-grid solar with 6 MW of short-term battery storage will be integrated with existing diesel fired power station and is expected to offset 20% of current diesel consumption and reduce CO$_2$ emissions by an estimated 12,000 metric tons per year (ARENA, 2015b). These projects have the potential to demonstrate the cost-effectiveness and reliability of high-penetration RE to the wider Australian mining sector.

The 2015 Northern Territory Solar Energy Transformation Program (SETuP) aims to provide solar power to over 30 remote communities over a three-year period, with the aim that most communities will be medium penetration sites with approximately 60% peak solar penetration. The Daly river site which also forms part of the SETuP program will be a 1000 kW high-penetration site achieving approximately 100% peak solar penetration which is about 50% average penetration (ARENA, 2014). This site will test a number of advanced technologies including:

- Tracking arrays to increase the solar penetration.
- Trial storage technologies such as lithium-ion batteries.
- Trial demand management to switch off non-essential loads such as air conditioners, when overall demand is high.
- Cloud forecasting technology to allow smaller diesel sets to operate without compromising the spinning reserve to ensure the reliability of supply.

5.2.2.6   *Growth in experience and capacity*
There is a growing technical and social engagement and capacity among several government, non-government and private organizations in the design, installation, operation and maintenance of solar energy technology in remote regions. These include: the Centre for Appropriate Technology, an organization which through its Bushlight program has delivered small-scale renewable energy systems to more than 130 remote indigenous communities and regularly undertakes maintenance services and training to community members (AECOM, 2013); the NT Power and Water Corporation which has two decades of experience in integrating solar technologies into remote community energy supply; and PowerCorp, Horizon and Origin and some private companies which are installing large-scale and high-penetration RE plants in remote towns (AECOM, 2013).

5.2.2.7   *Storage and integration*
Energy storage is one of the critical areas of research and development for improving the competitiveness and reliability of high level penetration or standalone large-scale RE supply systems. ARENA has been playing an important role in funding research and development, demonstration, deployment to commercialization and facilitating knowledge sharing on a variety of advanced RE storage technologies useful for solar energy supply in remote Australia. An example is a demonstration of CSIRO's (Commonwealth Scientific and Industrial Research Organisation) breakthrough UltraBattery® technology, which is noted to outperform commonly used storage batteries to deliver superior energy storage solutions in a variety of applications including in remote area power supply (RAPS) for off-grid renewable power solutions in remote areas of Australia (ARENA, 2012).

Though a long-term prospect, electric vehicles integrated with re-charging from RE could significantly transform domestic and commercial transport in remote regions. Electric vehicles have lower running costs, environmental benefits and can assist with managing demand on the electricity system if consumers are provided with an incentive to charge outside of peak energy use times. Charging electric vehicles at off-peak times improves the utilization of electricity infrastructure by increasing the demand on the electricity grid outside peak times. Batteries in electric vehicles could also be used as a storage device to meet household demand or demand on the electricity grid through feeding back to the grid at peak time (Li *et al.*, 2014).

However, in addition to the initial substantial cost of EVs and the current absence of charging stations in remote Australia, integrating them to RE supplied charging stations will take time. Syed *et al.* (2010) do not expect electric cars to make a significant contribution to Australian road transport before 2030 due to technological limitations and the requirement for an extensive re-charging network. Indeed Australia's targets for emissions reduction and vehicle efficiency are low when compared internationally (Climate Change Authority, 2014). To reach a market where second-hand vehicles play a central role in household motorization, strong incentives for a switch to electric vehicles would be required to accelerate the current slow rate of penetration of new technologies.

Table 5.3.   Key challenges and barriers of RE supply expansion in remote areas of Australia.

Key factors	Detail
Market	Market failure in competitive energy supply to remote regions because of small customer size, vast and remote locations with high transport and transaction costs and incumbent diesel fuel energy supply Market distortion due to non-consideration of externalities and diesel fuel subsidies with perverse outcomes of reinforcing the continued dominance of diesel powered energy generation in remote communities, leading to an unequal playing field that reduces competitiveness of RE
Financial and economic	High cost of living and proportion of expenditure on energy High upfront capital cost of RE installation Ability to source and access capital Need for investment certainty Financiers' inexperience and low confidence in RE technologies Costs associated with managing intermittent generation, grid and hybrid integration
Institutional	Uncertain and complex multi-layered policy environment Variable and inconsistent government commitment to renewable energy targets, GHG emission reductions and support to indigenous settlement in remote regions Dispersed and weak remote region voice and lobbying potential compared to coordinated and strong lobbying by incumbent non-renewable energy generators Regulatory framework that provide perverse subsidies and interpret challenges of existing and RE differently
Technical	Inappropriate RE technology and energy supply system design to social and cultural requirements and to extreme climate and geography Simple (non-technical) and visual informational needs A lack of standard codes and certification, lack of skilled personnel and training facilities, and unreliable or poorly designed systems Need for adequate accounting of resources needed for network hosting capacity, intermittent generation management, storage facilities, and system adaptability to rapid technological changes Guaranteed access to effective operation and maintenance services and supply of component parts
Social and cultural	Social acceptance and perception of product reliability Cultural appropriateness of RE technology and system design Customer awareness of technology and understanding of benefits Confidence on RE technologies Availability of land for deployment and native title and cultural heritage requirements Social capacity to follow through with RE installation and to maintain and service the systems
Community engagement	Lack of inclusive and effective consultation of community members Effective communication and information sharing Malicious damage and vandalism of RE installations

5.2.3   *Challenges for solar energy supply*

There have been several efforts to understand the barriers that lead to the currently very limited adoption of RE technology in remote off-grid locations (Byrnes *et al.*, 2015; Lowe and Lloyd, 2001; McKenzie and Howes, 2006; Painuly, 2001). Drawing from a comprehensive review of international literature, Painuly (2001) provided a framework for analyses of barriers to RE penetration. The article identified seven categories each having nested sets of multiple barriers. With slight modification and selection of element barriers, McKenzie and Howes (2006) have applied the categories to investigate the slow uptake of RE in remote Australia. A recent article by Byrnes *et al.* (2015) about indigenous communities in the remote Kimberly region in Western Australia confirms the validity of most of the barriers. Drawing from this literature, Table 5.3 summarizes broad categories of factors and detailes barriers to RE supply in remote communities.

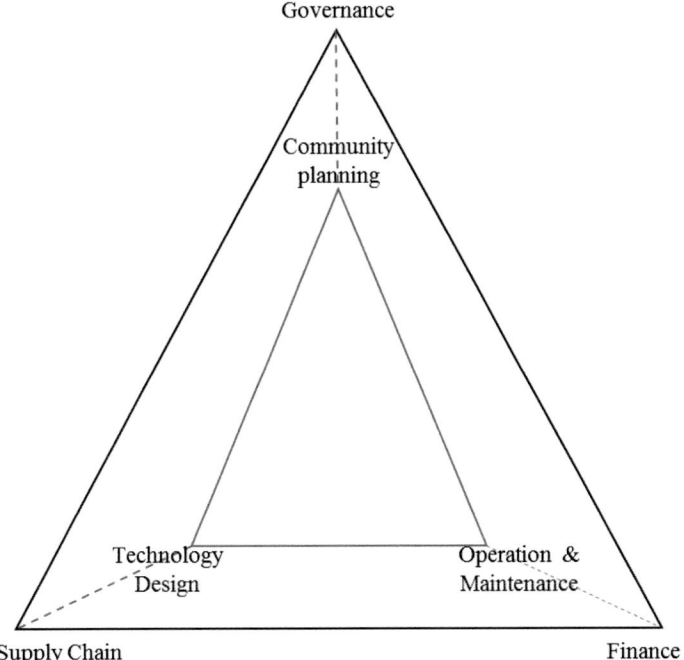

Figure 5.2.   Categories of barriers to RE supply in remote regions (source: Tuckwell *et al.*, 2011).

Tuckwell *et al.* (2011) developed a framework on challenges and barriers to RE supply in remote communities. This framework (Fig. 5.2) has been developed from experiences of successfully dealing with barriers to the supply of RE to small remote communities in northern Australia and to remote villages in India. The framework categorizes the barriers at different scales: one at an individual community level and the other at higher levels that need to be considered for successfully scaling out a model of RE supply across multiple remote communities.

At a local scale the framework identifies a triad of interacting challenges: (i) effective and inclusive community consultation and planning to understand actual energy needs and livelihood aspirations as well as to ensure community ownership of the planned RE supply system; (ii) appropriate technology selection and system design that is cost-effective, robust and capable of meeting demand over the technology lifetime without significant failure; and (iii) guaranteed access to effective locally developed and sourced operation and maintenance services to ensure reliable supply (Tuckwell *et al.*, 2011).

Tuckwell *et al.* (2011) note that addressing these key barriers at a community level is what led to the success of the Bushlight program in remote communities in Australia and villages in India. Despite this demonstrable success, the proportion of energy supply from renewable sources in off-grid areas in Australia still remains low. This is because effectively moving to scale brings a new set of barriers beyond the community level. These barriers can be broadly categorized into three areas: (i) access to adequate and suitable levels of finance; (ii) appropriate supply chain linkages to the remote locations that ensure standardized low operating cost and low failure technology at a scale where economies can be developed to provide meaningful capital cost reductions; and (iii) functional and sustained governance structures to safeguard and ensure revenue collection, to support and maintain local individual institutional structures, and to sustain system operation and maintenance processes.

The operations oriented framework is premised on the assumption that currently there are a variety of reliable small-scale RE technologies for standalone and hybrid energy supply systems to remote communities. While this is valid, further improvement in efficient energy conversion and storage are welcome as they reduce cost and improve reliability of supply and thus increase the level of penetration, making solar energy even more competitive than conventional fossil fuel sources. However, more cost-effective and reliable energy storage technology for high level penetration and large-scale RE supply is still a challenge (AECOM, 2014).

The following section provides detail on case studies of household energy conditions, needs and challenges in central Australian remote communities.

5.3   CASE STUDIES FROM INLAND REMOTE AUSTRALIA

5.3.1   *Description of Alice Springs and Lajamanu*

The two case study locations are Alice Springs and Lajamanu, in the Northern Territory. They are part of the central Australian region which is characterized by: remote to very remote locations; a low and highly variable rainfall, native vegetation adapted to semi-arid/arid conditions; a private sector dominated by mining and pastoral operations; a diverse public service sector and a low and highly dispersed heterogeneous population with a few larger towns (see Morton *et al.*, 2011; Stafford Smith and Cribb, 2009).

The two communities differ in geographical extent, socio-economic and demographic attributes (ABS, 2011) and degree of remoteness (ABS, 2013). Alice Springs is a remote service town with a population of 25,186 people of whom 19% are indigenous. Lajamanu is a very remote and small town at the northern boundary of central Australia and has a much lower population of 656 people with almost 90% of indigenous people. Only a minority of the population speak non-English languages in Alice Springs while most of the people in Lajamanu speak Warlpiri. The median weekly household income in Alice Springs (A\$ 1676) is considerably higher than in Lajamanu (A\$ 1199); yet when assessed for Aboriginal households only, the median weekly income drops markedly in both Alice Springs (A\$ 1073) and Lajamanu (A\$ 743). The annual rainfall and temperatures for Alice Springs and Lajamanu averaged over the past 10 years data from Bureau of Meteorology are 260 mm/29.5°C and 610 mm/33.7°C respectively (BoM, 2015). Although there is considerable year-to-year variation in the weather at both locations, the yearly climate variability is more pronounced for Lajamanu.

5.3.2   *Current state of energy supply mix in case studies*

The main sources of electrical power for remote settlements in the wider central Australia region are through localized diesel, or hybrid diesel-solar powered generators (McKenzie, 2013). However, the electrical power in Alice Springs is largely sourced from natural gas (58.7 MW), with a small amount from solar (1 MW), while in Lajamanu it is mainly from diesel (1.9 MW), with an additional input from solar (0.3 MW) (BREE (Bureau of Resources and Energy Economics), 2013).

In Alice Springs, the total fuel consumption of car traveling can be estimated to be between 60 and 80 million liters per year while in Lajamanu this figure decreases to a 450,000 to 550,000 liters range. This is the equivalent of around 3000 liters per person per year in Alice Springs and 1000 liters per person per year in Lajamanu.

5.3.3   *Opportunities and challenges for renewable energy in the case study locations*

Building on the preceding review, we examine a number of factors in the two case study locations that are likely to create opportunities and challenges for the adoption of RE.

In both locations, potential key local drivers for transition to RE are increasing energy use and associated costs. Currently, households in these remote desert towns require more energy for cooling than heating, with this situation expected to be accentuated under climate change (Wang *et al.*, 2010). Summer months in central Australia are characterized by the high level of use of conventional air conditioners and refrigerators (powered by electricity) for cooling, which is likely to be exacerbated under the climate change projected for this region. To counter the effects of warming temperatures, more energy is expected to be used to keep the existing buildings at the same level of comfort. A recent study by Race *et al.* (2016a) showed that adaptation to extreme weather by Aboriginal households has become more energy intensive and dependent on housing compared to a few decades ago.

Remote dwellings are commonly inefficient users of energy due to outdated or poor quality household appliances, variable housing quality, inability of tenants to modify rented houses, and a fluctuating number of residents that sometimes leads to lengthy periods of overcrowding (AIHW, 2014; Lea and Pholeros, 2010). A high proportion of electricity is used to power air conditioners and refrigeration, and to supply fresh water to maintain the livability of central Australia due to the hot summer season and semi-arid climate.

The excessive reliance on appliances powered by non-renewable energy sources during extended periods of hot weather may create a dependence on an adaptation pathway that contradicts mitigation objectives. Recent research of the household energy context in central Australia (Race *et al.*, 2016b) indicates 'low-income' households spent around 10% of their gross household income (before tax, so more than 10% of their disposable income) on energy costs, with about 20% of 'low-income' households facing difficulty paying their bills on time. The data reveals that many Aboriginal households in Alice Springs spend more than 10% of their income on electricity costs, and so can be categorized as 'energy poor' (energy bills >10% of the household income) households.

Further investigation of climate zone-specific data in the ABS energy consumption survey shows that the expenditure on household energy varied from A\$ 30 to A\$ 47 per week (excluding fuel for vehicle costs) across the designated climate zones, with most regions in central Australia spending around A\$ 37 to A\$ 40 per week. Around 43% of Alice Springs households live in rented homes with a median rent of A\$ 300 per week, and while the majority of Lajamanu residents live in rented homes where the median rent is just A\$ 20 per week (see ABS, 2011). The maximum rent in the Alice Springs town camps is in the range of A\$ 290 to A\$ 334 per week for a 2–4 bedroom house (Territory Housing, 2010), likely to be a considerable expense for families in the 'low-income' category.

Water supply in remote locations is also energy intensive with around 1100 kWh required per ML for Alice Springs. As the average house uses around 740 kL each year, that adds 814 kWh of electricity consumption for water supply per household – equating to an additional A\$ 200 per year. The cost of living in remote communities is thus generally higher. For example, the cost of a standard basket of goods (basket sufficient to provide food for a family of six people, including children and elderly for a fortnight) in 2012 in Alice Springs supermarkets was around A\$ 500, while a similar standard basket cost around A\$ 800 in remote community stores (see Department of Health, 2013).

Most households in the Alice Springs town camps and Lajamanu use pre-payment meters, which do not help in recording actual usage of power by the residents and actual amount spent on power cards for the household. The main disadvantage of using pre-payment meters in remote locations is that the purchase of power cards can be restricted to when the local shop is opened (e.g. absence of after-hours shops in remote locations) and availability of money among the residents (McKenzie, 2013) which in turn affects the reliability of power supply. This also means appliances such as fridges, air conditioners and heaters that are commonly used to store food and regulate housing temperatures are affected, creating more burdens on household income and the health of the residents. However, there are also important social benefits from pre-payment meters, such as providing an incentive for different members of the household (e.g. long-term residents, visitors) to contribute to the cost of electricity that they may use in the short term, and providing a more frequent indication of household energy use than less frequent billing.

In terms of electric vehicles, there is just one registered in Alice Springs and none in Lajamanu. There is one public solar powered recharge station for vehicles in the Northern Territory, located in the capital city of Darwin. There is significant initial capital cost outlay for charging stations and purchasing electric vehicles. Indeed first-generation battery (acid-lead and lithium-ion) prices are still high. Furthermore battery technology is under significant development which leads to potentially high depreciation rates. Finally, electric vehicles are currently limited to city transport. The long distance for commuting in remote Australia remains the biggest barrier to the use of electric vehicles, with regular long-distance travel in the range of 500 to 1000 km (Spandonide, 2014).

The uptake of affordable and suitable battery-fitted electric vehicles and charging facilities in remote communities will take a long time. Large organizations are expected to be the early adopters in the emerging electric vehicle market. However, the economic prospects of using electric vehicles in large remote communities are positive. In Alice Springs, the total annual vehicle fuel cost reaches A$ 100 million (or A$ 3500 per capita). In Lajamanu, the total annual vehicle fuel cost is over A$ 1 million (about A$ 2000 per capita). In Lajamanu, switching to RE-based electric vehicles would correspond to an average electricity need of 500 kW per day. Assuming that the vehicle charging would be equally spread between peak and off-peak times, it would be equivalent to an annual electricity bill of A$ 800, which would allow achieving considerable savings (around 150% of fuel expenses) for a population which is subject to high levels of financial and transport disadvantages: the vast majority of Aboriginal and Torres Strait Islander people living in very remote locations does not currently have access to appropriate transport (Spandonide, 2014).

5.4 DISCUSSION

The review of the literature highlights that there are significant opportunities for rapid transition to RE in remote areas of Australia. This pertains to the abundant high quality solar resource and growing demand for energy from the wider community and different industries, which is currently predominantly supplied by highly subsidized expensive diesel generation. Given this potential, a transition to RE sources in off-grid remote Australia is considered long overdue and still is a "low hanging fruit" in that it will be an easier, less complicated, more justifiable and attractive proposition for governments than is to transition to RE supply on-grid areas of the urban coastal regions (Byrnes *et al.*, 2015). However, at only 1% of the current total energy mix, RE in remote Australia remains low (AECOM, 2014).

Rapid scaling up of RE in remote Australia will require addressing several barriers and challenges at different scales. The theoretical and operational barriers to the supply of RE in remote Australia reviewed in the preceding section have several common challenges (e.g. finance, governance, technical, community engagement).

Community level barriers such as adequate and effective community engagement, appropriate RE technology and system design, reliable operation and maintenance still have to be considered in every new initiative for RE energy supply in remote communities. However, there is already experience, capacity and proven approaches to show that these are no longer the critical barriers for scaling up RE in remote Australia. The system level barriers, especially the financial and governance related issues, appear to be part of the critical set of challenges restricting the scaling out of RE in remote Australia. A synthesis of these barriers identifies what appear to be linked barriers at two additional scales – one at household scale and the other at higher national/regional scale.

The household scale relates to limited financial, physical and human capital of many indigenous households in remote areas, primarily underpinned by disadvantage and remoteness. While rooftop PV solar power has been affordable to many households in Australia with uptake hailed as a quiet revolution in solar energy supply (Flannery *et al.*, 2013), they are out of reach for most indigenous households in remote Australia. Such people have low average income, do not own houses and have limited incentive or literacy to understand alternative energy sources and rebates on offer and to make choices of RE supply at household scale. The case studies presented in this chapter also indicate that many indigenous households in remote regions are 'energy poor' in that they commonly spend more than 10% of their income on energy bills. Some suffer unrecognized power interruptions for various reasons including being unable to pay for power bills with significant food safety and health implications. For inland Australia, higher average temperatures and longer and more intense heat waves are projected with high certainty. This will increase demand for energy to cope with the impacts of climate change. Unless there is a rapid shift to RE sources, adaptation efforts will depend on existing diesel generated energy. Fuel costs for transport represent an equivalent proportion of household income.

A shift to RE is necessary to reduce long-term cost, to curb health and environmental externalities of use of diesel generation of electricity and support mitigation of greenhouse gases. However, the initial capital investment needed for RE installation is often beyond the reach of remote communities, and so requires external support. A feasibility study of new RE supply to three indigenous communities in the Kimberley region of Western Australia, all connected to isolated, diesel powered networks, found that RE can benefit both the power utility provider and the communities subject to grants being available to cover cost of capital (Byrnes *et al.*, 2015).

This leads us to the second additional but critical set of barriers at higher national or regional scale. This involves market failures and distortions. It also involves institutions in terms of ineffective regulatory frameworks and incentives, and uncertainty and inconsistency in policy commitment to support RE competiveness and investment.

In remote Australia, energy markets fail among other things because of small size, highly dispersed and socio-economically disadvantaged customers; and because of high transaction and transport costs to distant and ragged geography with limited and inadequate infrastructure. Governments intervene to correct these market failures through subsidies and incentives to incumbent fossil fuel powered energy supply to uphold universal service obligations of providing energy access to citizens irrespective of location. These subsidies hide the true cost of diesel generated energy supply, and by distorting the market they create an uneven playing field for new RE sources (Byrnes and Brown, 2015). The costs of these subsidies, and the unaccounted health, socio-economic and environmental externalities of diesel use, are borne by taxpayers and the environment.

Similarly, the monopolistic nature of transport services (public transport, fuel companies), implies that prices are less likely to reflect demand levels: the cost elasticity of transport is particularly low and public transport fare strategies tend to be less price sensitive than in urban environments.

The Australian Federal, State and Territory governments apply a multi-layer of policy instruments including exemptions, subsidies, grid connection costs and tariffs designed for different purposes but that affect energy prices and the rate of adoption of RE supply in off-grid regions. Significant fuel excise exemptions on fossil fuel industries as well as direct and indirect subsidies exist that partly shift the cost of diesel generation from utilities and state electricity consumers to the national tax payer (Byrnes and Brown, 2015). Tariff policies are responsibilities of the states and territories and different feed-in tariffs (FiT) across Australia have had an impact on rate of adoption of mainly small-scale solar systems (Byrnes *et al.*, 2013). However, most FiTs have been substantially reduced and/or phased out (Byrnes and Brown, 2015; OECD, 2013).

Recent government policies including the 2015 Energy White Paper (Commonwealth of Australia, 2015) promotes the development of the renewables industry on a market basis through a focus on competitive least-cost energy options. Byrnes and Brown (2015) note that these policies tend to downplay the inherent market distortions associated with tax exemptions, subsidies on fossil fuels and cost of connection to grids, which clearly make it difficult for emerging renewable technologies to properly compete and establish, especially in remote regions.

Australia wide, public awareness on greenhouse gas emission and pollution from fossil fuel energy is increasing and will continue to add political pressure to support a shift in value, preference and support towards clean and RE sources (Moss *et al.*, 2014).

Governments through a variety of regulatory incentive and policy instruments and partnerships need to support transition to RE supply. The carbon pricing scheme, Renewable Energy Targets (RET), ARENA and the Clean Energy Financing Corporation (CEFC – a government funded financier created to help overcome financing challenges associated with clean energy development) were all established to assist accelerating this transition to RE sources (Byrnes and Brown, 2015). SETuP is one recent example of how partnerships between different agencies may help this transition. Established in 2014, this A\$ 56 million equal-contribution partnership between ARENA and the Power and Water Corporation of the Northern Territory government is planned to produce a total of 10 MW. This type of partnership can assist the transition to RE supply in remote communities (ARENA, 2014). Power and Water Corporation notes that the solar energy produced through SETuP will replace up to four million liters of diesel fuel each year and these fuel savings will initially be used to finance the Northern Territory government's A\$ 27.5 million contribution to the program (NT Power and Water Corporation, 2014).

A faster transition to RE could be expected given the significant renewable resource potential in Australia and the fact that it has been among the first countries to innovate RE technologies. However, the uncertainty and lack of long-term bipartisan political commitment to the aforementioned institutions, and to broader reform of the fossil fuel based energy sector, has beset the rate of transition to RE supply. Implications of uncertainties generated by the repeal of the carbon pricing scheme, the intentions to close ARENA, the CEFC and the reduction of RET are discussed in detail elsewhere (Byrnes and Brown, 2015).

The cost of remoteness combined with the chronic disadvantage that drive market failure in remote Australia requires government intervention and commitment to facilitate a rapid transition to RE in remote regions. This transition can save taxpayers the cost of subsidizing and revenue lost from diesel fuel excise exemptions, reduce the energy supply cost to communities and contribute to reduce GHG emissions which can all support effective adaptation pathways.

Transformation pathways are needed to help indigenous communities not only in adapting to climate and other negative environmental changes and maintaining the current state of livelihoods and well-being, but also to escape welfare dependence and ill health for many (Maru *et al.*, 2014). Transition to RE supply to remote communities can be used as a catalyst for substantive change with possibilities for new lines of livelihood in the installation, operation and maintenance of solar power plants (Maru *et al.*, 2015).

5.5   CONCLUSION

Diesel energy generation in remote Australia is relatively expensive and depends on the volatile global prices of oil, and costly and risky transport to these remote locations. It also has often unaccounted environmental and health externalities due to pollutants and GHG emissions. Transitioning to RE in remote Australia is far from just being a financial or technical challenge. To make such a transition requires addressing a complex set of intertwined capital, cultural, institutional and technological challenges simultaneously. In many ways, creating a new energy system in remote Australia mirrors several other challenges facing remote Australia, such as building a relevant education system, providing adequate health and social services, and generating meaningful employment and sustainable industries – all critical challenges for remote Australian communities. Establishing a robust energy system based on renewables is central to building a prosperous and sustainable future for remote Australia. Successive governments have signaled their reluctance to continue to subsidize energy systems powered by conventional fuels, so transitioning to low-cost RE is more a case of when, rather than if.

Redesigning energy systems for remote Australia must consider in a holistic and integrated way stationary and transport energy requirements. The current energy use for transport in central

Australia amplifies the 'cost of living' pressures faced by many. Accessing appropriate housing, education, employment and government services will be increasingly unaffordable under the present energy *status quo*. This situation is typically more extreme in small very remote communities (e.g. Lajamanu) than in the larger regional towns (e.g. Alice Springs), but tends to occur throughout the vast and remote inland region of Australia. Any systemic change to energy generation and use throughout remote Australia will need to be widely embraced by its residents if it is to be enduring and successful. While advice, expertise and support will be necessary inputs from the mainstream, ultimately it will require the leadership of the communities that live and thrive in remote Australia to make the transition to RE. The lessons emerging from the grass-roots innovation and trials already underway in remote Australia illustrate that such a transition is possible and can mobilize widespread support. If such a transformation was to occur, and indeed our evidence suggests it must occur, then communities in remote Australia will be able to reinvent themselves as vibrant and sustainable places to live and work – offering an experience for others living in remote locations throughout the world.

ACKNOWLEDGEMENTS

We are grateful to an initial discussion and insightful suggestions by Lyndon Frearson. The authors also thank the informants who contributed to their understanding of energy use in Alice Springs and Lajamanu, Northern Territory, which in turn was supported by funding from the Cooperative Research Centre for Remote Economic Participation.

REFERENCES

ABS (2011) Census quick stats. Australian Bureau of Statistics (ABS), Canberra, ACT, Australia. Available from: http://www.abs.gov.au/websitedbs/censushome.nsf/home/quickstats [accessed March 2017].

ABS (2013) Remoteness structure. Australian Bureau of Statistics (ABS), Canberra, ACT, Australia. Available from: http://www.abs.gov.au/websitedbs/D3310114.nsf/home/remoteness+structure [accessed October 2016].

AECOM (2013) Australian remote renewables: opportunities for investment. Report, prepared by AECOM (Architecture, Engineering, Consulting, Operations, and Maintenance) for the Australian Trade Commission (Austrade), Canberra, ACT, Australia.

AECOM (2014) Australia's off grid clean energy market. Report prepared for ARENA (Australian Renewable Energy Agency), Canberra, ACT, Australia.

AIHW (2014) Housing circumstances of indigenous households: tenure and overcrowding. Cat. no. IHW 132, Australian Institute of Health and Welfare (AIHW), Canberra, ACT, Australia. Available from: http://www.aihw.gov.au/publication-detail/?id=60129548060 [accessed December 2016].

Altman, J. (2014) Culture and Society: Indigenous policy: Canberra consensus on a neoliberal project of improvement. Australian public policy: progressive ideas in the neoliberal ascendency In: Miller, C. & Orchard, L. (eds) *Australian Public Policy Progressive Ideas in the Neo-Liberal Ascendency*, 115. University of Bristol Press, Bristol.

ARENA (2012) UltraBattery® distributed PV support and UltraBattery® for remote area power supply. Australian Renewable Energy Agency (ARENA) Available from: http://arena.gov.au/project/ultrabattery-distributed-pv-support-and-ultrabattery-for-remote-area-power-supply/ [accessed December 2016].

ARENA (2014) Northern Territory Solar Energy Transformation Program (SETuP). Projects, Australian Renewable Energy Agency (ARENA). Available from: http://arena.gov.au/project/northern-territory-solar-energy-transformation-program/ [accessed December 2016].

ARENA (2015a) Weipa 6.7 MW solar photovoltaic (PV) solar farm. Australian Renewable Energy Agency (ARENA). Available from: http://arena.gov.au/project/weipa-solar-farm/ [accessed December 2016].

ARENA (2015b) Leading renewables-mining project pushes forward. Media releases, Australian Renewable Energy Agency (ARENA). Available from: http://arena.gov.au/media/leading-renewables-mining-project-pushes-forward/ [accessed December 2016].

Bahadori, A. & Nwaoha, C. (2013) A review on solar energy utilisation in Australia. *Renewable and Sustainable Energy Reviews*, 18, 1–5.

BoM (2015) Climate data online. Bureau of Meteorology Canberra, ACT, Australia. Available from: http://www.bom.gov.au/climate/data/ [accessed December 2016].

BREE (2013) Beyond the NEM and the SWIS: 2011–12 regional and remote electricity in Australia. Bureau of Resources and Energy Economics, Canberra, ACT, Australia. Available from: https://www.industry.gov.au/Office-of-the-Chief-Economist/Publications/Documents/rare/bree-regional-and-remote-electricity-201310.pdf [accessed March 2017].

Byrnes, L. (2014) The cost of failing to install renewable energy in regional western Australia. School of Economics, University of Queensland, Brisbane, QLD, Australia.

Byrnes, L. & Brown, C. (2015) Australia's renewable energy policy: the case for intervention. University Library of Munich, Munich, Germany.

Byrnes, L., Brown, C., Foster, J. & Wagner, L.D. (2013) Australian renewable energy policy: barriers and challenges. *Renewable Energy*, 60, 711–721.

Byrnes, L., Brown, C., Wagner, L. & Foster, J. (2015) Reviewing the viability of renewable energy in community electrification: the case of remote western Australian communities. *Renewable and Sustainable Energy Reviews*, 59, 470–481.

Clean Energy Council (2015) Clean Energy Australia report 2014. Melbourne, VIC, Australia.

Climate change Authority (2014) Light vehicle emissions standards for Australia research report Available from: http://climatechangeauthority.gov.au/files/files/Light%20Vehicle%20Report/Lightvehiclesreport.pdf [accessed February 2017].

Commonwealth of Australia (2015) 2015 Energy White Paper. Canberra, ACT, Australia.

Department of Health (2013) Market based survey 2012. Northern Territory government, Available from: http://digitallibrary.health.nt.gov.au/prodjspui/bitstream/10137/616/2/Northern%20Territory%20%20Market%20Basket%20Survey%20report%202014.pdf [accessed February 2017].

EC (2014) Energy prices and costs working document. European Commission, Brussels, Belgium. Available from: http://ec.europa.eu/energy/doc/2030/com_2014_15_en.pdf [accessed March 2017].

EIA (2011) International energy statistics. US Energy Information Agency (EIA), Washington, DC. Available from: https://www.eia.gov/cfapps/ipdbproject/iedindex3.cfm?tid=44&pid=45&aid=2&cid=AS,&syid=2008&eyid=2011&unit=MBTUPP [accessed December 2016].

Flannery, T.F., Sahajwalla, V. & Commission, C. (2013) The critical decade: Australia's future. Solar Energy. Climate Commission Secretariat, Department of Industry, Innovation, Climate Change, Science, Research and Tertiary Education, Canberra, ACT, Australia.

Frearson, L. (2014) What's stopping solar's assault on diesel? Business Spectator, Australian Independent Business Media, Melbourne, Australia. Available from: http://www.businessspectator.com.au/article/2014/7/18/solar-energy/whats-stopping-solars-assault-diesel [accessed March 2017].

Frischknecht, I. (2014) Increasing renewable energy off the grid. *Australian Remote Area Power Supply Conference*. Available from http://australien.ahk.de/fileadmin/ahk_australien/Dokumente/Events_neu_pg/Projekte/Erneuerbare_Energien_Bergbausektor/Praesentationen/Presentation_Ivor_Frischknecht.pdf [accessed February 2017].

Geoscience Australia n.d Available from: http://www.ga.gov.au/scientific-topics/energy/resources/other-renewable-energy-resources/solar-energy [accessed February 2017].

Green, D. & Minchin, L. (2014) Living on climate-changed country: indigenous health, well-being and climate change in remote Australian communities. *EcoHealth*, 11, 263–272.

Kaushik, C. (2011) Renewable energy for sustainable agriculture. *Agronomy for Sustainable Development*, 31(1), 91–118.

Lea, T. & Pholeros, T. (2010) This is not a pipe: the treacheries of indigenous housing. *Public Culture*, 22, 187–209.

Li, K., Lloyd, B., Liang, X. & Wei, Y. (2014) Energy poor or fuel poor: what are the differences? *Energy Policy*, 68, 476–481.

Lowe, D. & Lloyd, C. (2001) Renewable energy systems for remote areas in Australia. *Renewable Energy*, 22, 369–378.

Maru, Y.T., Chewings, V. & Sparrow, A. (2012) Climate change adaptation, energy futures and carbon economies in remote Australia: a review of the current literature, research and policy. CRC-REP Working Paper CW005, Ninti One Limited, Alice Springs, NT, Australia. Available from: http://www.crc-rep.com.au/resource/CW005_ClimateChangeAdaptationLitReview.pdf [accessed December 2016].

Maru, Y.T., Smith, M.S., Sparrow, A., Pinho, P.F. & Dube, O.P. (2014) A linked vulnerability and resilience framework for adaptation pathways in remote disadvantaged communities. *Global Environmental Change*, 28, 337–350.

Maru, Y.T., Race, D., Sparrow, A., Mathew, S. & Chewings, V. (2015) Adaptation as a trigger for transformation pathways in remote Indigenous communities. *Innovation in the Rangelands, Australian Rangeland Society 18th Biennial Conference*. Australian Rangeland Society, Alice Springs, NT, Australia.

Mathew, S., Lee, L.S. & Race, D. (2016) Conceptualising climate change adaption for native bushfood production in arid Australia. *International Journal of Learning in Social Contexts*, 19: 98–114.

McKenzie, M. (2013) Pre-payment meters and energy efficiency in indigenous households. Report by Bushlight, the Centre for Appropriate Technology, Alice Springs, NT, Australia.

McKenzie, M. & Howes, M. (2006) Remote renewable energy in Australia: barriers to uptake and the community engagement imperative. *Australasian Political Studies Association Annual Conference*. pp. 25–27. Available from: http://www98.griffith.edu.au/dspace/bitstream/handle/10072/12021/APSA_2006_Paper.pdf?sequence=1 [accessed February 2017].

Morton, S., Smith, D.S., Dickman, C., Dunkerley, D., Friedel, M., McAllister, R., Reid, J., Roshier, D., Smith, M. & Walsh, F. (2011) A fresh framework for the ecology of arid Australia. *Journal of Arid Environments*, 75, 313–329.

Moss, J., Coram, A. & Blashki, G. (2014) Solar energy in Australia. The Australian Institute, Canberra, ACT, Australia.

Mount, B. (2012) A report to the energy users association of Australia. CME report March 2012, Melbourne, VIC, Australia.

NT Power and Water Corporation (2014) Northern Territory Solar Energy Transformation Program (SETuP) factsheet. Available from: https://www.powerwater.com.au/sustainability_and_environment/setup [accessed February 2017].

OECD (2013) "Australia", in inventory of estimated budgetary support and tax expenditures for fossil fuels 2013. Organization for Economic Cooperation and Development (OECD Publishing. Available from: http://dx.doi.org/10.1787/9789264187610-5-en [accessed December 2016].

Painuly, J.P. (2001) Barriers to renewable energy penetration; a framework for analysis. *Renewable Energy*, 24, 73–89.

Pittock, B. (2011) Co-benefits of large-scale renewables in remote Australia: energy futures and climate change. *The Rangeland Journal*, 33, 315–325.

Race, D., Mathew, S., Campbell, M. & Hampton, K. (2016a) Understanding climate adaptation investments for communities living in desert Australia: experiences of indigenous communities. *Climatic Change*, 139(3), 461–475.

Race, D., Mathew, S., Campbell, M. & Hampton, K. (2016b) Are Australian aboriginal communities sustainably adapting to warmer climates? A study of communities living in semi-arid Australia. *Journal of Sustainable Development*, 9, 208–223.

Shafiullah, G., Amanullah, M., Ali, A.S., Jarvis, D. & Wolfs, P. (2012) Prospects of renewable energy – a feasibility study in the Australian context. *Renewable Energy*, 39, 183–197.

Spandonide, B. (2014) Transport systems in remote Australia: transport costs in remote communities. Technical Document. Ninti One Limited, Alice Springs, NT, Australia.

Stafford Smith, M. & Cribb, J. (2009) Dry Times. CSIRO publishing, Melbourne, VIC, Australia.

Stafford Smith (2008) The 'desert syndrome' – causally-linked factors that characterise outback Australia. *The Rangeland Journal*, 30(1), 3–14.

Syed, F., Fowler, M., Wan, D. & Maniyali, Y. (2010) An energy demand model for a fleet of plug-in fuel cell vehicles and commercial building interfaced with a clean energy hub. *International Journal of Hydrogen Energy*, 35(10), 5154–5163.

Territory Housing (2010) Rent in Alice Springs town camps. Northern Territory Government, Casuarina, Northern Territory, Australia. Available from: http://www.housing.nt.gov.au/__data/assets/pdf_file/0015/138210/rfs50_rent_astc_Sep12.pdf [accessed December 2016].

Tuckwell, M., Frearson, L., Behrendorff, G. (2011) Bushlight India, a new approach to rural electrification in India – Identifying and addressing structural barriers to remote village electrification. Centre for Appropriate Technology, Alice Springs, NT, Australia.

Wang, X., Chen, D. & Ren, Z. (2010) Assessment of climate change on residential heating and cooling requirement in Australia. *Building and Environment*, 45, 1663–1682.

Watterson, I., Abbs, D., Bhend, J., Chiew, F., Church, J., Ekström, M., Kirono, D., Lenton, A., Lucas, C., McInnes, K., Moise, A., Monselesan, D., Mpelasoka, F., Webb, L. & Whetton, P. (2015) Climate projections report: rangelands cluster report. In: Ekström, M., Whetton, P., Gerbing, C., Grose, M., Webb, L. & Risbe, J. (eds) Climate change in Australia – projections for Australia's natural resource management regions. CSIRO and Bureau of Meteorology, Australia. Available from: https://www.climatechangeinaustralia.gov.au/media/ccia/2.1.6/cms_page_media/172/RANGELANDS_CLUSTER_REPORT_1.pdf [accessed February 2017].

Wyld Group & MMA (2008) Solar energy. Chapter 10 in Australia's energy resource assessment report. Available from: http://arena.gov.au/files/2013/08/Chapter-10-Solar-Energy.pdf [accessed December 2016].

CHAPTER 6

Opportunities of adopting renewable energy for the nursery industry in Australia

Guangnan Chen, Erik Schmidt, Tek Maraseni, Jochen Bundschuh,
Thomas Banhazi & Diogenes L. Antille

6.1 INTRODUCTION

In Australia, the nursery and garden industry provides significant economic, cultural, social and environmental benefits to the community (NGIA, 2014). The production nurseries support a diverse array of industries and end users, including retail outlets, landscapers, cut-flower growers, orchardists, vegetable growers, interiorscapers, sustainable forestry and revegetation enterprises. Overall, the gross value of production (GVP) of the broad "nursery, flower and turf" industry in Australia is A$ 1271 million, which is 17% of the total GVP of Australian horticultural industry (ABS, 2013).

Amenity horticulture is currently one of the fastest growing industries in Australia (NGIA, 2009). Energy use efficiency has also become increasingly important due to the increasing cost and scarcity of energy sources and also the associated greenhouse gas (GHG) emissions causing global warming (Bundschuh and Chen, 2014; Chen and Baillie, 2009a). The horticultural sector contributes about 6% of the total agricultural GHG emissions in Australia (Deuter, 2008). Nursery operators are thus under increasing pressure to reduce their energy and carbon footprint. By improving energy efficiency and using clean energy sources, the nursery industry can drive down their carbon footprint (Abeliotis *et al.*, 2015; Beccaro *et al.*, 2014; Lazzerini *et al.*, 2014; Russo *et al.*, 2008), and also simultaneously increase their bottom line.

6.2 ENERGY USE IN NURSERY

To save energy, the first step is to conduct an energy audit to identify where the major energy uses are occurring, and which of these can be practically and economically changed (Chen and Baillie, 2009a; Chen and Maraseni, 2012). Quite often, conducting an energy audit may be one of the fastest and the most effective ways to save money and reduce energy demand. Furthermore, reducing energy demand also makes significant reductions in GHG emissions. This is important in maintaining the 'clean and green' of relevant products.

6.2.1 *Energy audits and assessments*

Energy audits refer to the systematic examination of an entity, such as a firm, organization, facility or site, to determine whether, and to what extent, it has used energy efficiently (Chen and Baillie, 2009a; Chen and Maraseni, 2012). They may also assess opportunities of potential energy savings through fuel switching, tariff negotiation and demand-side management.

Typically, an energy audit may involve the following steps:

- Gather data.
- Evaluate energy use.
- Recommendations.

- Report.
- Review.

Overall, the nursery is often a high-value and an energy intensive industry (Lazzerini *et al.*, 2014; Schmidt *et al.*, 2010). Energy uses in the nursery industry may include heating/cooling, irrigation and various machinery operations such as tillage, potting, fertilizing, and crop maintenance, etc. These operations may also be implemented by specific practices such as potting machines and trans-planters.

Nursery production is often divided into a number of phases relating to the age or type of plant under cultivation. These may include (Cameron and Emmett, 2003):

- Stock plant management or seed production.
- Propagation.
- A weaning and growing-on stage often carried out under protection.
- A growing-on period after plants have either been potted-on into larger containers (container stage) or planted in a field to reach marketable size (field-grown stage).

With the rising energy costs and increasing concern over GHG emissions, on-farm energy efficiency is becoming an increasingly important issue for many nursery operators. In many ways, on-farm energy inputs represent a major and one of the fastest growing cost inputs to the growers. In New Zealand, it was found that average energy use of flower greenhouse nursery ranged between 57 and 3768 MJ m^{-2} per annum, with an average 880 MJ m^{-2} or 8800 GJ ha^{-1} (Barber, 2004). Energy use was also found to be strongly influenced by management practice, regional location, the type and age of greenhouse, and the type of crop being grown. Generally, smaller operations were less energy intensive, possibly due to capital constraints. In comparison, greenhouse cultivation of cut flowers in Greece was found to require 37,343 kWh ha^{-1} of electricity consumption for refrigeration and 7900 kWh ha^{-1} for water pumping, equivalent to a total electricity energy use of 45,243 kWh ha^{-1} or 1628 GJ ha^{-1} (Abeliotis *et al.*, 2015). Lazzerini *et al.* (2014) also reported electricity use of 6.5–18.2 MWh ha^{-1} plus 600–806 kg of diesel use for both field- and container-grown plants in Italy. Kipp and Snodgrass (2006) found that using modern mechanical and lighting systems, energy use could be cut by up to 80% in Idaho, USA.

Despite the significant effort by the nursery industry, a number of barriers or challenges to the improvements of energy efficiency might still exist. These include (OAN, 2011):

- Energy use is decentralized across a large number of large, medium and small operations.
- There might not be a single entity synthesizing all the resources to support energy savings efforts.
- Inconsistent or unavailable data identifying projects with high degree of success, evaluating technologies, and real energy savings.
- Nurseries have limited time and staff to research and implement programs.
- The large capital investment necessary to fund efficiency projects is non-existent for many operations, and the paperwork for incentive programs is consistently mentioned as burdensome.
- The nursery industry currently has no energy baseline from which to measure the effectiveness of energy reduction activities.
- Education and awareness of new practices, programs and technologies is not focused on rural, small businesses.

6.2.2   *Case studies of nursery energy use in Australia*

Currently, there appears to be little data available regarding the energy use in the horticultural sector. In Australia, there also seems a total lack of energy data for the nursery sector outside Queensland (Chen *et al.*, 2015a).

There are evidences to suggest that most of the nurseries in Australia utilize greenhouse systems to maintain year-round production. In the following, very limited case studies of energy use of five small-to-medium nurseries around the Brisbane metropolitan area in Queensland will first

Table 6.1.   Summary of energy consumption in five different nurseries in Queensland.

Attributes	Nursery A	Nursery B	Nursery C	Nursery D	Nursery E
Used area for production [ha]	1.5	5	3.5	2.5	5
Annual turnover [million A$]	1.1	3	na	na	na
Energy consumed from electricity [GJ]	122.1	242.3	651.2	493.1	1558.5
Energy consumed from diesel [GJ]	0	92.6	204.6	na	867.5
Energy consumed from LPG gas [GJ]	0	488.3	586.0	na	119.4
Energy consumed from petrol [GJ]	0	0	0	0	165.4
Total direct energy consumption [GJ]	122.1	823.2	1441.8	493.1	2710.8
Total direct energy intensity [GJ ha^{-1}]	81.4	164.6	411.9	197.2	542.2
Total direct energy cost [A$]	4500	24650	41550	22000	73500
Energy cost per ha [A$ ha^{-1}]	3000	4930	11871	8800	14700

be presented (Schmidt *et al.*, 2010), noting that Brisbane is the state capital of Queensland, and it has a humid subtropical climate with hot, humid summers and dry, mild winters. The summer (December–February) average temperature is 20–28°C, while the winter (June–August) average temperature is 11–21°C (Guan and Bell, 2010).

In this study, the energy data was collected from these five farms over a period of one year (2010). For this purpose, the electricity consumption was obtained from the utility bills while the fuel (diesel, petrol and gas) consumption obtained from grower records. The machinery energy use data was based on either the figures provided by the growers or on-site observation. Diesel, electricity, petrol and LPG gas are used for various processes in these nurseries, such as preparation, establishment, in-season, irrigation, harvest, post-harvest, and general (Chen *et al.*, 2011). Electricity is mostly used for heating, pumping, refrigerating, running potting mixture, bag filler mixers and fertilizer agitator, and for general office use. LPG gas is mainly used for hot water heating and hot air burners, whereas diesel is mostly used for tractors and other vehicles.

The energy use data was then entered into an online energy management calculator (EnergyCalc), which was a process-based software for calculating and storing energy use and GHG emission data for the whole site or individual processes (Chen and Baillie, 2009b; Schmidt *et al.*, 2010). The software outputs are also grouped into four broad categories related to fuel use, electricity use, total on-farm energy use and carbon emissions. This software has been used for a systematic quantification of energy use across the nursery processes (e.g., potting machine, transferring/conveyor, trans-planter, lighting, irrigating, spraying and fertigation).

A summary of energy and production data for these five nurseries in Queensland is presented in Table 6.1, where the raw energy data from different sources has been converted into the standard energy unit [GJ].

It can be seen that in total, these five nurseries collectively used over 5591 GJ of energy. Out of them, 54.9% was consumed by electricity, 21.4% by LPG gas and 20.8% by diesel. Petrol used about 3% of the total energy.

From Table 6.1, it can also be seen that the overall energy consumption is generally not directly related to the production area. In Nursery A, energy consumption per hectare production area is 81.4 GJ, whereas in Nursery B it is 164.6 GJ, Nursery D 197.2 GJ, Nursery C 411.9 GJ and for Nursery E it is 542.2 GJ. Instead, energy consumption is more directly related to intensity of operations. These energy use figures may also be 20–1000 times higher than those of broad-acre cereal crops (0.4–4.4 GJ ha^{-1}) for Australia (Chen *et al.*, 2015a), but are still significantly lower than their New Zealand counterparts, which has a considerably cooler climate than subtropical Queensland.

Overall, it was also found that (Chen *et al.*, 2011; Schmidt *et al.*, 2010):

• Due to historical, personal and geographical reasons, a wide variety of growing methods and systems are used in different nurseries in Queensland.

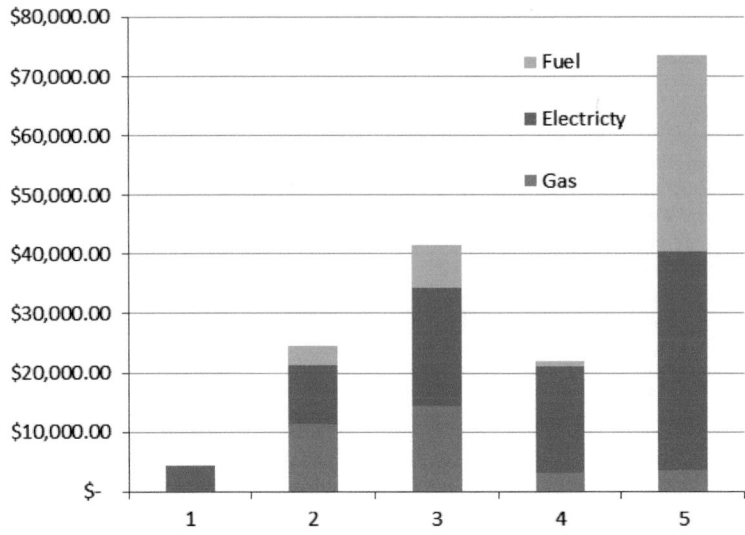

Figure 6.1. Total annual energy expenditures (A$) at the five nurseries in Queensland.

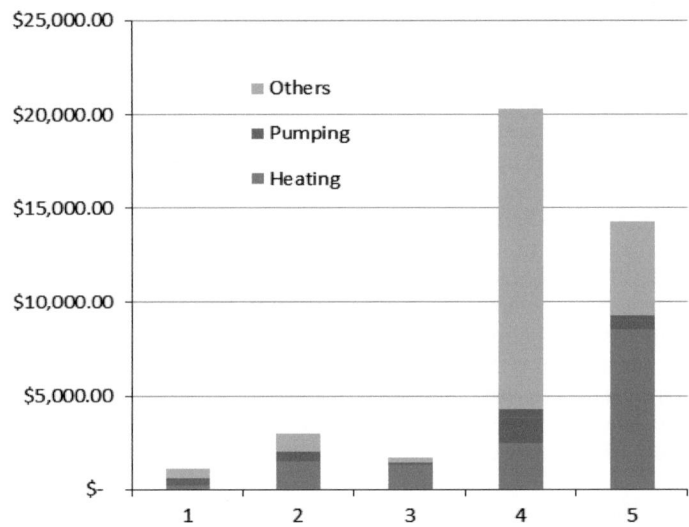

Figure 6.2. Annual heating energy costs (A$) per ha of total farm area at the five nurseries in Queensland.

- Energy costs vary significantly between different operators (Fig. 6.1), with their intensities varying between A$ 3000 to A$ 14,700 per hectare of production area (not the total farm area) (Table 6.1). Overall, the energy cost may consist of less than 0.5–1% of production value in these sites.
- Heating often forms a very significant component of the energy cost, particularly when this is supplied by electricity energy (Fig. 6.2). In this case, alternative heat sources such as LPG may be used to reduce the energy costs.
- There are few diesel pumps used in the nursery industry. The most common pumps used for irrigation in the nursery appear to be electric centrifugal pumps (both single and multistage) and are often run in excess of 12 hours per day.

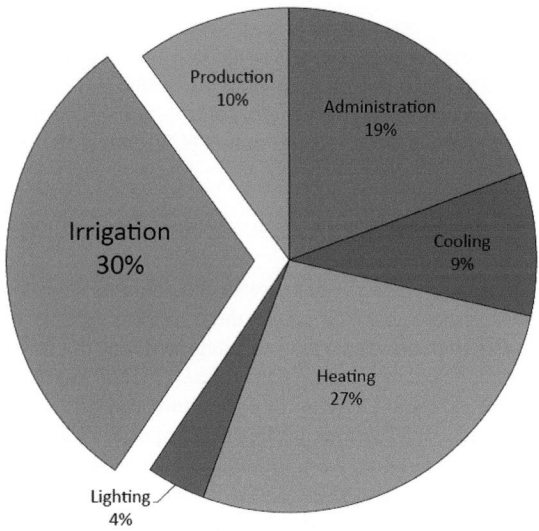

Figure 6.3.   Average annual energy use at the 16 nurseries in South East Queensland.

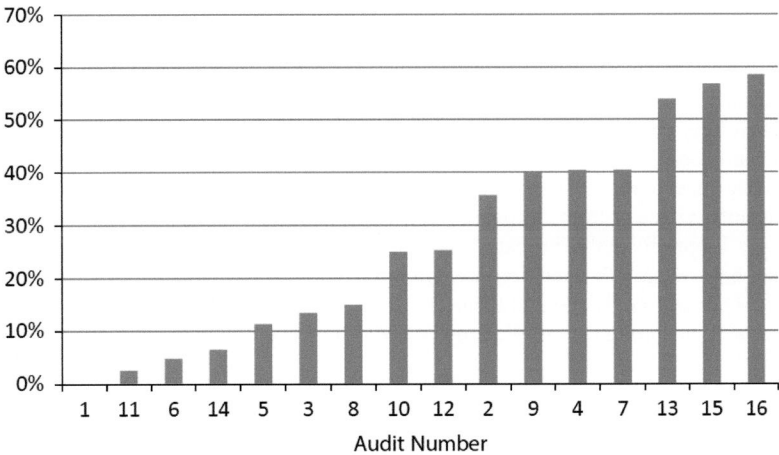

Figure 6.4.   Average irrigation energy use as a percentage of total electricity used at the 16 nurseries in South East Queensland.

- Growers are generally well aware of the need to reduce their energy usage. This may be evidenced by the widespread uses of vari-speed pumps to save pumping energy.
- It is observed that some motors used at a number of sites are significantly more powerful than the tasks they are required to perform.

In another extended study of irrigation energy use in nurseries, 16 further energy audits were carried out in South East Queensland (Eberhard *et al.*, 2013). It is shown that in these farms, an average 30% of the total electricity used on-site was for pumping water and irrigation (Fig. 6.3). This ranged from 6% (at a site where the heating dominated the usage) to 59% (Fig. 6.4). Irrigation is the single largest electricity using process in 7 out of 16 audits conducted. This further demonstrates that irrigation could be a significant user of energy in the nursery industry in this area.

McHugh *et al.* (2010) also suggested that for horticultural operations with considerable cooling requirements, heating and refrigeration systems could consume a large amount of electricity and thereby contribute significantly to the running costs of businesses. This is consistent with the findings of Figure 6.3. Improvements to the design, controls and operations of these systems have thus also great potential to reduce energy consumption (Cumming, 2015).

6.3   OPPORTUNITIES OF ADOPTING ALTERNATIVE ENERGY SOURCES

Fossil fuels are a limited resource and can also create considerable environmental problems such as GHG emissions, so it is desirable that where the opportunities are appropriate, renewable energy such as solar, wind, and bioenergy may be integrated into the nursery operations to save energy costs, and to reduce GHG emissions (Chen *et al.*, 2015b; Yusaf *et al.*, 2011). Examples of specific applications of renewable energy in agriculture and horticulture may include solar space and water heating, and using biomass for heating purposes (Sonneveld *et al.*, 2010). Other applications include lighting, irrigation, wastewater treatment pond aeration, communication and remote equipment operation, etc. (Gopal *et al.*, 2013; Mekhilef *et al.*, 2013; Vick and Almas, 2011). Biofuels may be used for tractors and road transport purposes (Basha, 2014).

6.3.1   *Opportunities of adopting solar energy*

Australia is a country which has a rich solar energy resource with the highest average solar radiation per square meter in the world (Geoscience Australia and ABARE, 2010; Yusaf *et al.*, 2011). The average annual solar radiation gathered in Australia is approximately 58 million PetaJoules (PJ), which is nearly 10,000 times the nation's annual energy consumption (Bahadori and Nwaoha, 2013). Therefore, solar energy in many cases is considered to be an attractive substitution of conventional energy. For example, electricity generation and solar hot water production can be used to either offset or replace the electricity used within a nursery operation and offers the ability to significantly reduce or even eliminate electricity costs to the nursery business.

Currently, it appears that the greatest interest exists in the nursery industry in the use of solar energy for photovoltaic (PV) electricity generation and thermal collectors for hot water production. A number of nursery and horticultural enterprises in Australia have already implemented some of these measures, including the installation of a PV system and upgrading to more energy efficient lighting such as LED lighting.

6.3.2   *Opportunities of adopting wind energy*

The idea of using wind to produce work is not a recent concept with traditional windmills being used to pump water and grind flour for centuries. With the advent of more efficient generators, the wind turbine has been used to generate electricity from the wind and is a promising source of renewable energy (Yusaf *et al.*, 2011).

Overall, wind power is seen as an effective alternative and renewable electrical energy source suitable to many areas of Australia, particularly along the coastline. Wind turbines can often have a shorter payback period and higher return on investment than solar power in areas with high average winds speeds and low wind turbulence. However, there is currently less interest in the wind energy in the nursery sector in Australia, owing to relatively few regions having average wind velocities which warrant wind power and also the potential issues for visual and noise pollution (Hall *et al.*, 2013).

6.3.3   *Opportunities of adopting bioenergy*

Biogas is a combustible gas produced as a result of anaerobic digestion (AD), a process in which organic matter is broken down by microbial activity in the absence of oxygen. Biogas normally consists of 50 to 60% methane. It is currently captured from landfill sites, sewage treatment plants, livestock feedlots and agricultural wastes. The typical conditions required are biomass combined with water in a lagoon or containment vessel, with the combined solution being maintained at between 16 and 60°C.

Overall, it appears that biogas production is a less suitable energy source for the nursery industry because nurseries generally do not produce enough suitable organic waste on-site to run an anaerobic digester. Biogas technology has not been tested in the nursery environment in Australia, and could pose a financial risk unless complete or major financial subsidization from external sources is provided. However, biogas could play a part in reducing energy and fertilizer costs of nurseries in the future, but at the moment other turnkey alternative energy sources exist that place themselves in a more suitable position for use.

6.4   DEVELOPMENT OF ONLINE CALCULATOR FOR ALTERNATIVE ENERGY SOURCES

Commissioned by Nursery & Garden Industry Australia (NGIA), an online calculator for renewable energy has recently been developed for the nursery industry by the National Centre for Engineering in Agriculture (NCEA), University of Southern Queensland. The overall framework for the calculator is shown in Figure 6.5. To run the program, the user will first need to capture energy demand information based on a utility statement of electricity usage or using calculations undertaken using EnergyCalc (Chen and Baillie, 2009b). A pre-determined percentage of energy demand to be met using wind or solar power is selected as is an appropriate local weather station for downloading of meteorological data (Fig. 6.5). The calculator is then able to size the appropriate solar or wind turbine system and provides a simple cost benefit based on energy demand, renewable energy generated, electricity costs, system capital and operating costs (Fig. 6.6).

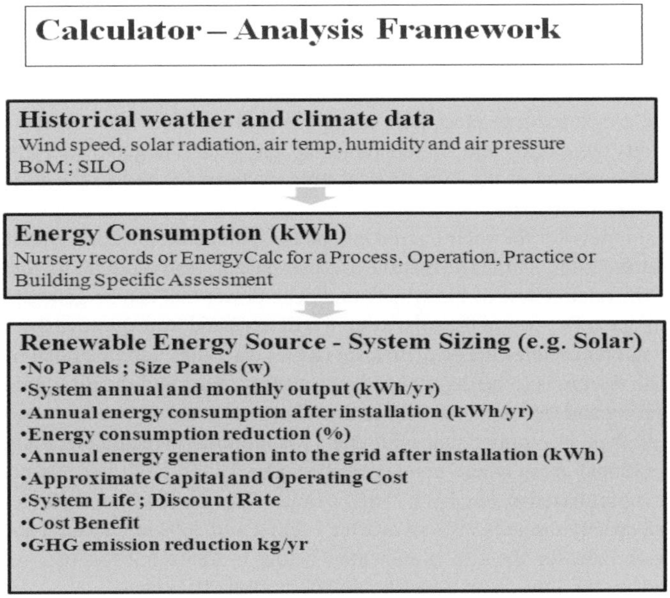

Calculator – Analysis Framework

Historical weather and climate data
Wind speed, solar radiation, air temp, humidity and air pressure
BoM ; SILO

Energy Consumption (kWh)
Nursery records or EnergyCalc for a Process, Operation, Practice or Building Specific Assessment

Renewable Energy Source - System Sizing (e.g. Solar)
•No Panels ; Size Panels (w)
•System annual and monthly output (kWh/yr)
•Annual energy consumption after installation (kWh/yr)
•Energy consumption reduction (%)
•Annual energy generation into the grid after installation (kWh)
•Approximate Capital and Operating Cost
•System Life ; Discount Rate
•Cost Benefit
•GHG emission reduction kg/yr

Figure 6.5.   The framework for the NGIA/NCEA online renewable energy calculator.

Items	Details
System Type	Solar
Rated Power	9.44 kW
System Lifetime	25 year
Capital Cost	A$ 36969
Operating Cost	A$ 739 per year
Electricity Sale Price*	A$ 0.44 per kWh
Electricity Purchase Price	A$ 0.22 per kWh
Panel Area	66.3 m^2
Energy Demand	18600 kWh per year
Energy Generated	18600 kWh per year
Greenhouse Gas Emissions Reduced**	19456 kg per year

Figure 6.6.   An example of NGIA/NCEA online renewable energy calculator output sheets.

6.5   CASE STUDIES

In the following, the above renewable energy calculator will be used to model and examine the generation and cost-effectiveness of solar and wind energy at a relatively large (25 ha) hypothetical nursery farm, noting the limitation of these case studies because both the assumed generation efficiency and costs of renewable energy could change significantly as the technology advances and government policy changes. The conclusion derived here is thus only applicable for the given energy tariff and also the particular location.

6.5.1   *Solar energy generation*

Sunlight can be converted into electricity using PV technologies. This can be used to replace existing electricity purchases and to sell power to the grid. The case study below illustrates generation of solar energy at the hypothetical nursery farm located in the inland township of Gatton, 90 km west of Brisbane.

Key technical inputs for the solar energy calculator include the location and intensity of solar radiation at that location. This information can be sourced from Bureau of Meteorology data. Others include the rated power of the system and peak power that can be generated during maximum solar irradiance. The cost of the solar system is determined from quotes to be A$ 4000.00 m^{-2} after subsidies and renewable energy certificate (REC) discounts administrated by the Australian government. The system is to be depreciated over twenty years and the discount rate is 5% with an annual operating and maintenance cost of 2% of capital cost.

In this case study, it is assumed that solar energy is required to match electricity demand from one of the operational areas where annual electricity is 10,000 kWh per annum. It is assumed that electricity is purchased at 18 c kWh^{-1} and can be sold into the grid at 40 c kWh^{-1}. It is also assumed that electricity usage is 20% in quarter 1 and 4 and 30% in quarter 2 and 3.

Solar radiation data for the site is indicated below in Table 6.2 together with solar power generation capacity for 1 m^2 panel based on a 13% system efficiency.

Figure 6.7 compares electricity demand for the production area (10,000 kWh) and generated electricity based on 40 m^2 of panels which is required to generate an equivalent annual power.

Table 6.2. Solar radiation for the modeled site and solar energy generation potential.

Solar Radiation Gatton MJ/m²												
Jan	Feb	Mar	Apr	May	Jun	Jul	Aug	Sep	Oct	Nov	Dec	Year
24.3	21.3	20.2	17.1	13.8	12.2	13.4	16.3	20	22.2	24.1	25.1	19.2

Solar Radiation Gatton kWh/m²/m												
202.5	177.5	168.3	142.5	115	101.7	111.7	135.8	166.7	185	200.8	209.2	1917

Energy Generated kWh/mth 1 square metre panel												
21.82	19.12	18.13	15.35	12.39	10.95	12.03	14.63	17.96	19.93	21.64	22.53	206.5

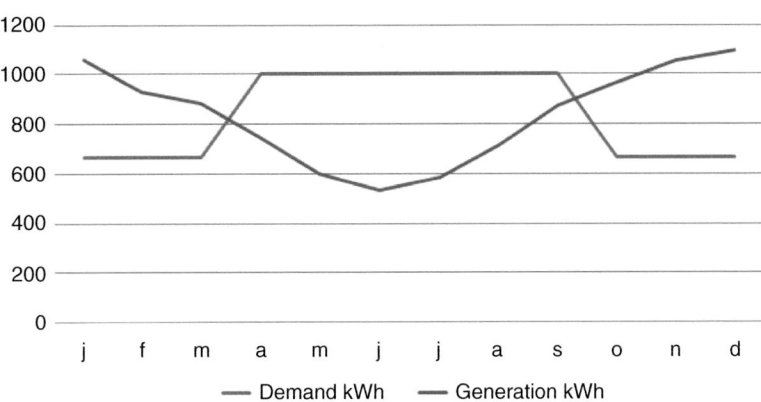

Figure 6.7.   Monthly energy demand and generation based on 40 m² solar panels.

It can be found that the annual energy generation is equal to energy consumption and the energy generation efficiency of the system is 100%. 1983 kWh of surplus power is generated in the summer months of October to March which can be sold into the grid at 40 c kWh^{-1} generating A\$ 793 of revenue. 1953 kWh of electricity has to be purchased during the winter months of April to September at 18 c kWh^{-1} costing A\$ 352. This is substantially less than the pre-solar annual cost of electricity (A\$ 1800).

Based on a discounted cash flow analysis over the 20-year life of the solar system, the annualized equivalent cost of purchasing and operating the solar system is A\$ 1162. The annualized equivalent cost of the current system is the cost of electricity purchased (A\$ 1800). The overall benefit-cost ratio is 1.55 and the payback period is 8.3 years.

6.5.2 *Wind energy generation*

Because of the inland location, the assumption for wind generation at the particular site of the hypothetical nursery farm is that the installed system would be designed to supplement current power demand and not to be a power plant. Thus, small-scale (less than 10 kW) wind turbines are chosen and installation would be on towers at 10 to 30 m high rather than 60 to 70 m for typical wind farms employing large turbines.

Analysis of the Bureau of Meteorology's wind data for the selected site provides an annual mean wind speed of only 1.96 m s^{-1}. This is quite low and results in an estimated annual production of only 343 kWh (Fig. 6.8).

Figure 6.9 provides a comparison of the monthly energy demand for the site versus the estimated energy production from a single small-scale wind turbine. Energy demand is read from the left-hand y-axis while energy production is read from the right-hand y-axis. The ratio of demand

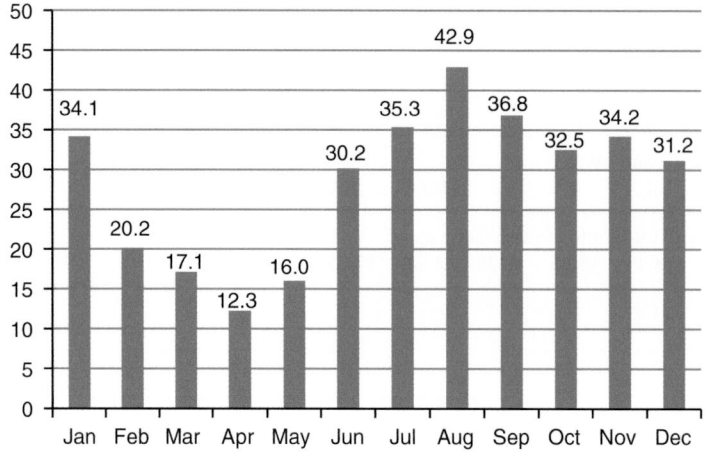

Figure 6.8.   Monthly estimated electricity production [kWh] for a wind turbine at the modeled site.

Figure 6.9.   Monthly energy demand vs. energy production for a single wind turbine at the modeled site.

to production is approximately 100:1. Therefore, energy generation from a single wind turbine can only produce 1% of the total energy demand for this particular site. No energy is able to be generated for sale to the grid, thus grid connection would not be feasible. 34,657 kWh of electricity also has to be purchased each year at 18 c kWh^{-1} costing A\$ 6238. The cost of a 6 kW wind turbine and installation is estimated to be approximately A\$ 55,000 after subsidies and REC discounts. The system is to be depreciated over 20 years and the discount rate is 5% with an annual operating and maintenance cost of 5% of capital cost. The benefit:cost ratio is 0.47 and the system is not viable at this location of very low wind speed.

6.6   CONCLUSION

Energy is increasingly becoming a key factor for the nursery industry to understand and manage in relation to sustainability and cost performance. Through case studies, this chapter has examined energy usage patterns in the nursery industry in Queensland, Australia. It has also identified potential renewable energy sources that could be utilized in the industry.

    It has been shown that energy uses in nurseries vary significantly between different operators. Energy use is strongly influenced by management practice, regional location, the type and age of greenhouse, and the type of crop being grown. The energy cost typically consists of 0.5–1% of production value. Irrigation and heating and cooling could be the major sources of energy use, each contributing to some 30% of total energy use.

The opportunity of adopting renewable energy in nursery operations has also been evaluated. An online renewable energy calculator has been used to provide an indication of potential for solar and wind systems to replace purchased electricity and feed electricity into the grid.

Future work should focus on developing a better understanding of energy usage patterns in the nursery industry across a representative range of nursery operations and sub-operations, regions, and scales. Opportunities of utilizing other potential renewable energy sources should also be explored. With the new technologies currently under development, renewable energy costs are falling rapidly. By improving energy efficiency and using clean energy sources, the nursery industry can not only drive down their carbon footprint but also simultaneously increase their production and profits.

REFERENCES

Abeliotis, K., Barla, S., Detsis, V. & Malindretos, G. (2015) Life cycle assessment of carnation production in Greece. *Journal of Cleaner Production*, 112(1), 32–38.

ABS (2013) Value of agricultural commodities produced, Australia 7503.0, 2011–12. Australian Bureau of Statistics, Canberra, ACT, Australia.

Bahadori, A. & Nwaoha, C. (2013) A review on solar energy utilisation in Australia. *Renewable and Sustainable Energy Reviews*, 18, 1–5.

Barber, A. (2004) Energy use and carbon dioxide emissions in the New Zealand vegetable & flower greenhouse industries. AgriLINK NZ Ltd., New Zealand.

Basha, S.A. (2014) Biodiesel emissions and performance. In: Bundschuh, J. & Chen, G. (eds.) *Sustainable Energy Solutions in Agriculture*. CRC Press, Boca Raton, FL. pp. 323–334.

Beccaro, G.L., Cerutti, A.K. Vandecasteele, I., Bonvegna, L., Donno, D. & Bounous, G. (2014) Assessing environmental impacts of nursery production: methodological issues and results from a case study in Italy. *Journal of Cleaner Production*, 80, 159–169.

Bundschuh, J. & Chen, G. (eds.) (2014) *Sustainable Energy Solutions in Agriculture*. CRC Press, Boca Raton, FL.

Cameron, R.W.F. & Emmett, M.R. (2003) Production systems and agronomy – Nursery stock and houseplant production. In: Brian, T. (ed.) *Encyclopedia of Applied Plant Sciences*. Elsevier, Oxford, UK.

Chen, G. & Baillie, C. (2009a) Agricultural applications: energy uses and audits. In: Capehart, B. (ed.) *Encyclopedia of Energy Engineering and Technology*, 1:1. Taylor & Francis Books, London, UK. pp. 1–5.

Chen, G. & Baillie, C. (2009b) Development of a framework and tool to assess on-farm energy uses of cotton production. *Energy Conversion & Management*, 50(5), 1256–1263.

Chen, G. & Maraseni, T. (2012) Agriculture: energy use and conservation. In: Jorgensen, S.E. (ed.) *Encyclopedia of Environmental Management*. Taylor & Francis Books, London, UK.

Chen, G., Maraseni, T., Banhazi, T. & Bundschuh, J. (2015a) Benchmarking energy use on farm. Report to Rural Industries Research and Development Corporation, National Centre for Engineering in Agriculture, University of Southern Queensland, Toowoomba, QLD, Australia.

Chen, G., Maraseni, T., Bundschuh, J. & Zare, D. (2015b) Agriculture: alternative energy sources. In: Anwar, S. (ed.) *Encyclopedia of Energy Engineering and Technology*. Taylor & Francis Books, London, UK.

Cumming, J. (2015) Economic evaluation of farm energy audits and benchmarking of energy use on vegetable farms. Horticulture Innovation Australia Ltd., Sydney, NSW, Australia.

Deuter, P. (2008) Vegetable industry carbon footprint scoping study – discussion paper 5: who will use the vegetable carbon tool? Horticulture Australia Ltd., Sydney, NSW, Australia.

Eberhard, J., McHugh, A., Scobie, M., Schmidt, E., McCarthy, A., Uddin, Md J., McKeering, L. & Poulter, R. (2013) Improving irrigation efficiency through precision irrigation in South East Queensland. National Centre for Engineering in Agriculture, University of Southern Queensland, Toowoomba, QLD, Australia.

Geoscience Australia and ABARE (2010) Australian energy resource assessment. Australian Government, Canberra, ACT, Australia.

Gopal, C., Mohanraj, M., Chandramohan, P. & Chandrasekar, P. (2013) Renewable energysource water pumping systems – a literature review. *Renewable and Sustainable Energy Review*, 25, 351–370.

Guan, L. & Bell, J.M. (2010) Monitoring energy performance of air conditioners in classrooms. In: Teng, J.G. (ed.) *Proceedings of the First International Conference on Sustainable Urbanization, 15–17 December, Hong Kong*. The Hong Kong Polytechnic University, Faculty of Construction and Land Use, Hong Kong, China. pp. 1915–1924.

Hall, N., Ashworth, P. & Devine-Wright, P. (2013) Societal acceptance of wind farms: analysis of four common themes across Australian case studies. *Energy Policy*, 58, 200–208.

Kipp, J.A. & Snodgrass, K. (2006) Modern systems cut energy use 80 percent at the coeur d'Alene nursery. Missoula, MT: U.S. Department of Agriculture, Forest Service, Missoula Technology and Development Center. Available from: http://www.fs.fed.us/t-d/php/library_card.php?p_num=0673%202326 [accessed December 2016].

Lazzerini, G., Lucchetti, S. & Nicese, F.P. (2014) Analysis of greenhouse gas emissions from ornamental plant production: a nursery level approach. *Urban Forestry and Urban Greening*, 13(3), 517–525.

McHugh, A.D., Erol, C. & Eberhard, J. (2010) SEQ irrigation futures R&D support: horticulture. Development of a framework and tools for packing shed water and energy audits. National Centre for Engineering in Agriculture, University of Southern Queensland, Toowoomba, QLD, Australia, 2010.

Mekhilef, S., Faramarzi, S.Z., Saidur, R. & Salam, Z. (2013) The application of solar technologies for sustainable development of agricultural sector. *Renewable and Sustainable Energy Reviews*, 18, 583–594.

NGIA (2009) Industrial overview. Nursery & Garden Industry Australia Ltd., 2009. Available from: http://www.ngia.com.au/Category?Action=View&Category_id=212 [accessed December 2016].

NGIA (2014) Environmental sustainability position statement. Nursery & Garden Industry Australia Ltd., Castle Hill, NSW, Australia.

OAN (2011) Nurseries for a sustainable world – Oregon Association of Nurseries sustainability roadmap. Oregon Association of Nurseries, Wilsonville, OR.

Russo, G., Scarascia Mugnozza, G. & De Lucia Zeller, B. (2008) Environmental improvements of greenhouse flower cultivation by means of LCA methodology. *Acta Horticulturae*, 801, 301–308.

Schmidt, E., Chen, G., Symes, T., Zhao, B. & Cameron, R. (201) Calculator for renewable energy resources for nursery production. National Centre for Engineering in Agriculture, University of Southern Queensland, Toowoomba, QLD, Australia.

Sonneveld, P.J., Swinkels, G., Campen, J., van Tuijl, B.A.J., Janssen, H.J.J. & Bot, G.P.A. (2010) Performance results of a solar greenhouse combining electrical and thermal energy production. *Biosystems Engineering*, 106(1), 48–57.

Vick, B.D. & Almas, L.K. (2011) Developing wind and/or solar powered crop irrigation systems for the Great Plains. *Applied Engineering in Agriculture*, 27(2), 235–245.

Yusaf, T., Goh, S.C. & Borserio, J.A. (2011) Potential of renewable energy alternatives in Australia. *Renewable and Sustainable Energy Reviews*, 15(5), 2214–2221.

CHAPTER 7

Fundamentals of solar energy

Maciej Klein, Kamil Łapiński, Katarzyna Siuzdak & Adam Cenian

7.1 INTRODUCTION

We are becoming more and more aware that fossil energy resources are limited – this leads to energy price increase even if sudden price variations sometimes hide such trends. It does not necessarily mean for us energy shortages but it does enforce changes in energy paradigm – photovoltaic cells, solar panels, wind turbines and bioenergy installations are being rapidly developed.

7.1.1 *Solar energy resources and potentials*

The amount of solar power reaching the earth's surface – 89 PW (Smil, 1991) or 2.8×10^{24} J per year – although most of it is reflected back, is more than enough to satisfy people's energy needs: 5–6×10^{20} J year^{-1} – see e.g. International Energy Statistics – EIA (www.eia.gov). It is estimated that photosynthesis processes alone can capture approximately 3×10^{21} J year^{-1} in produced biomass (Luo and Hong, 2012). Only a very small fraction of this energy will enable humans to survive.

Besides, the solar energy, absorbed in earth's atmosphere and ocean water, produces effects known as the water cycle and air convection resulting in water and wind energies. It has been estimated that solar energy (including wind, biomass and water flow energy) constitutes about 99.98% of all energy available to us; in comparison the heat from nuclear (geothermal) energy constitutes only 0.02%. The global exergy (the useful part of energy that allows us to do work and perform energy services) available on earth was estimated during the Global Climate & Energy Project at Stanford University (Fig. 7.1). The diagram summarizes the exergy reservoirs and flows in our sphere of influence including their interconnections, conversions, and eventual natural or anthropogenic destruction.

7.1.2 *Solar radiation and maps*

The sun emits electromagnetic (EM) radiation across most of the electromagnetic spectrum, in character close to a black body radiation with temperature around 5500°C, i.e. effective solar temperature (Fig. 7.2). The spectrum of solar light which approaches earth's surface changes due to interactions of light with earth's atmosphere. It is clearly seen that the interaction of radiation with the ozone sheath significantly reduces the amount of potentially hazardous UV radiation (with wavelength 200–300 nm).

The amount of incoming solar irradiance per unit area perpendicular to the rays (flux density), at a distance of one astronomical unit – roughly the mean distance from the sun to the earth – is called the solar constant. Actually, the direct solar irradiance at the top of the atmosphere fluctuates by about 6.9% during a year (from 1.412 kW m^{-2} in early January to 1.321 kW m^{-2} in early July) due to the varying distance from the sun. Its averaged value (total solar irradiance at the TOA) is reported by NASA (http://solarsystem.nasa.gov/planets/, 2014) and given as 1.365–1.369 kW m^{-2}.

The amount of solar radiation received at any location on the earth's surface depends on the state of the atmosphere, the location's latitude, and the time of day. Taking into account that the

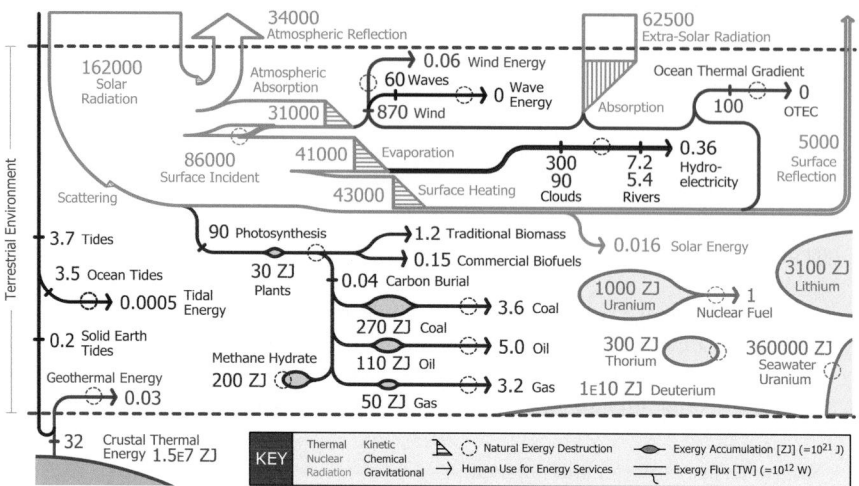

Figure 7.1.   Global exergy fluxes, reservoirs and use (after Global Climate and Energy Project, Stanford University (2007)).

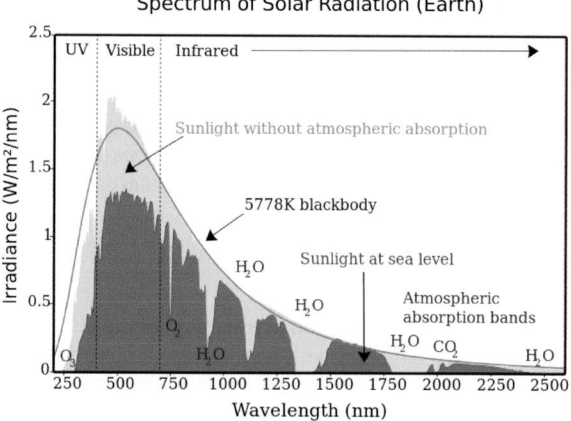

Figure 7.2.   Solar radiation spectrum at the top of the atmosphere (TOA) and sea level (after Wikimedia; https://commons.wikimedia.org/wiki/File:Solar_spectrum_en.svg (2016)).

angle at which solar radiation strikes the earth's surface changes and that at any one moment half the planet does not receive any solar radiation, averaged irradiation is close to one-fourth the solar constant (approximately $340 \, \text{W m}^{-2}$). Insolation for most people is from 150 to $300 \, \text{W m}^{-2}$ or 3.5 to $7.0 \, \text{kWh m}^{-2} \, \text{day}^{-1}$. Irradiance at the earth's surface perpendicular to the sun's rays at sea level on a clear day reaches approximately $1000 \, \text{W m}^{-2}$.

In order to enable better photovoltaic (PV) design and planning the global horizontal irradiance (GHI) has been defined which corresponds to the total amount of shortwave radiation received from the sun by a surface horizontal to the ground. This value includes two fluxes: direct normal irradiance (DNI) and diffuse horizontal irradiance (DIF). DNI represents the solar radiation that

Global Horizontal Irradiation   Australia

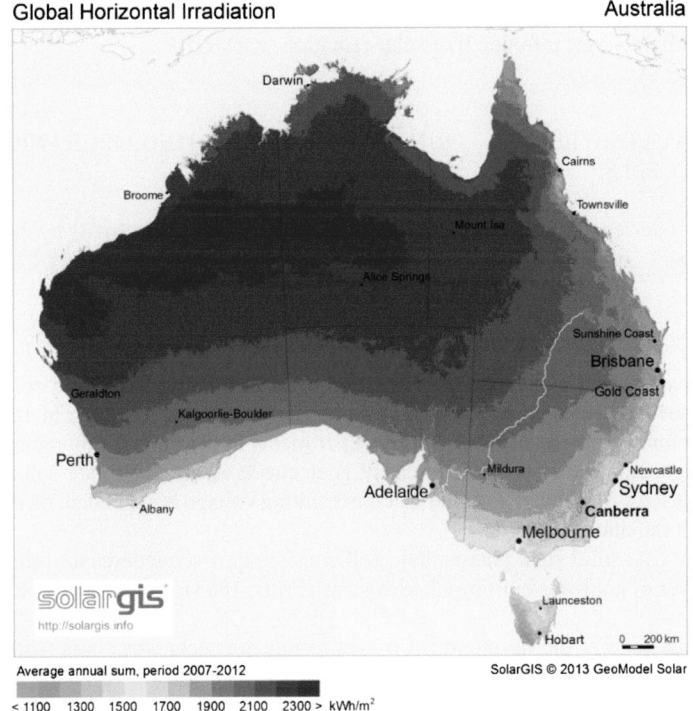

Global horizontal irradiation   Europe

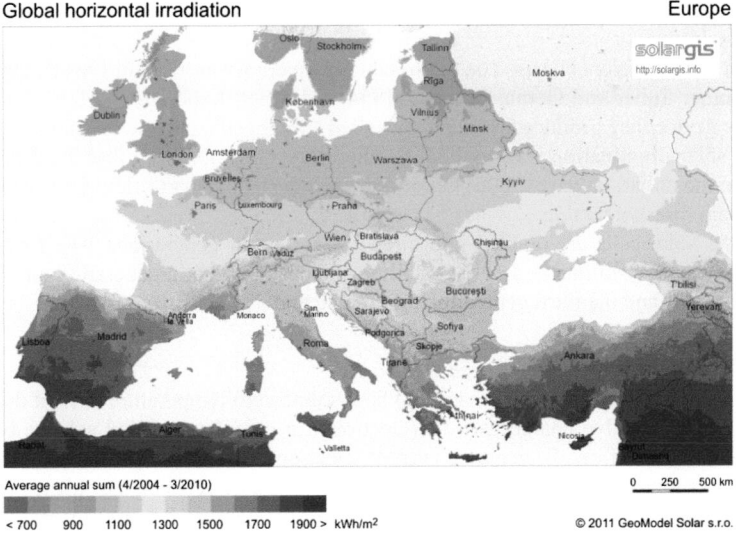

Figure 7.3.   Global horizontal irradiation for Australia and Europe (after SolarGIS (2014); http://solargis.
com/products/imaps/overview/).

comes in a straight line from the sun at its current position in the sky. DIF corresponds to the
scattered part of solar radiation – scattered by molecules and particles in the atmosphere. On a
clear day, most of the solar radiation received by a horizontal surface will be related to DNI, while
on a cloudy day most will be DIF.

Figure 7.3 presents the 6-years averaged annual global horizontal irradiation in Europe and
Australia. The northern part of Europe is characterized by 2–3 times smaller GHI than the southern

EU part and Australia. The global horizontal irradiations for various countries and regions are available at http://solargis.info/doc/free-solar-radiation-maps-GHI.

7.2   PHOTOVOLTAIC EFFECT – PRINCIPLE AND OPERATING MECHANISM OF SOLAR CELLS

A solar cell is a device that converts the solar radiation directly into electricity by the photovoltaic effect.

7.2.1   *History of photovoltaic effect discovery and PV-cell development*

This effect is a physical phenomenon of creation of electromotive force in a solid state under the influence of incident light radiation and was observed for the first time in 1839 by French physicist Edmond Becquerel (1839). During experiments in his father's laboratory he observed that the sunlight-exposed silver chloride (AgCl) electrode in an electrolyte solution may be a source of electricity. Moreover, he noted that the resulting voltage is dependent on the wavelength of the incident radiation.

In 1883 the first solid state photovoltaic cell consisting of semiconductor selenium covered with a thin layer of gold was built by Charles Fritts (Fritts, 1883). The efficiency of that cell was less than 1%.

Another milestone in the development of semiconductor technology was marked by Polish scientist Jan Czochralski, who invented a method of crystal growth to obtain single crystals of metals and semiconductor materials including monocrystalline silicon (Czochralski, 1918). Rod-shaped single crystals up to a few meters long and typically 0.2–0.3 m in diameter may be produced on a large scale. The Czochralski method is still widely used in the mass production of single crystals.

The first modern solar cell based on the silicon p-n junction was made in 1954 by Daryl Chapin, Calvin Souther Fuller and Gerald Pearson (1954) from Bell Laboratories, New Jersey, United States. The device they produced, which they called "photocell", had the efficiency of 6%, when $60 \, W \, m^{-2}$ solar illumination was applied. Furthermore, they observed that the electric current depends on the radiation wavelength; moreover, photocurrent starts to flow for wavelengths less than a certain threshold value.

Nowadays solar cells are divided into three categories, or generations. The first generation is based on monocrystalline and polycrystalline silicon, the second generation on thin films of various materials and the third generation on organic and dye-sensitized materials.

7.2.1.1   *First generation silicon-based solar cells*

Silicon is a semiconductor which means that at absolute zero temperature (0 K) it does not conduct electricity. Negative charge carriers (electrons) in completely filled valence band cannot participate in current flow. To generate current, electrons have to jump from the valence into a higher-lying completely empty conduction band. The distance between the valence and conduction band is called energy bandgap E_g. At temperatures greater than 0 K the probability of finding an electron in the conduction band is no longer zero. Electrons can be moved from the valence to the conduction band if the energy supplied (by any means) is greater than the energy bandgap E_g. This energy can be provided as a result of thermal activation or absorption of the incident radiation. The electron in the conduction band is a free electron or a free charge carrier. Moreover, it leaves in the valence band a vacant state called a hole which is a positive charge carrier (Kittel, 2005). The electron-hole generation process does not produce by itself the voltage (potential difference) in the semiconductor. To receive the potential difference, generated free charge carriers (electrons and holes) have to be separated. In solar cells the charge-separation process takes place on a so-called p-n junction.

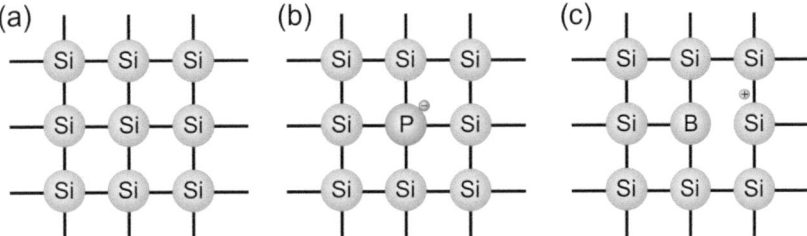

Figure 7.4.   Si crystal structure of (a) intrinsic semiconductor, (b) n-type semiconductor with P as a dopant atom and (c) p-type semiconductor with B as a dopant atom.

In the intrinsic semiconductor, also called undoped semiconductor, the number of electrons in the conduction band is equal to the number of holes in the valence band. To increase the electrical conductivity semiconductors are often doped with atoms of other elements. As a result semiconductors of n-type with a higher concentration of free electrons or p-type with a higher concentration of free holes are obtained (Kittel, 2005). To achieve an n-type semiconductor a Si atom (with four valence electrons) within the Si crystal structure should be substituted with a doped atom of five valence electrons (e.g. phosphorus, P), (Fig. 7.4). In a p-type Si semiconductor the atom of a dopant should have in turn three valence electrons (e.g. boron, B). In the p-n junction on the boundary separating p- and n-type areas an internal electric field is formed, due to charge diffusion process. The photogenerated electron-hole pairs are separated – internal electric field pulls electrons and holes toward the n- and p-type region, respectively. In this way spatially separated charges become a source of potential difference (photovoltage).

First generation solar cells consist of a thick layer (around 300 μm) of p-type silicon and a thin layer (a few μm) of n-type silicon on the top. Later, front and back metal contacts are added (Fig. 7.5b). A prepared cell is finally covered with an antireflective coating.

The operating principle of a solar cell is as follows:

- Incident solar radiation is absorbed in the solar cell.
- If the photon energy of absorbed radiation is higher than the energy bandgap, an electron-hole pair is created; the pair can recombine (contributing to losses) or be separated when it reaches the p-n junction.
- The carriers are separated by the electric field existing at the p-n junction (Figs. 7.5a and 7.5b); electrons are pulled into n-type region; holes are pulled into p-type region.
- When these carriers avoid recombination (as minority carrier or at junction) and reach electrodes, PV current and voltage are created (Luque and Hegedus, 2011).

Monocrystalline silicon (c-Si) solar cells – made of a single monolithic silicon crystal – reach efficiencies of up to 22% for module and visually have a dark color due to high value of absorption coefficient (Fig. 7.6a).

The polycrystalline or multi-crystalline silicon (poly-Si or mc-Si) solar cells – made from crystallized silicon or multiple interconnected monocrystals – reach efficiencies of up to 18% for module. They usually have a dark blue color and clearly defined borders of silicon crystals (Fig. 7.6b).

The manufacturing process of multi-crystalline solar modules consists of the following steps:

- Preparing of multi-crystalline silicon bricks.
- Wafer fabrication.
- Post-processing of solar cell.
- Assembly of solar module.

In the first step a base silicon material is melted in a furnace at 1673 K for 20 h. The next step is cutting the melted silicon block into bricks using a wire saw. The bricks obtained are washed in

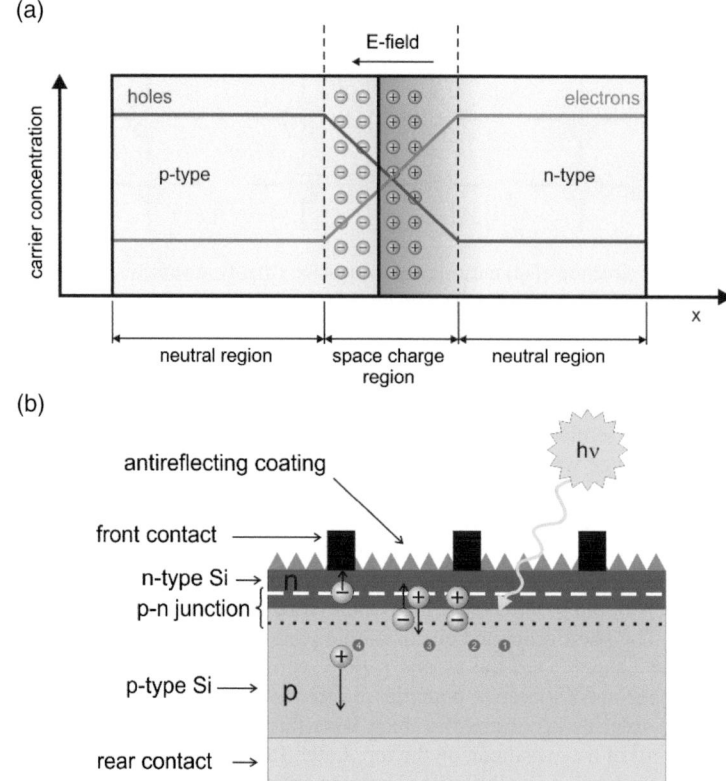

Figure 7.5.   Structure of (a) p-n junction and (b) operational mechanism of the Si solar cell.

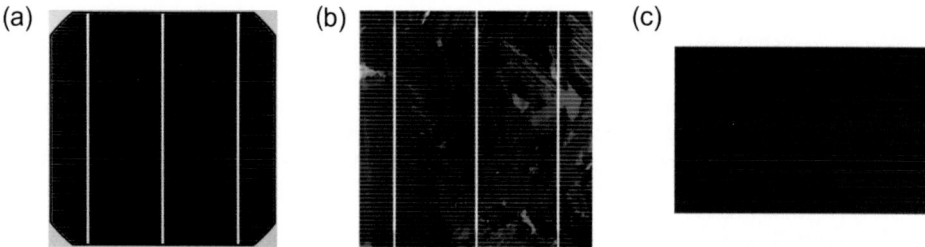

Figure 7.6.   Silicon solar cell: (a) monocrystalline, (b) polycrystalline and (c) amorphous (after MLsystem; http://mlsystem.pl/wp-content/uploads/2017/01/Katalog-wersja-ANGIELSKA.pdf).

a washing machine and then cut into thin discs, so-called wafers. The finishing process of silicon wafers fabrication consists of cleaning, polishing and edging (Fig. 7.7).

In order to fabricate the solar cell, the silicon wafer is cleaned once again and is then heated in a tube furnace at 1073 K and treated with a phosphoric gas. This gas penetrates the wafer which changes the electronic properties of the base material. As a result, on the surface of the p-type silicon the n-type layer is formed and p-n junction is formed. The resulting solar cells are cleaned and edged once again and then placed in an acid bath where only the bottom and edges are exposed to the acid. In effect, the top surface (n-type) is electrically isolated from the bottom side (p-type). In another step the solar cells are placed in a plasma chamber which gives them the typical blue color, reducing optical losses by preventing unwanted light reflections. Front and back

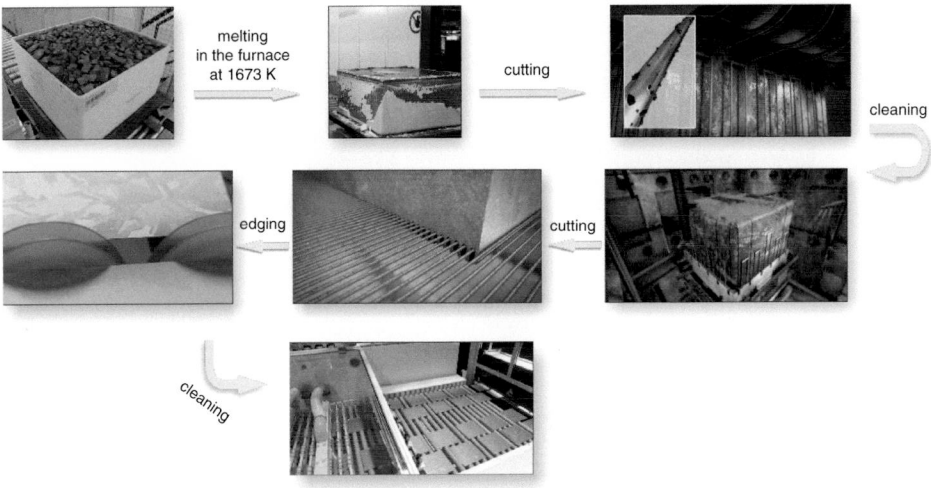

Figure 7.7.   Schematic diagram of the fabrication process of silicon wafers – after SolarWorld http://www. solarworld.de/en.

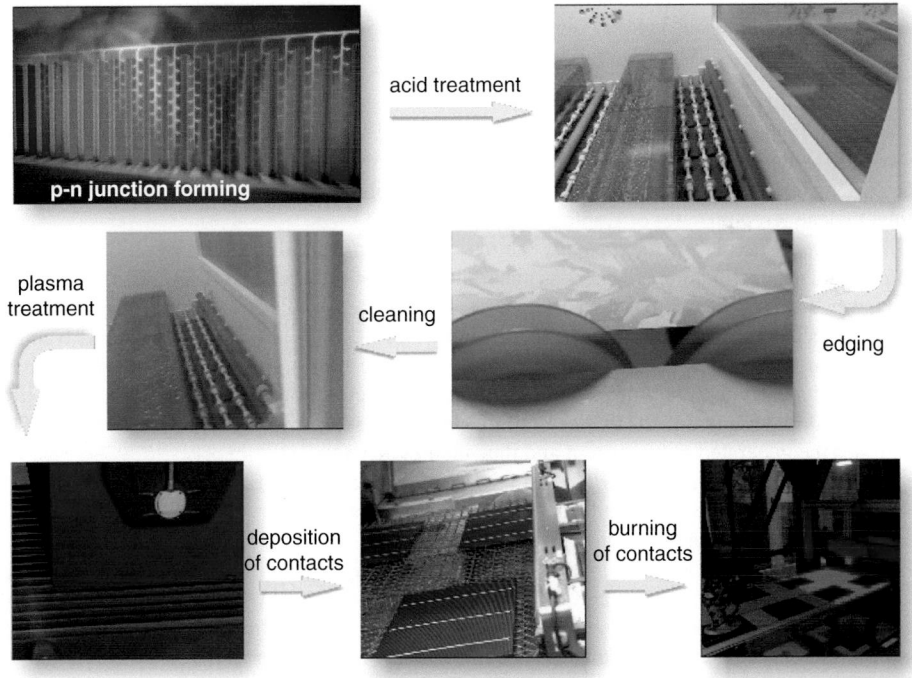

Figure 7.8.   Solar cells fabrication process from silicon wafers – after *SolarWorld.*

metal contacts made by the screen-printing technique are permanently connected with the cells by burning in the furnace (Fig. 7.8). The final steps are: connecting the fabricated solar cells in series and parallel, cells placement on the glass plate and encapsulation under vacuum conditions. Complete solar modules consist of the front cover glass plate, solar cells between two pieces of transparent encapsulant foil, back sheet, junction box and aluminum frame (Fig. 7.9).

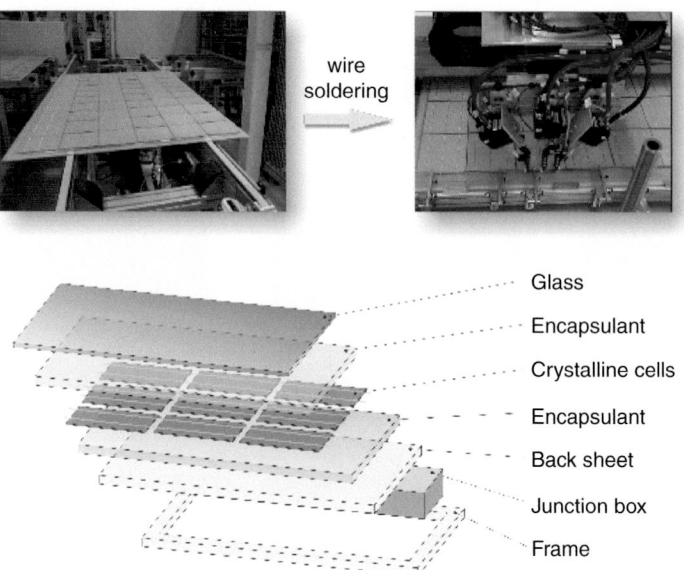

Figure 7.9.   Matrix of solar cells and soldering of wires to cells – top. Typical construction of the multi-crystalline solar module – top after *SolarWorld*.

7.2.1.2   *Second generation solar cells*

Thin-film solar cells constitute the second generation PVs in which the active semiconductor layer, due to its high absorption coefficient, can be thinner than 50 μm (Luque and Hegedus, 2011). This thin semiconductor layer can be deposited also on elastic substrates, such as a PET (polyethylene terephthalate) or Kapton foil.

Amorphous silicon (a-Si) solar cells are made of a non-crystalline form of silicon. The low temperature production process and less material consumption mean that the production costs are also lower. The a-Si solar cells reach efficiencies of up to 12% and have a slightly burgundy color and no visible crystals of silicon (Fig. 2.3c) (Luque and Hegedus, 2011). The main disadvantage of a-Si solar cells is a decrease in performance over time (up to 30%) as a result of long-term exposure to solar radiation which is related to the Staebler-Wronski effect (1980). Strong irradiation produces many defects in a crystalline structure caused by broken chemical bonds. This decreases the generated photocurrent and thus their efficiency.

Cadmium telluride solar cells (CdTe) are another type of thin-film PV. Construction is as follows: a p-type CdTe semiconductor creates an absorption layer which from the bottom is coated with a metal contact while on the top an n-type CdS semiconductor layer is deposited (Fig. 7.10a) (Luque and Hegedus, 2011). This configuration forms a p-n junction with an internal electric field. Moreover, the n-type CdS plays the role of an optical window layer which ensures better absorption of incident radiation by the p-type CdTe layer. A front contact layer is formed by a transparent conductive oxide (TCO) coated glass, which is characterized by low resistivity and high transparency to visible radiation. The CdTe modules reach efficiencies of above 10%. Cadmium telluride is a stable non-toxic chemical compound but elemental cadmium is one of the deadliest toxic materials known. This forms a big problem during recycling of used cells. Moreover, tellurium is an extremely rare element in the earth's crust. These two detrimental factors represent major obstacles to the further development of CdTe solar cells technology.

Solar cells with a *copper indium gallium selenide* (CuInGaSe or CIGS) active layer have a construction similar to CdTe solar cells (Fig. 7.10b). Composition of these four chemical components makes a material of high absorption and good physical properties (Luque and Hegedus, 2011). Electrical parameters depend on copper to indium volume ratio while energy band depends

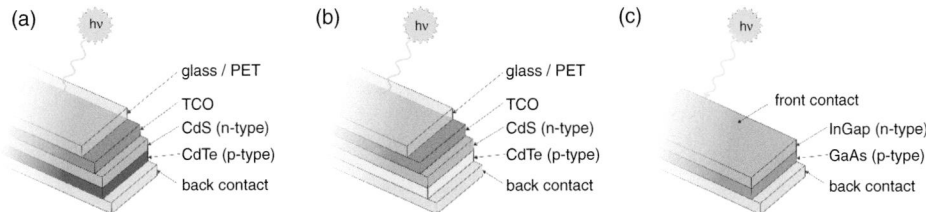

Figure 7.10.   Structures of (a) cadmium telluride, (b) CIGS and (c) gallium arsenide solar cell.

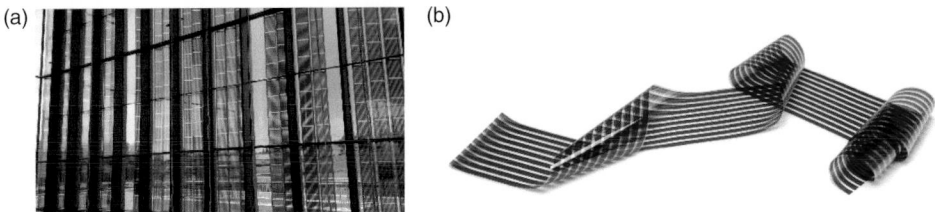

Figure 7.11.   (a) Dye solar cell façade at SwissTech Convention Center made by Solaronix (https://www.solaronix.com) and (b) flexible organic solar cell made by Konarka Technologies (after ECN (2014); http://www.ecn.nl).

on gallium amounts. The CIGS cells reach the efficiency around 15%, have excellent long-term stability and are environmentally friendly.

The next type of thin-film solar cells are *gallium arsenide solar cells* (GaAs) in which GaAs constitutes an absorption n-type layer and InGaP is a p-type layer forming an internal electric field and also an optical window (Fig. 7.10c). Since GaAs cells are prepared by crystal growth process they have to be deposited on an appropriate substrate with similar crystal structure and lattice constant (Luque and Hegedus, 2011). These conditions are fulfilled for example with a germanium (Ge) substrate. Gallium arsenide has an ideal energy bandgap (1.42 eV) which makes it an excellent material for solar cell application. Therefore, GaAs cells are the most efficient PV modules produced to date, achieving efficiencies up to 30%, and are widely used in space applications. However, GaAs is a very expensive and toxic chemical compound.

7.2.1.3   Third generation solar cell

The third generation solar cells are quite different from previous ones. There is no p-n junction in the cell structure; moreover, electron-hole pairs separate without internal electric field. Materials used in the fabrication process are non-toxic, mainly organic compounds. Physical and chemical properties of applied materials enable the manufacture of very thin structures deposited on glass or elastic substrates. Moreover, complete devices may be colorful and semitransparent which makes them very attractive from a practical and aesthetic point of view. They can be used in many smart appliances including semitransparent windows, special tent covers or even clothes. Figure 7.11a shows a dye solar cell multicolored façade at SwissTech Convention Center at École Polytechnique Fédérale de Lausanne campus made by Solaronix, and Figure 7.11b shows a flexible organic solar cell made by Konarka Technologies.

Dye-sensitized solar cells (DSSC) are devices with an operation mechanism similar to the process of photosynthesis in plants. The DSSC comprises of two electrodes separated by a liquid or *solid state electrolyte*. The first electrode, called a photoanode, is formed of mesoporous semiconductor (typically, titanium dioxide TiO_2) deposited on transparent conductive oxide coated glass substrate. Dye (monolayer) is adsorbed on the surface of the semiconductor. The real surface area of the semiconductor is 1000 times greater than geometric dimensions. The second electrode, called cathode or counter electrode, consists of a few nanometers thick (10^{-9} m) catalytic layer

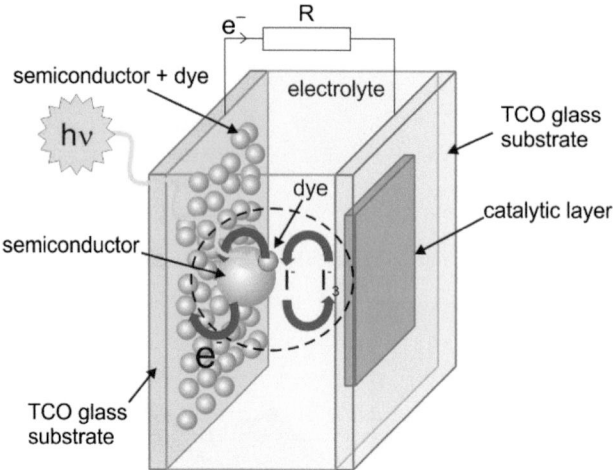

Figure 7.12.   Schematic structure and operating mechanism of DSSC.

(typically platinum, Pt) (Hagfeldt *et al.*, 2010). Figure 7.12 shows the schematic structure of the DSSC. For a dye, natural sensitizers can be used, including blackberry juice or hibiscus, but the highest known efficiencies are obtained with ruthenium-based complexes. All the materials used in DSSCs are cheap and most of them do not need very high purity. Moreover, these solar cells may be produced in typical laboratory conditions without special clean rooms and inert gas atmosphere.

The operational mechanism is as follows: incident solar radiation is absorbed by the dye molecule which is excited from the ground state S to the excited state S^* followed by electron transfer from S^* to the conduction band of semiconductor leaving the dye molecule in an oxidized state S^+. Then, injected electrons percolate through the semiconductor layer up to TCO coated glass, and next, through external load to the counter electrode. The working cycle of the cell is completed by reduction of the dye oxidized form with an electron transported from the cathode by special mediator pair (I^-/I_3^-) in the electrolyte.

The cell operation is associated with an appropriate adjustment of energy levels of each component, that is to say, the excited state of the dye molecule is located above the conduction band of the semiconductor and the mediator redox potential is lying higher than the ground state of the dye molecule. This means that there is no energy barrier for the electron path in the cell. Moreover, a very important role is played by reaction kinetics – processes leading to the generation of current in the external load occur much faster than the limiting processes (e.g. recombination or electron back-transfer). Dye-sensitized solar cells can achieve efficiencies of 14% in laboratory conditions, but less than 10% for commercial modules.

Organic solar cells (also called organic photovoltaic – OPV) are devices which convert sunlight directly into electricity using low molecular weight organic materials and polymers. In OPVs an active layer is situated between two electrodes – TCO coated glass substrate as a top semitransparent positive electrode and thermally evaporated metal as negative electrode at the OPV bottom. An active layer consists of two materials forming the heterojunction: a donor which is a good hole-transporting material and an acceptor which is a good electron-transporting material (Fig. 7.13). Donor and acceptor materials correspond to the p-type and n-type semiconductor, respectively, in silicon solar cells. One of the widely used materials as a donor is poly(3-hexylthiophene) (P3HT) and as an acceptor is [6,6]-phenyl-C61-butyric acid methyl ester (PCBM) (Brabec *et al.*, 2008).

In organic solar cells incident solar radiation is absorbed in the donor material. After that, molecular excitons (electron-hole pair on the same molecule) are created. These excitons move by diffusion to the donor/acceptor interface and then dissociate onto free charge carriers. Free

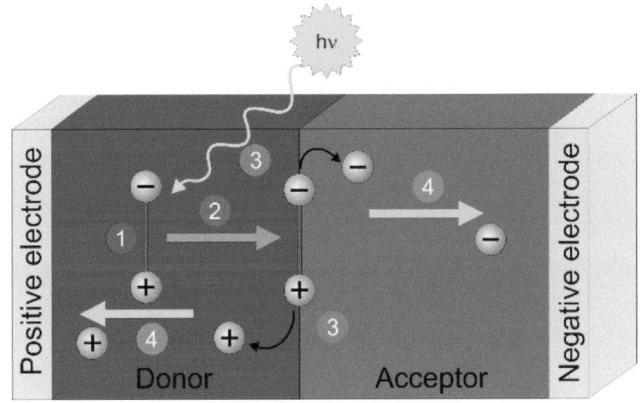

Figure 7.13.   Schematic structure and operating mechanism of OPV.

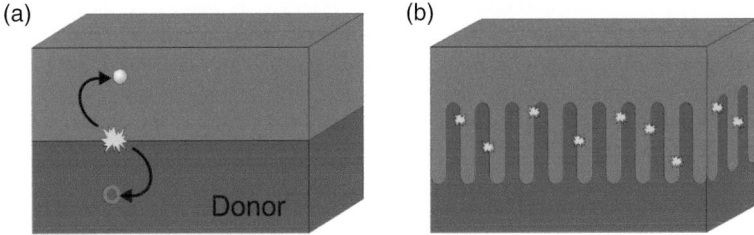

Figure 7.14.   Schematic diagram of (a) planar and (b) bulk heterojunction solar cell.

electrons diffuse through the acceptor layer to the negative electrode and free holes diffuse through the donor layer to the positive electrode. The photovoltage is created between these two electrodes.

Active layer may have mainly two different configurations: planar heterojunction and bulk heterojunction (compare Figs. 7.14a and 7.14b, respectively). In the planar configuration two layers placed one upon the other are well separated. In the bulk heterojunction solar cells donor and acceptor make interpenetrating composite material.

Besides their many advantages, organic solar cells have their drawbacks mainly connected with the physical properties of the active layer materials. They are very sensitive to water vapor and molecular oxygen. This introduces the necessity of the production of OPVs in special chambers (called gloveboxes) where the humidity is kept below 0.01% and in an inert gas atmosphere (i.e. nitrogen or argon). Therefore, before removing from a glovebox the finished cell should be encapsulated. The best organic solar cells reach an efficiency around 11% in laboratory applications and have a good long-term stability of electrical parameters.

7.2.2   *Main characteristics of solar cells*

In order to characterize the solar cell and to estimate its electrical parameters the current-voltage (I-V) characteristic under illumination should be determined. The I-V curve is a graphical representation of the current and voltage changes as different loads are applied. There are characteristic points on the I-V curve, these are: short-circuit current I_{SC} in amps [A], open-circuit voltage V_{OC} in volts [V] and maximum power point P_{MAX} in watts [W]. I_{SC} is the value of current in the short-circuit conditions i.e. when the voltage equals zero. V_{OC} is the value of voltage in the open-circuit conditions i.e. when the current does not flow. The power P produced by the solar cell can be easily calculated from the following equation $P = IV$, where I and V are current and voltage on the cell terminals. They can be estimated from the current-voltage characteristic.

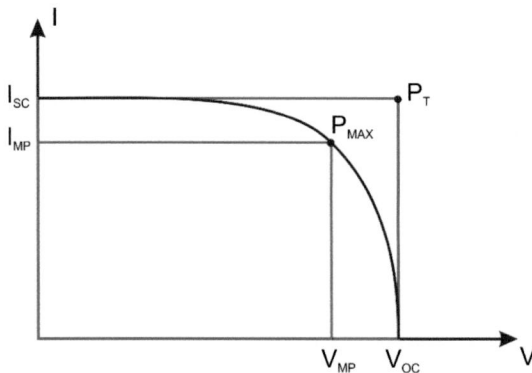

Figure 7.15.   Typical *I-V* curve with a graphical representation of the fill factor.

Maximum power point is the maximum value of P (marked as P_{MAX}). At this point the voltage and current are denoted V_{MP} and I_{MP}, respectively (Ginley and Cahen, 2012).

The basic quantity that characterizes the solar cell is the solar-to-electric power conversion efficiency η. This is the ratio of the electrical power output P_{MAX} measured in the maximum power point of the solar cell, to the solar power input P_{IN} which is irradiation intensity I_0, in $W\,m^{-2}$, related to the solar cell area S, in $m^2 - P_{IN} = I_0 S$:

$$\eta\,[\%] = \frac{P_{MAX}}{P_{IN}} = \frac{I_{MP}\,V_{MP}}{I_0\,S} \times 100$$

Another coefficient defining a quality of the solar cell is the fill factor (*FF*). It is calculated as a ratio between maximum power and the theoretical power P_T that would be generated if the cell could simultaneously generate the I_{SC} and V_{OC}:

$$FF = \frac{I_{MP}\,V_{MP}}{I_{SC}\,V_{OC}}$$

Fill factor can be also estimated from the *I-V* curve by comparing the area of the smaller (blue) and bigger (red) rectangles (Fig. 7.15).

Using the *FF* it is possible to calculate the solar-to-electric power conversion efficiency:

$$\eta\,[\%] = \frac{I_{SC}\,V_{OC}\,FF}{I_0\,S} \times 100$$

To compare various solar cells performance, standard test conditions (STC) were introduced with illumination corresponding to a summer sunny day. The STC conditions are given by temperature 298.15 K, using an $AM1.5$ spectrum under irradiation $1000\,W\,m^{-2}$, where air mass (*AM*) is the path length which light takes through the earth's atmosphere normalized to the shortest possible path length (that is, when the sun is directly overhead). The value "$AM1.5$", 1.5 atmosphere thickness, corresponds to a solar zenith angle of $z = 48.2°$ (Ginley and Cahen, 2012).

7.3   SYSTEM DESIGN: PV DIRECT, GRID-TIED, STAND-ALONE, GRID-TIED WITH BATTERY BACKUP, SOLAR THERMAL – PVT

Photovoltaic systems, known also as solar batteries, solar panels or solar modules are used to convert solar radiation energy into electric energy. There are two main types of photovoltaic systems. A photovoltaic system may be grid-tied (on-grid), or autonomous, utilized only for local (own) needs (off-grid), often installed in remote, rural locations without access to the grid. It is

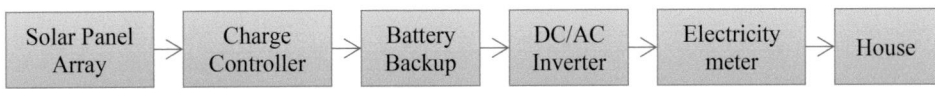

Figure 7.16. Block diagram of off-grid installation.

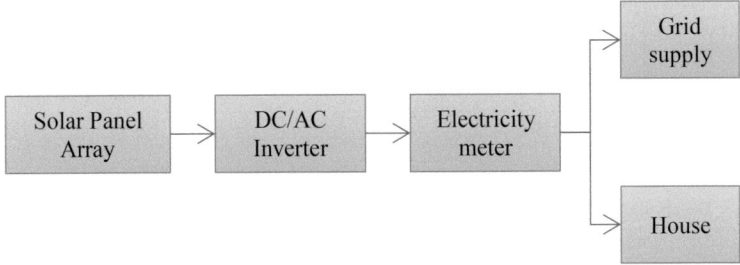

Figure 7.17. Block diagram of on-grid installation.

possible to build a hybrid grid, which can operate both stand-alone, and grid-tied with battery backup (Solanki, 2011, 2013).

7.3.1 *On-grid and off-grid systems and their applications*

Off-grid systems are photovoltaic systems, which operate independently of the public power distribution grid. Solar energy is converted into direct current (DC) by photovoltaic modules. First, the produced direct current passes through a voltage regulator and is then stored in accumulators. The next step is to convert the direct current stored in the power inverter into an alternating current consistent with the grid parameters, so it may be fully utilized by all domestic devices. This kind of installation is perfect in situations where the costs of grid connection are too high due to long distances to the closest grid.

The main components of an off-grid system are: photovoltaic modules, a power inverter, voltage regulators, accumulators, the wiring system and mounting systems (Fig. 7.16). Off-grid systems are suited to situations where the energy produced can be fully utilized on site (Hankins, 2010).

The construction of an autonomous system does not imply that the building will not be connected to the grid. Stand-alone installations may also operate in buildings connected to the grid. However, in such cases the energy produced by the photovoltaic modules will not be returned to the grid, but will be utilized for the building's own needs in real-time or stored in batteries (accumulators). In the case of stand-alone installations the power inverter generates its own microgrid and supplies separate circuits in the building or the entire building. The power inverter may be connected to the grid directly to enable emergency accumulator charging.

On-grid systems are photovoltaic systems connected to the local grid. One of the characteristics of an on-grid system is its considerable flexibility. The surplus of the energy produced by the system on a sunny day is sold to the grid, while in case of energy shortages caused by a cloudy day or high power consumption, the power will be drawn from the grid in a way unnoticed by the user.

The main components of an on-grid system (Fig. 7.17) are: photovoltaic modules, a power inverter, safety devices that turn off the installation in cases of grid inefficiency (a blackout), smart energy meter, the wiring system and mounting systems (Balfour *et al.*, 2013).

On-grid systems are currently the most economically viable systems because they do not require the produced energy to be stored. These systems are usually fitted with two electric energy meters. One measures the surplus energy which was unused for own needs and resold to the grid, while the other one measures the energy drawn from the grid. Sometimes one two-way meter is used.

On-grid systems with battery backup are photovoltaic systems connected to the grid, which allow independence from electric energy supplied from an external grid. The surplus of electric

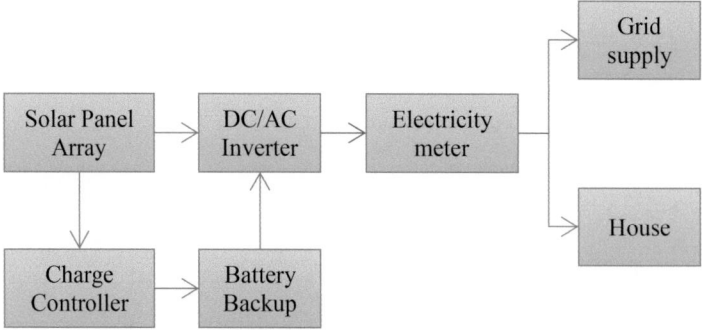

Figure 7.18.   Block diagram of on-grid with battery backup installation.

energy on sunny days is not resold to the grid but instead stored in accumulators and utilized in situations of sudden energy shortage (blackout). The energy stored in accumulators may be utilized on cloudy days or during the night, reducing the amount of energy drawn from the grid.

The main components of an on-grid system with battery backup (Fig. 7.18) are: photovoltaic modules, a power inverter, voltage regulators, safety devices, accumulators, the wiring system and mounting systems. Due to the energy storage equipment, these systems are more expensive than the traditional on-grid systems.

In stand-alone installations (off-grid) and grid-tied installations (on-grid) the direct current produced by the photovoltaic modules is converted by the power inverter to alternating current, and in this form is utilized by the end user. Photovoltaic systems co-operating with a low voltage direct current installation are rare due to the dominance of electrical devices being designed for alternating current.

Photovoltaic systems supplying private-owned buildings with electric energy are installed mainly on roofs or front elevations. The power output of such systems usually fits in the 2–10 kW_e range. The amount of space required to install such a system depends on the type of photovoltaic modules. For the most popular currently available silica-based panels, installation of 1 kW_e requires around 7 m^2.

Changing sunlight conditions during the day and temporal shifts in energy production in regards to energy demand in the system make a fixed connection of the installation to the grid an advantageous solution. Connecting the installation to the grid results in stabilization of the source operation. Autonomous off-grid systems are appropriate mainly in locations without the possibility of connection to the grid, or where such a connection would be very expensive.

7.3.2   *PVT systems and their application*

Photovoltaic thermal (PVT) are liquid-cooled photovoltaic modules, that produce both thermal energy and electricity. Due to that, instead of two separate installations – solar thermal collectors for water heating and photovoltaic panels for electric energy production – only one is installed, which makes it a perfect solution for limited spaces. The cooling of the photovoltaic panels with liquid contributes also to an increase in their efficiency particularly on hot days.

During the summer, when modules in an installation will heat up to high temperatures exceeding 320 or even 350 K, the use of cooling in PVT modules enables the temperature to be kept at a lower level so the efficiency of the cell remains relatively high (Fig. 7.19) (Szymański, 2013). For the PVT module to operate efficiently, it should remain at a stable, low operating temperature of 300–310 K; that is why it is important to search for solutions which decrease the operating temperature of the module.

There are two different designs among hybrid systems. First are hybrid systems based on the construction of a flat liquid collector, in which underneath the photovoltaic cells there is an

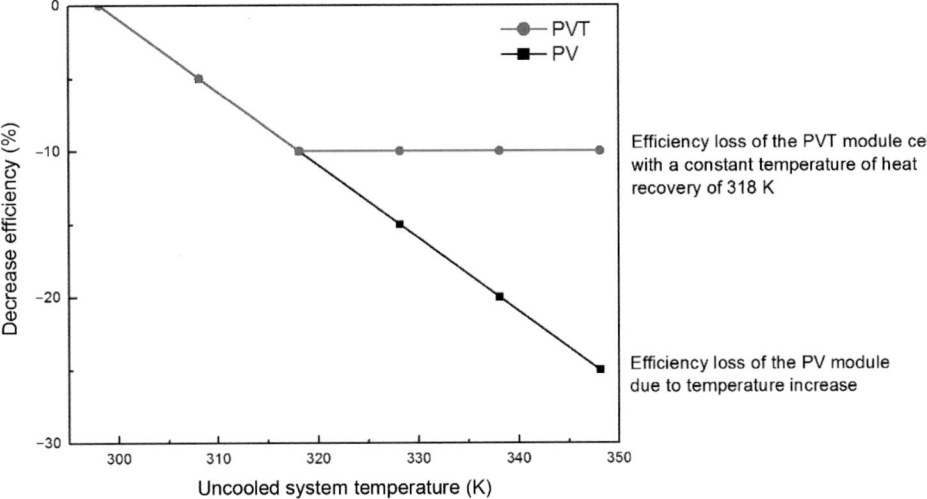

Efficiency loss of the PVT module cell with a constant temperature of heat recovery of 318 K

Efficiency loss of the PV module due to temperature increase

Figure 7.19. Comparison of efficiency drop of regular PV module with PVT module.

Figure 7.20. The frame of the module with an installed prototype of the hybrid device.

absorber and a heat exchanger, through which a flowing heating medium recovers the heat. This type of hybrid PVT module is connected to a heating installation in the same way as in a regular solar thermal panel installation. The second type of system is the hybrid system based on the concept of solar air collectors. In that case the air is the heating medium, which flows underneath the photovoltaic cells. Hybrid systems based on the solar air collector are cheaper and have a simpler structure, but their application becomes justified only when there is a demand for warm air heating.

An interesting solution that has appeared recently on the PVT market is the direct lamination of photovoltaic cells on the absorber plate, which makes the heat exchange between the cell and the heating medium much more effective.

The PVT system proposed by IMP PAN, designed to lower the operating temperature of the photovoltaic module, consists of a photovoltaic module with a micro heat exchanger, mounted on a common frame. Between the boundaries of the photovoltaic module and the micro heat exchanger, there is a layer of thermal interface material. A stable link between the photovoltaic module and the micro heat exchanger is provided by an adhesive and clamps.

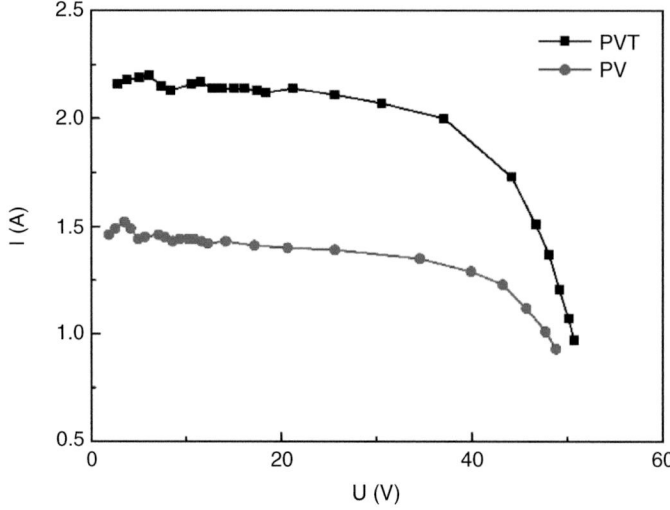

Figure 7.21. Voltage-current characteristic for PVT and PV.

The micro heat exchanger contains two aluminum plates, and between them another plate with cuts forming flow canals for the exchanger. The goal of the research is to obtain a module of higher efficiency. Moreover, the configuration enables a temperature lower than the equilibrium temperature to be achieved without such cooling.

Figure 7.21 shows results of a comparison for an Avancis 180 photovoltaic panel, facing south at a 30° angle without a heat exchanger, and with a mounted micro heat exchanger prototype. It is clear that the PVT system achieved a much higher efficiency for the PV panel itself (12.6%) in comparison to the same PV panel without a heat exchanger (11.8%) (Łapiński and Cenian, 2013):

$$FF = \frac{I_{MP}\,V_{MP}}{I_{SC}\,V_{OC}}$$

$$FF_{PV} = 0.69$$

$$FF_{PVT} = 0.72$$

The next stage of research conducted at IMP PAN will be the application of phase change materials (PCM) in order to maintain low operating temperature of a photovoltaic cell.

7.4   STORAGE: BATTERIES, CAPACITORS AND SUPERCAPACITORS, OPERATION PRINCIPLE, NEW DEVELOPMENT (GRAPHEN ETC.)

Due to their instability, incorporation of multiple renewable power sources into one electrical system requires appropriate energy storage facilities. They include systems related to short term (up to 30 s), medium (from 15 min up to a few days) and long-term (seasonal) stability and (energy) quality. For each of these storage periods different technologies are applied, starting from batteries and supercapacitors, through pumped-storage hydroelectricity and compressed air energy storage, up to power to gas solutions.

Here we focus on small scale installations applicable in the case of rural regions, including batteries and galvanic cells, so-called redox flow cells (RFCs) and electrochemical capacitors.

7.4.1   *Batteries*

Batteries accumulate energy that can be stored in one or more units. In principle we consider two main categories: primary and secondary batteries (Botte, 2007). In contrast to primary, secondary-type batteries can be charged. The charging and discharging process can be repeated many times, the number of recharging cycles (cyclability) being strongly related to the battery type. Alkaline or zinc-carbon technology are examples of primary batteries whereas lead-based, nickel-based, lithium-based sodium nickel chloride technology are secondary-type batteries and nowadays provide distinctive and important functions to many household and industry applications. Every secondary-type technology has the potential for significant technological advancement and cost reduction in the near future.

7.4.1.1   *Primary batteries*
Alkaline-based batteries
Alkaline-based technology is characterized by high energy density and long lifetime. The typical values of voltage and current are 1.5 V and 700 mA, respectively. Alkaline batteries are commonly used in remote controls, clocks, cameras or MP3 players.

Zinc-carbon batteries
Zinc-carbon based batteries are also known as dry batteries. They consist of a carbon rod immersed in a mixture of carbon powder and manganese dioxide. Their usual voltage is about 1.5 V. Zinc-carbon batteries are characterized by long lifetime and efficient work but only at moderate thermal conditions. These types of batteries are used in low power drain applications, e.g. flash lights, remote control or clocks.

7.4.1.2   *Secondary (rechargeable) batteries*
Lead-based batteries
Batteries with lead electrode were invented in the 1880s and until now they have been successfully applied in many off-grid systems. Currently, almost 80% of all installed capacity of industrial batteries are based on lead-acid technology.

Lead-based batteries are characterized by:

- Capacitance: 1–16,000 Ah.
- Energy density: 25–50 Wh kg^{-1}.
- Energy efficiency: >85%.
- Cyclability: >2000 cycles at 80% of discharge.
- Temperature conditions: −30 to +50°C.

Lead-based batteries can be arranged in large systems without any special management. These large and quite heavy types of batteries are mostly applied in starting, lighting and ignition systems present in vehicles. Because of environmental protection policy, more than 95% of lead-based batteries are recycled in a closed loop.

Nickel-based batteries
The Ni-based technologies (nickel-cadmium, nickel-metal hydride, nickel-hydrogen, nickel-zinc) are the second most popular, just after the Pb-based batteries. They are used under severe conditions: extreme temperatures, cycling or charging velocity. Nickel batteries offer different sizes and forms: pocked, fiber and foam electrodes. Cells can be arranged in series or in parallel in order to build systems with determined voltage, power and energy value.

Nickel-based batteries are characterized by:

- Capacity: 0.5–2000 Ah.
- Energy density: 20–80 Wh kg^{-1}.
- Energy efficiency: >90%.
- Life time: 25 years.

- Cyclability: 3000 cycles.
- Operating conditions: −40 to +60°C.

Nickel-based batteries can be connected in large configurations without any special management systems. Recycling rates of these industrial batteries in European countries is almost 100%.

Development of the Ni-based technologies focuses on increase in cyclability and temperature range as well as cost reduction. Thanks to their performance, operational safety and operation at critical conditions, nickel-based batteries can hardly be replaced by other technologies.

Lithium-based batteries

Lithium-based batteries were commercialized in the 1990s and after a few years of operation they now cover over 50% of the small portable market.

Lithium-based batteries are characterized by:

- High energy density: 150–200 Wh kg^{-1}.
- High efficiency.
- Long cycle life: over 5000 cycles at 80% depth of discharge.
- Maintenance-free design.

The most important advantage of Li-based batteries is their versatility – they can be easily adjusted to desired voltage, power or energy. The main disadvantage concerns demand for complicated control systems. Nonetheless, since 2010 the application of Li-ion batteries has significantly increased because of the development of the electric and hybrid vehicles market.

Li-based batteries are used in many portable electronic devices, e.g. calculators, iPad, watches, toys or digital notepads. Nowadays, improvements in Li-ion technology are seen in increased energy density, cyclability and lifetime.

Sodium nickel chloride batteries

Sodium nickel chloride batteries were recently introduced to the electric and hybrid motors market. Various NaNiCl-based technologies are available for railway backup or on-grid/off-grid energy storage systems. Many international projects have been devoted to increase application of these batteries. A complete turnkey energy storage system with a power up to MW range has been developed recently and it is under test now.

Sodium nickel chloride batteries are characterized by:

- Capacity: 380 V, 40 Ah.
- Energy density: 120 Wh kg^{-1}.
- Energy efficiency: 92%.
- Lifetime: over 10 years.
- Cyclability: 4500 cycles.
- Thermal conditions: −30 to +60°C.

Sodium-based batteries have existed on the market for only 15 years and their further development is necessary. The research is focused on improvement of specific power and cyclability, and cost reduction.

7.4.2 Capacitors and supercapacitors

The concept of electrochemical capacitors (ECs) has been known for many years. In 1957 a capacitor design (Becker, General Electric) was patented; 13 years later, in 1970, the company SOHIO introduced these devices into the market. Since the 1990s, intensification of scientific and technical research has taken place in the area of electrochemical capacitors. That is related to the application of ECs in vehicles equipped with electric or hybrid motors.

The electrochemical capacitors, commonly known as supercapacitors, are regarded as a promising alternative to primary electric charge storage devices. In this type of capacitor, a charge is collected through the electrical double layer present at the electrode/electrolyte interface.

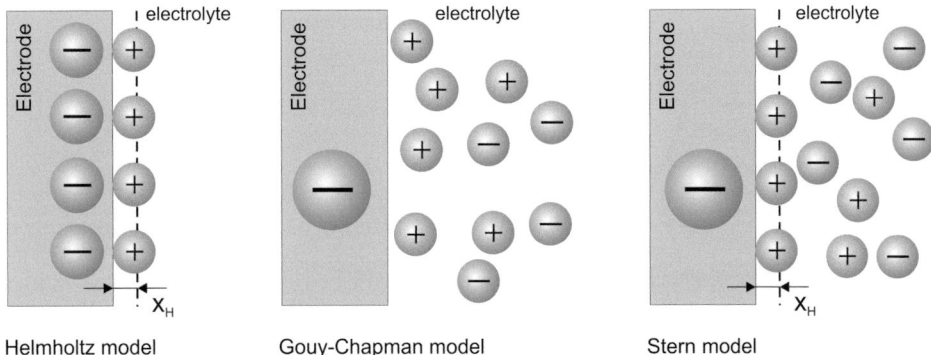

Figure 7.22.   Models of the electric double layer present at the interface of first type conductor (metal) and second type conductor (electrolyte).

The concept of the electric double layer has a centuries-old history. In modern times it was reoffered by Helmholtz (1853) concerning the dispersed phase. The development of scientific thought in this field is depicted in Figure 7.22. The first model takes into account the rigid charge arrangement (Helmholtz, 1853). Independently of one another Gouy (1910) and Chapman (1913) developed a diffuse model (see middle part of Fig. 7.22), where the applied potential and the concentration of the electrolyte influence the double layer capacitance. This leads to the diffuse effect of layer due to thermal motion. It means that the thickness of the layer varies and at some distance from the electrode ions can move freely. Stern (1924) linked the rigid model for the Helmholtz potential far from E_z (potential at minimum load) with a diffusion Gouy-Chapman model for the near E_z. A double layer is formed here by a layer adjacent to the electrode ions, followed by a diffusion layer that is extending to the bulk of the solution. As a result, we are dealing with two capacitors connected in series: the capacitor of the rigid layer with capacitance C_H and the capacitor of the diffuse layer with the capacitance C_{dif}. The total capacity of the electric double layer is: $C_{dl}^{-1} = C_H^{-1} + C_{dif}^{-1}$.

Therefore, the charge accumulated near the contact of two electrically conductive phases builds a capacitor. Its capacitance C depends on the accumulated charge under the influence of potential $C = dQ/dV$, and depends on its geometry (area S and distance between electrodes d). The thickness of the interface capacitor depends on the size of solvent molecules, so, d is the diameter of particles or their clusters. The works of Graham (1947) and the model proposed by Parsons (1981) take into account the presence of the solvent dipoles at the capacitor interface (Fig. 7.23).

The capacitance of smooth surface metal electrode C_{dl} reaches $20–40\,\mu\text{F cm}^{-2}$. However, the electrochemical capacitors utilize materials with very high, developed surface area. The term "developed surface" means a much larger effective surface area as compared to the geometric area of a smooth surface. Thanks to e.g. developed surface of the activated carbons it is possible to increase the double layer capacitance 1000 times in comparison to the flat film with the same geometrical area. Accumulation of electrostatic charge at the electrode/electrolyte interface using appropriate materials enables the achievement of a capacitance of several hundred $F\,g^{-1}$ (Brett and Brett, 1993; Conway, 1999).

The group of electrochemical capacitors described above is known under the name of EDLC (electrical double layer capacitors). The term "double" indicates the fact that we are dealing with two types of layers: rigid and diffuse, as shown in Figure 7.22 and 7.23.

General Motors Company obtained its first patent for electrochemical supercapacitors in 1957. Further improvement was proposed by Trassati *et al.* (1975). It concerns the application of ruthenium oxide; in this case, besides electrostatic charge, on the surface appears another charge attributed to the charge transfer reactions. This effect, leading to so-called redox pseudocapacitance, also characterizes electroactive polymers, including: polyaniline, polythiophene and polypyrrole.

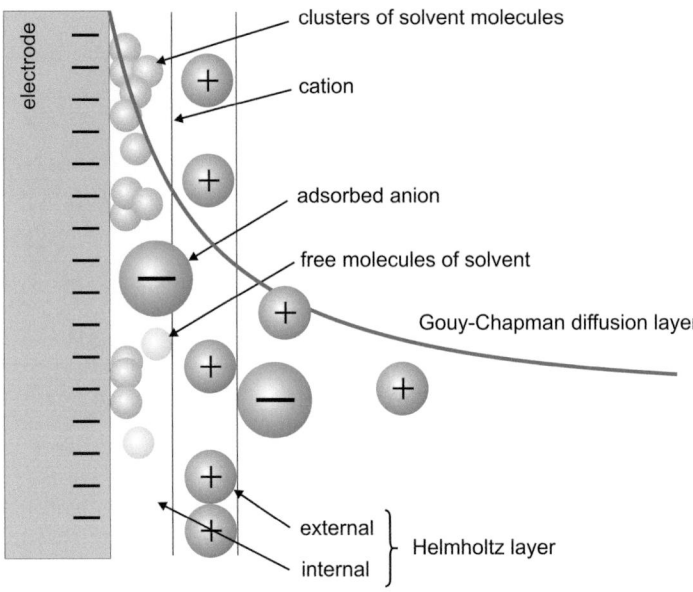

Figure 7.23.   Electric double layer at the electrode/electrolyte according to the Parsons's model (1981).

Systems of electrochemical supercapacitors (called "redox flow cells" in the UK) are currently the main solution for electric energy accumulation in power plants around the world. Generally, electrochemical capacitors are used in UPS (uninterruptible power supply) systems, electronic devices that protect against the discontinuity in power supply.

Another important application of electrochemical capacitors is related to a power supply for electric and hybrid vehicles. In these applications, the capacitor is often co-working with fuel cells or galvanic cells. The most important feature of supercapacitors in this application is a very short charging period. This feature distinguishes the electrochemical capacitors from batteries and fuel cells. Figure 7.24 presents the charge and discharge curves of a galvanic cell (battery) and a supercapacitor without considering the time constants of the process. Discharge-curve voltage of a galvanic cell maintains a stable value over a long period of time, whilst discharging the supercapacitor is characterized by a constant voltage drop during the discharge process (Fig. 7.24).

The energy stored in the battery is two times higher than the energy available from the capacitor at the same level of charge Q and the potential V:

$$E_B = \int V_B dQ; \quad E_C = \int V_C dQ; \quad E_C = (1/2)E_B$$
$$V_C = q/C; \quad E_C = QV = 1/2CV^2$$

In the symmetric capacitor (where both electrodes – the positive and negative one – are made of the same material) as soon as discharge level achieves half of discharge $1/2Q$ at the initial state, no further discharge is possible, because there is no potential difference between the electrodes.

7.4.2.1   *Supercapacitor construction*

In general, supercapacitor as typical capacitor consists of electrodes that are separated by electrolyte.

Electrode materials commonly used are:

- Various carbon materials with high active surface area that allows the electrode to accommodate much more electrolyte, and this determines the amount of charge the supercapacitor can hold. Usually carbon nanotubes are used, modified with functional groups of ionic character,

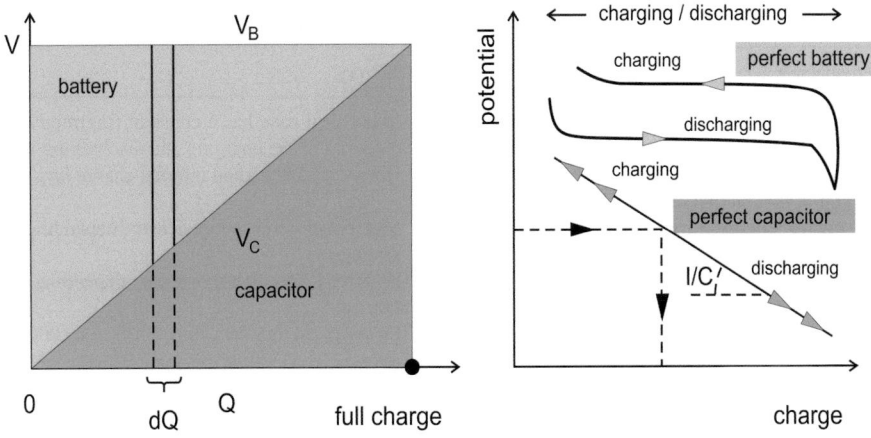

Figure 7.24.   Comparison between charge and discharge process in a battery and capacitor.

activated carbon or mesoporous carbon (Frackowiak, 2007). Recently, much attention has been focused on graphene as a promising electrode material. This material consists of graphite monolayers with a surface area of about $2600\,m^2\,g^{-1}$ and large gravimetric energy densities. For example, graphene-based supercapacitors exhibit a specific energy density of $86\,Wh\,kg^{-1}$ at a current density of $1\,A\,g^{-1}$ (Liu *et al.*, 2010). This is comparable with that of the Ni-metal hydride-battery but its charging or discharging process lasts only seconds or minutes. Furthermore, as the production process of graphene is improved and the price of the material is going down gradually, that favors graphene application for large-scale installations.

- Electroactive polymers (Irvin *et al.*, 2007; Snook *et al.*, 2011) – commonly described as synthetic metals – are mixed electron-ion conductors. The most commonly used polymers are: polypyrrole, thiophene derivatives, and polyaniline. These materials are characterized by fast oxidation and reduction processes during charging/discharging, ease of production (chemical or electrochemical synthesis) and high charge density.
- Metal oxides – often used in electrochemical capacitors utilizing redox reaction. Commonly investigated metal oxides are: ruthenium oxide, titanium oxide, iridium oxide, manganese oxide, cobalt oxide or ferrites (Lokhande *et al.*, 2011).
- Hybrid systems, where – in order to obtain increase in the charge density per unit area of the electrode – additional centers capable of redox reactions are introduced (Lisowska-Oleksiak and Nowak, 2007).

In order to isolate two electrodes, the porous separator, filled by electrolyte is used. The electrolyte is typically one or more solvents containing ionic species. Usually, the physical and electrochemical properties of the electrolyte employed determine the power properties of a super-capacitor. The thickness d of electrical double layer at the electrode/electrolyte depends strongly on the dielectric permittivity of the electrolyte used and therefore shows a simple relationship between the applied electrolyte and the capacitance of electrode materials.

As electrolyte, aqueous and non-aqueous solvents are applied. Aqueous solvents narrow operational voltage to 1 V, because at higher potential molecules undergo decomposition: at the anode and cathode, oxygen and hydrogen are generated respectively. However, aqueous electrolytes are characterized by high conductivity (i.e. $0.8\,S\,cm^{-1}$ for sulfuric acid), low price and environmental friendliness. With non-aqueous solutions, which do not contain chemically active hydrogen atoms in their molecules, we can achieve enhancement in the stability potential range. Because decomposition of the solvent is not experienced, the operational potential window can be widened up to 3 V. The only drawback is their high specific resistance that lowers the supercapacitor's power.

Table 7.1.   Comparison of the characteristics of an electrochemical capacitor and a battery.

Supercapacitor	Battery
Charging and discharging curve is sloped straight	In an ideal case has a constant (thermodynamic) potential of the charging and discharging process
Charging curve due to straight slope is a good indication of the internal charge level	Does not have a good internal charge level
Relatively low energy density	Medium or good energy density, depending on the kinetics
Good power density	Relatively low power density, depending on the kinetics
Has an excellent cycling and cycle time performance due to simple addition and withdrawal charge (double layer type)	Lower cycle time by about $1/100 \sim 1/1000$ due to redox of irreversibility and phase change processes in three dimensions
Internal resistance results from a highly-developed area of the matrix and electrolyte	Internal resistance results from electrolyte and active materials
Little or no activation of C polarization but may be dependent on temperature	Significant temperature dependence of the activation polarization (faradic resistance)
Long life (except corrosion of current collectors)	Low life time due to degradation or reconstruction of active materials
Conductivity of the electrolyte can be reduced during charging process due to the ionic adsorption	Conductivity of the electrolyte can increase or decrease depending on the chemistry of faradaic processes

7.4.2.2 *Supercapacitor vs. other energy storage devices*

Electrochemical capacitors as energy storage and conversion devices can be placed between electrolytic capacitors and batteries. Table 7.1 lists the information that allows the comparison of electrochemical capacitors with batteries (secondary galvanic cells).

Comparison of characteristics of supercapacitors and batteries shows that these devices are complementary but often not interchangeable. The diagram presented in Figure 7.25 illustrates a comparison between energy and power densities for various types of primary cells and typical electrochemical capacitors. The characteristics of an internal combustion engine with spark ignition are also indicated. Presented data concerns the application of supercapacitors in mechanical vehicles with electric engines.

Batteries – depending on the materials used – store different amounts of energy. Most popular and cost-competitive are lead-acid batteries. However, they are characterized by low energy density in comparison to other systems. A typical energy requirement of a modern electric vehicle is only about $200\,Wh\,km^{-1}$; therefore storage systems should be capable to provide energy of $500\,Wh\,kg^{-1}$. Systems of batteries, fuel cells or electrochemical supercapacitors are able to cope with such requirements.

In electric vehicles supercapacitors can be applied together with fuel cells or batteries which act as the energy source. Supercapacitors do not store as much energy as batteries, but are capable of fast charge and discharge. Parallel connection of supercapacitors with chemical power sources, such as fuel cells or batteries, offers the following advantages (Faggioli *et al.*, 1999):

- Improves efficiency and reduces energy consumption under varying operating conditions.
- Provides high efficiency irrespective of the type and condition of the energy source.
- Increases the strength of the current sources across a highly variable rate of energy consumption.
- Increases the maximum possible range that the vehicle can cover thanks to the efficient recovery of braking energy.
- Increases the stability of the electricity transmission system.

Figure 7.26 presents a supply system for vehicles with electric engines.

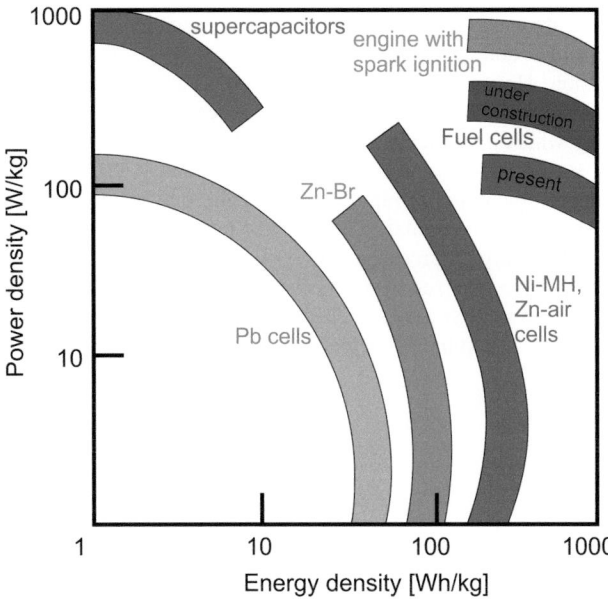

Figure 7.25.   Comparison of power and energy densities for supercapacitors, batteries, fuel cells and gasoline engines (after Shukla *et al.*, 2001).

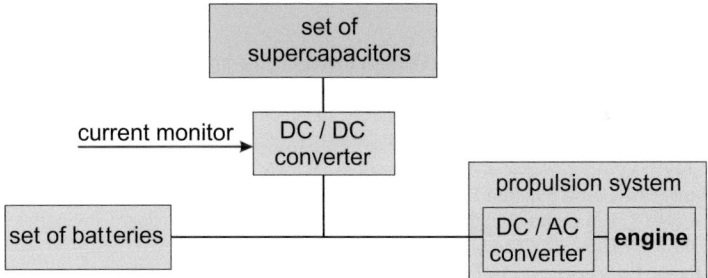

Figure 7.26.   Schematic of supply system (after Faggioli *et al.*, 1999).

The DC/DC converter is responsible for fast power flow to and from the system of supercapacitors. This makes possible dynamic acceleration and energy recovery during braking. The voltage of supercapacitor systems could be lower than the voltage of the accumulator. Lower operating voltage enables reduction in number of cell units and simplification of the supercapacitor's electric module. The presented scheme, without its modification, could be enhanced by an additional generator (as an auxiliary power element). Apart from their application in the drive system, supercapacitors could have additional functions, e.g. as power steering, electric heating, break energy regeneration or engine start-up in hybrid systems.

Furthermore, supercapacitors are increasingly being used in photovoltaic and wind-turbine systems (Li, 2012). The stochastic nature (fluctuation in solar irradiance and wind) of RES output and power demand leads to frequent charge/discharge actions of batteries and a rapid battery ageing. Thus, supercapacitors are regarded as the best solution for short time energy storage and enable fast dynamic regulation of a power-grid. Usually, the PV array is connected to the DC-bus via a PV converter and the outputs of the photovoltaic panels are sent to a bank of ultracapacitors for energy storage. In modern PV systems a combination of supercapacitors with batteries or fuel cells are used (Thounthong *et al.*, 2011). Such hybrid systems provide efficient energy transfer: from PV converter into the storage media and from storage toward the energy consuming application. However it should be taken into account that the voltage output of a supercapacitor series may differ (be much lower or higher) from the operational voltage of the circuit it is powering, which necessitates a sophisticated circuit design to harness all of the stored energy.

Summarizing, supercapacitors are rapidly entering the power engineering market. Legislative solutions concerning environmental protection and sustainable development strongly support installation of renewable energy sources that incorporate reliable energy storage systems. Electrochemical capacitors are characterized by a very fast charge/discharge process and long lifetimes, though they are not able to store volumes of electricity as high as classic batteries. For that reason it is a very good solution to combine supercapacitors with other chemical sources of electricity storage.

REFERENCES

Balfour, J.R., Shaw, M. & Jarosek, S. (2013) *Introduction to Photovoltaics*. Jones & Bartlett Learning, Burlington, MA.

Becquerel, E. (1839) Memoires sur les effects electriques produits sous l'influence des rayons, *Compotes Rendues*, 9, 561–567.

Botte, G.G. (2007) Batteries: basic principles, technologies, and modeling. In: Bard, A.J. & Stratmann, M. (eds) *Encyclopedia of Electrochemistry*. Wiley-VCH Verlag, Weinheim, Germany. pp. 377–424.

Brabec, C., Dzakonv, V. & Scherf, U. (2008) *Organic Photovoltaics*. Wiley-VCH Verlag, Weinheim, Germany, 2008.

Brett, C.M.A. & Brett, A.M.O. (1993) *Electrochemistry. Principles, methods and applications*. Oxford University Press, New York, NY.

Chapin, D.M., Fuller, C.S. & Pearson, G.L. (1954) A new silicon p-n junction photocell for converting solar radiation into electrical power. *Journal of Applied Physics*, 25, 676–677.

Chapman, D.L. (1913) A contribution to the theory of electrocapillarity. *Philosophical Magazine*, 25, 475–481.

Conway, B.E. (1999) *Electrochemical Supercapacitors. Scientific Fundamentals and Technological Applications*. Kluwer Academic/Plenum Publishers, New York, NY.

Czochralski, J. (1918) Ein neues Verfahren zur Messung der Kristallisationsgeschwindigkeit der Metalle. *Zeitschrift für Physikalische Chemie*, 92, 219–221.

Faggioli, E., Rena, P., Daniel, V., Andrieu, X., Mallant, R. & Kahlen, H. (1999) Supercapacitors for the energy management of electric vehicles. *Journal of Power Sources*, 84, 261–269.

Frackowiak, E. (2007) Carbon materials for supercapacitors. *Physical Chemistry Chemical Physics*, 9, 1774–1785.

Fritts, C. (1883) On a new form of selenium photocell. *Proceedings of the American Association for the Advancement of Science*, 33, 97.

Ginley, D.S. & Cahen, D. (2012) *Fundamentals of Materials for Energy and Environmental Sustainability*. Cambridge University Press, New York, NY.

Gouy, G. (1910) Constitution of the electric charge at the surface of an electrolyte. *Journal de Physique*, 9, 457–467.

Graham, D.C. (1947) The electrical double layer and the theory of electrocapillarity. *Chemical Reviews*, 41, 441–501.

Hagfeldt, A., Boschloo, G., Sun, L., Kloo, L. & Pettersson, H. (2010) Dye-sensitized solar cells. *Chemical Reviews*, 110, 6595–6663.

Hankins, M. (2010) Stand-alone Solar Electric Systems. The Earthscan Expert Handbook for Planning, Design and Installation. *Earthscan's Expert Series*. Routledge, Abingdon-on-Thames. pp. 117–140.

Helmholtz, von H.L.F. (1853) Über einige Gesetze der Vertheilung elektrischer Ströme in körperlichen Leitern mit Anwendung auf die thierisch-elektrischen Versuche. *Annals of Physics*, 89, 211–233.

Irvin, J.A., Irvin, D.J. & Stenger-Smith, J.D. (2007) Electroactive polymers for batteries and supercapacitors. Chapter 9, in: Skotheim, T.A. & Reynolds, J.R. (eds.) *Conjugated Polymers: Processing and Applications*. CRC Press, Boca Raton, FL. pp. 9.1–9.29.

Kittel, C. (2005) *Introduction to Solid State Physics*. Wiley, Hoboken, NJ.

Łapiński, K. & Cenian, A. (2013) Optimization of heat exchange processes in hybrid photovoltaic-thermal (PVT) appliances with and without glass. Project final report. Institute of Fluid-Flow Machinery Polish Academy of Sciences, Gdańsk, Poland.

Li, J., Chen, Y. & Liu, Y. (2012) Research on a stand-alone photovoltaic system with a supercapacitor as the energy storage device. *Energy Procedia*, 16, 1693–1700.

Lisowska-Oleksiak, A. & Nowak, A.P. (2007) Metal hexacyanoferrate Network synthesized inside polymer matrix for electrochemical capacitors. *Journal of Power Sources*, 173, 829–836.

Liu, C., Yu, Z., Neff, D., Zhamu, A. & Jang, B.Z. (2010) Graphene-based supercapacitor with an ultrahigh energy density. *Nano Letters*, 10, 4863–4858.

Lokhande, C.D., Dubal, D.P. & Joo, O. (2011) Metal oxide thin film based supercapacitors. *Current Applied Physics*, 11, 255–270.

Luo, F.L. & Hong, Y. (2012) *Renewable Energy Systems: Advanced Conversion Technologies and Applications*. CRC Press, Boca Raton, FL.

Luque, A. & Hegedus, S. (2011) *Handbook of Photovoltaic Science and Engineering*, 2nd edn. Wiley, Chichester, UK.

Parsons, R. (1981) The electrical double layer at solid/liquid interfaces. *Journal of Electroanalytical Chemistry*, 118, 3–18.

Shukla, A.K., Arico, A.S. & Antonucci, V. (2001) An appraisal of electric automobile power sources. *Renewable and Sustainable Energy Reviews*, 2, 137–155.

Smil, V. (1991) *General Energetics: Energy in the Biosphere and Civilization*. Wiley, New York, NY.

Snook, G.A., Kao, P. & Best, A.S. (2011) Conducting-polymer-based supercapacitor devices and electrodes. *Journal of Power Sources*, 196, 1–12.

Solanki, C.S. (2011) *Solar Photovoltaics: Fundamental, Technologies and Applications*, 2nd edn. Prentice Hall of India, New Delhi, India.

Solanki, C.S. (2013) *Solar Photovoltaic Technology and Systems: A Manual for Technicians, Trainers and Engineers*. PHI Learning Private Limited, Delhi, India. pp. 219–248.

Staebler, D.L. & Wronski, C.R. (1980) Optically induced conductivity changes in discharge-produced hydrogenated amorphous silicon. *Journal of Applied Physics*, 51, 3262–3268.

Stern, O. (1924) Zur Theorie der elektrolytischen Doppelschicht. *Zeitschrift für Elektrochemie*, 30, 508–516.

Szymński, B. (2013) Photovoltaic installation. GLOBEnergia, Kraków, Poland.

Thounthong, P., Chunkag, V. Sethakul, P., Sikkabut, S., Pierfederici, S. & Davat, B. (2011) Energy management of fuel cell/solar cell/supercapacitor hybrid power source. *Journal of Power Sources*, 196, 313–324.

Trassati, S., Galizzioli, D. & Tantardini, F. (1975) Ruthenium dioxide: A new electrode material. II. Non-stoichiometry and energetics of electrode reactions in acid solutions. *Journal of Applied Electrochemistry*, 203, 203–214.

CHAPTER 8

Renewable energy technologies for greenhouses in semi-arid climates

Francisco Javier Cabrera, Jorge Antonio Sánchez-Molina, Guillermo Zaragoza, Manuel Pérez-García & Francisco Rodríguez-Díaz

8.1 INTRODUCTION

Modern agricultural systems are characterized by the intensive and optimal use of land and water, turning agricultural exploitation into a semi-industrial concept. Greenhouses are systems suitable both for zones with unfavorable climatic conditions – allowing crop growth regardless of the ambient temperature – and for regions with less restrictive weather – with the aim of increasing crop productivity and improving fruit quality. In this context, a secure and environmentally friendly energy supply must be considered, for any power range or circumstance, including for stand-alone installations.

Crop growth is primarily determined by climate and the amount of water and fertilizers applied through irrigation. Therefore, greenhouses are ideal for farming because they allow one to optimize these physical parameters, via the photosynthetic process (Ramírez-Arias et al., 2012), to enhance biomass production. This manipulation requires energy consumption, depending on the crop's physiological requirements, and the production patterns adopted for yield quantity and timing.

The present general concern on the development of more efficient and sustainable productive activities has increased interest in the evaluation of alternatives to the conventional energy sources in the sector; presently many are involved in processes for assessing the environmental feasibility and technological development of such alternatives (Antón et al., 2007; Bojacá et al., 2014; Martínez-Blanco et al., 2011; Page et al., 2012; Romero-Gámez et al., 2012; Torrellas et al., 2013; van der Werf et al., 2014). This work aims to provide an overview of the existing options for the integration of renewable sources in greenhouses located in semi-arid regions where, for example, the high availability of solar radiation facilitates its use for the fulfillment of a certain percentage of the heat and/or electricity loads of greenhouses. Accordingly, this overview contains the basic technological aspects of the main renewable technologies applicable in greenhouses, some simplified design tools and criteria for their selection. Finally, it includes a summary of selected experiences in this field. The main objective is to contribute to a better understanding of the technologies that should support the promotion and development of projects for the implementation of renewable energies in agriculture.

With respect to the integration of renewable energy sources, for modern greenhouses the overall starting point, that is, orientation, shape and means of heat distribution, is favorable, thanks to the energy optimization implicit in the basic design of any present-day greenhouse system (von Zabeltitz, 2011). However, excluding costs and generic maturity constraints present in any renewable project, the most important limiting factor for alternative energy source integration in greenhouses is the competition for ground use between intensive cropping and the energy systems themselves. This results in a reduction of the area cultivated or the blocking of photosynthetic active radiation (PAR), in cases where a fraction of the greenhouse roofs are used for systems installation. This limiting factor jeopardizes the implementation of low-density renewable energy sources in greenhouses, such as low temperature solar thermal and solar photovoltaic (PV),

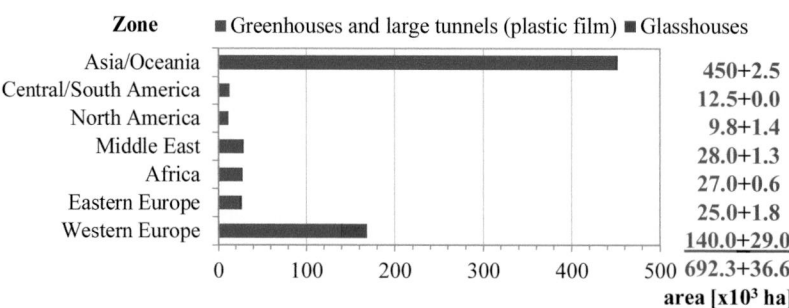

Figure 8.1.   Distribution of permanent crop protection areas around the world (compiled from Giacomelli *et al.*, 2008).

yet favors high-density sources, such as biomass, geothermal or wind energy. Furthermore, from a structural point of view, systems integration has to take into account that any additional mechanical loads should not involve changes in the support frames and elements, which are generally designed for light envelopes (plastic or thin glass).

Additionally, it must also be pointed out that, contrary to the traditional application of renewable energy in agriculture, greenhouses are usually located in areas where electricity demands are fully covered by the pre-existing network. Consequently, in these cases, self-sufficiency is not the objective because the installation will operate by net metering of the instantaneous and/or aggregated consumption that is directed into the production network. Occasionally, local energy production can be incorporated into nearby distribution lines, subject to the corresponding national access regulations, e.g., to the existing national feed-in tariff schemes aimed at encouraging electricity produced from renewable energy sources; or better still, integrated into smart-grid schemes at the district level, including other nearby agricultural facilities, stores, cold rooms, and buildings, among others.

Conversely, there are farms in isolated locations where the costs of accessing conventional mains energy are prohibitive, making a classic approach based on alternative on-site sources necessary. In these situations, the sizing of the system elements is crucial, given the need to optimally balance the unpredictable nature of some renewable resources, such as solar and wind, with the daily and seasonal greenhouse energy consumption patterns; these determine storage requirements and, in an eventuality, the use of conventional back-up systems (not necessary for farms in the previous case). Stand-alone greenhouses in this category are no different from other stand-alone farm or agricultural facilities, provided the specific electricity loads are known.

Worldwide, the crop surface covered by some type of protection is around five million hectares (ha). Of this, the surface occupied by permanently protected crops is more than 700,000 ha, 95% being protected by plastic greenhouses (Fig. 8.1). Most of the farms are concentrated in Southeast Asia (primarily China, South Korea and Japan), followed by the Mediterranean basin, which includes about 215,000 ha of permanent greenhouses and macro-tunnels (Tüzel and Leonardi, 2009).

Spain is the country with the largest area of greenhouses in the Mediterranean (52,000 ha, mainly concentrated in the southeast, a semi-arid zone). Along with Italy, Turkey and Morocco, these countries have more than 70% of the overall area of greenhouses in the Mediterranean. Glass greenhouses are concentrated in Northern Europe (von Elsner *et al.*, 2000). The Southern European greenhouses are characterized by plastic cladding and, due to the climatic conditions, less restrictive specifications, leading to lower costs (Castilla and Hernández, 2007; Giacomelli *et al.*, 2008; von Zabeltitz, 2011).

According to the number of modules making up their structure, greenhouses can be classified as either single- or multi-span. The replication of a basic shape in multi-span greenhouses serves for larger cultivation areas, keeps the necessary slope to avoid condensation on the interiors of the walls and roof, facilitates rainwater removal, and improves the capture of solar radiation and

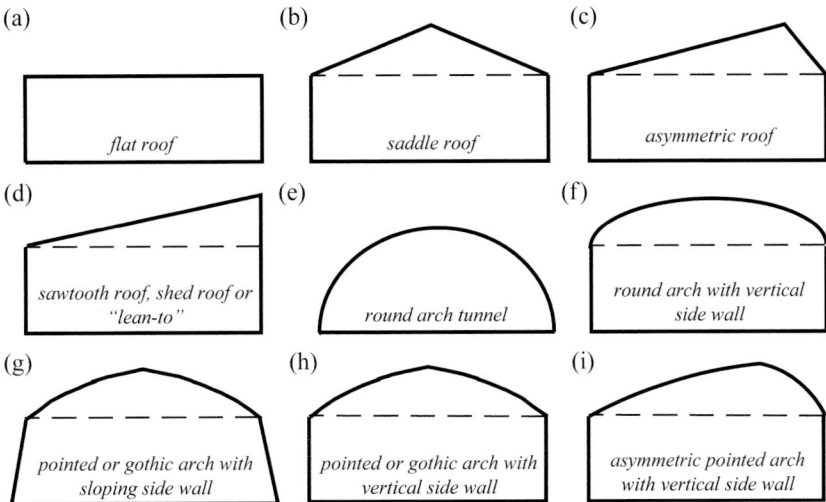

Figure 8.2.   Most frequently used greenhouse shapes (adapted and modified from Baudoin *et al.*, 2002; von Zabeltitz, 1999).

the indoor climate. The most commonly used shapes in greenhouse construction are shown in Figure 8.2. Usually the span width varies between 4 and 12 m, with a total height not exceeding 5–8 m.

This chapter aims to give an overview of renewable energy use in greenhouses, how to achieve the energy demand and how this can be obtained from different renewable energy sources, such as wind, solar thermal, PV, geothermal, and biomass energy. The chapter ends with the development of energy saving measures in greenhouses using optimal controllers.

8.2   GREENHOUSES IN SEMI-ARID CLIMATES

The semi-arid climate, which exists in various locations around the world, is characterized by low annual precipitation (200–400 mm year^{-1}) and consequently high solar radiation availability (Verheye, 2006). This solar radiation availability for photosynthesis, together with warm summers and mild temperatures in winter, means greenhouses in these regions generally perform well, provided that crop water requirements are met by local underground resources[1] and using optimized irrigation systems. In these zones, low thermal loads are required during the winter to reach optimal temperatures inside the greenhouse; this is thanks to the confinement effect, resulting from the decrease in air exchange with the outside environment, and the low transparency to far infrared radiation (emitted by the crop, the soil and the inner greenhouse elements) and high transparency to sunlight (Castilla and Baeza, 2013). During the summer, especially in the Mediterranean and tropical areas where the inside temperature can exceed the recommended maximum threshold levels, indoor conditions can be controlled by passive means, such as the use of openings in the cladding for natural ventilation.

The most common crops in semi-arid regions are ornamental plants, flowers, fruits and vegetables. Figure 8.3 shows an overview of existing climatic conditions for different semi-arid locations all over the world and, as a reference, the data for a hot arid and a cold location. The figure also includes the greenhouse climate control methods for the cultivation of thermophilic plants (Castilla, 2007); it shows the advanced suitability of greenhouses in these locations, with

---

[1] or using desalination systems (Bundschuh and Hoinkis, 2012).

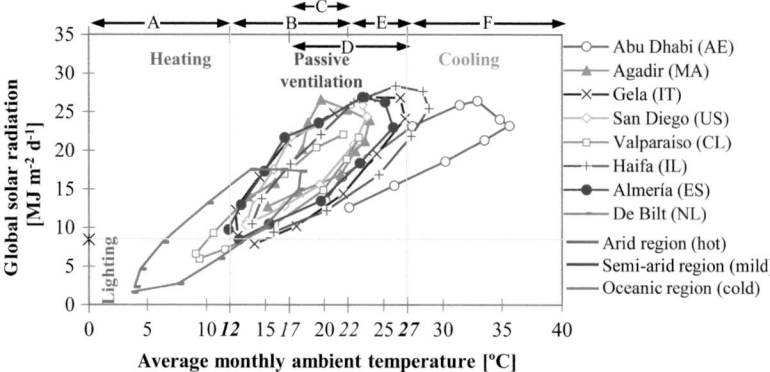

A.   Heating necessary [<12°C]
B.   Greenhouse without active climate control (necessary passive ventilation) [12-27°C]
C.   Cultivation is possible in open air (inland areas) [17-22°C]
D.   Cultivation is possible in open air (coastal areas) [17-27°C]
E.   Active cooling system [22-27°C]
F.   Excessive temperatures [>27°C]
x.   Minimum daily radiation requirement [8.5 MJ m⁻² d⁻¹]

Figure 8.3.   Climate suitability for cultivation of thermophilic vegetable species inside greenhouses in semi-arid, arid and oceanic regions.

regard to those needing active heating and/or cooling, which would result in higher fossil fuel consumption.

Nonetheless, these averaged climatic conditions and strategies are the result of a dynamic interaction between the greenhouse structures and the varying daily values of outdoor temperature, solar irradiance, relative humidity and wind velocity. In spite of the overall good performance shown above, it is possible that in a relevant number of hours over the year, active measures are required to maintain crop growth and increase greenhouse productivity (Baudoin, 2013; Castilla, 2007; von Zabeltitz, 2011).

In addition to this, unlike locations where there is only a warm or cold climate, in semi-arid regions, especially in the Mediterranean area, the passive, and eventually active, climate control strategies must address both heating and cooling demands, introducing an added complexity factor to the system's selection and operation. An obvious example of this is the passive cooling technique consisting of whitening the greenhouse roofs and walls used to decrease heat gain during summer; this must be removed to allow cladding transparency during the winter months. Another example is temperature and humidity control using ventilation, which are factors that often conflict, e.g., when a greenhouse is vented to lower the temperature, the humidity is also lowered, or when vented to lower humidity, the temperature likewise lowers.

8.3   OVERVIEW OF ENERGY DEMANDS IN GREENHOUSES

In greenhouses, the following energy demands can be identified:

- Demands related to indoor climate control.
- Demands related to irrigation, fertigation and sanitation.
- Demands associated with the use of computers and devices for measuring and control, lighting and security.
- Demands for the systems supporting harvest and post-harvest tasks.

The first two demands must be considered basic, given that they are related to the physiological needs of the crops. Both demands are determined by an elementary process of mass and heat

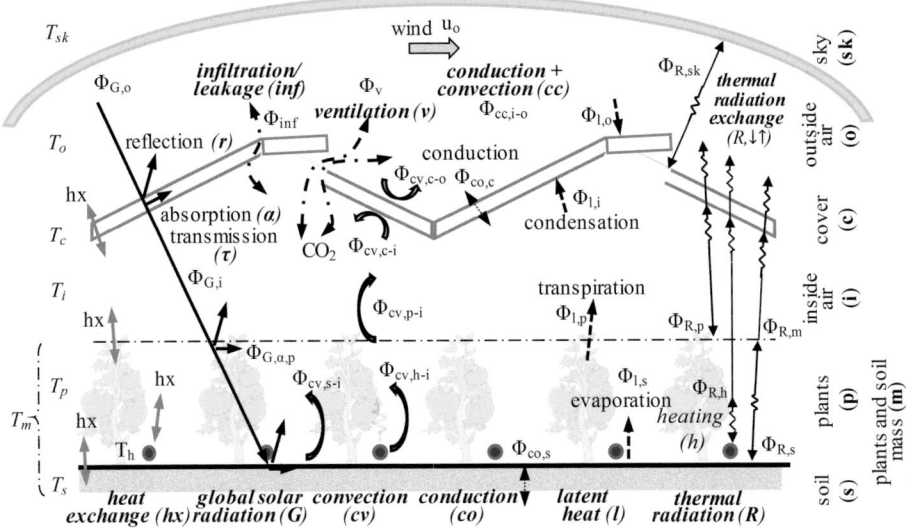

Figure 8.4.   Physical processes and energy fluxes in a greenhouse that need to be considered in the calculation of the total energy balance (see nomenclature table for definitions).

transfer involving the plants, the air inside, the cladding surfaces and the meteorological variables. As a result of the corresponding balances, a certain amount of energy must be added or extracted to fulfill the established temperature set points. A precise knowledge of this amount, either on an instantaneous or an accumulated basis, is key to the design of greenhouse energy systems, because it allows the assessment of:

- The energy system device specifications, including their size, their interconnections and distribution lines and the way in which they interact with the greenhouse environment.
- An estimation of the yearly operating costs related to these basic energy demands as well as the specific daily and seasonal patterns.
- The eventual need for energy storage, both for back-up purposes overnight and during low radiation or low wind periods; as well as for renewable sources and for systems performance optimization.
- The options for installation and operational improvements, either by means of systems control or through efficiency measures, such as heat recovery, among others.

In the case of climate control energy demands, the load must be provided in the form of sensible or latent heat and, correspondingly, a specific source (solar radiation, geothermal well or biomass, etc.) or sink (ground, sky or air etc.) must be considered in the case of cooling. Thermal balance in greenhouses can be addressed by thinking of the greenhouse as a solar collector (Boulard and Baille, 1987) or a control volume (González-Real and Baille, 2010). Figure 8.4 summarizes the main energy and mass fluxes to be taken into account according to the above (Baille, 1999, 2006; Bot and Van de Braak, 1995; Day and Bailey, 1999; González-Real and Baille, 2010; Hanan, 1998; Montero *et al.*, 1998; Rodríguez, 2002; von Zabeltitz, 1999).

An accurate thermal load calculation must consider the overall specifications of the materials and elements that determine the fluxes in Figure 8.4 and the means for a fast, accurate and precise physical and mathematical representation of the processes, including their dynamic nature and as high a spatial resolution as possible. However, in a first stage design, there are some alternative formulations that have a reliable level of accuracy, based on the approximation of the actual behavior to steady-state conditions and a lumped parameter representation.

In addition to these well characterized thermal demands, a relevant quantity of energy in the form of electricity must also be considered for climate control in greenhouses. Besides the eventual use of electrically driven heat pumps or chillers, the elements required for heat distribution inside a greenhouse are water and air pumps, as well as fans and motors for roof windows. Also, irrigation pumps are fed by electricity and, as in the case of heat pumps; the corresponding energy load presents a climate dependent pattern, because of the relationship between evapotranspiration, and the inside and outside variables.

8.3.1   *Greenhouse heating and cooling loads*

As put forward in a first stage design phase, the heating and cooling loads can be estimated using simplified formulations based on a steady-state approach and a lumped-parameters representation. One of the more accepted is expressed as follows (modified from ASABE, 2003):

$$Q_h = Q_{cc,i-o} + Q_{inf} + Q_v + Q_s + Q_{R,\downarrow\uparrow m-sk} - H_{sen} \tag{8.1}$$

$$Q_h = U_c A_c (T_i - T_o) + \left[\rho_i V_i \frac{N}{3600} [c_{p,i}(T_i - T_o) + L_i(\chi_i - \chi_o)] \right] + [A_s \rho_{ex} V_v [c_{p,ex}(T_{ex}$$
$$- T_{inlet}) + L_{ex}(\chi_{ex} - \chi_{inlet})]] + Q_s + Q_{R,\downarrow\uparrow m-sk} - [A_s \tau_{G,c} G_o(1-f)] \tag{8.2}$$

where H_{sen} [W] corresponds to global solar radiation not used for transpiration that heats the greenhouse; Q_s [W] is the soil energy flux; $Q_{R,\downarrow\uparrow m-sk}$ is the thermal radiative exchange between the mass (plants and soil) and the sky; Q_h [W] is the heating supplement; Q_{inf} [W] is the infiltration or leakage losses; Q_v [W] is the ventilation losses; $Q_{cc,i-o}$ [W] is the conduction-convection losses; U_c [W m^{-2} °C^{-1}] is the overall heat loss coefficient; A_c [m^2] is the cover surface; T_i [°C] is the inside temperature; T_o [°C] is the outside temperature; V_i [m^3] is the greenhouse volume; N [h^{-1}] is the air renewal rate by infiltration; ρ_i [kg m^{-3}] is the inside air density; $c_{p,i}$ [J kg^{-1} °C^{-1}] is the specific heat of the inside air; L_i [J kg^{-1}] is the latent heat of water vaporization at T_i; χ_i and χ_o [kg$_{water}$ kg$_{air}^{-1}$] are the inside and outside absolute humidities, respectively; V_v [m^3 m^{-2} s^{-1}] is the floor area ventilation rate required to maintain a given temperature rise between the temperature of exhaust air ($T_{ex} \approx T_i$) and the temperature of air entering the greenhouse ($T_{inlet} \approx T_o$); ρ_{ex} [kg m^{-3}] is the exhaust air density; $c_{p,ex}$ [J kg^{-1} °C^{-1}] is the specific heat of the exhaust air; L_{ex} [J kg^{-1}] is the latent heat of water vaporization at T_{ex}; χ_{ex} and χ_{inlet} [kg$_{water}$ kg$_{air}^{-1}$] are the exhaust and inlet absolute humidities, respectively; $\tau_{G,c}$ [–] is the transmission coefficient of the cover for outside global radiation, G_o [W m^{-2}], received by the greenhouse surface A_s [m^2]; and f [–] is the radiation fraction used for transpiration.

Using standard values for greenhouse cladding material properties (Table 8.1) and air infiltration rates (Table 8.2), and with the set points of inside relative humidity $RH_i \leq 80\%$; heating temperature, $T_{i,hot}$, 12°C; and refrigeration temperature, $T_{i,cool}$, 27°C), Table 8.3 shows an overall estimation of the energy and mass balances. It provides the order of magnitude of the loads, to be considered for one semi-arid, one warm and one cold location, for different growing cycles.

Along with this approach, which provides a yearly overview of energy demands, an adequate renewable design requires knowledge of the daily and seasonal energy demand patterns, to design the systems to the actual resources available. Unlike conventional systems, they have a certain level of unpredictability and their own daily and seasonal patterns. For example, using a conservative approach by adapting the systems configuration and size to low-solar availability conditions will result in higher costs along with oversizing and wasted energy for the rest of the year.

Figure 8.5 illustrates this, showing the estimation of the complete annual thermal load for a tomato-growing cycle (September–August, 365 days) in Almería (southeast Spain), for a plastic greenhouse (Table 8.3). Not only can the seasonal patterns be seen, but also the daily ones, facilitating an advance in the existing requirements; for example, for better design of the dimensions and heating systems control (both conventional and renewable). Concerning the use of solar radiation, the figure also shows clearly the seasonal variations in the incidence patterns for a horizontal and a vertical surface; these should be taken into account for better selection of the tilting angle of the collector systems to adapt renewable energy harvesting to the thermal or electrical requirements.

Table 8.1. Characteristics of different cladding materials: thickness, e_c; absorption, α, reflection, r, and transmission, τ, for global solar radiation, G, and thermal radiation, R; conduction heat transfer coefficient, U_c (mean wind speed 4 [m s^{-1}], mixed heating system and standard conditions), and density, ρ_c (compiled from ASABE, 2003; Castilla, 2007; Feuilloley and Issanchou, 1996; Nijskens et al., 1985; Papadakis et al., 2000; Valera et al. 2008; von Zabeltitz, 2011).

Material	Thickness e_c [mm]	Global solar radiation[1] G (0.3 < λ < 3 μm) [–]				Thermal radiation[1] R (λ > 3 μm) [–]			$U_c^{[2]}$ [W m^{-2} C^{-1}]	ρ_c [kg m^{-3}]
		$\alpha = \varepsilon$	r	τ_{Gb}	τ_{Gd}	$\alpha = \varepsilon$	r	τ		
Single glass	4	0.03	0.08	0.89	0.82	0.9	0.1	0	6.0–8.8	2400
Polymethyl methacrylate (PMMA)	8	0.06	0.12	0.82	0.82	0.98	0.02	0	3.4	1190
Polycarbonate (PC)	4	0.08–0.11	0.14–0.15	0.78	0.78	0.89–0.98	0.09	0.02–0.03	3.2–3.6	170–200
Single thermal polyethylene film (PE–th)	0.18	0.03	0.08	0.89	0.81	0.7–0.77	0.03	0.2–0.27	6.0–13.0	910
Copolymer ethylene vinyl acetate (EVA)	0.1	0.02	0.07–0.09	0.89–0.91	0.82	0.42–0.58	0.03	0.39–0.55	7.8	940
Co-extrusion PE–EVA–PE	0.2	0.02–0.04	0.09–0.14	0.82–0.89		0.59	0.03	0.38	6.0–12.1	930
Polyvinyl chloride (PVC)	0.1	0.02	0.07	0.91		0.62	0.32	0.06	7.7	1300
Double glass	8	0.15	0.13	0.72		0.83	0.17	0	3.2–5.2	
Double PE	0.2	0.03	0.14	0.83		0.28	0.06	0.66	4.0–6.0	

[1] Optical properties for normal incidence radiation. [2] Varies with the wind velocity, the heating system and cladding material (e.g., *multi-span* type, PE–EVA–PE–covered: water heating, $U_c = 0.56u_o + 4.8$; *parral* type, PE–th–covered; air heating, $U_c = 0.75u_o + 7.5$ and air heating, $U_c = 2.7u_o + 3.11$) (López, 2003).

Table 8.2.   Estimated infiltration rates for greenhouses by type and construction (modified[1] from ASABE, 2003).

| | Infiltration rate, N [h^{-1}][1] | |
| | Low leakage or good maintenance greenhouse ("*multi-span*" or "*venlo*") | High leakage or poor maintenance greenhouse ("*parral*") |
Type and construction		
New construction:		
plastic film	min (0.75; 0.5 $N_{airtight}$)	min (1.5; 0.5 $N_{non\text{-}airtight}$)
glass or fiberglass	min (0.5; 0.5 $N_{airtight}$)	min (1; 0.5 $N_{airtight}$)
Old construction:		
plastic film	min (2; $N_{airtight}$)	min (4; $N_{non\text{-}airtight}$)
glass or fiberglass	min (2; 2 $N_{airtight}$)	min (4; 2 $N_{airtight}$)

[1]Internal air volume exchanges per unit time [h^{-1}]. High winds or direct exposure to wind will increase infiltration rates; $N_{airtight} = 0.075u_o + 0.25$, for low leakage greenhouses (e.g., *multi-span* or *venlo*) and $N_{non\text{-}airtight} = 0.29u_o + 0.76$ (López *et al.*, 2001) for high leakage greenhouses (e.g., *parral*). (u_o [m s^{-1}], outside wind velocity).

Table 8.3.   Estimated availability of solar radiation, G_o; heating requirements, Φ_h; ventilation rate, V_v; and net irrigation water requirement, *NIWR*, for greenhouses in different locations, types of greenhouse and growing cycles.

Growing cycle[1] (days)	Location	Type	Cladding material	A_c/A_s[2] [–]	G_o [GJ m^{-2}]	Φ_h[3] [GJ m^{-2}]	V_v[4] [hm^3 m^{-2}]	*NIWR* [m^3 m^{-2}]
Tomato S.-A. (365)	Abu Dhabi (AE)	*gothic arch*	PE–EVA– PE	1.34	7.26	0.01	13.73	1.61
Tomato S.-A. (365)	Almería (ES)	*gothic arch*	PE–EVA– PE	1.34	6.79	0.17	6.28	0.97
Tomato S.-A. (365)	Almería (ES)	*venlo*	glass	1.29	6.79	0.17	6.53	1.05
Tomato S.-A. (365)	De Bilt (NL)	*venlo*	glass	1.29	3.54	0.82	0.75	0.61
Tomato S.-M. (272)	Abu Dhabi (AE)	*gothic arch*	PE–EVA– PE	1.34	5.04	0.01	11.51	0.95
Tomato S.-M. (272)	Almería (ES)	*gothic arch*	PE–EVA– PE	1.34	4.36	0.17	2.45	0.55
Tomato S.-M. (272)	Almería (ES)	*venlo*	glass	1.29	4.36	0.17	2.55	0.61
Tomato S.-M. (272)	De Bilt (NL)	*venlo*	glass	1.29	2.01	0.79	0.26	0.36

[1]S.-A.: September to August, S.-M.: September to May. [2]Greenhouse area: $A_s = 10,000$ m^2. [3]Heating temperature: $T_{i,hot} = 12°C$. [4]Refrigeration temperature: $T_{i,cool} = 27°C$.

8.3.2   *Greenhouse electricity requirements*

As already mentioned, the greenhouse electricity requirements relate to supporting thermal processes using climate control and supplying irrigation pumps and other appliances for specific agricultural tasks. The electrical power P_j [kW] consumed can be estimated using the relationship between the electro-mechanical specifications of the motors and pumps to the water and air

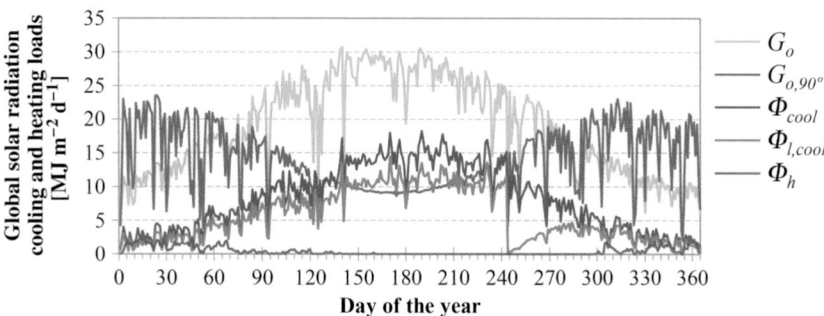

Figure 8.5. Evolution of daily global solar radiation, G_o; global solar radiation for vertical surface (Pérez *et al.*, 1986), $G_{o,90°}$; total cooling load, Φ_{cool}; latent cooling load, $\Phi_{l,cool}$; and total heating load, Φ_h, for a tomato-growing cycle (September–August) in Almería for a plastic greenhouse [heating temperature $T_{i,hot} = 12°C$ and refrigeration temperature $T_{i,cool} = 27°C$].

movement requirements. Consequently, as for the thermal requirements, the electricity demand is directly related to climate. The number and size of all the electrical elements is, additionally, a key parameter for the design of stand-alone systems because it determines the capacity for instantaneous self-generation of the corresponding grids.

The energy requirements of each appliance, $E_{L,j}$ [kWh], will be approximated by the time integration during the considered operation period of the electrical power consumed P_j [kW]:

$$E_{L,j} = \int_{t1}^{t2} P_j(t)dt \tag{8.3}$$

The operation time for each appliance is, in turn, dependent on the assigned appliance function. There is an extremely wide range of options ranging from 24-h cycles for computers and measuring-and-control systems to occasional switch-ons of actuators to cover natural ventilation demands and scheduled irrigation periods. As for the thermal loads (in this case more relevant because of the non-intermittency requirements of some of the performed tasks), daily pattern estimation is necessary to specify both the renewable generation and the storage systems.

For example, it is estimated that the annual electrical energy requirements for forced greenhouse ventilation $E_{L,FV}$ [kWh m^{-2}] in the Mediterranean area is between 7 and 10 kWh m^{-2} year^{-1} (Kittas *et al.*, 2012, 2013). This value can be estimated on a daily basis by knowing the eventual daily forced ventilation demand, according to the estimated daily thermal loads, if the greenhouse air volume to be exchanged by exhaust fan operation is known $V_{v,FV}$ [m^3 m^{-2}], both for cooling during the summer and for humidity control during the winter:

$$E_{L,FV} = k_F P_F V_{v,FV}/c_F \tag{8.4}$$

where P_F [kW] is the power of the fan, c_F [m^3 h^{-1}] is the discharge capacity of the fan and k_F [–] is a corrective factor that accounts for non-ideal fan performance and the eventual inductive losses of the power unit or engine. Due to different daily ventilation demands, a daily variation of $E_{L,FV}$ must also be expected (Fig. 8.6a).

An equivalent approach can be carried out to estimate the electricity required for the irrigation pumps. There are models (Allen *et al.*, 1998; Seginer, 2002) capable of estimating irrigation water requirements (*IWR*) [m^3 m^{-2}] for the crops, as a function of the inside climate along with plant physiological performance and evapotranspiration processes. Taking into account the required pressure expressed in height of an equivalent column of water H [m] to deal with the well depth, the drops in pipeline friction pressure and the pressure requirements for drip-irrigation systems, the estimation of electrical energy required for irrigation purposes $E_{L,I}$ [kWh m^{-2}] is:

$$E_{L,I} = 2.778 \times 10^{-7} k_I \rho_{water} g \, H \, IWR/\eta_P \tag{8.5}$$

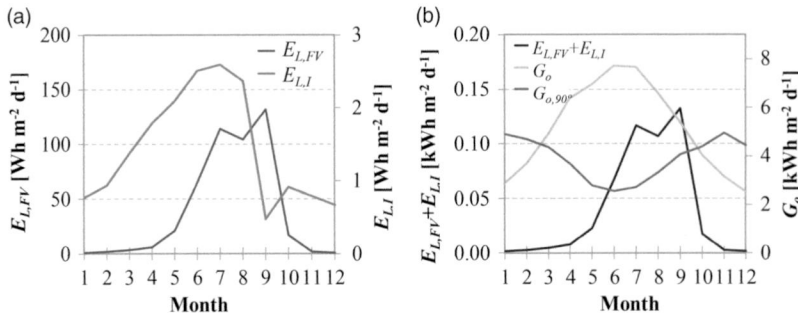

Figure 8.6.   Evolution of mean daily (a) electrical energy requirements for forced ventilation, $E_{L,FV}$, and for irrigation, $E_{L,I}$, and (b) sum, $E_{L,FV} + E_{L,I}$, and global solar radiation, horizontal G_o, and vertical (Pérez *et al.*, 1986) $G_{o,90°}$, for a tomato-growing cycle (September–August) in Almería inside a plastic greenhouse [$P_F/A_s = 0.005\,\text{kW m}^{-2}$, $c_F/A_s = 220\,\text{m}^3\,\text{h}^{-1}\,\text{m}^{-2}$, $H = 100\,\text{m}$].

where k_I [–] is a corrective term that takes into account the inductive losses of the pump motors and other hydraulic and electrical system inefficiencies; 2.778×10^{-7} is the conversion factor from J to kWh; ρ_{water} [kg m^3] is the water density; g [m s^{-2}] is the acceleration from gravity; and η_P [–] is the average electro-mechanical performance of the pump, ranging from 30% to 60%. In the case of drip irrigation, additional interest must be paid to the feeding controllers and valves that allow efficient systems performance. An example of this estimation is included in Figure 8.6.

8.4   RENEWABLE ENERGIES APPLICABLE TO GREENHOUSES

The concept of renewable energy covers a wide range of energy conversion technologies, of which the common link is the use of permanent and naturally occurring primary sources. Their potential use in agriculture has been considered since the 1970s, when different networks were created and promoted by international organizations (FAO, 1999); these have made it possible to find reference works on solar and renewable heating (von Zabeltitz, 1986, 1988; von Zabeltitz and Popovski, 1989), biomass (Strehler, 1988), geothermal (Popovski, 1988; von Zabeltitz and Popovski, 1989), and waste heat (Cornet d'Elzius *et al.*, 1989) greenhouse applications.

At first glance, the following renewable technologies can be identified, which are applicable to greenhouses:

- Micro and mini wind energy.
- Low and medium temperature solar thermal energy.
- PV solar energy.
- Geothermal energy.
- Biomass energy.

Regardless of the systems' final production, whether it is for heat or for electricity, the variations and options for their implementation are related to site-dependent factors. As regards resource availability, solar energy (both PV and thermal) can be considered at the regional level, while conventional geothermal or wind energy is only available at specific sites. Biomass use is similarly suitable only if it is available near agricultural sites. In terms of intensity, while high solar fractions require considerable surface occupation, other renewable resources, such as geothermal or biomass, are capable of providing sizable power yields from compact devices.

With independence of the above overall criteria, the integration of renewable energy sources in greenhouses allows multiple design options; in the literature, it is even possible to find configurations in which the renewable technologies are combined, aiming to maximize the global energy performance. That is the case for example of the prototype studied by Esen and Yuksel (2013) for greenhouse heating, which includes biogas, solar collectors and a ground source heat pump (GSHP) (Esen and Yuksel, 2013; Esen *et al.*, 2015).

8.4.1 *Wind energy*

Applying wind energy to greenhouses is more than just a case of adapting to the power scale of a particular wind turbine. Turbine operation is contingent on the existence of a minimum number of wind hours, with speeds above the threshold value for each device along with the possibility of spatial integration among them, especially in areas of high greenhouse concentration.

The usable conversion power into mechanical energy, and subsequently into electrical energy, contained in the wind, can be measured by the following expression:

$$P_{\rm w} = \frac{1}{2}\rho_{\rm o}A_{\rm w}u_{\rm o}^3 \tag{8.6}$$

where $P_{\rm w}$ [W] is the power; $\rho_{\rm o}$ [kg m^{-3}] is the air density; $A_{\rm w}$ [m^2] is the area swept by the wind turbine rotor; and $u_{\rm o}$ [m s^{-1}] is the wind velocity. The non-linear nature of this expression shows the importance of adapting the conversion systems to a greater wind speed range. The energy conversion is, in any case, limited by the Betz limit, which establishes a 16/27 ($\sim$59%) ratio.

Current wind turbine technology, whether on horizontal or vertical axes, allows for more intensive electricity production than other renewable technologies, such as solar power. This makes it particularly interesting in terms of its footprint, which is a critical aspect in greenhouse applications. To approximate, there is a 1:10 relationship in the power per unit of active area in each case (blades area/module area). This determines that, under favorable operating conditions, the energy [kWh] produced by a wind turbine occupying the same space as a PV module will be substantially higher. However, wind resource availability is strongly influenced by climatic and spatial factors (height, obstacles and natural intensification pathways, etc.), which determine the implementation potential and in this case makes it more restrictive.

8.4.2 *Low and medium temperature solar thermal energy*

The analogy between a greenhouse and a solar thermal collector, although not strictly accurate, is apposite in justifying the potential use of renewable energy in this sector. During the 1970s and 1980s, after the first oil crisis, many experiments took place in Europe (Floris and Parodi, 1986; von Zabeltitz, 1986, 1988; von Zabeltitz and Popovski, 1989) and USA (Mears *et al.*, 1980) to evaluate different solar collector configurations applied to greenhouses, which basically detailed the following three categories:

- Thermal collectors external to the greenhouse (Bargach *et al.*, 1999; Castilla *et al.*, 1985; Montero *et al.*, 1985; Xu *et al.*, 2014).
- Thermal collectors integrated in the greenhouse cover (Vox *et al.*, 2008).
- Using the greenhouse as the solar collector (Boulard and Baille, 1987; Zaragoza *et al.*, 2007).

The starting points were conventional flat plate collectors (FPCs) (more specifically, the development of unglazed "low cost" collectors made of EPDM rubber) and aluminum roll-bond-type collectors (Montero *et al.*, 1985). Subsequent experiments have confirmed their potential, expanding the options to unglazed "low cost" collectors using transparent plates of alveolar polyethylene inside the greenhouse (Bargach *et al.*, 2004), water-filled plastics tubes or sleeves (Baille (González-Real) *et al.*, 1977; Ntinas *et al.*, 2014), or air-heating systems that consist of plastic coatings on adsorbent sand beds, placed on the ground (Ghosal *et al.*, 2005). Solar air collectors that use air as a thermal fluid have also been tested (Alkilani *et al.*, 2011).

Table 8.4 summarizes the different types of solar thermal collectors, accounting for their operating temperature range, along with other operational features, such as concentration ratio.

The capacity of a solar collector for thermal generation is determined by its characteristic curve; this relates to the performance of the collector, $\eta_{\rm a|Gt}$ [–], which is defined as the useful heat

Table 8.4.   Types of solar collectors (compiled from Kalogirou, 2009).

Sun-tracking	Collector type	Absorber type	Concentration ratio[1]	Temperature range [°C]
None	Unglazed ("low cost") collector	Tube/bag	1	<30
	Flat plate collector (FPC)	Flat	1	30–80
	Evacuated tube collector (ETC)	Flat	1	50–200
	Compound parabolic collector (CPC)	Tubular	1–5	60–240
Single-axis tracking			5–10	60–300
	Linear Fresnel reflector (LFR)	Tubular	10–40	60–250
	Parabolic trough collector (PTC)	Tubular	15–45	60–300
	Cylindrical trough collector (CTC)	Tubular	10–50	60–300
Two-axes tracking	Parabolic dish reflector (PDR)	Point	100–1000	100–500
	Heliostat field collector (HFC)	Point	100–1500	150–2000

[1]Concentration ratio is defined as the aperture area divided by the receiver/absorber area of the collector.

produced Q_u [W] divided by the incident solar energy. The incident solar energy is determined by the global solar irradiance G_t [W m^{-2}] received in the collector aperture area A_a [m^2], with characteristic parameters $\eta_o = F_R \overline{\tau\alpha}$ [–], a_1 [W m^{-2} °C^{-1}] and a_2 [W m^{-2} °C^{-2}] obtained according to standardized tests and operating conditions EN 12975-2 (2006), replaced by ISO 9086 (2013):

$$\eta_{a|G_t} = \frac{Q_u}{G_t A_a} = F_R \overline{\tau\alpha} \frac{K_{\theta b}(\theta) G_{b,t} + K_{\theta d} G_{d,t}}{G_t} - a_1 \frac{(T_{col} - T_o)}{G_t} - a_2 \frac{(T_{col} - T_o)^2}{G_t} \tag{8.7}$$

where T_o [°C] is the outside air temperature; T_{col} [°C] is the average fluid temperature in the collector; and $K_{\theta b}$ [–] and $K_{\theta d}$ [–] are the incidence angle modifiers for direct $G_{b,t}$ [W m^{-2}] and diffuse $G_{d,t}$ [W m^{-2}] solar irradiation, respectively. Performance is expressed as a fixed coefficient associated with the collector's solar absorption characteristics η_o (optical performance); and two variable thermal loss coefficients, a_1 and a_2, which increase with decreasing isolation or increasing operating temperature values. For example, "low cost" collectors without insulation can achieve optimal operation at low temperatures, but they suffer a significant performance loss at higher temperatures (Fig. 8.7).

Conventional FPCs provide η_o values between 0.7 and 0.8, and a_1 values between 4 and 7 W m^{-2} °C^{-1}; the coefficient a_2 is not commonly used to characterize these types of collectors, so the performance equation fits a straight line. Evacuated tube collectors (ETCs) are collectors where the absorber, instead of occupying a rectangular cavity, is placed in a glass tube that has been subjected to a vacuum to improve insulation. It has worse optical performance, but presents lower a_1 values and therefore has a better energy performance at higher temperatures. The same applies to other types of collectors (Fig. 8.7a), such as parabolic trough collectors (PTCs). A thermal range with acceptable performance can usually be established, depending on the demand, of about <30°C for unglazed collectors, 40–80°C for conventional FPCs, 70–150°C for ETCs, 70–250°C for linear Fresnel reflectors (LFCs), and 70–300°C for PTCs (Table 8.4 and Fig. 8.7a).

The average cost of solar collectors is between 400 and 700 US$ m^{-2}, and the unglazed collectors may cost 90–95% less. In terms of thermal power, an estimation of 0.7 kW$_{th}$ m^{-2} can be assumed (IEA, 2012). Overall, systems primary investments are extremely case-dependent because of the costs of auxiliary elements such as storage, pumps, pipes and insulation, among others.

8.4.2.1   *Components and configurations of solar thermal energy systems for greenhouses*
Performance is the main limiting factor for the application of this technology in greenhouses. Heat demand occurs in winter, when available solar radiation is lower. This, together with the

(a)   (b)

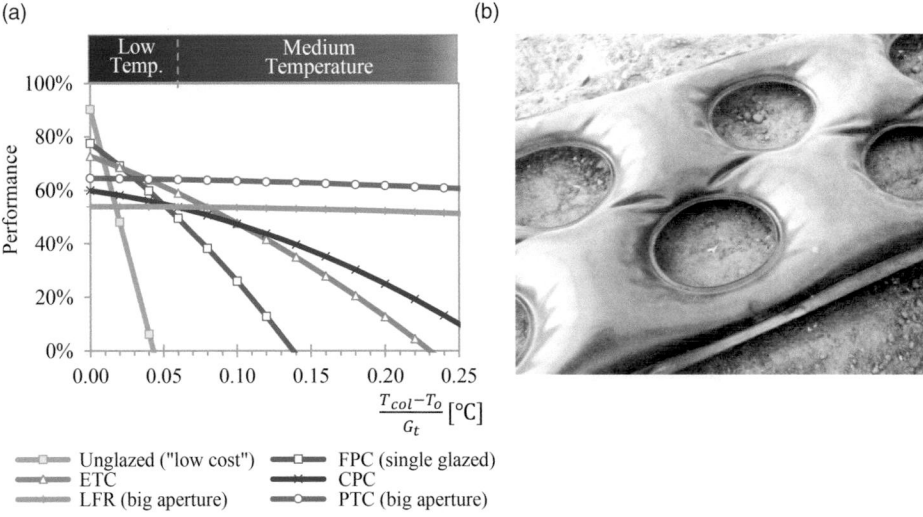

Figure 8.7.   (a) Performance of different solar collectors related to the global radiation in the aperture area, G_t (normal incidence, diffuse fraction $G_{d,o}/G_o = 0.15$; T_{col} is the average fluid temperature in the collector), and (b) sensible storage in black PE bags used as "low cost" solar collectors (picture from Baille (González-Real) *et al.*, 1977).

losses from the conversion process, means that the solar field area necessary for the generation of high solar fractions involves considerable land occupation.

This inconvenience can be mitigated as follows:

- Through the use of solar energy during the months of maximum availability when heat is not needed to generate cooling by absorption machines (Sethi and Sharma, 2007).
- Through integrated approaches that include centralization and thermal distribution networks, where other energy sources other than strictly solar (biomass, waste heat from auxiliary industries, etc.) can be implemented and storage mechanisms included as already done in certain urban environments (Bot *et al.*, 2005; Pinel *et al.*, 2011; Voulgaraki and Papadakis, 2008).
- By increasing the insulation value of the greenhouse cover to reduce thermal energy demand (Bot *et al.*, 2005).

Inside the greenhouses, alongside the solar collection methods mentioned, the following types of thermal storage (Sethi and Sharma, 2008; Sethi *et al.*, 2013) have been identified; all are geared to combating the daily thermal cycle caused by an excess of energy built up during daylight hours (Fig. 8.8):

- Bags or pipes.
- Rigid deposits inside the greenhouse.
- Conventional external storage tanks.
- Thermal storage in bedrock (pebbles, concrete, etc.).
- Thermal storage underground.
- Thermal storage with phase change materials (PCMs) (Boulard *et al.*, 1990).

Energy storage can be performed in the short-term (day/night) with a reasonable storage volume, $0.2\,\mathrm{m^3\,m^{-2}}$ of greenhouse soil area, or long-term storage from summer to winter (seasonal), where $V_{sto}/A_s = 12\text{–}18\,\mathrm{m^3\,m^{-2}}$ could be necessary (von Zabeltitz, 2011). Lazaar *et al.* (2015) elaborated a comprehensive review of experiences on thermal storage for greenhouses heating in a comparative study of the performance two heating sources that were intended to increase

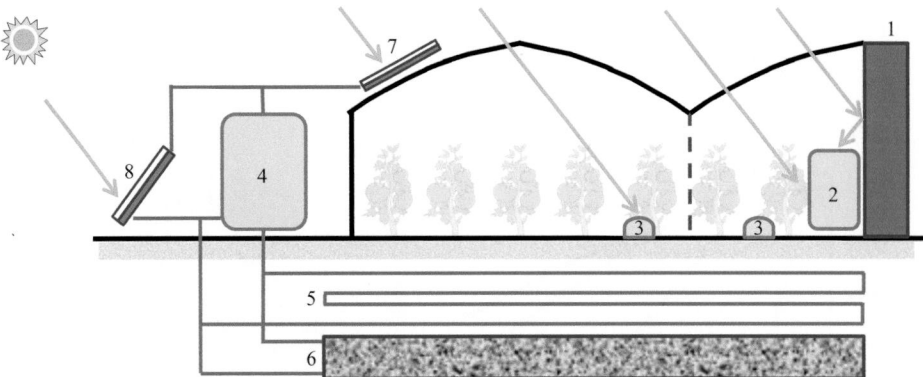

Figure 8.8.   System layout configurations for solar collectors and thermal storage in greenhouses. [1: thermal storage in north wall; 2: storage in inner tanks (water/phase change materials); 3: storage in bags or pipes; 4: conventional external storage tanks; 5: underground storage with water distribution; 6: storage in bedrock; 7: integrated solar collector; and 8: external solar collector].

the nocturnal internal air temperature of two 100 m^2 tunnel greenhouses in Tunisia: an electrical heating system and a solar ETC coupled to a storage water tank with a volume of 0.2 m^3.

The energy distribution for heating and/or cooling applications in greenhouses can be realized with heat exchangers, *water*-air or *air*-air.

8.4.2.2   *Experience in the use of solar thermal energy in greenhouses*

Several experiments with solar heating in greenhouses using black PE bags (Baille (González-Real) *et al.*, 1977) (Fig. 8.7b) and other solar collectors (Castilla, 1980, 1981; Solís and Castilla, 1980; Castilla *et al.*, 1985; Montero *et al.*, 1985) (Fig. 8.9), were carried out during the 1970s–1980s, leading to the use of transparent polyethylene tubes (250 microns) covering 30–40% of the ground in Almería greenhouses. The thermal water storage per unit of greenhouse area was 0.07–0.1 m^3 m^{-2}, the demonstrated solar collector efficiency was 30–50%, and the achieved minimum temperature increases were 3°C on average and 5°C on very favorable days. Despite being considered a good frost protection system, the resultant high soil occupancy inside the greenhouses meant that using these systems was limited, due to the difficulty in realizing the necessary growing practices. The limiting factor of high soil occupancy inside greenhouses from transparent polyethylene tubes can be avoided as shown by Ntinas *et al.* (2014), by placing the crop on top of the polyethylene sleeves, making all of the greenhouse ground available for cultivation and carrying out necessary growing practices. Another facility in Bari, southern Italy, has an integrated solar collector in the south-facing greenhouse wall surface (90° elevation angle), with a storage capacity per unit of collector area of 0.08 m^3 m^{-2} and 12 m^2 of total collector area. It supplies around 64% of the crop's heating requirements, with a collector performance of ~34% and a relative collector surface of 0.15 m^2 m^{-2} of greenhouse area; heating requirements of the greenhouse are fully covered ($T_{i,hot} = 14$°C and using a thermal screen) (Vox *et al.*, 2008, 2010).

A solar heating system with seasonal storage (0.61 m^3 m^{-2}) was installed in a greenhouse in Thessaloniki (Greece) with a solar collector area of 0.9 m^2 m^{-2}. The solar fraction obtained in the simulations was 0.4 with FPC (Voulgaraki and Papadakis, 2008) (heating temperature $T_{i,hot} = 12$°C, $\Phi_h = 0.705$ GJ m^{-2}; storage temperature up to $T_{sto} = 90$°C in the warmer months with lower demand and a minimum of 20°C). A solar fraction of 1 could be obtained for $A_{col}/A_s = 3$ m^2 m^{-2} and $V_{sto}/A_{col} = 0.4$ m^3 m^{-2}.

Calculations done in Almería for unglazed and FPC with a day/night storage size of $V_{sto}/A_{col} = 0.05$ m^3 m^{-2}, show that a solar fraction close to 1 can be obtained using a collector

Figure 8.9.   Experimentation with solar heating in Almería greenhouses using "low cost" solar collectors during the 1980s (source: Research Center Cajamar "Las Palmerillas").

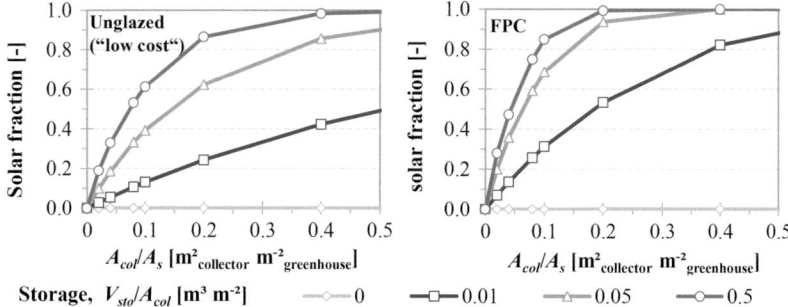

Figure 8.10.   Solar fraction for heating a greenhouse with different solar field areas, A_{col}/A_s, and day/night storage sizes, V_{sto}/A_{col}, for (a) unglazed ("low cost") and (b) flat plate collector (FPC). [Site located in Almería; tomato crop; heating temperature $T_{i,hot} = 12°C$; $\Phi_h = 0.17\,GJ\,m^{-2}$; generation, storage and distribution temperature, $T_{sto} = T_h = 30°C$; heat transfer coefficient of storage tank $U_{sto} = 0.5\,W\,m^{-2}\,°C^{-1}$; collector elevation angle = latitude, TRNSYS simulations].

area of $A_{col}/A_s = 0.2$ and $0.4\,m^2\,m^{-2}$, to cover the heating needs of a greenhouse with a minimum temperature of $T_{i,hot} = 12$ (Fig. 8.10) and 16°C, respectively.

Other research has been carried out on the use of solar collectors for cooling greenhouses by absorption devices (Vox *et al.*, 2014), the disinfection of hydroponic water (Tripanagnostopoulos and Rocamora, 2008) and greenhouse soil (Pérez, 2007) with FPC and PTC, desalination modules in roofs (Chaibi and Jilar, 2005), and the installation of LFR and PTC solar collectors to supply thermal desalination systems used for irrigation of crop greenhouses (Bundschuh and Hoinkis, 2012), e.g., The Sahara forest project.

8.4.3   *PV solar energy*

PV systems appear to be very well suited to provide electricity for stand-alone agricultural and livestock farms (Qoaider and Steinbrecht, 2010). In addition, due to its maturity and modular nature, PV technology is being considered as an option for direct on-grid renewable energy integration at the utility and consumer levels. Accordingly, it might be possible to consider the so-called greenhouse integrated PV Installation (GHIPV) as a facility for simultaneous production of renewable electricity and food, provided the blocking of PAR caused by the modules in indoor greenhouse areas is not significant for crop-growing cycles. This approach can be added to the more general concept of *agrovoltaic* farms (Dupraz *et al.*, 2011). This optimized mixed land use

Table 8.5.   Main characteristics of commercial PV technologies for greenhouses.

Type	Performance [%]	Weight, cells $[\mathrm{g\,m^{-2}}]$	Weight, modules $[\mathrm{kg\,m^{-2}}]$	Specific yield $[\mathrm{Wp\,m^{-2}}]$
Crystalline	14–22	1200–1500	10–20	120–150
Thin film	4–12	4–5	0.5–4	80–100

configuration is proposed for producing food and solar PV energy in a similar way to that used for agroforestry systems, which combine trees and food crops.

8.4.3.1   Components and configurations of PV systems for greenhouses

For descriptive purposes, a PV cell can be considered as an electrical generator with an output that is directly related to the amount of incident solar radiation on it. This amount of electricity is, to a lesser extent and depending on the cell type, a function of the cell temperature. This capability of solar cells comes from the layers of semiconductor material coated both with electron-donor and electron-acceptor atoms. These allow the absorbed light photons of adequate frequency to make the electrons rise to higher energy states and thus move from the solar cell into an external circuit. Because only certain frequencies are able to produce charge-carrying movement, depending on the semiconductor material band gap, the efficiency of the energy conversion is not great; a significant percentage of the solar spectrum has no effect on the process. For example, the so-called theoretical Shockley-Queisser limit of a single-junction cell with an energy band gap of crystalline silicon has 31% energy conversion efficiency. The cells' characteristics relate to their actual electrical performance, physical specifications (weight, color, rigidity, strength, durability, etc.) and costs acquired during their manufacturing process. High-purity silicon-based crystalline cells in wafer form are, so far, the most advanced technologically. In addition to this, amorphous thin film cells can be manufactured by evaporating silicon on polymeric, and other, substrates (Table 8.5). These types of cells are well suited to large-area roll-to-roll processing and, consequently, they are very suitable for use in plastic greenhouses. This cell manufacturing procedure requires very little active material, leading to production cost savings but, on the other hand, results in high defect densities and less efficiency than crystalline solar cells. New materials and principles must also be considered, such as those used in organic and dye-sensitized cells that can even result in translucent PV elements and a significant cost reduction. Although they are far from being available for mass application, due to their present stability constraints and active lifespan, the ongoing research for this technology is very promising; there have even been some pilot greenhouse application studies (Marucci et al., 2012).

Solar cells are arranged and connected to form larger units called modules that also include weather protection sheets and frames for enhanced mechanical strength and durability. The modules are connected to each other in series (arrays) to increase the total voltage produced by the system. The arrays are connected in parallel to increase the system current. The typical power generated by PV modules ranges from a few tens of watts up to 300–350 W, depending on the module size and technology. The module's performance is defined as the ratio between the corresponding electrical power delivered and the power of the solar radiation contributed as irradiance. For standardization purposes, PV modules are rated on the basis of the power delivered under the so-called standard test conditions (STCs): 1 kW m^{-2} of irradiance, 25°C PV cell temperature and the spectrum corresponding to the air mass (AM 1.5), the subscript p (peak) being used in this case to fix these conditions into corresponding specifications.

PV systems also include (Fig. 8.11):

- Batteries and charge controllers (for stand-alone applications), consisting of a set of lead-acid electrochemical accumulators connected to the solar generator and the load by an electronic controller, the aim of which is to protect the battery from overcharging or discharging. Although

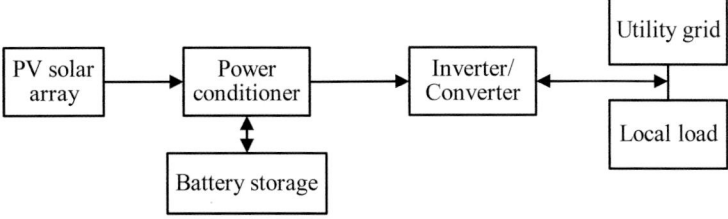

Figure 8.11.   Diagram of a typical PV system.

lead-acid accumulators are well established, new developments specifically designed for renewable systems and accounting for environmental protection, life cycle, costs and energy range requirements are also becoming available (Poullikkas, 2013). The sizing of the storage systems must be determined by the trade-off between the daily electricity demand, the system's generation capability (i.e., the installed PV power) and the probability for a certain, expected loss of load (Lucio *et al.*, 2012).

- Inverters, to convert the direct current (DC) power generated by the PV system and that extracted from the batteries to alternating current (AC) power. Inverters allow systems to be compatible with the conventional electricity distribution networks and to feed the most common electrical appliances. Inverter power ranges from a few hundred watts (normally for stand-alone systems), to several kilowatts (the most frequently used range, and the most suitable for greenhouse installations), and even up to a hundred kilowatts for large-scale utility systems.

The investment costs of PV systems are still high, although they are in an evident decreasing trend, as a result of cells manufacturing technology improvements and, especially, the existence of an emerging market of large volume and scale, especially at the utility level. Costs of PV installations are also related to their nature (stand-alone or grid connected), because of the weight of batteries on initial investment. According to the International Energy Agency (IEA, 2010), it is expected that turnkey system prices and electricity generation costs will drop more than two-thirds by 2030, from current figures between US$ 4000 and US$ 6000 per kW.

Regarding the possible structural arrangement of the PV modules over a greenhouse roof, two possible options have been identified in the literature:

- Cladding-integrated opaque modules (Fig. 8.12). In this case, seasonal and daily variability of shading effects must be carefully analyzed to avoid the generation of permanently shadowed areas and microclimate changes (Fatnassi *et al.*, 2015) inside the greenhouse. According to Yano *et al.* (2010), a checkerboard arrangement reduces this effect. In addition to this, the modules can be placed internally, under the greenhouse cover (as in the above-mentioned work) or externally, fixed to the main supporting greenhouse lines (Pérez-Alonso *et al.*, 2012). In both cases, light, flexible thin film modules were used.
- Modules arranged along the focal lines of solar concentrators conforming to the cover structure of the greenhouses (Fig. 8.13), both in cylindrical and in Fresnel-lens geometries (Sonneveld *et al.*, 2010, 2011; Wenger and Teitel, 2012). This configuration has the advantage of using concentrated radiation, so the electricity generation per unit area used is higher; therefore, the number of elements causing radiation blockage are reduced. The main inconvenience is that the required optics to produce the concentration effect involve a critical intervention on the greenhouse structure. This configuration allows for simultaneous thermal use because the heat dissipated by the PV modules under higher radiation fluxes can be collected by water pipes, which improves their performance and allows an energy surplus in the form of heat.

In addition to this, some configurations are based on conventional cell technologies used in buildings, which can be applied to glass greenhouses (Cerón *et al.*, 2013). These designs

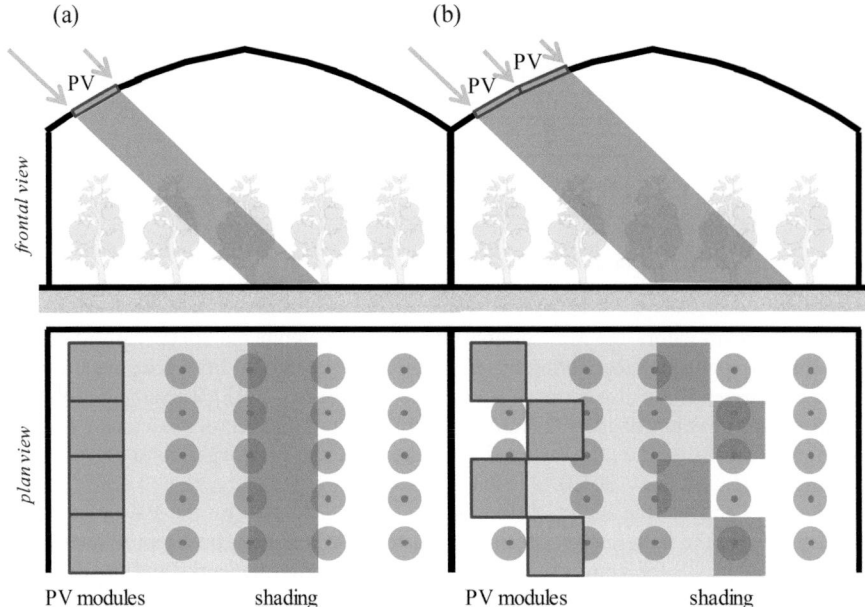

Figure 8.12.   A greenhouse with PV modules integrated into the cover with (a) linear and (b) non-linear (quincuncial) alignments.

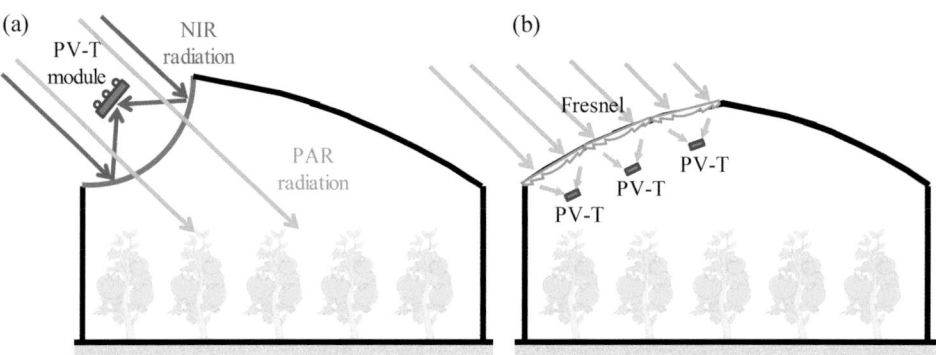

Figure 8.13.   A greenhouse with PV concentrators in the cover with (a) cylindrical and (b) Fresnel lenses.

are made up of modules with a reduced number of opaque cells built between two transparent covers, normally glass/glass (the arrangement and separation of the cells permits varying levels of transparency), and continuous and semi-transparent modules formed by thin film cells textured by laser, producing a high-density mesh of pointy or linear openings laid over the cell allowing light to pass through. Also, in aiming to obtain solar PV semi-transparent materials intended for greenhouse roof applications, Yano *et al.* (2014) have studied the use of 1.8 mm diameter crystalline silicon spherical solar microcells embedded in transparent modules with different levels of densities; the higher the density of cells, the higher the production of electricity and the lower the transparency. The estimations of the annual electrical energy production per unit greenhouse land area have shown a reliable energy performance for greenhouses in high-irradiation regions.

Finally, it must also be noted that this general approach, aimed at providing greenhouse cladding with additional energy capabilities, includes more advanced concepts, such as fluorescent solar concentrators or photoselective materials for different purposes. Lamnatou and

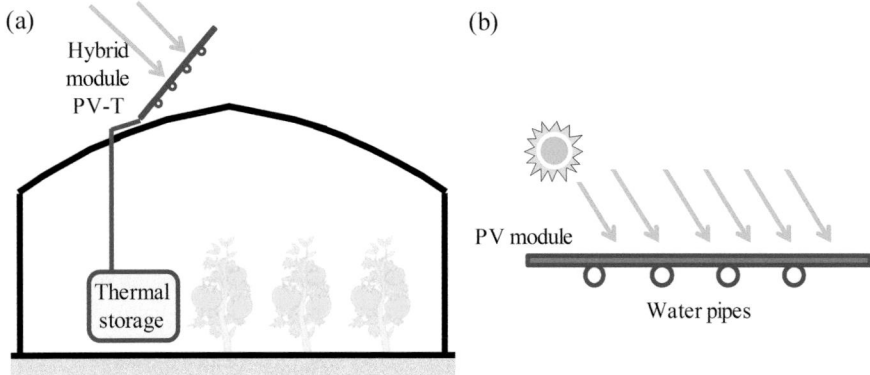

Figure 8.14. (a) A greenhouse with PV-T hybrid modules in the cover; and (b) the detail of a hybrid module.

Chemisana (2013a, 2013b) carried out a comprehensive overview of all these current options, including, among others, some of the above-mentioned, such as Fresnel and near infrared (NIR) concentration covers. Further new developments are favoring new light and transparent PV materials, appropriate for use in greenhouses (Emmott *et al.*, 2015; Zhao *et al.*, 2014).

8.4.3.2 *Experience in the use of PV energy in greenhouses*

A literature review shows many examples of greenhouses equipped with PV modules to feed ventilation systems (Davies *et al.*, 2008; Janjai *et al.*, 2009; Nayak and Tiwari, 2008; Yano *et al.*, 2007) or in combined systems including thermal collectors and geothermal heat pumps (Nayak and Tiwari, 2009; Ozgener and Hepbasli, 2007). Al-Ibrahim *et al.* (2006) and Al-Helal *et al.* (2006), together with a comprehensive review of previous PV experiments in greenhouses, demonstrated the performance of a pilot stand-alone installation situated in a hot climate location in Saudi Arabia. It feeds, among other electrical loads, a fan and pad evaporative cooling system with an installed PV power of 14.72 kWp and a 3000 A battery storage system for a 9 m × 39 m fiberglass greenhouse. Considering electricity storage, alternatives can be found to conventional electrochemical batteries, such as the use of electrolyzers combined with fuel-cells (Ganguly *et al.*, 2010).

For the so-called greenhouse integrated PV installations (GHIPV), some proposals and research also exist. This approach is the most suitable for zones with a high greenhouse concentration, because of the lack of free space. Rocamora and Tripanagnostopoulos (2006) proposed an integrated scheme based on hybrid thermal-PV modules (Fig. 8.14). Structural integration is performed in the ventilation areas and a thermal storage tank can be placed inside the greenhouse. As a case study, a multi-tunnel greenhouse (22,500 m²) is considered in the Murcia region (Spain) with a yearly electricity consumption of 150,000 kWh to feed the fans, irrigation and fertilization equipment, the compressor for the pesticides application, motors for the screens, and climate control actuators. In total, 10% of the south cover is occupied, meaning a 5% occupation of the total cover area. Furthermore, an interior shading percentage of 8% was estimated on clear days, as an effect of the solar geometry. By applying the typical climatic conditions for this location and considering the cells' performance of 10%, one can determine the electricity production under these conditions exceeding 12% of the greenhouse's own consumption.

As mentioned before, Yano *et al.* (2010) studied the spatial distribution of radiation inside an east–west oriented greenhouse, with a curved roof having 12.9% of its area covered by PV modules in the city of Matsue in Japan (35°30′N, 133°E). In this work, the electricity generation from two configurations of 30 flexible modules (0.90 m² × 0.46 m²) located inside the greenhouse was assessed for both a continuous and a checkerboard alignment. The results showed more uniform interior radiation distribution with the quincuncial alignment (Fig. 8.12) and an estimated yearly PV production of approximately $8\,kWh\,m^{-2}$, with no relevant yield differences between the two.

Sonneveld *et al.* (2010) designed a new type of greenhouse (Fig. 8.13a) which uses a curved cover as a reflecting surface to direct solar radiation in the NIR towards a hybrid module of the type previously described. In addition to the generation of heat and electricity, this design decreased the shaded area because of the concentration, in the order of $\times 30$; this produced higher energy fluxes over the module and therefore made higher occupation unnecessary. The yearly electricity generation of this greenhouse, normalized to a greenhouse surface, is in the order of $20\,kWh\,m^{-2}$ with a thermal production of $576\,MJ\,m^{-2}$. In addition, Sonneveld *et al.* (2011), based on the concept of Souliotis *et al.* (2006), proposed the integration of Fresnel-type solar concentrator lenses in the greenhouse cover in such a way that hybrid thermal-PV modules occupied their focal line. According to their proposal, re-directing the radiation coming from the sun in summer decreases the sun contribution to the greenhouse interior by 75%, therefore decreasing the required cooling load by a factor of four. The concentration factor in this prototype is $\times 25$, with electricity generation of $29\,kWh\,m^{-2}$ and thermal production of $518\,MJ\,m^{-2}$, which, according to the authors, indicates that the system could provide the total energy demand for a well-insulated greenhouse in Northern Europe.

Pérez-Alonso *et al.* (2012) tested a $1024\,m^2$ plastic greenhouse in Almería ($36°52'N$, $2°17'W$), in which three zones were demarcated, aiming to study the influence on the crop yield of tomato, of different shading levels provoked by PV external modules installed in the greenhouse cover. One zone was shadow-free for control purposes and the other two were covered with 12 flexible thin film modules over a $3.399 \times 0.461\,m^2$ surface. Each active cover had a 9.79% roof occupation ratio, the first consisted of 12 individual modules uniformly distributed in a quincuncial alignment and the second of 6 module pairs. The results obtained indicated that, for the conditions undertaken in the experiment, the yearly electricity production, normalized to the greenhouse ground surface, was $8.25\,kWh\,m^{-2}$.

Cossu *et al.* (2014) assessed the climate conditions, temperature and solar radiation, inside a $960\,m^2$ east–west pitched-roof greenhouse in which the south-oriented roof of each span was replaced by multi-crystalline silicon PV modules to reach 50% roof coverage, with peak rated power of the PV system at $68\,kWp$. The prototype was placed at Decimomannu (Sardinia, Italy, $39.333°N$, $8.989°E$). The annual electricity production was $112\,kWh\,m^{-2}$ of greenhouse area. The higher modules roof occupation and the use of crystalline cells justified the higher electricity production compared with previous experiences. In addition to this, the electricity produced was used to feed a supplementary lighting system and an air-heating system, to evaluate its agronomic and economic convenience.

Regarding crop yield in GHIPV greenhouses, there are few published results[2] showing the effects on the crops, which in general terms are not extreme; however, the trade-off between investment and revenues is still not well established. Ureña-Sánchez *et al.* (2012), in analyzing the tomato cultivation of Pérez-Alonso *et al.* (2012), observed differences between the control and treatments with respect to mean fruit mass and maximum fruit diameter; however, since all the fruit produced fell into the same commercial class, the final production price was not affected. Kadowaki *et al.* (2012), on the basis of the systems tested in Yano *et al.* (2010), found that for Welsh Onion cultivation, the checkerboard arrangement diminished the inhibitory effects of PV-array shading, compared with a straight alignment that produced a 25% decrease in fresh weight. They also carried out an economic analysis, obtaining payback times greater than 20 years for the technology used. Costs and payback periods can be drastically reduced with lower product

---

[2]Considering those with roof occupation ratios of less than 0.15 or, in general terms, those with a relevant light transmitting function; unlike the existing commercial GHIPV that have higher opacity levels because of the large number of solar modules on the roof – present mainly for electricity production on a utility scale, as exists in some Southern European countries as a result of extremely favorable (but time limited) feed-in tariff schemes.

prices and better performance, which are feasible when using crystalline cells instead of flexible thin layer cells.

As an example of agronomic performance of a PV greenhouse with a high fraction of occupied roof area, the above-mentioned work of Cossu *et al.* (2014) observed that the inhomogeneous shading inside the greenhouse produced an 18% reduction in tomato harvesting on the plant rows farthest from the PV cover of the span. The supplementary lighting, powered without exceeding the energy produced by the PV-array, was not enough to affect the crop production, the revenue of which was lower than the cost for heating and lighting.

8.4.3.3 *Sizing of PV systems for greenhouse applications*

Starting from knowledge of the electricity demand on a daily basis, $E_{\mathrm{L}}^{\mathrm{d}}$ [kWh day^{-1}], there are different approaches to estimate specifications for elements in PV systems. A correct and balanced estimation might include a dynamic simulation, accounting for all the cell dependences and system parameters including storage, along with further optimization of the process, from both the cost and performance points of view. It is clear, for example, that for the stand-alone systems, the fulfillment of electrical demands could be reached by combining different values of module sizes and storage capability. Higher instantaneous generation could reduce the need for accumulation, while large batteries would allow a lesser number of PV modules in the installation. The trade-off between these two aspects for certain value loss of load probability (LLP), defined as the ratio between the energy deficit and the energy demands on the load, over a long period of time, determines the size of the elements. LLP is always greater than zero due to the random nature of the solar radiation. It has different values according to the installation use. Higher reliability and a low LLP value are required for installations where faults might result in critical load malfunctioning.

For a preliminary design, a simplified approach can be adopted. The daily energy produced by a solar module can be estimated as follows:

$$E_{\mathrm{PV}}^{\mathrm{d}} = A_{\mathrm{PV}}\eta^{*}G_{\mathrm{t}}^{\mathrm{d}} \tag{8.8}$$

where A_{PV} [m^2] is the module surface, η^{*} [–] is its averaged performance and $G_{\mathrm{t}}^{\mathrm{d}}$ [kWh m^{-2} day^{-1}] is the daily solar energy received on the plane of the module. In fact, accounting for STCs (1000 W m^{-2}), the above expression can be transformed into:

$$E_{\mathrm{PV}}^{\mathrm{d}} = P_{\mathrm{PV}} H_{\mathrm{P(s)}} \tag{8.9}$$

in which P_{PV} [kW] is the peak power of the module and $H_{\mathrm{P(s)}}$ [h day^{-1}] is the so-called daily sun hours peak – the value of irradiation expressed in hours in which peak conditions are reached. This approach has the advantage of providing a rapid interpretation of PV yield as the product of the power, in terms of hours, of PV electricity production at peak conditions. The PV system's final yield will be affected by different losses due to cabling, interconnections and converters, among others – all of them integrated in the performance ratio parameter, *PR* [–], which normally has values between 0.7 and 0.8 for grid connected systems and between 0.6 and 0.7 for stand-alone systems. Accordingly, matching between electricity demands and PV generation is obtained simply by:

$$E_{\mathrm{L}}^{\mathrm{d}} = PR\,P_{\mathrm{PV}}\,H_{\mathrm{P(s)}} \tag{8.10}$$

Consequently, knowing the electricity demand, the performance ratio and the value of daily solar radiation, it is possible to assess the P_{PV} value, which will determine the number of required modules to fulfill the electricity demand. As energy demand and solar radiation vary throughout the year, of all possible combinations, the size adopted is that of the so-called worst day, which corresponds to a day in the month with the least radiation according to the tilting angle, with the aim of guaranteeing the required production under those conditions. This approach overestimates the size of the generation elements for all-year-round performance, but is widely accepted as a first estimation during the early design stages.

Something similar occurs in the case of storage requirements, which can also be estimated at an early stage of the design process as follows:

$$C_B = (N_d E_L^d 1000)/(\Delta V_{DC} DOD \, \eta_B) \qquad (8.11)$$

where C_B [Ah] is the estimated charge of the battery; N_d [day] is an estimated number of days of fully autonomous system performance; DOD [–] is the maximum depth of battery discharge; η_B [–] is the performance of the battery regarding specific losses (self-discharge, temperature, etc.); and ΔV_{DC} [V] is the DC installation voltage.

As an example, and as mentioned before in the electricity demand section, it is estimated that the annual requirement for electrical energy for greenhouse ventilation in the Mediterranean area is between 7 and 10 kWh m^{-2} year^{-1} (Kittas *et al.*, 2012, 2013). In this case, if the installation were located on a greenhouse cover tilted at 15° (east–west-oriented) in Almería (Spain), a grid connected installation of 60 kWp thin film modules, occupying 10% of the greenhouse roof surface would produce the required electricity to move the fans throughout the year (Fig. 8.6). If the installation were located under stand-alone conditions, the size of the PV generator would rise to more than 200 kWp, which, together with a 24,000 Ah battery bank, would make this option very expensive. However, if considering electricity demand for water pumping (Fig. 8.6), an easily affordable installation of only 4 kWp would balance greenhouse electricity consumption on a yearly basis, whereas a 6 kWp installation and a 1000 Ah battery could cover electricity demand in a secure way, if the greenhouse were in a remote area.

Those gross figures can be further refined through detailed calculations taking into account the specific pumping demands (specific head pressure) and, especially, the irrigation strategies used by the irrigation sectors and programmed irrigation campaigns. These detailed estimations allow more feasible configurations, as demonstrated by Reca *et al.* (2016), who assessed the reliability of a PV pumping installation of 1.6 kWp to irrigate 1 ha of a Mediterranean greenhouse with only one irrigation sector; they found it possible to reduce the maximum power value by the optimization of the irrigation sectors number.

The sizing methods of PV systems in greenhouses must not only deal with final electricity production. As demonstrated by previously mentioned existing experiences, other aspects are also very important, such as shading levels, which influence plant growing processes by reducing photosynthetic rates. Sizing of PV systems must account for the trade-off between energy demands and costs, and crop production rates and their revenues.

8.4.4   *Geothermal energy*

Geothermal energy is growing in interest as a result of the significant consideration given to promoting the use of renewable energy sources under Directive 2009/28/EC of the European Parliament and of the Council on 23 April 2009. Although geothermal energy has great potential for greenhouse application, it is only possible at sites with suitable geothermal resources that can be developed (Adaro *et al.*, 1999; Bakos *et al.*, 1999; Popovski, 1988). Their effective utilization in greenhouses is no different, from the integration point of view, compared with any other thermal, renewable source, but others do not have the restrictions in resource availability.

To directly use geothermal water from the subsoil, the following should be considered: if the water is available at a shallow depth in the subsoil; if the water temperature is suitable; and if the water is not too corrosive, it can be used both for heating and for irrigation (Vasilevska *et al.*, 2011; von Zabeltitz, 2011).

Geothermal outflows are compatible with a wide range of greenhouse heating systems, especially with a water temperature in the 40–60°C range (low temperature). Different approaches and schemes use geothermal energy in greenhouses: using a geothermal well as an energy source; purchasing geothermal heat from an Energy Service Company (ESCO); using a GSHP; or using systems providing both space heating and cooling with underground heat exchangers to extract heat from (usually) shallow depths (Campiotti *et al.*, 2009).

Table 8.6. Distribution of biomass consumption according to origin and applications in 2006, as stated in the PER 2011–2020 (IDAE, 2011).

Resources	Electrical [ktep]	Thermal [ktep]
Workshop wood	0	950
Pruning wood	0	250
Agricultural wood	0	400
Cereal straw	80	20
Black liquor	600	0
Sawdust and chips	0	450
Peel	170	380
Olive pomace	250	450
Others	0	200
Total	1100	3100

The latter scheme type, also called enhanced geothermal system (EGS), allows the exploitation of geothermal energy in locations other than hydrothermal reservoirs, such as using the subsoil for thermal storage or as a thermal source for electric heat pumps coupled to the ground (for heating and cooling), either directly or by way of water circulation wells. This type of application takes advantage of the thermal stability of underground layers (depending on the composition of the ground, a few meters down, the temperature is constant and the external annual thermal wave is cancelled) and the use of efficient systems, such as heat pumps for heat generation, which offer good results, reducing consumption from air-conditioning (Milenić *et al.*, 2010; Ozgener and Hepbasli, 2005).

A rough estimate for the climatic conditions in Almería, with an average outside nocturnal temperature of 11°C during the winter months, demonstrates that using the ground as a thermal reservoir could produce savings in electricity consumption of about 25% for heat pumps operating at 45°C (based on Carnot-cycle operation and a ground temperature equal to the location's mean annual temperature).

For use of geothermal heat in greenhouses, investment costs range from US$ 500 to US$ 1000 per kW$_{th}$ (IEA, 2011; IPCC, 2011).

8.4.5 *Biomass*

The use of biomass as an energy source has undergone significant developments over recent years due to the various initiatives implemented by organizations and institutions operating renewable energy policy. The 2011–2020 Spanish Renewable Energy Plan (PER) (IDAE, 2011) highlighted biomass energy use as one of its priorities; it included an initial assessment in which the incorporation of biomass linked to intensive agriculture would be expected (Table 8.6).

Biomass-based energy has a carbon-neutral cycle and, therefore, does not contribute to the atmospheric greenhouse effect – this is its main advantage over fossil fuels (Werther *et al.*, 2000). Other advantages include less particle emissions and gaseous pollutants, such as carbon monoxide, hydrocarbons, and nitrogen oxides (IDAE, 2007). Biomass use would alleviate an important environmental problem and could help to control nocturnal greenhouse temperatures during cold periods.

In the case of greenhouses, biomass consumption can achieve negative carbon dioxide (CO_2) balances by re-using flue gases from combustion emitted from crop-enrichment processes.

8.4.5.1 *Biomass preparation*
In general, the use of vegetable waste as an energy source presents a series of practical problems concerning low density, high transport and storage costs, and high humidity content, promoting biological degradation and fermentation processes that are difficult to control. The thermal

Figure 8.15.   Different types of vegetable waste used in biomass boilers; from left to right and top to bottom: almond shells, crushed olive stones, greenhouse tomato vegetable residues (*agri-pellets*) and pine waste pellets.

exploitation of the vegetable residues requires good knowledge of their characteristics (ash and moisture content, calorific power, etc.) and the processing methods necessary for their use in biomass boilers. A densification process is usually employed, during which the vegetable mass is exposed to high pressures and turned into elements of greater volumetric density; this makes it easier to store and transport, as well as more homogeneous, cleaner and far more manageable (Ortíz *et al.*, 2003; Werther *et al.*, 2000). The most commonly used densification techniques are pelleting and briquetting.

Briquettes and pellets have similar calorific values, humidity and chemical properties. The main difference is their size and density. Pellets are cylindrical, high-density elements with a diameter of between 6 and 8 mm and lengths of between 6 and 12 mm. Briquettes are usually cylindrical as well, but are larger than pellets (with diameters of between 5 and 10 cm and lengths of between 15 and 50 cm) and of lower density. Pellets have certain advantages over briquettes: they are smaller and twice as dense, it is easier to automate their movement, and they occupy less space in transportation and storage.

Currently the most commonly used treated biomass products are pellets made with forest waste (sawdust, wood pruning, etc.). However, the increased demand for pellets and briquettes has resulted in countries such as Denmark and Norway reaching their production limit. Coupled to this is the importance of agriculture in some countries, which has led many researchers to study the potential of agricultural waste being used to manufacture these high-density elements, mainly pellets (*agri-pellets*, Fig. 8.15).

The use of agricultural waste presents certain disadvantages, given that it is very diverse in origin: cotton thistle stems (Abasaeed, 1992), wheat residues mixed with paper (Demirbaş and Şahin, 1998), tea residues (Demirbaş, 1999), olive residues (Yaman *et al.*, 2000), cardoon (*artichoke thistle*) (Alonso, 2004), esparto grass (Debdoubi *et al.*, 2005), sugar cane (Erlich *et al.*, 2005) and rice husks (Maiti *et al.*, 2006) are examples. The generation of waste is not uniform throughout the year, reaching high levels at the end of the various agricultural seasons. High dispersion means more expensive transport and there may be pollution from toxic products following phytosanitary treatments performed during the crop's development (Callejón-Ferre *et al.*, 2011) – elevated levels of up to 70 toxic substances in vegetable waste from green beans, watermelon and melon crops have been registered (Garrido-Frenich *et al.*, 2003). There might also be a high resulting ash content (14–45%) (Fernández *et al.*, 2013) or a significant moisture content (50–90%).

Table 8.7. Calorific biomass power (adapted from Callejón-Ferre *et al.*, 2011; Valera *et al.*, 2008).

Kind of biomass	$LCV^{1)}$ [MJ kg^{-1}]
Wood pellets	>16.75
Olive stones	>15.91
Almond shells	>15.49
Grape pomace	>15.91
Woodchips	~6.70–13.82
Greenhouse waste:	14.24–20.10
Zucchini plant (*Cucurbita pepo* L.)	14.95
Cucumber plant (*Cucumis sativus* L.)	14.65
Eggplant (*Solanum melongena* L.)	19.22
Tomato plant (*Solanum lycopersicum* L.)	17.25
Bean plant (*Phaseolus vulgaris* L.)	19.80
Pepper plant (*Capsicum annuum* L.)	17.75
Watermelon plant (*Citrullus vulgaris* Schrad.)	16.58
Melon plant (*Cucumis melo* L.)	15.70

[1]The lower calorific value is the amount of energy that the unit mass of matter can detach to produce a chemical oxidation reaction.

In addition, their extraction and treatment is an important outlay for the farmer, which takes place at different stages, starting with the plant cutting (not including the root), drying in the greenhouse for 3–4 days, and then dropping to the ground. Finally, the waste material has to be gathered for transportation to the storage site. During the storage period, the waste material has to be periodically turned to facilitate drying and prevent anaerobic fermentation. Extraction techniques and natural drying can improve the waste's properties, such as reducing the ash content and thus increasing its calorific value (Fernández *et al.*, 2013). In general, various studies have indicated a higher crumbling tendency and slightly lower calorific power with agri-pellets compared with wood pellets (Table 8.7).

In addition to the above, agri-pellet combustion compared with that of wood pellets produces higher emissions, greater waste generation and more substantial corrosion problems, because agri-pellets have a higher nitrogen, sulfur, chlorine and potassium contents than forest waste. However, these problems can be resolved using gas-filtering technologies and adequate boilers (Pastre, 2002).

8.4.5.2 *Biomass heaters*

There are a wide variety of biomass boilers currently on the market; each is designed for one or several types of biomass. It is important not to use another biomass with different conditions as the boiler's energy efficiency will be lower and problems caused by ash might appear. Additionally, it is important to choose the boiler type according to the heating power required from the system, based on the environment and the desired thermal conditions.

The heating power calculation must be made taking into account the different heat distribution possibilities throughout the greenhouse to optimize the energy use. In this sense, the energy requirements depend on the type of structure. Heat dissipation for each type of structure is different and will depend on the thermal jump desired – where thermal jump is the temperature difference in the greenhouse with or without heating. The system energy balance is the heat dissipated and the resultant heat that must be provided by the boiler (the heat demand).

Once the heat demand is defined, it is necessary to know the power transferred by the fuel to the fluid transporter, the useful power P_h [kW]. This depends on the daily heat demand, Φ_h [kWh m^{-2}], the heating surface, A_s [m^2], the boiler performance (given by the manufacturer,

η_h [–]), and the daily time use N_h [h]:

$$P_h = \Phi_h A_s / (N_h \eta_h) \tag{8.12}$$

During combustion, the boilers are going to generate gases. These gases have a high CO_2 concentration and a low waste content, making it easier to use for the CO_2 enrichment processes of greenhouse crops. Therefore, a biomass boiler should be able to generate the necessary heat for heating as well as sufficient CO_2 for enrichment in greenhouses. This is especially important when photosynthetic rates are higher, because the CO_2 concentration falls below the atmospheric concentration, producing a deficit. It intensifies when the crop nears maximum development (Sánchez-Guerrero *et al.*, 2005). Assays performed in Mediterranean greenhouses indicate that the CO_2 concentration decreases by up to 20% compared with atmospheric CO_2 concentration (Lorenzo *et al.*, 1990), even when the windows are opened. As a means of addressing this deficit, certain authors propose various carbon enrichment techniques to achieve increased crop production (Edwards *et al.*, 2008; Hao *et al.*, 2008). Conversely, many authors assert that the photosynthesis rate is enhanced by increasing CO_2 concentration above the atmospheric value (0.0350–0.0380% volume fraction) up to 0.1%. Above this level, the photosynthesis rate cannot be surpassed by increasing the CO_2 concentration (Edwards *et al.*, 2008; Hao *et al.*, 2008; Nederhoff, 1990; Rodríguez, 2002). This rate increase translates into a significant rise in production in accordance with that shown in Bailey (2002), Kläring *et al.* (2007), Linker *et al.* (1999), Nederhoff (1990), Portree (1996), Sánchez-Guerrero *et al.* (2005), Sánchez-Guerrero *et al.* (2009), Schmidt *et al.* (2008), Tremblay and Gosselin (1998), and Zabri and Burrage (1997). It is also essential to optimize carbon fertilization strategy responses, performing them in the absence of other limiting factors – the best results are obtained when radiation levels are high; that is, the photosynthetic rate is higher when temperature and CO_2 concentrations rise along with an increase in available solar radiation (Aikman, 1996; Schmidt *et al.*, 2008). Carbon enrichment not only affects the photosynthetic rate, but also increases water consumption efficiency by gradually reducing the stomatal opening rate, decreasing water loss caused by transpiration (Sánchez-Guerrero *et al.*, 2009).

To achieve the desired goals, the system can be operated in two alternate ways:

- Burning at night – the heat would be supplied directly to the greenhouse when the crop requires it. Moreover, CO_2 is captured and stored to supply the crops during the day (Sánchez-Molina *et al.*, 2014). The biomass combustion provides heat (about 17 MJ kg^{-1}) (Callejón-Ferre *et al.*, 2011), which is transferred to the water, and CO_2, which is generated from the combustion flue gases. The system is responsible for gas purification through CO_2 capture, releasing the rest into the atmosphere. The use of the capture system allows for a reduction in the gas volume stored under pressure, obtaining more concentrated CO_2 gas.
- Burning during daylight hours, CO_2 is generated when there is most demand. In this case, the heated water would be stored for use at night.

8.5   ENERGY SAVINGS FROM CROP GROWTH CONTROL IN GREENHOUSES USING OPTIMAL CONTROLLERS

Apart from using optimal control algorithms as a method to obtain energy savings in greenhouses, these algorithms are also used for the efficient management of the energy-consuming systems, such as the climate and irrigation actuators. This has been made possible by the introduction of computers. However, in most cases, computers are used to operate actuators; control strategies are primarily empirical and reflect classical methods used by farmers to manage the actuators (Bakker and Challa, 1995).

In the most advanced control systems, control is based on mathematical models, as is normal in the process industry (van Straten *et al.*, 2010). From an objective function, with the use of climate, fertigation and growing models, along with optimization techniques, the optimal trajectories are determined to follow the set points during the crop cycle (Ioslovich and Seginer, 2002; López, 2003; Rodríguez, 2002; Tap, 2000; van Henten, 1994; van Henten *et al.*, 2006; van Straten *et al.*, 2010). There are a few published works with experimental results in which the optimal trajectories for climatic variables have been obtained (Ioslovich and Seginer, 2002; Rodríguez, 2002; Tap, 2000; van Henten, 1994; van Straten *et al.*, 2010). In research works in greenhouses, important savings in energy costs have been achieved by applying these production techniques (Tap, 2000). Although these preliminary experiments have obtained acceptable results, the economic and energy-related effects are difficult to assess with so few cases, and their validity can only be measured by applying these techniques in real commercial greenhouses (van Straten *et al.*, 2010).

The greenhouse production agro-system has commonly been approached using a hierarchical control architecture (Challa and van Straten, 1993; Rodríguez *et al.*, 2003, 2008; Tantau, 1993), where the system is supposed to be divided into different time scales and the control system is divided into different layers to achieve optimal crop growth. As a first approach, the production control system could be split into the following levels (Rodríguez *et al.*, 2003, 2008; Tantau, 1993):

- Top layer: Market (time scale of weeks/months) – taking into account the long-term objectives (market prices, harvesting dates, and required quality) and the long-term predictions of the growth using a growth model (for the estimation of yield and profits) and the energy consumption models; the optimization is performed to calculate the set-point trajectories of the crop growth variables along the considered control horizon.
- Intermediate layer: Crop (with a time scale of days) – based on instructions received from the top layer, the system determines the paths that the water and nutrient supply should follow and the climatic variables involved in crop growth.
- Bottom layer: Greenhouse (time scale of minutes) – this includes the controllers that try to cancel set-point tracking errors (these set points are calculated by the upper layer), minimizing the energy cost.

However, most research on optimal greenhouse control considers only one optimization problem objective, mainly focused on increasing the grower's profit (Challa and van Straten, 1993; Rodríguez *et al.*, 2003, 2008; Tantau, 1993), by simply minimizing the CO_2 supply (van Henten and Bontsema, 2009) or including the thermal integral in the control decision maker (Körner and Van Straten, 2008). Other contributions have been published on optimal climate control for greenhouse crops where a scheme for seasonal optimal control based on the Hamilton-Jacobi-Bellman formalism was produced, showing promising simulation results (Ioslovich *et al.*, 2009).

Even if control methods have focused only on a single objective solution, many issues in nature need the optimization of several objectives at the same time, often in conflict with one another (Coello, 2003). Even though maximizing the benefits criteria have dominated the greenhouse production system (Rodríguez, 2002; Tap, 2000; van Henten, 1994), other imperatives are emerging related to a need for quality (CAJAMAR, 2004; van Uffelen *et al.*, 2000), a reduction of pollutants into the environment (Siddiqi *et al.*, 1998; Stanghellini *et al.*, 2003), and energy savings or water-use efficiency, mainly in regions where water is a limited resource (de Pascale and Maggio, 2005; Stanghellini *et al.*, 2003).

The different greenhouse elements and the multiple relationships that exist make it a complex system in which energy, mass, and information flows have different dynamics and magnitudes. The crop is the centerpiece, about which variables such as weather (temperature, atmospheric humidity, CO_2 and PAR), nutrition (water and nutrients), biotics (pests, diseases, virus, bacteria and weeds) and cultural management (pruning) revolve. These variables establish interactions between themselves, meaning the layered complexity has to be studied in depth and identified within subsystems. The problem often has no optimal solution, which simultaneously optimizes all elements, but it has a set of suboptimal or non-dominated alternative solutions known as a Pareto optimal set (Liu *et al.*, 2003); here, a compromise solution may be selected from that set by a decision process. Different criteria, such as physical yield, crop quality, product quality,

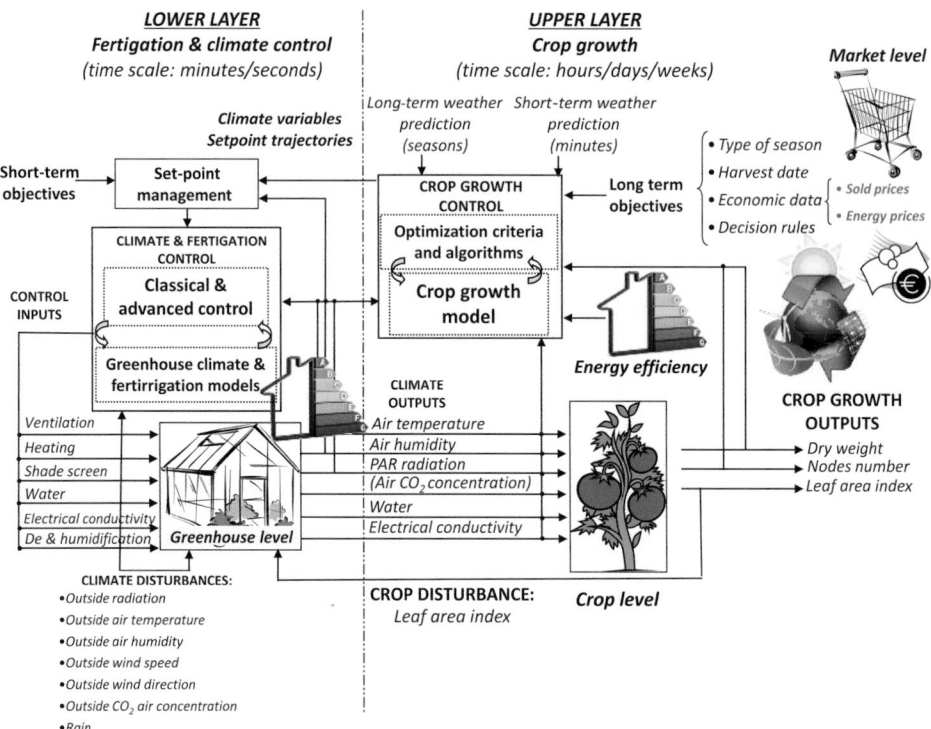

Figure 8.16.   Multilayer hierarchical system proposed to control crop growth.

production process timing, production costs and risks, energy consumption or prioritizing the use of renewable energies, can be formulated within greenhouse crop management. These criteria will often give rise to controversial climate and fertigation requirements, which have to be solved explicitly or implicitly at the so-called tactical level, where the grower has to make decisions about several conflicting objectives.

In this approach, a multi-objective, hierarchical control architecture has been developed by the Automatic Control, Robotic and Mechatronic Research Group at the University of Almería. It is described in Figure 8.16, and, as can be observed, is composed of two layers, the upper layer and the lower layer.

The upper layer level solves an optimization problem as a function of the expected production and associate costs or the desired date of harvest, including the energy costs. This optimization problem maximizes an objective function that represents the profit obtained, based on the climatic variables that affect the plant's growth, providing the set points that must follow these climatic variables throughout the season. Based on determined criteria, the climatic conditions calculated are: temperature, humidity and CO_2 concentration, along with irrigation conditions (water supply and electrical conductivity (EC) directly to the lower layer (the controller) that must achieve it). In a preliminary phase, the following objectives have been selected (Ramírez-Arias *et al.*, 2012): economic benefits, energy efficiency (decreasing energy consumption for the production of a unit of agricultural product), the fruit quality and the use of water efficiency, related to CO_2 control inside the greenhouse to achieve concentrations between that required by the crop and that which is harmful to operators who have to work in the greenhouse interior. An important phase is the characterization and quantification of these objectives; for example, considering the energy costs (fuel and electricity), the water and fertilizer costs. The quality of the fruit, including elements such as flavor (characterized by the amount of sugar and acidity), firmness and an organoleptic appraisal of the fruit:

$$J_1 = Y_{\text{inc}}(t_2)\eta_{\text{FFT}}(t_2) - \int_{t_1}^{t_2} Y_{\text{cos}}(t)dt \tag{8.13}$$

$$J_2 = \int_{t_1}^{t_2} (w_{\text{ssol}}Y_{\text{ssol},n} + w_{\text{av}}Y_{\text{av},n} + w_{\text{fi}}Y_{\text{fi},n} + w_{\text{tf}}Y_{\text{tf},n}) \tag{8.14}$$

$$J_3 = \eta_{\text{FFT}}(t_2)\Big/ \int_{t_1}^{t_2} IWR(t)dt \tag{8.15}$$

In this initial approach, the above objectives can be raised in the first instance, using the following objective functions: where J_1 is the objective function to maximize benefits, J_2 is the objective function of fruit quality maximization, and J_3 is the maximization of water-use efficiency. Y_{inc} is the income from selling the product, Y_{cos} is the crop production costs (energy, water and fertilizers), η_{FFT} is the fresh fruit performance, t_1 is the initial time, t_2 is the final time, $Y_{\text{ssol},n}$ is the sugar content in the fruit, $Y_{\text{av},n}$ is the degree of appreciable acidity in the fruit, $Y_{\text{fi},n}$ is the firmness of the fruit, $Y_{\text{tf},n}$ is the size of the fruit, w_{ssol}, w_{av}, w_{fi} and w_{tf} are weighting factors, and IWR is the water provided.

In principle, the factors involved in the indicated cost functions are a function of air temperature and EC, as well as disturbances, such as the PAR or the CO_2 concentration, assuming that the crop is well irrigated. The multi-objective optimization must provide optimal daylight and nighttime set points for EC in drainage water, and for temperature and CO_2 concentration for the rest of the season. Furthermore, the multi-objective problem cost function and the resolution method that suits it must be studied by analyzing techniques of simple weighting, weighted sums, ε-restrictions, goal settings, games theory and evolutionary algorithms (Coello, 2003; Laabidi and Bouani, 2004; Liu *et al.*, 2003), as well as considering aspects of uncertainty systems (Hu *et al.*, 2007). Alternatively, using a receding horizon strategy, when a night–day transition or vice versa occurs, the optimization problem is again solved by using new, real measured data for climatic variables and crop growth, trying to reduce errors coming from a plant-model mismatch, deviations in the weather forecast or the errors produced when climatic variables are not able to achieve the climatic set points, because of actuator disturbances or limitations.

Several simulation tests have been performed to study the response of the system under different conditions. All of them start with the same initial conditions:

- In southeastern Spain, heating is usually necessary in the autumn/winter seasons, so all the studies refer to these months (October/February), although the proposed algorithm is generic and applicable to any season.
- The initial crop state is the same for all the tests with 10.8 leaves, $60\,\text{g m}^{-2}$ of total dry matter and a density of 3 plants m^{-2} of soil surface.
- The crop cycle length is 90 days.

From the point of view of energy savings, the response of the system to changes in the energy prices is studied. As the energy for the heating systems is more expensive than the electricity, four cases are considered: the energy price is constant throughout the entire season; the fuel price diminishes in the middle of the season; the energy price increases in the middle of the season; and the energy is free (renewable). When the price is cheaper or free, the system tends to use the heating system more (Table 8.8). As can be observed, there is a significant difference between the final dry matter obtained in the different simulations.

The lower layer level in Figure 8.16 is responsible for achieving the set points of climatic and irrigation variables, which are calculated in the upper layer. There are plenty of set-point drivers (Ramírez-Arias, 2005; Rodríguez, 2002). This layer may also be used as an efficiency criterion, using optimal energy controllers. As an example, in Ramírez-Arias *et al.* (2005), the use of aero-thermal heating appears in the night temperature control, where a generalized predictive control (GPC) control approach has been used (Clarke *et al.*, 1987). With these kinds of controllers, the

Table 8.8.   Results based on fuel prices.

	Final dry matter [g m^{-2}]	Number of nights with heating
Constant prices	576.87	23
High prices at the beginning	580.76	24
High prices at the end	581.61	24
Renewable energy	650.45	36

aim is to optimize the following cost function:

$$J = EV \left[\sum_{j=N_1}^{N_2} \delta(j)[\hat{y}(t+j|t) - w(t+j)]^2 + \sum_{j=1}^{N_u} \beta(j)[\Delta u(t+j-1)]^2 \right] \qquad (8.16)$$

where EV is the mathematical expectation; $\hat{y}(t+j|t)$ is an optimal system output prediction sequence performed with data known up to instant t; $\Delta u(t+j-1)$ is a sequence of future control increments, obtained from cost function minimization; N_1 and N_2 are the minimum and maximum prediction horizons; N_u is the control horizon; and $\beta(j)$ and $\delta(j)$ are weighting sequences that penalize future tracking and control efforts, respectively, along the horizons (here δ equals 1 and β is a user-chosen constant). The reference trajectory, $w(t+j)$, can be the set point or a smooth approximation from the present output value $y(t)$ to the set point, usually implemented as a first-order filter. If no constraints are taken into account, as the model is linear and the optimization criterion is quadratic, an explicit solution can be found. Otherwise, a quadratic programming optimization algorithm is used.

Subsequently, there is an example of the introduction of advanced control algorithms compared to classic controllers: the use of different '*tuning knobs*' in the GPC algorithm and different sample times and dead-zones in the on/off control. The GPC controller produces more commutations and less consumption than the on/off controller; the number of commutations is within the ranges recommended by the supplier. This assay was tried out in a research greenhouse in Almería (southeastern Spain) during a typical night in which the heating system was on using the on/off control for 221 min (11.27 € cost), but only for 164 min with the GPC controller (8.36 € cost). There might also have been effects on the crop, but this was not evident given the relatively short timescale used here. In conclusion, significant economical savings can be achieved using energy efficiency criteria in the greenhouse controllers.

8.6   CONCLUSIONS

A survey of the renewable energy technologies applied in greenhouses has been done, emphasizing those applications related to the use of solar radiation and biomass resources for the fulfillment of the thermal and electrical loads in farms located in semi-arid regions. Together with the basic description of the energy conversion devices and systems, some simplified approaches have been presented for the estimation of greenhouse energy demands at these locations, with the aim of using them as reference for sizing the main elements of renewable installations.

In general terms, renewable energies are considered as very suitable for their integration in greenhouses located in regions with semi-arid climates, provided the corresponding systems do not compete for ground occupation or do not interfere with radiation catching by the crops. This is a result of the patterns of basic energy demands of greenhouses at these locations, usually low-cost structures, where large energy consumption may not be required, due to the existence of favorable outdoor disturbances, but some energy use can still optimize the performance in terms of crop production.

Of the existing renewable technologies, those related to the utilization of solar energy, both in the way of heat and/or electricity must be considered as the primary option. In these semi-arid regions, the high values of solar radiation and its overall yearly availability, make PV cells and solar thermal collectors well suited for fulfilling the energy demands of greenhouses.

Biomass is also considered a suitable option, provided enough supply is available. Two major advantages of biomass, with respect to solar technologies, are to be considered: its intensive nature, which reduces its impact on the farms in terms of land occupation, and, especially, the possibility of re-using the CO_2 produced during combustion for increasing plant production, through re-injection under controlled indoor climate conditions in the greenhouse.

For integrating the different technologies discussed, it is important to use optimal control algorithms as a method to obtain energy savings in greenhouses. Moreover, these algorithms are also used for an efficient management of the energy-consuming systems.

New developments as, for example, nanotechnology-produced light and transparent PV materials that can be used for greenhouse cladding and, especially, smart systems for the integration and management of the energy and mass exchange processes present in greenhouses, will contribute to increasing the penetration of the systems presented in this work.

ACKNOWLEDGMENTS

This work is part of Controlcrop projects P10-TEP-6174 and P10-RNM-5927, supported by the Andalusian Ministry of Economy, Innovation and Science (Andalusia, Spain) and the National Plan Project DPI2011-27818-C02-01 of the Spanish Ministry of Science and Innovation, as well as the European Regional Development Fund (EU-ERDF). This work was also supported by EU-ERDF funds and the project ENERPRO DPI2014-56364-C2-1-R "Control and energy management strategies in production environments with support of renewable energy", financed by the Spanish Ministry of Economy and Competitiveness. The authors are grateful for the invaluable contributions of the Cajamar Foundation Experimental Station at "Las Palmerillas".

NOMENCLATURE

AC   alternating current
CPC   compound parabolic collector
CTC   cylindrical trough collector
DC   direct current
EC   electrical conductivity
EGS   enhanced geothermal Systems
EPDM   ethylene propylene diene monomer
ESCO   energy service company
ETC   evacuated tube collector
FPC   flat plate collector
GHIPV   greenhouse integrated photovoltaic installations
GPC   generalized predictive control
GSHP   ground source heat pump
HFC   heliostat field collector
LFR   linear Fresnel reflector
NIR   near infrared radiation
PAR   photosynthetic Active radiation
PDR   parabolic dish reflector
PE   polyethylene

PER	renewable energy plan
PTC	parabolic trough collector
PV	photovoltaic
PV-T	photovoltaic and thermal
toe	ton of oil equivalent
η	performance [–]
η_o	optical efficiency [–]
$\Delta u(t+j-1)$	dequence of future control increments, obtained from cost function minimization
A	area [m^2]
a_1	first-order heat loss coefficient of the solar collector [W m^{-2} °C^{-1}]
a_2	second order heat loss coefficient of the solar collector [W m^{-2} °C^{-2}]
c	discharge capacity [m^3 h^{-1}]
C_B	battery charge [Ah]
c_p	specific heat [J kg^{-1} °C^{-1}]
DOD	maximum depth of battery discharge [–]
e	thickness [mm]
E_L	energy requirements [kWh]
EV	mathematical expectation or expected value
ε	emission coefficient [–]
f	fraction of the radiation used for transpiration [–]
G	global solar radiation [W m^{-2}]
g	gravity acceleration [g m^{-2}]
H	required pressure, height of equivalent column of water [m]
$H_{P(s)}$	daily peak sun hours [h d^{-1}]
H_{sen}	global solar radiation not used for transpiration that heats the greenhouse [W]
IWR	irrigation water requirement [m^3 m^{-2}]
J	objective function
k	corrective coefficient [–]
K_θ	incidence angle modifier [–]
L	latent heat of water vaporization [J kg^{-1}]
N	air renovation rate by infiltration [h^{-1}]
N_1, N_2	minimum and maximum prediction horizons
N_d	number of days [d]
N_h	daily-use time [h]
$NIWR$	net irrigation water requirement [m^3 m^{-2}]
N_u	control horizon
P	electrical power [kW]
PR	performance ratio [–]
Q	energy flux [W]
R	thermal radiation [W m^{-2}]
r	reflection coefficient [–]
RH	relative humidity [%]
T	temperature [°C]
t	time [s]
$T_{i,cool}$	refrigeration temperature [°C]
$T_{i,hot}$	heating temperature [°C]
u	wind velocity [m s^{-1}]
U_c	overall heat loss coefficient [W m^{-2} °C^{-1}]
V	volume [m^3]
V_v	natural ventilation rate [m^3 s^{-1} m^{-2}]
w	weighting factors [–]

$w(t+j)$	reference trajectory
Y	variables of the objective function
$\hat{y}(t+j|t)$	optimal system output prediction sequence performed with data known up to instant t

Greek symbols

α	absorption coefficient [–]
$\beta(j), \delta(j)$	weighting sequences that penalize the future
ΔV	voltage [V]
λ	wavelength [μm]
ρ	density [kg m^{-3}]
τ	transmission coefficient [–]
Φ	energy flux density [W m^{-2}]
χ	absolute humidity [kg$_{\text{water}}$ kg$_{\text{air}}^{-1}$]

Subscripts

a	aperture
av,n	degree of appreciable acidity in the fruit n
b	beam radiation
B	battery
c	cover
cc	conduction + convection
co	conduction
col	solar collector
cool	cooling
cos	production costs
cv	convection
d	diffuse radiation
ex	exhaust air
F	fan
FFT	fresh fruit
fi,n	firmness of the fruit n
FV	forced ventilation
h	heating
hx	heat exchange
i	inside
I	irrigation
inc	product income
inf	infiltration
inlet	inlet air
l	latent heat
m	plant and soil mass
o	outside
p	plant
P	pump
R,$\downarrow\uparrow$	thermal radiative exchange
s	soil
sk	sky
ssol,n	sugar content of the fruit n
sto	storage
t	titled
tf,n	size of the fruit n
u	useful

v   ventilation
w   wind

Superscripts
*   averaged, normalized
d   daily

REFERENCES

Abasaeed, A.E. (1992) Briquetting of carbonized cotton stalk. *Energy*, 17(9), 877–882.

Adaro, J.A., Galimberti, P.D., Lema, A.I., Fasulo, A.I. & Barral, J.R. (1999) Geothermal contribution to greenhouse heating. *Applied Energy*, 64(1–4), 241–249.

Aikman, D.P. (1996) A procedure for optimizing carbon dioxide enrichment of a glasshouse tomato crop. *Journal of Agricultural Engineering Research*, 63(2), 171–183.

Al-Helal, I., Al-Abbadi, N. & Al-Ibrahim, A. (2006) A study of evaporative cooling pad performance for a photovoltaic powered greenhouse. *Acta Horticulturae* (ISHS), 710, pp. 153–164.

Al-Ibrahim, A., Al-Abbadi, N. & Al-Helal, I. (2006) PV greenhouse system: system description, performance and lesson learned. *Acta Horticulturae* (ISHS), 710, 251–264.

Alkilani, M.M., Sopian, K., Alghoul, M.A., Sohif, M. & Ruslan, M.H. (2011) Review of solar air collectors with thermal storage units. *Renewable and Sustainable Energy Reviews*, 15(3), 1476–1490.

Alonso, J.J. (2004) Dinamización de los municipios Segovianos de Abades, Fuentemilanos y Valdeprados, mediante la instalación de una industria de pelletizado de cardo (*Cynara cardunculus* L.) para su posterior uso como combustible. Aplicación al caso particular de la sustitución de gasóleo C por pellets en la residencia de Válidos de Segovia, para la obtención de calefacción y agua caliente sanitaria [Dynamization of the Segovian municipalities of Abades, Fuentemilanos y Valdeprados through the installation of a pelletizing industry of cardoon (*Cynara cardunculus* L.) for later use as fuel. Application to the particular case of the substitution of gasoil C by pellets in the residence of Válidos de Segovia, for heating and domestic hot water]. Final thesis, Technical University of Madrid, Madrid, Spain.

Allen, R.G., Pereira, L.S., Raes, D. & Smith, M. (1998) Crop evapotranspiration: guidelines for computing crop water requirements. FAO Irrigation and Drainage, Paper 56, Food and Agriculture Organization of the United Nations, Rome, Italy.

Antón, A., Castells, F. & Montero, J.I. (2007) Land use indicators in life cycle assessment. Case study: the environmental impact of Mediterranean greenhouses. *Journal of Cleaner Production*, 15(5), 432–438.

ASABE (2003) Heating, ventilating and cooling greenhouses. ASAE EP 406.4, American Society of Agricultural and Biological Engineers, St. Joseph, MI.

Bailey, B.J. (2002) Optimal control of carbon dioxide enrichment in tomato greenhouses. *Acta Horticulturae* (ISHS), 578, 63–69.

Baille (González-Real), M., Nicolas, B., Rassent, E. & Brun, R. (1977) The heating of the greenhouses by water circulation in plastics film tubes laid out on the soil: microclimatic study and agronomic results. *Plasticulture*, 36, 13–29.

Baille, A. (1999) Energy cycle. In: Enoch, H.Z. & Stanhill, G. (eds) *Ecosystems of the World* 20, *Greenhouse Ecosystems*. Elsevier, Amsterdam, The Netherlands. pp. 265–286.

Baille, A., López, J.C., Bonachela, S., González-Real, M.M. & Montero, J.I. (2006) Night energy balance in a heated low-cost plastic greenhouse. *Agricultural and Forest Meteorology*, 137(1–2), 107–118.

Bakker, J.C. & Challa, H. (1995) Aim of the greenhouse climate control. In: Bakker, J.C., Bot, G.P.A., Challa, H. & Van de Braak, N.J. (eds) *Greenhouse Climate Control: An Integrated Approach*. Wageningen Academic Publishers, Wageningen, The Netherlands. pp. 1–3.

Bakos, G.C., Fidanidis, D. & Tsagas, N.F. (1999) Greenhouse heating using geothermal energy. *Geothermics*, 28(6), 759–765.

Bargach, M.N., Dahman, A.S. & Boukallouch, M. (1999) A heating system using flat plate collectors to improve the inside greenhouse microclimate in Morocco. *Renewable Energy*, 18(3), 367–381.

Bargach, M.N., Tadili, R., Dahman, A.S. & Boukallouch, M. (2004) Comparison of the performance of two solar heating systems used to improve the microclimate of agricultural greenhouses in Morocco. *Renewable Energy*, 29(7), 1073–1083.

Baudoin, W. (ed.) (2013) Good agricultural practices for greenhouse vegetable crops: principles for Mediterranean climate areas. FAO Plant Production and Protection, Paper 217. Food and Agriculture Organization of the United Nations, Rome, Italy.

Baudoin, W., Grafiadellis, M., Jiménez, R., La Malfa, G., Martínez-García, P.F., Monteiro, A.A., Nisen, A., Verlodt, H., de Villele, O., von Zabeltitz, C. & Garnaud, J.C. (2002) El cultivo protegido en clima Mediterráneo [Protected crops in Mediterranean climate]. FAO Plant Production and Protection Paper 90. Food and Agriculture Organization of the United Nations, Rome, Italy.

Bojacá, C.R., Wyckhuys, K.A.G. & Schrevens, E. (2014) Life cycle assessment of Colombian greenhouse tomato production based on farmer-level survey data. *Journal of Cleaner Production*, 69, 26–33.

Bot, G.P.A., van de Braak, N.J., Challa, H., Hemming, S., Rieswijk, T., van de Straten, G. & Verlodt, I. (2005) The solar greenhouse: state of the art in energy saving and sustainable energy supply. *Acta Horticulturae* (ISHS), 691(2), 501–508.

Bot, G.P.A. & van de Braak, N.J. (1995) Physics of greenhouse climate. Energy balance. In: Bakker, J.C., Bot, G.P.A., Challa, H. & van de Braak, N.J. (eds) *Greenhouse Climate Control: An Integrated Approach.* Wageningen Academic Publishers, Wageningen, The Netherlands. pp. 135–141.

Boulard, T. & Baille, A. (1987) Analysis of thermal performance of a greenhouse as a solar collector. *Energy in Agriculture*, 6(1), 17–26.

Boulard, T., Razafinjohany, E., Baille, A., Jaffrin, A. & Fabre, B. (1990) Performance of a greenhouse heating system with a phase change material. *Agricultural and Forest Meteorology*, 52(3–4), 303–318.

Bundschuh, J. & Hoinkis, J. (eds) (2012) *Renewable Energy Applications for Freshwater Production.* CRC Press, Boca Raton, FL.

CAJAMAR (2004) Analysis of horticultural season in Almería. Season 2003/2004. Informes y Monografía 5, Cajamar Caja Rural, Almería, Spain.

Callejón-Ferre, A.J., Velázquez-Martí, B., López-Martínez, J.A. & Manzano-Agugliaro, F. (2011) Greenhouse crop residues: energy potential and models for the prediction of their higher heating value. *Renewable and Sustainable Energy Reviews*, 15(2), 948–955.

Campiotti, C., Alonzo, G., Belmonte, A., Bibbiani, C., di Carlo, F., Dondi, F. & Scoccianti, M. (2009) Renewable energy and innovation for sustainable greenhouse districts. *Fascicula de Energetica*, 15, 196–201.

Castilla, N. (1980) Ensayo de apoyo energético solar en abrigo de polietileno [Test of solar energy support in polyethylene wrap]. Cajamar Caja Rural, Almería, Spain.

Castilla, N. (1981) Experiencias sobre aprovechamiento de energía solar en invernaderos en Almería (España) [Experiences on the use of solar energy in greenhouses in Almería (Spain)]. Cajamar Caja Rural, Almería, Spain.

Castilla, N. (2007) *Invernaderos de Plástico: Tecnología y Manejo* [*Plastic Greenhouses: Technology and Management*]. Mundi-Prensa, Madrid, Spain.

Castilla, N. & Baeza, E. (2013) Greenhouse site selection. In: Baudoin, W. (ed.) Good agricultural practices for greenhouse vegetable crops: principles for Mediterranean climate areas. FAO Plant Production and Protection Paper 217. Food and Agriculture Organization of the United Nations, Rome, Italy. pp. 21–34.

Castilla, N. & Hernández, J. (2007) Greenhouse technological packages for high quality production. *Acta Horticulturae* (ISHS), 761, 285–297.

Castilla, N., Montero, J. I., Bretones, F., Gálvez, J. L., Jímenez, M. & Sevilla, A. (1985) Essays on solar heating of greenhouses in Almería, Spain. *CNRE Workshop on Solar Heating of Greenhouses, 18 February 1985, Nicosia, Cyprus.*

Cerón, I., Caamaño-Martín, E. & Neila, F.J. (2013) 'State-of-the-art' of building integrated photovoltaic products. *Renewable Energy*, 58(0), 127–133.

Clarke, D.W., Mohtadi, C. & Tuffs, P.S. (1987) Generalized predictive control – part i. The basic algorithm, and part ii. Extensions and interpretations. *Automatica*, 23(2), 137–160.

Coello, C.A. (2003): Guest editorial: Special issue on evolutionary multiobjective optimization. *IEEE Transactions on Evolutionary Computation*, 7(2), 97–99.

Cornet d'Elzius, C., Ferrero, G.L. & Kocsis, K. (eds) (1989) Utilization of waste heat from power stations. Proceedings of an international seminar. Commission of the European Communities, CD-NA-12423-2A-C, Brussels, Belgium.

Cossu, M., Murgia, L., Ledda, L., Deligios, P.A., Sirigu, A., Chessa, F. & Pazzona, A. (2014) Solar radiation distribution inside a greenhouse with south-oriented photovoltaic roofs and effects on crop productivity. *Applied Energy*, 133, 89–100.

Chaibi, M.T. & Jilar, T. (2005) Effects of a solar desalination module integrated in a greenhouse roof on light transmission and crop growth. *Biosystems Engineering*, 90(3), 319–330.

Challa, H. & van Straten, G. (1993) Optimal diurnal climate control in greenhouses as related to greenhose management and crop requirements. In: Hashimoto, Y., Bot, G., Tantau, H.J., Day, W. & Nonami, H. (eds) *The Computerized Greenhouse.* Academic Press, London. pp. 119–137.

Davies, P.A., Hossain, A.K., Lychnos, G. & Paton, C. (2008) Energy saving and solar electricity in fan-ventilated greenhouses. *Acta Horticulturae* (ISHS), 797, 339–346.

Day, W. & Bailey, B.J. (1999) Physical principles of microclimate modification. In: Stanhill, G. & Enoch, H.Z. (eds) *Ecosystems of the World* 20. *Greenhouse Ecosystems*. Elsevier Science B.V., Amsterdam, The Netherlands. pp. 71–99.

de Pascale, S. & Maggio, A. (2005) Sustainable protected cultivation at a Mediterranean climate. Perspectives and challenges. *Acta Horticulturae* (ISHS), 691, 29–42.

Debdoubi, A., El amarti, A. & Colacio, E. (2005) Production of fuel briquettes from esparto partially pyrolyzed. *Energy Conversion and Management*, 46(11–12), 1877–1884.

Demirbaş, A. (1999) Evaluation of biomass materials as energy sources: upgrading of tea waste by briquetting process. *Energy Sources*, 21(3), 215–220.

Demirbaş, A. & Şahin, A. (1998) Evaluation of biomass residue. 1. Briquetting waste paper and wheat straw mixtures. *Fuel Processing Technology*, 55(2), 175–183.

Dupraz, C., Marrou, H., Talbot, G., Dufour, L., Nogier, A. & Ferard, Y. (2011) Combining solar photovoltaic panels and food crops for optimising land use: towards new agrivoltaic schemes. *Renewable Energy*, 36(10), 2725–2732.

Edwards, D., Jolliffe, P., Baylis, K. & Ehret, D. (2008) Towards a plant-based method of CO_2 management. *Acta Horticulturae* (ISHS), 797, 273–278.

Emmott, C.J.M., Rohr, J.A., Campoy-Quiles, M., Kirchartz, T., Urbina, A., Ekins-Daukes, N.J. & Nelson, J. (2015) Organic photovoltaic greenhouses: a unique application for semi-transparent PV? *Energy and Environmental Science*, 8(4), 1317–1328.

Erlich, C., Öhman, M., Björnbom, E. & Fransson, T.H. (2005) Thermochemical characteristics of sugar cane bagasse pellets. *Fuel*, 84(5), 569–575.

Esen, H., Esen, M. & Yuksel, T. (2015) Modelling of biogas, solar and a ground source heat pump greenhouse heating system by using ensemble learning. In: Mastorakis, N.E. & Solomon To, C.W. (eds) *New Developments in Mechanics and Mechanical Engineering*. WSEAS Press, Vienna, Austria. pp. 74–81.

Esen, M. & Yuksel, T. (2013) Experimental evaluation of using various renewable energy sources for heating a greenhouse. *Energy and Buildings*, 65(0), 340–351.

FAO (1999) European System of Cooperative Research Networks in Agriculture (ESCORENA). Progress report 1996–1998. Food and Agriculture Organization of the United Nations. Regional Office for Europe.

Fatnassi, H., Poncet, C., Bazzano, M.M., Brun, R. & Bertin, N. (2015) A numerical simulation of the photovoltaic greenhouse microclimate. *Solar Energy*, 120, 575–584.

Fernández, M.D., Reinoso, J.V., Pérez, C., López, J.C., Acién, F.G., Sanchez, J.A., Meca, E. & Gázquez, J.C. (2013) Influencia de las técnicas de recogida sobre las propiedades de residuos vegetales de invernaderos para su aprovechamiento energético [Influence of collection techniques on the properties of vegetable residues of greenhouses for their energy use]. *VII Congreso Ibérico de Agroingeniería y Ciencias Hortícolas, 26–29 August 2013, Madrid, Spain.* pp. 2051–2055.

Feuilloley, P. & Issanchou, G. (1996) Greenhouse covering materials measurement and modelling of thermal properties using the hot box method, and condensation effects. *Journal of Agricultural Engineering Research*, 65(2), 129–142.

Floris, R. & Parodi, G. (1986) Application of solar energy in greenhouses and solar crop drying. United Nations Environment Programme (UNEP). Mediterranean Action Plan. Priority Actions Programme (PAP-11/ME-5102-83-05). Mediteranean Cooperative Network in Renewable Sources of Energy, UNEP, Split, Croatia.

Ganguly, A., Misra, D. & Ghosh, S. (2010) Modeling and analysis of solar photovoltaic-electrolyzer-fuel cell hybrid power system integrated with a floriculture greenhouse. *Energy and Buildings*, 42(11), 2036–2043.

Garrido-Frenich, A., Arrebola, F.J., González-Rodríguez, M.J., Vidal, J.L.M. & Díez, N.M. (2003) Rapid pesticide analysis, in post-harvest plants used as animal feed, by low-pressure gas chromatography-tandem mass spectrometry. *Analytical and Bioanalytical Chemistry*, 377(6), 1038–1046.

Ghosal, M.K., Tiwari, G.N., Das, D.K. & Pandey, K.P. (2005) Modeling and comparative thermal performance of ground air collector and earth air heat exchanger for heating of greenhouse. *Energy and Buildings*, 37(6), 613–621.

Giacomelli, G.A., Castilla, N., van Henten, E.J., Mears, D.R. & Sase, S. (2008) Innovation in greenhouse engineering. *Acta Horticulturae* (ISHS), 801, 75–88.

González-Real, M.M. & Baille, A. (2010) Tecnología de invernaderos [Greenhouse technology]. Polytechnic University of Cartagena, Cartagena, Murcia, Spain.

Hanan, J.J. (1998) *Greenhouses: Advanced Technology for Protected Horticulture*. CRC Press, Boca Raton, FL.

Hao, X., Wang, Q. & Khosla, S. (2008) Responses of greenhouse tomatoes to summer CO_2 enrichment. *Acta Horticulturae* (ISHS), 797, 241–246.

Hu, Z., Chan, C.W. & Huang, G.H. (2007) Multi-objective optimization for process control of the in-situ bioremediation system under uncertainty. *Engineering Applications of Artificial Intelligence*, 20(2), 225–237.

IDAE (2007) Energía de la biomasa [Biomass energy]. Spanish Ministry of Industry, Energy and Turism – IDAE – APIA, Madrid, Spain.

IDAE (2011) Plan energías renovables de España 2011–2020 [Plan for renewable energies in Spain 2011–2020]. Spanish Ministry of Industry, Energy and Turism – IDAE, Madrid, Spain.

IEA (2010) Technology roadmap. Solar photovoltaic energy. International Energy Agency, Paris, France.

IEA (2011) Technology roadmap. Geothermal heat and power. International Energy Agency, Paris, France.

IEA (2012) Technology roadmap. Solar heating and cooling. International Energy Agency, Paris, France.

Ioslovich, I., Gutman, P.-O. & Linker, R. (2009) Hamilton-Jacobi-Bellman formalism for optimal climate control of greenhouse crop. *Automatica*, 45(5), 1227–1231.

Ioslovich, I. & Seginer, I. (2002) SE-structures and environment: acceptable nitrate concentration of greenhouse lettuce: two optimal control policies. *Biosystems Engineering*, 83(2), 199–215.

IPCC (2011) Special report on renewable energy sources and climate change mitigation. Edenhofer, O., Pichs-Madruga, R., Sokona, Y., Seyboth, K., Kadner, S., Zwickel, T., Eickemeier, P., Hansen, G., Schlömer, S., von Stechow, C. & P. Matschoss (eds). Cambridge University Press, UK and New York, NY. pp. 1075.

Janjai, S., Lamlert, N., Intawee, P., Mahayothee, B., Bala, B.K., Nagle, M. & Müller, J. (2009) Experimental and simulated performance of a PV-ventilated solar greenhouse dryer for drying of peeled longan and banana. *Solar Energy*, 83(9), 1550–1565.

Kadowaki, M., Yano, A., Ishizu, F., Tanaka, T. & Noda, S. (2012) Effects of greenhouse photovoltaic array shading on Welsh Onion growth. *Biosystems Engineering*, 111(3), 290–297.

Kalogirou, S. (2009) *Solar Energy Engineering: Processes and Systems*. Elsevier/Academic Press, London.

Kittas, C., Katsoulas, N. & Bartzanas, T. (2012) Greenhouse climate control in Mediterranean greenhouses. *Cuadernos de Estudios Agroalimentarios. Innovación en Estructuras Productivas y Manejo de Cultivos en Agricultura Protegida*, 3, 89–113.

Kittas, C., Katsoulas, N., Bartzanas, T. & Bakker, S. (2013) Greenhouse climate control and energy use. In: Baudoin, W. (ed.) Good agricultural practices for greenhouse vegetable crops: principles for Mediterranean climate areas. FAO Plant Production and Protection Paper 217. Food and Agriculture Organization of the United Nations, Rome. pp. 63–96.

Kläring, H.P., Hauschild, C., Heißner, A. & Bar-Yosef, B. (2013) Model-based control of CO_2 concentration in greenhouses at ambient levels increases cucumber yield. *Agricultural and Forest Meteorology*, 143(3–4), 208–216.

Körner, O. & Van Straten, G. (2008) Decision support for dynamic greenhouse climate control strategies. *Computers and Electronics in Agriculture*, 60(1), 18–30.

Laabidi, K. & Bouani, F. (2004) Genetic algorithms for multiobjective predictive control. *First International Symposium on Control, Communications and Signal Processing 21–24 March, Hammamet, Tunisia*. pp. 149–152.

Lamnatou, C. & Chemisana, D. (2013a) Solar radiation manipulations and their role in greenhouse claddings: fluorescent solar concentrators, photoselective and other materials. *Renewable and Sustainable Energy Reviews*, 27(0), 175–190.

Lamnatou, C. & Chemisana, D. (2013b) Solar radiation manipulations and their role in greenhouse claddings: Fresnel lenses, NIR- and UV-blocking materials. *Renewable and Sustainable Energy Reviews*, 18(0), 271–287.

Lazaar, M., Bouadila, S., Kooli, S. & Farhat, A. (2015) Comparative study of conventional and solar heating systems under tunnel Tunisian greenhouses: thermal performance and economic analysis. *Solar Energy*, 120, 620–635.

Linker, R., Gutman, P.O. & Seginer, I. (1999) Robust controllers for simultaneous control of temperature and CO_2 concentration in greenhouses. *Control Engineering Practice*, 7(7), 851–862.

Liu, G.P., Yang, J.B. & Whidborne, J.F. (eds) (2003) Multiobjective optimisation and control. *Engineering Systems Modelling and Control Series*. Research Studies Press Limited, Baldock, Hertfordshire, UK.

López, J.C. (2003) *Sistemas de Calefacción en Invernaderos Cultivados de Judía en el Litoral Mediterráneo* [*Heating Systems in Greenhouses cultivating Beans on the Mediterranean Coast*]. PhD Thesis, University of Almería, Almería, Spain.

López, J.C., Pérez, J., Montero, J.I. & Antón, A. (2001) Air infiltration rate of Almería "parral type" greenhouses. *Acta Horticulturae* (ISHS), 559, 229–232.

Lorenzo, P., Maroto, C. & Castilla, N. (1990) CO_2 in plastic greenhouse in Almeria (Spain). *Acta Horticulturae* (ISHS), 268, 165–170.

Lucio, J.H., Valdés, R. & Rodríguez, L.R. (2012) Loss-of-load probability model for stand-alone photovoltaic systems in Europe. *Solar Energy*, 86(9), 2515–2535.

Maiti, S., Dey, S., Purakayastha, S. & Ghosh, B. (2006) Physical and thermochemical characterization of rice husk char as a potential biomass energy source. *Bioresource Technology*, 97(16), 2065–2070.

Martínez-Blanco, J., Muñoz, P., Antón, A. & Rieradevall, J. (2011) Assessment of tomato Mediterranean production in open-field and standard multi-tunnel greenhouse, with compost or mineral fertilizers, from an agricultural and environmental standpoint. *Journal of Cleaner Production*, 19(9–10), 985–997.

Marucci, A., Monarca, D., Cecchini, M., Colantoni, A., Manzo, A. & Cappuccini, A. (2012) The semitransparent photovoltaic films for Mediterranean greenhouse: a new sustainable technology. *Mathematical Problems in Engineering*, Article ID 451934.

Mears, D.R., Roberts, W.J. & Cipolletti, J.P. (1980) Solar heating of commercial greenhouses. New Jersey Agricultural Experiment Station. Rutgers University, New Brunswick, NJ.

Milenić, D., Vasiljević, P. & Vranješ, A. (2010) Criteria for use of groundwater as renewable energy source in geothermal heat pump systems for building heating/cooling purposes. *Energy and Buildings*, 42(5), 649–657.

Montero, J.I., Antón, A. & Muñoz, P. (1998) Fundamentos [Fundamentals]. In: Pérez-Parra, J. & Cuadrado, I. (eds) Tecnología de invernaderos [Greenhouse technology]. Consejería de Agricultura y Pesca, FIAPA, Caja Rural de Almería, Almería, Spain. pp. 253–266.

Montero, J.I., Castilla, N. & Bretones, F. (1985) Evaluación de colectores solares térmicos de bajo costo [Evaluation of low-cost solar thermal collectors]. *X Reunión Bioclimática [Bioclimatic meeting]*. Estcaction Experimental Zonas Áridas, CSIC, Almería, 1–4 October 1985. Las Palmerillas, Cajamar, Almería, Spain.

Nayak, S. & Tiwari, G.N. (2008) Energy and exergy analysis of photovoltaic/thermal integrated with a solar greenhouse. *Energy and Buildings*, 40(11), 2015–2021.

Nayak, S. & Tiwari, G.N. (2009) Theoretical performance assessment of an integrated photovoltaic and earth air heat exchanger greenhouse using energy and exergy analysis methods. *Energy and Buildings*, 41(8), 888–896.

Nederhoff, E.M. (1990) Technical aspects, management and control of CO_2 enrichment in greenhouses. *Acta Horticulturae* (ISHS), 268, 127–138.

Nijskens, J., Deltour, J., Coutisse, S. & Nisen, A. (1985) Radiation transfer through covering materials, solar and thermal screens of greenhouses. *Agricultural and Forest Meteorology*, 35(1–4), 229–242.

Ntinas, G.K., Fragos, V.P. & Nikita-Martzopoulou, C. (2014) Thermal analysis of a hybrid solar energy saving system inside a greenhouse. *Energy Conversion and Management*, 81(0), 428–439.

Ortíz, L., Tejada, A., Vázquez, A. & Piñero, G. (2003) Aprovechamiento de la biomasa forestal producida por la cadena monte-industria. III. Producción de elementos densificados [Use of forest biomass produced by the mountain-industry chain. III. Production of densified elements]. *Revista CIS-Madeira*, 11, 13–32.

Ozgener, O. & Hepbasli, A. (2005) Performance analysis of a solar-assisted ground-source heat pump system for greenhouse heating: an experimental study. *Building and Environment*, 40(8), 1040–1050.

Ozgener, O. & Hepbasli, A. (2007) A parametrical study on the energetic and exergetic assessment of a solar-assisted vertical ground-source heat pump system used for heating a greenhouse. *Building and Environment*, 42(1), 11–24.

Page, G., Ridoutt, B. & Bellotti, B. (2012) Carbon and water footprint tradeoffs in fresh tomato production. *Journal of Cleaner Production*, 32(0), 219–226.

Papadakis, G., Briassoulis, D., Scarascia Mugnozza, G., Vox, G., Feuilloley, P. & Stoffers, J.A. (2000) SE-structures and environment: radiometric and thermal properties of, and testing methods for, greenhouse covering materials. *Journal of Agricultural Engineering Research*, 77(1), 7–38.

Pastre, O. (2002) Analysis of the technical obstacles related to the production and utilisation of fuel pellets made from agricultural residues. ALTENER 2002-012-137-160. European Pellet Centre, European Biomass Industry Association (EUBIA), Brussels, Belgium.

Pérez-Alonso, J., Pérez-García, M., Pasamontes-Romera, M. & Callejón-Ferre, A.J. (2012) Performance analysis and neural modelling of a greenhouse integrated photovoltaic system. *Renewable and Sustainable Energy Reviews*, 16(7), 4675–4685.

Pérez, M.C. (2007) *Applications of Solar Energy to thermal Treatment of Soils in Greenhouses*. PhD Thesis, Universidad de Cordoba, Cordoba, Spain.

Perez, R., Stewart, R., Arbogast, C., Seals, R. & Scott, J. (1986) An anisotropic hourly diffuse radiation model for sloping surfaces: description, performance validation, site dependency evaluation. *Solar Energy*, 36(6), 481–497.

Pinel, P., Cruickshank, C.A., Beausoleil-Morrison, I. & Wills, A. (2011) A review of available methods for seasonal storage of solar thermal energy in residential applications. *Renewable and Sustainable Energy Reviews*, 15(7), 3341–3359.

Popovski, K. (ed.) (1988) Geothermal energy resources and their use in European agriculture. European Cooperative Networks on Rural Energy (CNRE), Study no. 2. REUR Technical Series, n°4, Food and Agriculture Organization of the United Nations (FAO), Rome, Italy.

Portree, J. (1996) Greenhouse vegetable production guide for commercial growers. Province of British Columbia Ministry of Agriculture, Fisheries and Food, Victoria, BC, Canada.

Poullikkas, A. (2013) A comparative overview of large-scale battery systems for electricity storage. *Renewable and Sustainable Energy Reviews*, 27(0), 778–788.

Qoaider, L. & Steinbrecht, D. (2010) Photovoltaic systems: a cost competitive option to supply energy to off-grid agricultural communities in arid regions. *Applied Energy*, 87(2), 427–435.

Ramirez-Arias, A., Rodriguez, F., Guzman, J.L., Arahal, M.R., Berenguel, M. & Lopez, J.C. (2005) *Improving efficiency of greenhouse heating systems using model predictive control*. Proceedings of the 16th IFAC World Congress, 3–8 July 2005, Prague, Czech Republic. IFAC Proceedings Volumes, 38(1), 40–45.

Ramírez-Arias, A., Rodríguez, F., Guzmán, J.L. & Berenguel, M. (2012) Multiobjective hierarchical control architecture for greenhouse crop growth. *Automatica*, 48(3), 490–498.

Ramírez-Arias, J.A. (2005) *Multiobjective hierarchical Control of Greenhouse Crop Growth*. PhD Thesis, Editorial Universidad de Almeria, Almeria, Spain.

Reca, J., Torrente, C., López-Luque, R. & Martínez, J. (2016) Feasibility analysis of a standalone direct pumping photovoltaic system for irrigation in Mediterranean greenhouses. *Renewable Energy*, 85, 1143–1154.

Rocamora, M.C. & Tripanagnostopoulos, Y. (2006) Aspects of PV/T solar system application for ventilation needs in greenhouses. *Acta Horticulturae*, 719, 239–245.

Rodríguez, F. (2002) *Modelling and hierarchical Control of Greenhouse Crop Production*. PhD Thesis, Universidad de Almeria, Almeria, Spain.

Rodríguez, F., Berenguel, M. & Arahal, M.R. (2003) A hierarchical control system for maximizing profit in greenhouse crop production. *European Control Conference ECC 2003, 1–4 September 2003, Cambridge, UK*. pp. 2753–2758.

Rodríguez, F., Guzmán, J.L., Berenguel, M. & Arahal, M.R. (2008) Adaptive hierarchical control of greenhouse crop production. *International Journal of Adaptive Control and Signal Processing*, 22(2), 180–197.

Romero-Gámez, M., Suárez-Rey, E.M., Antón, A., Castilla, N. & Soriano, T. (2012) Environmental impact of screenhouse and open-field cultivation using a life cycle analysis: the case study of green bean production. *Journal of Cleaner Production*, 28(0), 63–69.

Sánchez-Guerrero, M.C., Lorenzo, P., Medrano, E., Baille, A. & Castilla, N. (2009) Effects of EC-based irrigation scheduling and CO_2 enrichment on water use efficiency of a greenhouse cucumber crop. *Agricultural Water Management*, 96(3), 429–436.

Sánchez-Guerrero, M.C., Lorenzo, P., Medrano, E., Castilla, N., Soriano, T. & Baille, A. (2005) Effect of variable CO_2 enrichment on greenhouse production in mild winter climates. *Agricultural and Forest Meteorology*, 132(3–4), 244–252.

Sánchez-Molina, J.A., Reinoso, J.V., Acién, F.G., Rodríguez, F. & López, J.C. (2014) Development of a biomass-based system for nocturnal temperature and diurnal CO_2 concentration control in greenhouses. *Biomass and Bioenergy*, 67, 60–71.

Schmidt, U., Huber, C., Rocksch, T., Salazar Moreno, R. & Rojano Aguilar, A. (2008) Greenhouse cooling and carbon dioxide fixation by using high pressure fog systems and phytocontrol strategy. *Acta Horticulturae* (ISHS), 797, 279–284.

Seginer, I. (2002) SE-structures and environment: the Penman-Monteith evapotranspiration equation as an element in greenhouse ventilation design. *Biosystems Engineering*, 82(4), 423–439.

Sethi, V.P. & Sharma, S.K. (2007) Greenhouse heating and cooling using aquifer water. *Energy*, 32(8), 1414–1421.

Sethi, V.P. & Sharma, S.K. (2008) Survey and evaluation of heating technologies for worldwide agricultural greenhouse applications. *Solar Energy*, 82(9), 832–859.

Sethi, V.P., Sumathy, K., Lee, C. & Pal, D. S. (2013) Thermal modeling aspects of solar greenhouse microclimate control: A review on heating technologies. *Solar Energy*, 96(0), 56–82.

Siddiqi, M.Y., Kronzucker, H.J., Britto, D.T. & Glass, A.D.M. (1998) Growth of a tomato crop at reduced nutrient concentrations as a strategy to limit eutrophication. *Journal of Plant Nutrition*, 21(9), 1879–1895.

Solís, A. & Castilla, N. (1980) Calefacción solar de invernaderos. Aplicaciones de la Energia, S.A. (APLESA), Madrid, Spain.

Sonneveld, P.J., Swinkels, G.L.A.M., Campen, J., van Tuijl, B.A.J., Janssen, H.J.J. & Bot, G.P.A. (2010) Performance results of a solar greenhouse combining electrical and thermal energy production. *Biosystems Engineering*, 106(1), 48–57.

Sonneveld, P.J., Swinkels, G.L.A.M., van Tuijl, B.A.J., Janssen, H.J.J., Campen, J. & Bot, G.P.A. (2011) Performance of a concentrated photovoltaic energy system with static linear Fresnel lenses. *Solar Energy*, 85(3), 432–442.

Souliotis, M., Tripanagnostopoulos, Y. & Kavga, A. (2006) The use of Fresnel lenses to reduce the ventilation needs of greenhouses. *Acta Horticulturae* (ISHS), 719, 107–114.

Stanghellini, C., Kempkes, F.L.K. & Knies, P. (2003) Enhancing environmental quality in agricultural systems. *Acta Horticulturae* (ISHS), 609, 277–283.

Strehler, A. (ed.) (1988) Biomass combustion technologies. European Cooperative Networks on Rural Energy (CNRE). Guideline nº 1. REUR Technical Series nº 2, Food and Agriculture Organization of the United Nations, Rome, Italy.

Tantau, H.J. (1993) Optimal control for plant production in greenhouse. In: Hashimoto, Y. (ed.) *The Computerized Greenhouse: Automatic Control Application in Plant Production*. Academic Press, San Diego, CA. pp. 139–152.

Tap, F. (2000) *Economised-based optimal Control of Greenhouse Tomato Crop Production*. PhD Thesis, Agricultural University of Wageningen, Wageningen, The Netherlands.

Torrellas, M., Antón, A. & Montero, J.I. (2013) An environmental impact calculator for greenhouse production systems. *Journal of Environmental Management*, 118(0), 186–195.

Tremblay, N. & Gosselin, A. (1998) Effect of carbon dioxide enrichment and light. *HortTechnology*, 8(4), 524–528.

Tripanagnostopoulos, Y. & Rocamora, M.C. (2008) Use of solar thermal collectors for disinfection of greenhouse hydroponic water. *Acta Horticulturae* (ISHS), 801, 749–756.

Tüzel, Y. & Leonardi, C. (2009) Protected cultivation in Mediterranean region: trends and needs. *Ege Üniversitesi Ziraat Fakültesi Dergisi*, 46(3), 215–223.

Ureña-Sánchez, R., Callejón-Ferre, A.J., Pérez-Alonso, J. & Carreño-Ortega, A. (2012) Greenhouse tomato production with electricity generation by roof-mounted flexible solar panels. *Scientia Agricola*, 69(4), 233–239.

Valera, D.L., Molina, F.D. & Álvarez, A.J. (2008) Ahorro y eficiencia energética en invernaderos [Energy savings and efficiency in greenhouses]. Instituto para la Diversificación y Ahorro de la Energía. IDAE, Madrid, Spain.

van der Werf, H.M.G., Garnett, T., Corson, M.S., Hayashi, K., Huisingh, D. & Cederberg, C. (2014) Towards eco-efficient agriculture and food systems: theory, praxis and future challenges. *Journal of Cleaner Production*, 73(0), 1–9.

van Henten, E.J. (1994) *Greenhouse Climate Management: An optimal Control Approach*. PhD Thesis, Agricultural University of Wageningen, Wageningen, The Netherlands.

van Henten, E.J. & Bontsema, J. (2009) Time-scale decomposition of an optimal control problem in greenhouse climate management. *Control Engineering Practice*, 17(1), 88–96.

van Henten, E.J., Buwalda, F., de Zwart, H.F., de Gelder, A., Hemming, J. & Bontsema, J. (2006) Toward an optimal control strategy for sweet pepper cultivation. 2. Optimization of the yield pattern and energy efficiency. *Acta Horticulturae* (ISHS), 718, 391–398.

van Straten, G., van Willigenburg, G., van Henten, E. & van Ooteghem, R. (2010) *Optimal Control of Greenhouse Cultivation*. CRC Press, Boca Raton, FL.

van Uffelen, R.L.M., van der Mass, A.A., Vermeulen, P.C.M. & Ammerlaan, J.C.J. (2000) TQM applied to the dutch glasshouse industry: state of the art in 2000. *Acta Horticulturae* (ISHS), 536, 679–686.

Vasilevska, S.P., Gecevska, V. & Popovski, K. (2011) Geothermal energy-convenient heat source for renewal and new development of protected crop cultivation in Macedonia. *Renewable and Sustainable Energy Reviews*, 15(6), 2909–2920.

Verheye, W. (2006) Dry lands and desertification. In: Verheye, W.H. (ed.) *Land Use, Land Cover and Soil Sciences. Encyclopedia of Life Support Systems (EOLSS)*. UNESCO-EOLSS Publishers, Oxford, UK.

von Elsner, B., Briassoulis, B., Waaijenberg, D., Mistriotis, A., von Zabeltitz, C., Gratraud, J., Russo, G. & Suay-Cortes, R. (2000) Review of structural and functional characteristics of greenhouses in European Union countries. I. Design requirements. *Journal of Agricultural Engineering Research*, 75(1), 1–16.

von Zabeltitz, C. (ed.) (1986) Greenhouse heating with solar energy: draft state-of-the-art-study to be discussed at the second European Cooperative Networks on Rural Energy (CNRE) workshop. REUR Technical Series, n°1, Food and Agriculture Organization of the United Nations, Rome, Italy.

von Zabeltitz, C. (ed.) (1988) Energy conservation and renewable energies for greenhouse heating. European Cooperative Networks on Rural Energy (CNRE). Guideline n°2. REUR Technical Series, n°3, Food and Agriculture Organization of the United Nations, Rome, Italy.

von Zabeltitz, C. (1999) Greenhouse structures. In: Stanhill, G. & Enoch, H.Z. (eds) *Ecosystems of the World* 20. *Greenhouse Ecosystems*. Elsevier Science B.V., Amsterdam, The Netherlands. pp. 17–69.

von Zabeltitz, C. (2011) *Integrated Greenhouse Systems for Mild Climates: Climate Conditions, Design, Construction, Maintenance, Climate Control*. Springer Berlin Heidelberg, Germany.

von Zabeltitz, C. & Popovski, K. (eds) (1989) Utilization of solar and geothermal energy for heating greenhouses. Report and proceedings of joint CNRE workshop. European Cooperative Networks on Rural Energy (CNRE). Bulletin (FAO) n°21, Food and Agriculture Organization of the United Nations, Rome, Italy.

Voulgaraki, S.I. & Papadakis, G. (2008) Simulation of a greenhouse solar heating system with seasonal storage in Greece. *Acta Horticulturae* (ISHS), 801, 757–764.

Vox, G., Blanco, I., Scarascia Mugnozza, G., Schettini, E., Bibbiani, C., Viola, C. & Campiotti, C.A. (2014) Solar absorption cooling system for greenhouse climate control: technical evaluation. *Acta Horticulturae* (ISHS), 1037, 533–538.

Vox, G., Schettini, E., Lisi Cervone, A. & Anifantis, A. (2008) Solar thermal collectors for greenhouse heating. *Acta Horticulturae* (ISHS), 801, 787–794.

Vox, G., Teitel, M., Pardossi, A., Minuto, A., Tinivella, F. & Schettini, E. (2010) Sustainable greenhouse systems. In: Salazar, A. & Rios, I. (eds) *Sustainable Agriculture: Technology, Planning and Management*. Nova Science Publishers Inc. New York. pp. 1–80.

Wenger, E. & Teitel, M. (2012) Design of a concentrated photovoltaic system for application in high tunnels. *Acta Horticulturae* (ISHS), 952, 401–407.

Werther, J., Saenger, M., Hartge, E.U., Ogada, T. & Siagi, Z. (2000) Combustion of agricultural residues. *Progress in Energy and Combustion Science*, 26(1), 1–27.

Xu, J., Li, Y., Wang, R.Z. & Liu, W. (2014) Performance investigation of a solar heating system with underground seasonal energy storage for greenhouse application. *Energy*, 67(0), 63–73.

Yaman, S., Şahan, M., Haykiri-açma, H., Şeşen, K. & Küçükbayrak, S. (2000) Production of fuel briquettes from olive refuse and paper mill waste. *Fuel Processing Technology*, 68(1), 23–31.

Yano, A., Kadowaki, M., Furue, A., Tamaki, N., Tanaka, T., Hiraki, E., Kato, Y., Ishizu, F. & Noda, S. (2010) Shading and electrical features of a photovoltaic array mounted inside the roof of an east–west oriented greenhouse. *Biosystems Engineering*, 106(4), 367–377.

Yano, A., Onoe, M. & Nakata, J. (2014) Prototype semi-transparent photovoltaic modules for greenhouse roof applications. *Biosystems Engineering*, 122, 62–73.

Yano, A., Tsuchiya, K., Nishi, K., Moriyama, T. & Ide, O. (2007) Development of a greenhouse side-ventilation controller driven by photovoltaic energy. *Biosystems Engineering*, 96(4), 633–641.

Zabri, A.W. & Burrage, S.W. (1997) The effects of vapour pressure deficit (VPD) and enrichment with CO_2 on water relations, photosynthesis, stomatal conductance and plant growth of sweet pepper (*Capsicum annum* L.) grown by NFT. *Acta Horticulturae* (ISHS), 449, 561–568.

Zaragoza, G., Buchholz, M., Jochum, P. & Pérez-Parra, J. (2007) Watergy project: towards a rational use of water in greenhouse agriculture and sustainable architecture. *Desalination*, 211(1–3), 296–303.

Zhao, Y., Meek, G.A., Levine, B.G. & Lunt, R.R. (2014) Near-infrared harvesting transparent luminescent solar concentrators. *Advanced Optical Materials*, 2(7), 606–611.

CHAPTER 9

Solar photocatalytic disinfection of water for reuse in irrigation

Pilar Fernández-Ibáñez, María Inmaculada Polo-López, Sixto Malato,
Alba Ruiz-Aguirre & Guillermo Zaragoza

9.1 INTRODUCTION

The close relationship between fresh water and human beings makes this element a globally vital resource. The planet's fresh water resources are mainly wells, reservoirs, rivers and lakes. These water resources are currently becoming scarce, and their limited accessibility and chemical and microbial pollution are affecting agriculture, industry and domestic activities.

In recent decades, alternative fresh water sources, such as the effluents of municipal wastewater treatment plants (MWTPs), have been sought. However, the wide diversity of microorganisms (some of them human pathogens) present in this type of water makes it necessary for the microbial load to be removed from wastewater before its reuse to ensure that water is safe for human consumption and to keep disease from spreading. The spectrum of microorganisms in wastewater effluents is very wide, from fungi, such as *Fusarium* spp. and *Phytophthora* spp. (associated with pests in agriculture), bacteria, such as *Escherichia coli* and *Enterococcus faecalis* (associated with risk of human illness from consumption of polluted water), and viruses (e.g. enteroviruses, rotavirus, hepatitis A) to protozoa (e.g. *Giardia lamblia*, *Cryptosporidium parvum*) and parasites (e.g. *Taenia* spp. and *Ancylostoma* spp.), among others.

There are many techniques for water disinfection. Physical processes, such as filtration and UV-C radiation, and chemicals, such as chlorine and ozone, are the most common large-scale treatments. All such techniques have some negative aspect, be it the high cost of maintenance and facilities, clogging of filters and/or lack of residual effect. The main disadvantage of chlorination is the formation of potentially mutagenic disinfection byproducts, such as trihalomethanes, when there is natural organic matter in the water. In agriculture, chemical treatments are typically used to remove pathogens; however, most such treatments are phytotoxic for plants, they may cause the accumulation of toxic compounds in the subsurface, and sometimes they are not effective due to microbial resistance. For these reasons, huge efforts are being made to find alternative techniques that can meet all the needs for removing microorganisms and toxic chemical compounds from water.

The success of advanced oxidation processes (AOPs) in removing hazardous chemical compounds and pathogens from water has been widely demonstrated in recent decades. AOPs are characterized by their generation of highly oxidizing species, especially the hydroxyl radical (OH$^\bullet$). The low specificity of this radical favors its reaction with any organic compound, resulting in complete mineralization to H_2O and CO_2. Among the AOPs available, those that can make use of natural sunlight as a source of photons, such as heterogeneous photocatalysis with TiO_2 and homogeneous photo-Fenton processes, are of special interest because of their low cost and low environmental impact. Other solar processes, such as hydrogen peroxide and sunlight (H_2O_2/solar), are also very attractive for water disinfection due to their high efficiency and low cost. The use of solar radiation for drinking water disinfection is also well known. Solar disinfection (SODIS) has the advantage of inactivating pathogens in water by means of the synergistic effect of mild temperatures and solar photons. Nevertheless, its efficiency is low and it has mainly been applied in isolated communities in developing countries with high solar irradiance and scarce fresh water resources because of its low cost.

In view of the above, the use of devices to enhance the collection of solar photons has also become an important natural solar radiation treatment issue. Among the different reactor configurations reported in the literature, the compound parabolic collector (CPC) has been shown to be the best option for solar water treatment (Malato *et al.*, 2009).

Concentration of pollutants facilitates the treatment of wastewater with AOPs. In general, low pollutant concentrations follow pseudo-first-order ($r = k_{ap}C$) degradation kinetics. This means that higher pollutant concentrations increase the reaction rate, thereby improving process efficiency. For water disinfection, use of a pathogen concentration method based on membrane technologies could also be of interest, as high-quality water can be produced at the same time as a stream of concentrate. Membrane distillation (MD) is an attractive option for increasing disinfection efficiency, as it combines mild temperature and selective membranes to guarantee the absence of pathogens in fresh water by producing a retentate containing a high concentration of microorganisms, which is easier to treat.

9.2   COMPOUND PARABOLIC COLLECTORS FOR WATER DISINFECTION

CPCs are stationary solar collectors with a non-imaging reflective surface. They collect all incident solar radiation on a given surface area and reflect it towards a receiver at their focus. In water treatment applications, the CPC mirror geometry is an ordinary involute of two truncated parabolas, so all the solar radiation below the acceptance angle reaches a tubular photoreactor (Fig. 9.1a). This geometry enables complete illumination of the whole tubular circumference. The photoreactor is made of borosilicate glass with 90% UV-A transmittance.

CPC reactors have the advantage of making highly efficient use of direct and diffuse solar radiation without solar tracking and, therefore, are a lower-cost system than solar-concentrating collectors. Water temperatures are low, favoring photochemical reaction efficiency (recommended <50°C for TiO_2 or photo-Fenton). A number of experimental studies have demonstrated that CPC photoreactors are a good option for solar photochemical treatment, removing both hazardous chemical compounds and pathogens from water (Malato *et al.*, 2009).

The development of CPC pilot plants for water disinfection led to the construction of the 60 L solar CPC photoreactor pilot plant shown in Figure 9.1b. This photoreactor consists of two CPC mirror modules placed on an anodized-aluminum platform. To enhance the solar radiation collection, CPC reactors may be tilted at the same angle as the local latitude. This inclination favors collection of solar radiation throughout the year (Malato *et al.*, 2009). At Plataforma Solar de Almería (PSA) (southeast Spain), CPC reactors are inclined at an angle of 37° relative to the horizontal plane (local latitude of PSA: 37°84′N). The total collector surface of the photoreactor is 4.5 m^2. Each module is made up of ten borosilicate glass tubes, each 1.5 m long and 50 mm outer diameter. The CPC mirror is made of highly reflective anodized aluminum with a concentration factor of one. The ratio of irradiated water to total water is 75%, and the total treatment volume is 60 L. Water is circulated through the tubes to a tank by a centrifugal pump (150 W). Dissolved oxygen (DO), pH and temperature sensors are installed in the pipes to continuously monitor these parameters, which are recorded with data-acquisition software (Polo-López *et al.*, 2010).

Measurement of the solar UV radiation is essential for proper evaluation of the results of solar photochemical water treatment because, under natural sunlight, the incident radiation during the experiment varies. UV radiation is measured with a global UV-A pyranometer (295–385 nm) placed near the reactor and at the same inclination angle (37°). The pyranometer provides incident UV-A irradiance data in W m^{-2}. Inactivation kinetics during disinfection depend not only on elapsed time but also on the total amount of UV-A photons received in the system during solar tests. The cumulative UV-A energy per unit of volume (Q_{UV}) received in the photoreactor enables

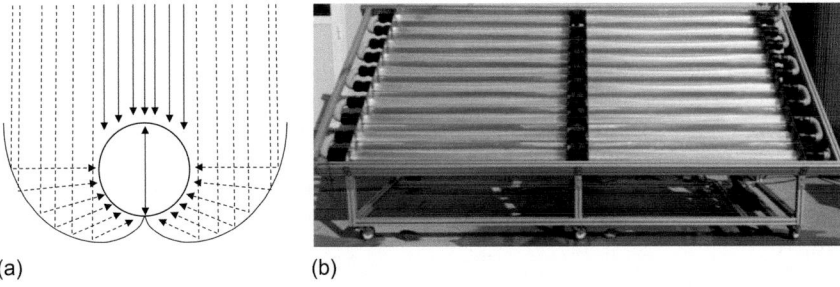

Figure 9.1. (a) Diagram of solar photon flux incident on the aperture area of a CPC mirror with tubular photoreactor in the focus of the mirror; (b) 60 L recirculating CPC reactors for water disinfection at the PSA facility.

comparison of results under different irradiation conditions, on different days for example, as calculated by:

$$Q_{UV,n} = Q_{UV,n-1} + \Delta t_n \overline{UV}_{G,n} A_r / V_t;$$

$$\Delta t_n = t_n - t_{n-1} \tag{9.1}$$

where $Q_{UV,n}$ and $Q_{UV,n-1}$ are the UV-A energies received per liter $[\text{kJ L}^{-1}]$ at times n and $n-1$, $UV_{G,n}$ is the average incident radiation on the irradiated area, t_n is the sample experimental time n [s], A_r is the illuminated area of the solar collector $[\text{m}^2]$ and V_t is the total volume [L] (Fernández-Ibáñez *et al.*, 2009).

9.3 SOLAR WATER DISINFECTION (SODIS)

The first successful application of solar radiation for water disinfection was demonstrated by Acra *et al.* (1980). They used sunlight for disinfection of oral rehydration solutions taken to developing countries as part of a World Health Organization (WHO) disease-control program. The full potential of SODIS for inactivation in water of microorganisms such as *E. coli* (and other coliforms), *Cryptosporidium* spp. and cysts of *Giardia muris* has since been well reported in the literature (McGuigan *et al.*, 2012), and has been recommended by WHO as one of the best options for disinfecting household water (WHO, 2007). The SODIS treatment consists of filling a 1.5 L poly(ethylene) terephthalate (PET) bottle with polluted water and exposing it to direct sunlight for 6 h (or two consecutive days in the case of cloudy conditions) (SODIS, 2011).

SODIS inactivates microorganisms in water mainly by means of the short UV-B (280–320 nm) and the total UV-A (320–400 nm) wavelengths of the solar spectrum. This radiation generates oxidative radicals – e.g. hydroxyl radical (OH$^\bullet$), superoxide (O$_2^{\bullet-}$), hydrogen peroxide (H$_2$O$_2$) – that produce alterations in DNA strands, proteins and other compounds, causing cell death after prolonged exposure (Wegelin *et al.*, 1994). Moreover, temperature is also important for SODIS treatment. Differences in bacterial inactivation between 12 and 40°C have been found to be negligible. However, when the temperature rises to 50°C, the bactericidal action is doubled, due to the strong synergy between UV radiation and the thermal effect (McGuigan *et al.*, 1998).

The impact of SODIS for household drinking water on community health in developing countries has recently been studied. Randomized controlled trials have evaluated the incidence of diseases such as dysentery and non-dysentery diarrhoea among children from six months to five years of age living in peri-urban and rural communities in Kenya (du Preez *et al.*, 2011) and Cambodia (McGuigan *et al.*, 2011). These trials demonstrated that children drinking water treated by SODIS had a lower incidence of dysentery and non-dysentery diarrhoea than control groups (water not treated by SODIS). Moreover, du Preez *et al.* (2011) found that the SODIS treatment influenced height and weight in the same children in Kenya, where median weight- and height-for-age increased significantly in the SODIS group, corresponding to an average of 0.8 cm in height over a one-year period for the group as a whole, with median weight-for-age 0.23 kg higher over the same period.

Nevertheless, the SODIS treatment also has several disadvantages. It is limited by the availability of solar radiation reaching the earth's surface, the long exposure required, the volume of water treated (1.5 L per bottle), and the resistance of some pathogens to inactivation by solar radiation and mild temperature alone. Therefore, other treatments are being studied to solve these limitations.

9.4   ADVANCED OXIDATION PROCESSES (AOPs)

9.4.1   *Photocatalysis with titanium dioxide (TiO₂/solar)*

Semiconductor photocatalysis by TiO_2 has been widely studied for 50 years. When irradiated at wavelengths below 380 nm, an electron/hole pair is generated on the surface of this semiconductor catalyst. The O_2 electron acceptor generates a superoxide radical, and the H_2O electron donor generates a hydroxyl radical (Carey *et al.*, 1976). Heterogeneous photocatalysis with TiO_2 is one of the technologies most studied for removal of both organic chemical compounds and microorganisms from water. Numerous studies have demonstrated the effectiveness of TiO_2 in the inactivation of microorganisms of different natures (Markowska-Szczupak *et al.*, 2011).

The TiO_2 inactivation mechanism is based on OH• attack on the cell wall, which disrupts cell wall functionality and eventually causes cell death (Blanco-Gálvez *et al.*, 2007). Internal damage can also be caused by very small particles of TiO_2 that affect vital cell functions, culminating in cell death. Figure 9.2a shows the TiO_2/solar inactivation mechanisms. In addition, it is well known that adsorption of the catalyst on the cell surface enhances the inactivation mechanism (Polo-López *et al.*, 2010). Adsorption is based on electrostatic attraction between the catalyst and cell surfaces, because the cell surface has a negative charge, while the surface charge of the TiO_2 particle depends on the pH, being positive at acid pHs (favoring adsorption) and negative when the pH is basic. Adsorption enables OH• radicals to be generated directly on the microorganism surface, thus enhancing inactivation efficiency. Figure 9.2b shows *Fusarium* spp. (Polo-López *et al.*, 2010) and *Phytophthora* sp. spores in water and surrounded by TiO_2 particles.

The photocatalytic efficiency of this treatment is strongly dependent on catalyst concentration. To reach the maximum photocatalytic yield, catalyst concentrations must be tested to find the best photoreactor conditions. Figure 9.3a shows inactivation of *F. solani* spores in a 200 mL solar

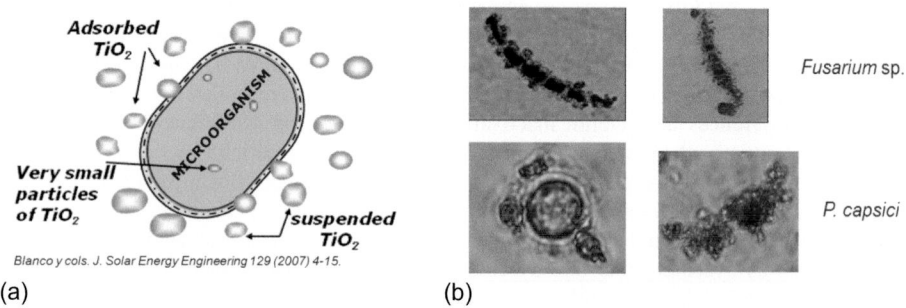

(a)   (b)

Figure 9.2.   (a) TiO_2 disinfection mechanisms (modified from Blanco-Gálvez *et al.*, 2007); (b) optical images of *Fusarium* (Polo-López *et al.*, 2010) and *Phytophthora capsici* spores surrounded by TiO_2 particles during photocatalytic treatment.

stirred batch reactor (SSBR) at several TiO$_2$ concentrations. It was observed that the higher the catalyst concentration, the higher the photocatalytic inactivation efficiency, up to a concentration of 35 mg L^{-1}, where inactivation results were best; higher concentrations did not lead to better inactivation.

A similar trend was observed in a 14 L CPC reactor with borosilicate tubes of 50 mm outer diameter. However, in this case, the best inactivation efficiency was found with a catalyst concentration of 100 mg L^{-1} (Fig. 9.3b). In both kinds of photoreactor, inactivation efficiency decreased as catalyst concentration increased or decreased from the optimum. This is because, at low catalyst concentrations, OH$^{\bullet}$ generation is insufficient to inactivate the pathogens, and some of the solar radiation is not used in the treatment. At high catalyst loads, part of the radiation is scattered and adsorbed by the catalyst, and not all of the sample is illuminated. This may prevent damage to microorganisms from the action of direct sunlight and OH$^{\bullet}$ generation in unlit areas. In both cases, the complete inactivation of microorganisms cannot be attained and microbial recovery and regrowth during the post-treatment period (i.e. storage) may occur. On the other hand, these results highlight the importance of the photoreactor diameter, which determines the path length of the solar radiation through the photoreactor, and affects the optimal catalyst concentration for the best photocatalytic inactivation results. Hence, to achieve the best inactivation efficiency, a lower catalyst load is required with a longer SSBR light-path length (11 cm internal diameter), and vice versa (Fernández-Ibáñez *et al.*, 2009).

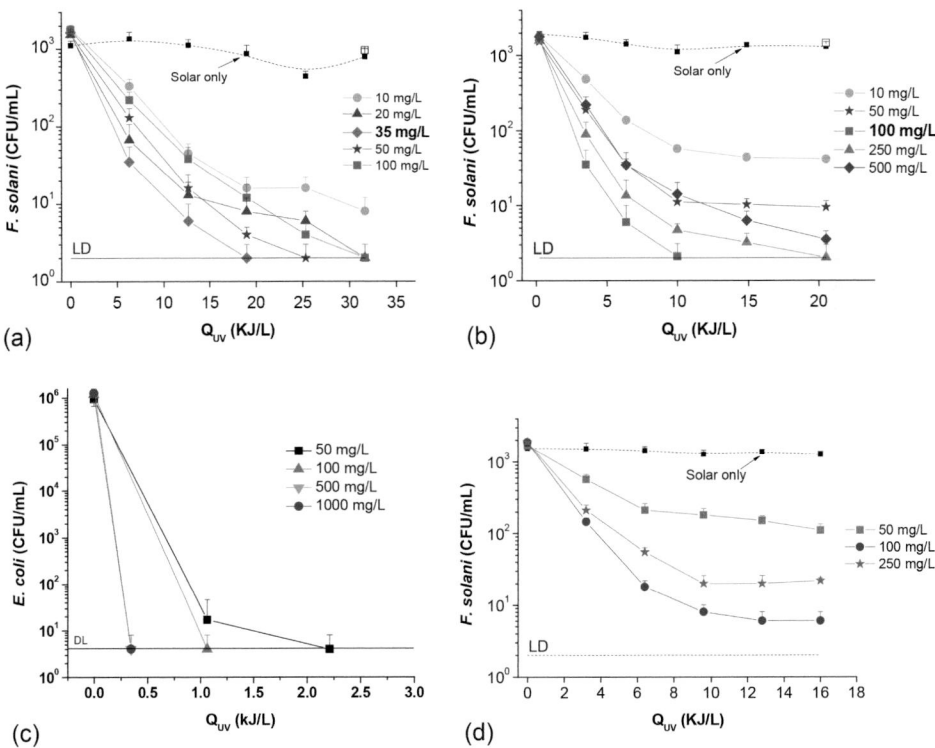

Figure 9.3. Inactivation of *F. solani* at several concentrations of TiO$_2$ catalyst in: (a) 200 mL solar stirred reactor; (b) 14 L CPC photoreactor in distilled water (modified from Fernández-Ibáñez *et al.*, 2009). (c) Inactivation of *E. coli* at several concentrations of TiO$_2$ catalyst in 200 mL solar stirred photoreactor (modified from Helali *et al.*, 2013). (d) Inactivation of *F. solani* at several concentrations of TiO$_2$ catalyst in well water in 14 L CPC photoreactor (modified from Fernández-Ibáñez *et al.*, 2009).

Another important efficiency parameter is the type of target, because it may also influence the optimal catalyst concentration. Figure 9.3c shows inactivation of *E. coli* K12 in distilled water in a 200 mL SSBR. Several TiO_2 catalyst (Evonik P25) concentrations (from 0.05 to $1 g L^{-1}$) were tested under sunlight. The best inactivation results in these experiments were obtained with $500 mg L^{-1}$ of catalyst (Helali *et al.*, 2013). Other authors have studied the influence under constant illumination of TiO_2 concentration on the *E. coli* inactivation rate and similar results, that is, an optimal TiO_2 concentration for bacterial inactivation of 0.25 to $1 g L^{-1}$, have been reported (Benabbou *et al.*, 2007; Rincón and Pulgarín, 2003; Wei *et al.*, 1994). These results do not agree with those found by Fernández-Ibáñez *et al.* (2009) for fungi, that is, *Fusarium* spores ($35 mg L^{-1}$ in a 200 mL SSBR), which highlights the importance of the type of pathogen in determining the catalyst concentration for the best photocatalytic efficiency.

Besides the above, factors such as chemical compounds in the water matrix can affect the photocatalytic inactivation efficiency. It is well known that inorganic salts, such as carbonates/bicarbonates, in the water matrix are a limiting factor in photocatalytic disinfection, because HCO_3^- reacts with the hydroxyl radicals, producing the less reactive anion radical $CO_3^{\bullet-}$ and scavenging $OH^\bullet$ radicals (Buxton and Elliot, 1986). This effect may be observed in Figure 9.3d, which shows the inactivation kinetics in a 14 L CPC reactor of *Fusarium solani* in well water, which contains a higher concentration of carbonates/bicarbonates. The inactivation efficiency in well water is observed to be much lower (Fig. 9.3d) than in distilled water (Fig. 9.3b) under similar operating conditions. On the other hand, the presence of organic matter in the water matrix also reduces the efficiency of the photocatalytic treatment. Recent studies on photocatalytic disinfection treatments carried out with municipal wastewater effluent (MWWE) containing concentrations of dissolved organic carbon (DOC) in the range of 10–20 mg/L have demonstrated a reduction in the photocatalytic efficiency compared with water matrices without organic matter. This effect is attributed to competition between organic compounds and microorganisms for generated $OH^\bullet$ radicals (García-Fernández *et al.*, 2015; Polo-López *et al.*, 2014).

9.4.2 *Photocatalysis with the photo-Fenton process (Fe/H₂O₂/solar)*

The photocatalytic Fenton-based treatment essentially consists of a catalytic cycle in which $OH^\bullet$ radicals are generated by Fe^{2+} and H_2O_2. The main reaction involved in this catalytic cycle is shown in Equation (9.2). However, the efficiency of the treatment increases when the system is irradiated at wavelengths below 580 nm (photo-Fenton) because Fe^{3+} aqua complexes are reduced to Fe^{2+}, generating an extra $OH^\bullet$ (Equation (9.3)) and enabling the cycle to restart. This is called the photo-Fenton treatment (Malato *et al.*, 2009).

$$Fe^{2+} + H_2O_2 \rightarrow Fe^{3+} + OH^- + OH^\bullet \qquad (K \approx 70 M^{-1} s^{-1}) \qquad (9.2)$$

$$Fe(OH)^{2+} + h\nu \rightarrow Fe^{2+} + OH^\bullet \qquad (9.3)$$

The first study using a photo-Fenton process for water disinfection was conducted in 2005 by Rincón and Pulgarín, who investigated the efficiency of photo-Fenton treatment in inactivating *E. coli* cells with low reagent concentrations ($0.3 mg L^{-1}$ Fe; $10 mg L^{-1}$ H_2O_2) with very successful results (Rincón and Pulgarín, 2005). Since then, the treatment of other pathogens, and the chemical parameters related to the iron, have been investigated.

One of the most important parameters for photo-Fenton treatment efficiency is the pH, which is optimal at 2.8. This parameter affects the solubility of the iron in aqueous solution and the type of aqua complexes formed. At near-neutral pH, most of the active iron is lost from the solution due to precipitation, thereby reducing the efficiency of the treatment (Pignatello *et al.*, 2006), and at pH close to 7, the complexes formed are not photoactive. The influence of pH on the inactivation efficiency of *Fusarium* sp. spores by photo-Fenton treatment is shown in Figure 9.4. These experiments were carried out in a 200 mL SSBR using $5 mg L^{-1}$ of Fe^{2+} and $10 mg L^{-1}$

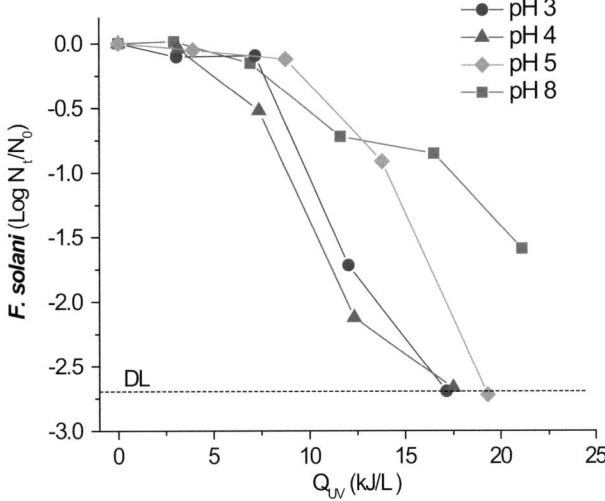

Figure 9.4.   *Fusarium* sp spore inactivation by photo-Fenton treatment in simulated wastewater effluent at different pH values (modified from Polo-López *et al.*, 2012).

of H_2O_2 at pH values of 3, 4, 5 and 8 in simulated wastewater effluent (SWWE) characterized by $25 \, mg \, L^{-1}$ of DOC (Polo-López *et al.*, 2012). As expected, inactivation efficiency increased at acidic pH values, but no significant differences were observed between pH values 3 and 4. These results show that the amount of $OH^•$ generated at pH 4 is enough to cause a strong negative effect on microbial viability. On the other hand, DOC concentration, which was also monitored, decreased significantly more at pH 3 than at pH 4. This result demonstrates the difference between removal from water of chemical organic compounds and of pathogens.

The use of photo-Fenton treatment for disinfection at the optimal pH (2.8) has several disadvantages: (i) the acidic pH has a very negative effect on the viability of some pathogens (such as *E. coli* and related bacteria), and some microorganism inactivation may be due to the low pH rather than the solar treatment; (ii) acidification and neutralization of the pH before and after the treatment increases both the cost and the water salinity. Therefore, research is currently focusing on the operating conditions for water disinfection treatment at near-neutral pH values. The solar photo-Fenton process at near-neutral pH has recently been investigated and has proved to be a promising treatment for the disinfection of water containing different kinds of microorganisms, such as bacteria, fungi, viruses, etc. (García-Fernández *et al.*, 2012; Ortega-Gómez *et al.*, 2013, 2014; Polo-López *et al.*, 2013; Rodríguez-Chueca *et al.*, 2014; Spuhler *et al.*, 2010).

Another parameter evaluated in water disinfection is the influence of the iron salt used for the treatment. The effect on photo-Fenton treatment efficiency in removing *Phytophthora capsici* zoospores in distilled water in a 200 mL SSBR under natural solar radiation was recently studied using two different iron sources, ferrous sulfate (Fe^{2+}) and ferric nitrate (Fe^{3+}) (Polo-López *et al.*, 2013). The highest inactivation rate was measured with $5 \, mg \, L^{-1}$ of Fe^{3+} and $10 \, mg \, L^{-1}$ of H_2O_2, which required a solar UV dose of $2.5 \, kJ \, L^{-1}$. Inactivation kinetics using $5 \, mg \, L^{-1}$ of Fe^{2+} and $10 \, mg \, L^{-1}$ of H_2O_2 required a solar UV dose of $6 \, kJ \, L^{-1}$ (Polo-López *et al.*, 2013). The effect of Fe^{2+} and Fe^{3+} ($0.6 \, mg \, L^{-1}$) with UV/vis radiation on *E. coli* inactivation has also been investigated (Spuhler *et al.*, 2010): in contrast to the results described above for *P. capsici*, this research demonstrated better inactivation in demineralized water with Fe^{2+} than with Fe^{3+}.

The different results reported with these two iron salts may be due to the nature of the microorganisms evaluated and the different roles of the iron speciation in the cells. Fe^{2+} may become diffused within the cells and, when the internal iron concentration increases, the possibility of $OH^•$ generation via intracellular Fenton reactions also increases, as per Equations (9.2) and (9.3)

(Spuhler *et al.*, 2010). On the other hand, Fe^{3+} is likely to be adsorbed onto cell walls, generating exciplexes on the bacterial membrane. This is promoted by the transport of Fe^{3+} into the cells via its binding to specific proteins; photosensitization of these exciplexes may lead to direct oxidation of the membrane and/or the generation of Fe^{2+} and $OH^{\bullet}$ immediately adjacent to the microorganism (Spuhler *et al.*, 2010). The high presence of iron exchange-specific sites on the zoospore cell wall and the influence of osmotic force may cause a high Fe^{3+} adsorption rate in this kind of microorganism (Morris and Phuntumart, 2009; Polo-López *et al.*, 2013), which could enhance inactivation under solar radiation. This also explains why higher Fe^{3+} concentrations achieve complete inactivation of *P. capsici* spores more quickly than lower ones.

9.4.3   *Solar-driven process with H_2O_2*

Hydrogen peroxide photolysis occurs when it is irradiated by photons at wavelengths below 300 nm, yielding two $OH^{\bullet}$ radicals (Goldstein *et al.*, 2007). However, solar radiation at the earth's surface contains only a few photons below 300 nm (down to 280 nm, UV-B), and almost all of them are absorbed by the vessel wall (plastic, glass) through which the water is illuminated. Therefore, solar energy is inefficient for the generation of hydroxyl radicals by this pathway, and the combined effect of solar energy and H_2O_2 had not been investigated as a disinfection method until recently. The lethal synergy of H_2O_2 and near-UV light was first reported for phage T7 in 1977 (Anathaswamy and Eisenstark, 1977). Nevertheless, the use of H_2O_2 with sunlight has latterly awakened great interest in solar water disinfection, due to its strong germicidal effect on several microorganisms. Sichel *et al.* (2009) observed very good inactivation of *Fusarium* sp. spores under natural sunlight and various concentrations of H_2O_2 (from 0 to 500 mg L^{-1}). Since then, other studies have also reported the strong synergistic effect of H_2O_2 and sunlight with different reactor configurations and microorganisms, but all using very low H_2O_2 concentrations (<50 mg L^{-1}).

The efficiency of the H_2O_2/solar treatment at a low reagent concentration (10 mg L^{-1}) in inactivating *Fusarium* sp. (Polo-López *et al.*, 2011) and *Phytophthora capsici* (Polo-López *et al.*, 2013) in a 200 mL SSBR is shown in Figures 9.5a and 9.5b, respectively. In both cases, complete inactivation of pathogens was achieved in all three water matrices: distilled water, well water and SWWE. Better inactivation was found when H_2O_2 was added to the sample than with solar radiation alone. In addition, more accumulated solar UV-A energy was required in the distilled water sample than in SWWE (Polo-López *et al.*, 2011). Spuhler *et al.* (2010) also showed the benefit of using H_2O_2 to inactivate *E. coli* in different water solutions, with and without the

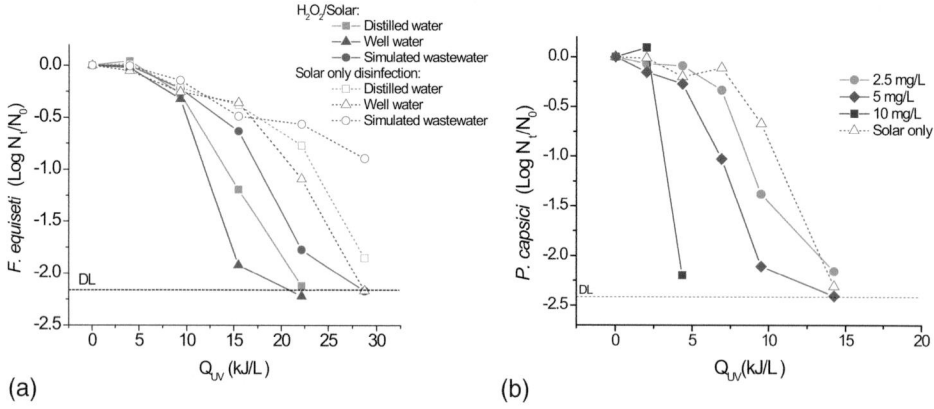

Figure 9.5.   (a) *Fusarium equiseti* inactivation by sunlight and H_2O_2/sunlight (10 mg L^{-1}) in different water matrices in a 200 mL SSBR (modified from Polo-López *et al.*, 2011); (b) *Phytophthora capsici* inactivation by sunlight and H_2O_2/solar with 2.5, 5 and 10 mg L^{-1} in a 200 mL SSBR (modified from Polo-López *et al.*, 2013).

addition of chemical inorganic compounds and natural organic matter. García-Fernández *et al.* (2012) reported similar inactivation results for *F. solani* spores and *E. coli* in distilled water with several low concentrations of H_2O_2.

The inactivation mechanisms of this solar treatment are still unknown; however, cell biology could be of help in explaining them. The inactivation hypothesis is supported by two fundamental facts: (i) diffusion of free H_2O_2 inside cells as an uncharged molecule; (ii) the presence of iron and other metals such as copper inside cells as part of their molecular and biological structure. Therefore, the possibility of H_2O_2 diffusion inside cells and its reaction with iron inside cells may result in the formation of reactive oxygen species, such as $OH^\bullet$, by a photo-Fenton reaction (Equations (9.2) and (9.3)). The radicals generated inside the cells attack the internal components and structures, which impairs cell functionality, ending in cell death (Polo-López *et al.*, 2011; Sichel *et al.*, 2009; Spuhler *et al.*, 2010).

The main advantages of this solar treatment are: (i) the potentially low cost due to the use of small amounts of hydrogen peroxide and sunlight as the photon source; (ii) the effluent is safe for irrigation since concentrations of H_2O_2 below $50 \, mg \, L^{-1}$ are non-toxic for crops (Sichel *et al.*, 2009); (iii) decomposition of H_2O_2 into water and oxygen means the reagent is consumed by the treatment, so there is no concern for secondary pollution or pH correction. Other AOPs, such as titanium dioxide or photo-Fenton treatments, have the disadvantages of post-treatment requirements, which involve TiO_2 catalyst or iron removal, or pH modification.

9.5 DISINFECTION OF WASTEWATER EFFLUENTS BY SOLAR TREATMENTS

The use of TiO_2/solar, photo-Fenton and H_2O_2/solar treatments have successfully removed pathogens from MWWEs. The main challenge for disinfection of wastewater matrices is the high chemical and pathogen load. Table 9.1 shows the characteristics of wastewater effluent from the El Bobar MWTP, located in south-east Spain. The high concentrations of carbonates/bicarbonates, the presence of organic matter and a high load of microorganisms ($10^4 \, CFU \, mL^{-1}$ of *E. coli*) reduced or limited photocatalytic inactivation efficiency.

Figure 9.6 shows the inactivation of naturally occurring fungi (*Fusarium* spp., *Aspergillus* spp., *Candida* sp., etc.) in MWWE by solar treatment using the 60 L CPC photoreactor described above. Fungi were completely removed in all the solar treatments evaluated. The relative efficiency of inactivation was as follows: H_2O_2/solar ($10 \, mg \, L^{-1}$) > photo-Fenton ($5 \, mg \, L^{-1}$ of Fe^{2+}, $10 \, mg \, L^{-1}$ of H_2O_2 at pH 3) > TiO_2/solar ($100 \, mg \, L^{-1}$).

Other studies on microorganism removal from MWWEs by AOPs under natural sunlight have been performed. Agulló-Barceló *et al.* (2013) reported inactivation of naturally occurring *E. coli*, somatic coliphages, F-specific RNA bacteriophages and *Clostridium* sp. in a 10 L

Table 9.1. Physicochemical characteristics of municipal wastewater treatment plant effluent (modified from Rodríguez-Chueca *et al.*, 2014).

Characteristic [unit]		Characteristic [unit]	
Na^+ [mg L^{-1}]	211.40 ± 22.80	SO_4^{2-} [mg L^{-1}]	102.60 ± 28.80
NH_4^+ [mg L^{-1}]	32.00 ± 11.70	Cl^- [mg L^{-1}]	337.50 ± 10.80
K^+ [mg L^{-1}]	33.10 ± 5.80	NO_3^- [mg L^{-1}]	23.50 ± 16.00
Mg^{2+} [mg L^{-1}]	48.00 ± 5.90	PO_4^{3-} [mg L^{-1}]	17.10 ± 29.80
Ca^{2+} [mg L^{-1}]	117.00 ± 10.30	Conductivity [µS cm^{-1}]	1458 ± 89.80
pH	7.53 ± 0.10	Turbidity [NTU]	14.60 ± 6.62
DOC [mg L^{-1}]	17.00 ± 3.00	DIC [mg L^{-1}]	56.50 ± 6.60
E. coli [CFU mL^{-1}]	10^3	*E. faecalis* [CFU mL^{-1}]	10^3

DOC, dissolved organic carbon; DIC, dissolved inorganic carbon.

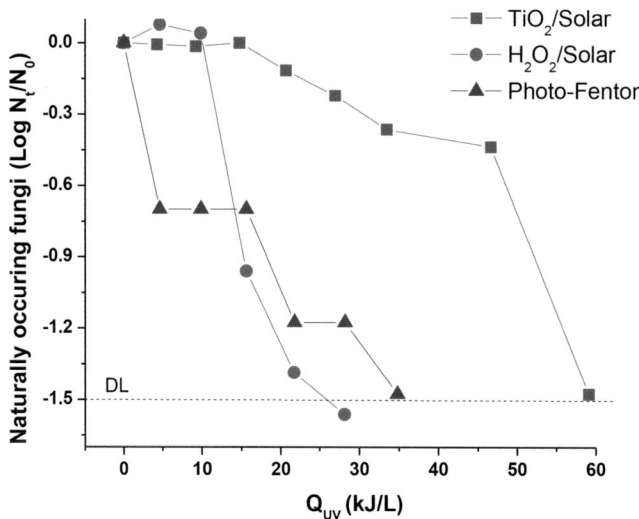

Figure 9.6.   Inactivation of naturally occurring fungi in MWWE in a 60 L CPC reactor using TiO_2/solar $(100\,mg\,L^{-1})$, H_2O_2/solar $(10\,mg\,L^{-1})$ and photo-Fenton $(5\,mg\,L^{-1}$ of Fe^{2+}, $10\,mg\,L^{-1}$ of H_2O_2 at pH 3).

CPC reactor. Efficiencies were different for each group of pathogens. Inactivation efficiency for *E. coli* was highest with photo-Fenton treatment at pH 3, followed by H_2O_2 $(20\,mg\,L^{-1})$/solar, then TiO_2/solar and, finally, solar photo-inactivation. For viral indicators, the order was photo-Fenton at pH 3 > TiO_2/solar > H_2O_2 $(20\,mg\,L^{-1})$/solar > solar photo-inactivation. Sulfite-reducing clostridia (SRC) was the most resistant microorganism indicator of all the processes evaluated (Agulló-Barceló *et al.*, 2013). In a recent publication, Rodríguez-Chueca *et al.* (2014) reported very good inactivation efficiency in the removal of naturally occurring *E. coli* and *E. faecalis* from wastewater effluents in a 10 L CPC reactor using photo-Fenton treatment at pH 5.

Reuse of treated wastewater for agriculture is one of the most obvious final uses. An experimental study on wastewater disinfection with H_2O_2/solar for reuse in irrigating lettuce was conducted by Bichai *et al.* (2012). Lettuce is grown worldwide for raw consumption and, therefore, its irrigation with polluted water increases the risk to human health. The impact of pathogens in the water on H_2O_2/solar water disinfection efficiency was investigated in distilled water, well water, SWWE and MWWE from the El Bobar (Almería) MWTP using 25 L static CPC batch reactors and 2.5 L PET bottles. *E. coli* inactivation was complete in all cases, reducing the bacterial load to the limit of detection $(2\,CFU\,mL^{-1})$ from $10^4\,CFU\,mL^{-1}$ in the real wastewater effluent, and from 10^6 seeded *E. coli* (ATCC 23631) in the distilled water, well water and SWWE. The H_2O_2 concentrations tested were 5 and $10\,mg\,L^{-1}$. Figure 9.7 shows the results with H_2O_2 in the CPC reactor. In all cases, it was observed that the addition of a small concentration of H_2O_2 led to very good inactivation.

The treated wastewater effluent was used for lettuce irrigation, and the presence/absence of *E. coli* in the lettuce leaves was evaluated. Positive (mineral water spiked with $10^6\,CFU\,mL^{-1}$ of *E. coli*) and negative (mineral water only) controls were also evaluated. Results showed that, out of a total of 28 crop samples irrigated with solar- or solar/H_2O_2-disinfected wastewater effluent, 26 showed no presence of *E. coli*. These results were the first experimental observations of the potential for low-cost solar disinfection to enhance the microbial safety of wastewater reuse in agricultural irrigation (Bichai *et al.*, 2012).

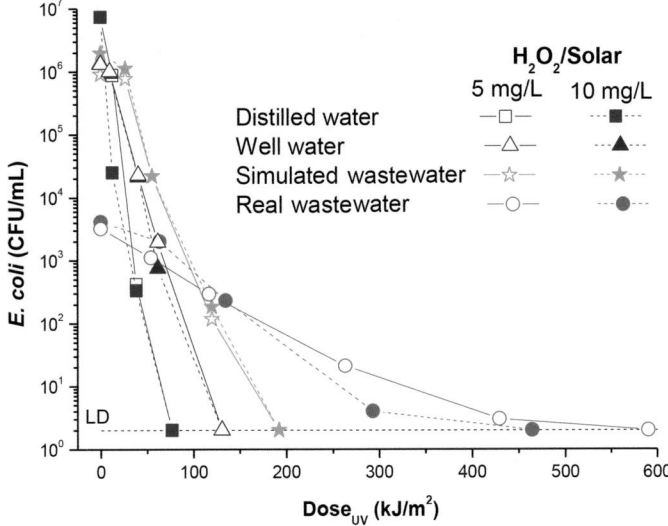

Figure 9.7.   *E. coli* inactivation in different types of water using 25 L static batch solar reactor with H_2O_2/solar (5 and $10\,mg\,L^{-1}$) (modified from Bichai *et al.*, 2012).

9.6   IMPROVEMENTS IN WATER DISINFECTION BY MEMBRANE DISTILLATION PRETREATMENT

Membrane distillation (MD) is a non-isothermal separation process, in which a hydrophobic microporous membrane allows vapor or other volatile compounds to pass through it while retaining the liquid and the solutes. A temperature difference between the two sides of the membrane creates a difference in vapor pressure that, in turn, drives the process (Khayet, 2011). Because the membrane is hydrophobic, liquids do not penetrate it unless the hydrostatic pressure exceeds a certain limit (the liquid entry pressure, LEP).

MD is used to concentrate ions, colloids or other non-volatile compounds in aqueous dilutions. As a wastewater pretreatment for AOPs, not only does it improve the downstream AOP by concentrating the microorganisms, but the raised temperature in MD also contributes to their inactivation or weakening. Furthermore, the distillate is free of microorganisms and can be added to the clean water supply.

The most important advantages of MD over other filtration mechanisms are (Alkhudhiri *et al.*, 2012):

- The process is not driven by absolute pressure; therefore the size of the pores is larger (from 0.2 to 1 μm), the risk of clogging is much lower, and water of much worse quality, with higher solute concentration, can be treated.
- The feedwater does not usually require chemical pretreatment before entering the modules, just a simple pre-filtration.
- It is a low-energy-demand process, which can be run at atmospheric pressure and at temperatures ranging from 60 to 90°C. Therefore low-grade heat or renewable energies, such as solar with non-concentrating collectors, can be used.
- Since it is an evaporative process, the quality of the distillate is very good, regardless of the feedwater.

Different configurations are used to establish the vapor pressure difference across the membrane which drives the MD process. The simplest uses a coolant in direct contact with the permeate side of the membrane (direct contact membrane distillation, DCMD). The vapor condenses in

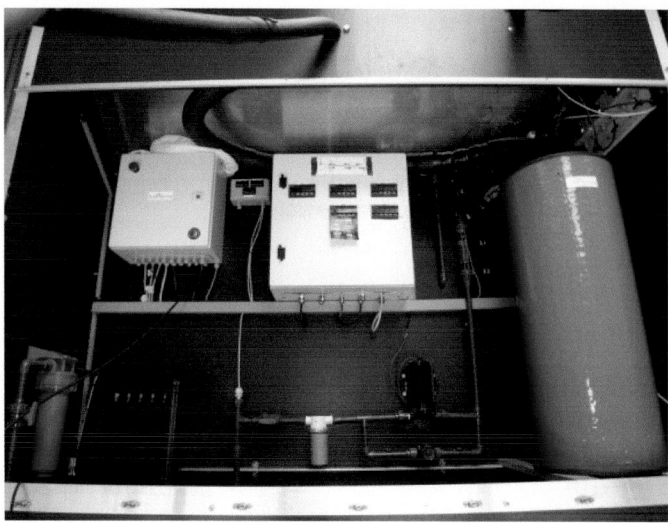

Figure 9.8.   Oryx 150 module at the PSA facility.

the liquid-vapor interface created on the other side of the membrane by the cooling solution. In air-gap membrane distillation (AGMD), a layer of stagnant air is introduced between the permeate side of the membrane and a condensing surface in contact with the cooling solution. This improves the energy efficiency by decreasing conductive heat losses across the membrane but reduces the distillate production, because the air gap increases mass transfer resistance. This can be mitigated if the distillate is left inside the gap as it is produced, until it finally leaves the module by overflowing (liquid-gap membrane distillation, LGMD). In other configurations, the vapor is extracted from the module to an external condenser using a cold inert gas (sweeping gas membrane distillation, SGMD) or a vacuum (vacuum membrane distillation, VMD).

Most current MD research is at laboratory scale, and most results are satisfactory, with distillate fluxes larger than those of mechanical filtration processes such as reverse osmosis. However, pilot-size energy efficiency and distillate flux are not yet optimal (Zaragoza *et al.*, 2014).

Figure 9.8 shows the Oryx 150 MD system designed by the Fraunhofer Institute for Solar Energy Systems, Germany, and marketed by Solar Spring GmbH. The Oryx 150 concentrates pathogens in wastewater. It has a spiral-wound module with an LGMD configuration and a 70 µm thick polytetrafluoroethylene (PTFE) membrane, with a 0.2 µm average pore size, 80% porosity and 10 m^2 effective area. The module is made up of three channels: the evaporator, the condenser and the distillate channel, which is located between the first two and fills up during the operation, releasing the distilled water by overflow. The module is built entirely of plastic. In addition to the PTFE membrane, the condenser is an ethylene tetrafluoroethylene (ETFE) foil, the spacers separating the channels are low-density polyethylene, and the shell is made of glass-fibre-reinforced plastic (Winter *et al.*, 2011). The system is designed to maximize internal heat recovery, so feedwater is pumped into the condenser channel, where its temperature is raised by the latent heat of condensation, and then flows out of the module through a heat exchanger in a countercurrent to the flow from the heat source, re-entering the module through the evaporator channel. The maximum temperature in the evaporator channel is 85°C. In these experiments, heat was supplied by a solar thermal field of stationary collectors.

This system has been used successfully to produce clean water from feedwater contaminated by *F. solani* spores and *E. coli*, which were inactivated by the thermal effect. During the MD procedure, water temperatures can be as high as 80°C, which is detrimental to thermophiles, such as *Fusarium* spp. and *E. coli*. Using combined MD (as pretreatment) and solar AOP technologies,

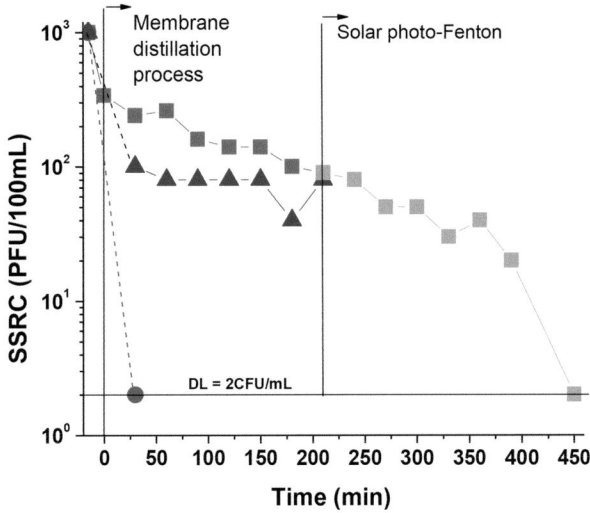

Figure 9.9. Removal of SSRC by combined MD and photo-Fenton processes. Concentration of *Clostridium* sp. spores in feed tank (-■-), in 60 L CPC photoreactor (-■-), concentrate outlet (-▲-) and distillate outlet (-●-) (modified from Ruiz-Aguirre *et al.*, 2015).

spores of SRC (SSRC) were evaluated because: (i) they are found in wastewater (at concentrations of ~10^3 CFU mL^{-1}); (ii) they resist temperatures up to 80°C; (iii) they are not easily removed with traditional disinfection techniques (Ruiz-Aguirre *et al.*, 2015). This genus has been used as an indicator of the presence of *Cryptosporidium* in water, and also of pathogenic microorganisms in drinking water by European Community legislators (Committee on Indicators for Waterborne Pathogens, 2004).

Tests were carried out using effluent from the El Bobar (Almería) MWTP, which is naturally contaminated by SSRC. The feed was diluted with demineralized water in a proportion of 1:2 to a concentration of 340 SSRC CFU per 100 mL. This feedwater was pumped into the MD module. Then the distillate produced was collected in a tank while the concentrate was returned to the feed tank. Recirculation lasted 210 minutes. Throughout the test, samples were taken every 30 minutes from the feed tank, concentrate outlet and distillate outlet. The temperature in the evaporator channel increased from 50 to 75°C and the feed flow rate was 600 L h^{-1}, producing a total of 100 L of distillate (Ruiz-Aguirre *et al.*, 2015).

The results showed that the distillate was completely free of microorganisms and, in the concentrate, the population of *Clostridium* sp. spores fell to 90 CFU per 100 mL (Fig. 9.9), rather than increasing. Although *Clostridium* sp. spores resist high temperatures, some of them may have been more sensitive to the thermal stress and been killed.

Then 60 L of the concentrate feed were subjected to solar photo-Fenton treatment for four hours. The pH was adjusted to 4, dissolving 60% of the total iron. This pH favors the removal of carbonates and bicarbonates, increasing process efficiency. The starting concentrations of Fe^{2+} and H$_2$O$_2$ were 10 mg L^{-1} and 20 mg L^{-1}, respectively. However, several 20 mg L^{-1} doses of H$_2$O$_2$ had to be added due to the consumption of peroxide, up to a total amount of 140 mg L^{-1}. The spore concentration fell below the limit of detection (1 CFU per 100 mL) during the treatment, achieving complete removal of SSRC for the first time (Fig. 9.9) (Ruiz-Aguirre *et al.*, 2015). Previous experiments without MD pretreatment (Fig. 9.10) had only managed a 1.3-log decrease in 4 h using a total of 100 mg L^{-1} of H$_2$O$_2$, supplied in 20 mg L^{-1} doses, and 11 mg L^{-1} of dissolved iron. In other tests carried out at pH values of 3 and 8, spores also could not be completely inactivated (Agulló-Barceló *et al.*, 2013).

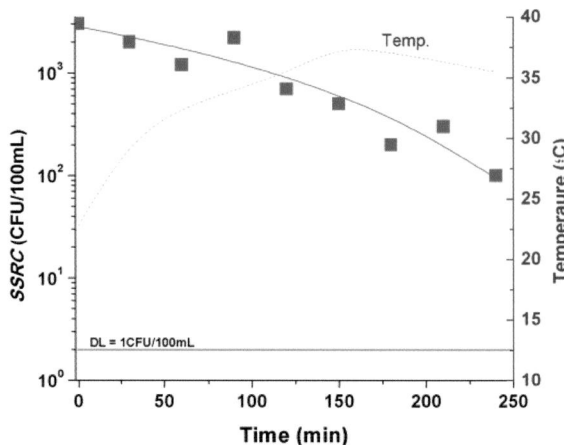

Figure 9.10.   Removal of SSRC using photo-Fenton treatment without MD pretreatment (modified from Ruiz-Aguirre *et al.*, 2015).

Therefore, wastewater pretreatment with MD is necessary for complete removal of *Clostridium* sp. with solar photo-Fenton treatment. The reason may not be so much the preconcentration of microorganisms by MD as the thermal stress experienced during MD, which seems to affect the physiological system of the spores, leaving them more vulnerable to attack by hydroxyl radicals. However, the MD process alone was also able to produce 100 L of distillate completely free of *Clostridium* sp. spores.

9.7   CONCLUDING REMARKS AND FUTURE PERSPECTIVES

In general, wastewater has a range of constituents, from dissolved metals and trace organic compounds to solids and a number of microorganisms. Treating wastewater for recovery requires a secondary treatment, which removes large suspended solids, dissolved organic matter, nutrients, and inorganic contents. The main objective in treated wastewater reuse is decreasing the health risk associated with hazardous pathogens. International and national regulations and guidelines concerning the restrictions on wastewater reuse establish limits for a number of biological parameters to control this biological risk. The challenge for future technologies that introduce solutions for feasible wastewater reuse is to efficiently remove pathogens and chemical contaminants in the same processes. This chapter has illustrated a number of solar-driven processes, and approaches coupled with other new technologies such as membrane distillation, that are highly efficient in the removal of pathogens from different water matrices. Their disinfection capabilities open up new opportunities for tertiary treatments of wastewater, which may be considered after conventional and advanced secondary biological treatments, to improve the quality of treated water for restricted reuse, such as irrigation among others. There are still, however, scientific limitations and knowledge gaps regarding their effectiveness in complex scenarios, such as brackish wastewater or very resistant water pathogens, that must be investigated prior to their implementation in the marketplace.

ACKNOWLEDGMENTS

The financial support given by the EU Switch-Asia Programme under the ZCR-2 project (DCI-ASIE/2013/334 140) and by the Spanish Ministry of Economy and Competitiveness under the WATER4CROP project (CTQ2014-54563-C3-3) are acknowledged.

REFERENCES

Acra, A., Karahagopian, Y., Raffoul, Z. & Dajani, R. (1980) Disinfection of oral rehydration solutions by sunlight. *The Lancet*, 2(8206), 1257–1258.

Agulló-Barceló, M., Polo-López, M.I., Lucena, F., Jofre, J. & Fernández-Ibáñez, P. (2013) Solar advanced oxidation processes as disinfection tertiary treatments for real wastewater: implications for water reclamation. *Applied Catalysis B: Environmental*, 136–137, 341–350.

Alkhudhiri, A., Darwish, N. & Hilal, N. (2012) Membrane distillation: a comprehensive review. *Desalination*, 287, 2–18.

Anathaswamy, H.N. & Eisenstark, A. (1977) Repair of hydrogen peroxide induced single strand breaks in *Escherichia coli* deoxyribonucleic acid. *Journal of Bacteriology*, 130, 187–191.

Benabbou, A.K., Derriche, Z., Felix, C., Lejeune, P. & Guillard, C. (2007) Photocatalytic inactivation of *Escherichia coli*: effect of concentration of TiO_2 and microorganism, nature, and intensity of UV irradiation. *Applied Catalysis* B: *Environmental*, 76, 257–263.

Bichai, F., Polo-López, M.I. & Fernández-Ibáñez, P. (2012) Solar disinfection of wastewater to reduce contamination of lettuce crops by *Escherichia coli* in reclaimed water irrigation. *Water Research*, 46, 6040–6050.

Blanco-Gálvez, J., Fernández-Ibáñez, P. & Malato-Rodríguez, S. (2007) Solar photocatalytic detoxification and disinfection of water: recent overview. *Journal of Solar Energy Engineering*, (ASME), 129, 1–12.

Buxton, G.V. & Elliot, A.J. (1986) Rate constant for reaction of hydroxyl radicals with bicarbonate ions. *Radiation Physics and Chemistry*, 27(3), 241–243.

Carey, J.H., Lawrence, J. & Tosine, H.M. (1976) Photodechlorination of PCBs in the presence of TiO_2 in aqueous suspensions. *Bulletin of Environmental Contamination and Toxicology*, 16(6), 697–701.

Committee on Indicators for Waterborne Pathogens, National Research Council (2004) *Indicators for Waterborne Pathogens*. The National Academies Press, Washington, DC.

du Preez, M., Conroy, R.M., Ligondo, S., Hennessy, J., Elmore-Meegan, M., Soita, A. & McGuigan, K.G. (2011) Randomized intervention study of solar disinfection of drinking water in the prevention of dysentery in Kenyan children aged under 5 years. *Environmental Science and Technology*, 45, 9315–9323.

Fernández-Ibáñez, P., Sichel, C., Polo-López, M.I., de Cara-García, M. & Tello, J.C. (2009) Photocatalytic disinfection of natural well water contaminated by *Fusarium solani* using TiO_2 slurry in solar CPC photo-reactors. *Catalysis Today*, 144, 62–68.

García-Fernández, I., Polo-López, M.I., Oller, I. & Fernández-Ibáñez, P. (2012) Bacteria and fungi inactivation using Fe^{3+}/sunlight, H_2O_2/sunlight and near neutral photo-Fenton: a comparative study. *Applied Catalysis* B: *Environmental*, 121–122, 20–29.

García-Fernández, I., Fernández-Calderero, I., Polo-López, M.I. & Fernández-Ibáñez, P. (2015) Disinfection of urban effluents using solar TiO_2 photocatalysis: a study of significance of dissolved oxygen, temperature, type of microorganism and water matrix. *Catalysis Today*, 240, 30–38.

Goldstein, S., Aschengrau, D., Diamant, Y. & Rabani, Y. (2007) Photolysis of aqueous H_2O_2: quantum yield and applications for polychromatic UV actinometry in photoreactors. *Environmental Science and Technology*, 41, 7486–7490.

Helali, S., Polo-López, M.I., Fernández-Ibáñez, P., Ohtani, B., Amano, F., Malato, S. & Guillard, C. (2013) Solar photocatalysis: a green technology for *E. coli* contaminated water disinfection. Effect of concentration and different types of suspended catalyst. *Journal of Photochemistry and Photobiology A: Chemistry*, 276, 31–40.

Khayet, M. (2011) Membranes and theoretical modeling of membrane distillation: a review. *Advances in Colloid and Interface Science*, 164, 56–88.

McGuigan, K.G., Joyce, T.M., Conroy, R.M., Gillespie, J.B. & Elmore-Meegan, M. (1998) Solar disinfection of drinking water contained in transparent plastic bottles: characterizing the bacterial inactivation process. *Journal of Applied Microbiology*, 84, 1138–1148.

McGuigan, K.G., Samaiyar, P., du Preez, M. & Conroy, R.M. (2011) High compliance randomized controlled field trial of solar disinfection of drinking water and its impact on childhood diarrhea in rural Cambodia. *Environmental Science and Technology*, 45, 7862–7867.

McGuigan, K.G., Conroy, R.M., Mosler, H.J., du Preez, M., Ubomba-Jaswa, E. & Fernández-Ibáñez, P. (2012) Solar water disinfection (SODIS): a review from bench-top to roof-top. *Journal of Hazardous Materials*, 235–236, 29–46.

Malato, S., Fernández-Ibáñez, P., Maldonado, M.I., Blanco, J. & Gernjak, W. (2009) Decontamination and disinfection of water by solar photocatalysis: recent overview and trends. *Catalysis Today*, 147, 1–59.

Markowska-Szczupak, A., Ulfig, K. & Morawski, A.W. (2011) The application of titanium dioxide for deactivation of bioparticulates: an overview. *Catalysis Today*, 169, 249–257.

Morris, P.F. & Phuntumart, V. (2009) Inventory and comparative evolution of the ABC superfamily in the genomes of *Phytophthora ramorum* and *Phytophthora sojae*. *Journal of Molecular Evolution*, 68(5), 563–575.

Ortega-Gómez, E., Esteban García, B., Ballesteros Martín, M.M., Fernández-Ibáñez, P. & Sánchez Pérez, J.A. (2013) Inactivation of *Enterococcus faecalis* in simulated wastewater treatment plant effluent by solar photo-Fenton at initial neutral pH. *Catalysis Today*, 209, 195–200.

Ortega-Gómez, E., Ballesteros Martín, M.M., Esteban García, B., Sánchez Pérez, J.A. & Fernández-Ibáñez, P. (2014) Solar photo-Fenton for water disinfection: an investigation of the competitive role of model organic matter for oxidative species. *Applied Catalysis B: Environmental*, 148–149, 484–489.

Ortega-Gómez, E., Ballesteros Martín, M.M., Carratalà, A., Fernández-Ibáñez, P., Sánchez Pérez, J.A. & Pulgarín, C. (2015) Principal parameters affecting virus inactivation by the solar photo-Fenton process at neutral pH and μM concentrations of H_2O_2 and $Fe^{2+/3+}$. *Applied Catalysis B: Environmental*, 174–175, 395–402.

Ortega-Gómez, E., Ballesteros Martín, M.M., Esteban García, B., Sánchez Pérez, J.A. & Fernández-Ibáñez, P. (2016) Wastewater disinfection by neutral pH photo-Fenton: the role of solar radiation intensity. *Applied Catalysis B: Environmental*, 181, 1–6.

Pignatello, J.J., Oliveros, E. & MacKay, A. (2006) Advanced oxidation processes for organic contaminant destruction based on the Fenton reaction and related chemistry. *Environmental Science and Technology*, 36, 1–84.

Polo-López, M.I., Fernández-Ibáñez, P., García-Fernández, I., Oller, I., Salgado-Tránsito, I. & Sichel, C. (2010) Resistance of *Fusarium* sp. spores to solar TiO_2 photocatalysis: influence of spore type and water (scaling-up results). *Journal of Chemical Technology and Biotechnology*, 85, 1038–1048.

Polo-López, M.I., García-Fernández, I., Oller, I. & Fernández-Ibáñez, P. (2011) Solar disinfection of fungal spores in water aided by low concentrations of hydrogen peroxide. *Photochemical & Photobiological Sciences*, 10, 381–388.

Polo-López, M.I., García-Fernández, I., Velegraki, T., Katsoni, A., Oller, I., Mantzavinos, D. & Fernández-Ibáñez, P. (2012) Mild solar photo-Fenton: an effective tool for the removal of *Fusarium* from simulated municipal effluents. *Applied Catalysis B: Environmental*, 111–112, 545–554.

Polo-López, M.I., Oller, I. & Fernández-Ibáñez, P. (2013) Benefits of photo-Fenton at low concentrations for solar disinfection of distilled water. A case study: *Phytophthora capsici*. *Catalysis Today*, 209, 181–187.

Polo-López, M.I., Castro-Alférez, M., Oller, I. & Fernández-Ibáñez, P. (2014) Assessment of solar photo-Fenton, photocatalysis, and H_2O_2 for removal of phytopathogen fungi spores in synthetic and real effluents of urban wastewater. *Chemical Engineering Journal*, 257, 122–130.

Rincón, A.G. & Pulgarín, C. (2003) Photocatalytical inactivation of *E. coli*: effect of (continuous-intermittent) light intensity and of (suspended-fixed) TiO_2 concentration. *Applied Catalysis B: Environmental*, 44, 263–284.

Rincón, A.G. & Pulgarín, C. (2005) Comparative evaluation of Fe^{3+} and TiO_2 photoassisted processes in solar photocatalytic disinfection of water. *Applied Catalysis B: Environmental*, 63, 222–231.

Rodríguez-Chueca, J., Polo-López, M.I., Mosteo, R., Ormad, M.P. & Fernández-Ibáñez, P. (2014) Disinfection of urban wastewater effluents (real and simulated) using mild solar photo-Fenton. *Applied Catalysis B: Environmental*, 150–151, 619–629.

Ruiz-Aguirre, A., Polo-López, M.I., Fernández-Ibáñez, P. & Zaragoza, G. (2015) Assessing the validity of solar membrane distillation for disinfection of contaminated water. *Desalination and Water Treatment*, 55(10), 2792–2799.

Sichel, C., Fernández-Ibáñez, P., De Cara, M. & Tello, J. (2009) Lethal synergy of solar UV-radiation and H_2O_2 on wild *Fusarium solani* spores in distilled and natural well water. *Water Research*, 43, 1841–1850.

SODIS. (2011) SODIS Method: How does it work? Available from: http://www.sodis.ch/methode/anwendung/index_EN [accessed October 2016].

Spuhler, D., Rengifo-Herrera, J.A. & Pulgarín, C. (2010) The effect of Fe^{2+}, Fe^{3+}, H_2O_2 and the photo-Fenton reagent at near neutral pH on the solar disinfection (SODIS) at low temperatures of water containing *Escherichia coli* K12. *Applied Catalysis B: Environmental*, 96, 126–141.

Wegelin, M., Canonica, S., Mechsner, K., Fleischmann, T., Pesaro, F. & Metzler, A. (1994) Solar water disinfection: scope of the process and analysis of radiation experiments. *Journal of Water Supply: Research and Technology – AQUA*, 43, 154–169.

Wei, C., Lin, W.Y., Zainal, Z., Williams, N.E., Zhu, K., Kruzic, A.P., Smith, R.L. & Rajeshwar, K. (1994) Bactericidal activity of TiO_2 photocatalyst in aqueous media: toward a solar-assisted water disinfection system. *Environmental Science and Technology*, 28, 934–938.

WHO. (2007) Combating waterborne disease at the household level. World Health Organization, Geneva.

Winter, D., Koschikowski, J. & Wieghaus, M. (2011) Desalination using membrane distillation: experimental studies on full scale spiral wound modules. *Journal of Membrane Science*, 375, 104–112.

Zaragoza, G., Ruiz-Aguirre, A. & Guillén-Burrieza, E. (2014) Efficiency in the use of solar thermal energy of small membrane desalination systems for decentralized water production. *Applied Energy*, 130, 491–499.

CHAPTER 10

Solar PV for water pumping and irrigation

István Patay, István Seres & Jochen Bundschuh

10.1 INTRODUCTION

Everywhere in the world, water supply and irrigation are fundamental to agricultural production. The main aim of water supply in arable land and plantations is to:

- Ensure the elementary conditions of production using water
- Increase the quantity of yield
- Increase the quality of yield
- Realize other technological elements through irrigation (e.g. fertilization, plant protection).

The quantity of water needed for irrigation of a given land area depends on a number of factors, the most important of which are (Goswami, 1986):

- Nature of the crop and crop growth cycle
- Climatic conditions
- Type and condition of soil
- Land topography
- Field application efficiency
- Water quality.

The added water demand usually changes over time depending on rainfall, evapotranspiration and the actual water demand of the crop.

Water pumping, from groundwater and surface water bodies, is a principal energy consumer – and therefore a significant cost factor – in agriculture. The cost of any irrigation has three elements: water, energy and human resource. The costs can be decreased by water- and energy-saving irrigation methods and the automation of the applied systems. One possible way of reducing energy costs is the utilization of locally available renewable energy sources such as solar and wind. The implementation of renewable energy technologies, as a substitute for the electricity produced from the fossil fuels commonly used for water pumping, can further reduce costs and improves the chances of achieving sustainable irrigation, as well as contributing to climate protection.

In the agricultural sector, water pumping is required in many situations, the most important being irrigation, livestock support (for fodder production, livestock watering), and other on-farm operations such as cleaning. It can represent a significant, or even the largest, share of energy consumption in any given agricultural sector. In this chapter, we discuss the technical, and especially the economic, feasibility of implementing solar-powered (photovoltaic, PV) water pumping in agriculture and interpret this as far as possible, given the limited data currently available from the industry (on-farm only).

In the case of irrigation, which is ahead of livestock watering as the most important application of water pumping, the share of total energy input depends on crop species, climate, soil type, water demand, energy sources used (e.g. diesel, grid electricity), type of irrigation system used, and pumping height (e.g. depth to groundwater table). This means that total production costs are highly variable between and within different commodities according to geographical location, and therefore require case-specific assessments.

Solar energy is a form of radiation energy with a broad frequency spectrum. The total (or global) irradiation consists of both direct and diffuse radiation. Solar energy techniques (e.g. solar thermal and solar electricity) mainly use direct radiation with good levels of efficiency. Because of the virtual movement of the sun, the radiation value varies with time and according to geographical situation. However, the sum of the radiation at any point of the earth over a long-term period is nearly constant. In Figure 9.1, the annual surface radiation values can be observed, based on meteorological and energetical measurements. The solar energy map shows that the average annual energetical sums $(1–2.5\,\mathrm{MWh\,m^{-2}})$ are high enough in the main agricultural areas for utilization of the solar radiation.

During the growth period of vegetation, when irrigation is often needed, there is a high volume of radiation energy coming from the sun. This solar energy can be converted into electric energy by PV applications to power irrigation pumps.

As the prices of solar PV panels continue to reduce and the additional elements required (converters, controllers, pumps) are also increasingly available, the use of solar electric pumping systems has quickly spread around the world. Many types of installation have been tested for different applications, and a lot of practical experience has thus been acquired. Electricity shortages and high diesel costs affect the pumping available for community water supplies and irrigation; therefore, using solar energy for water pumping is a promising alternative to conventional electricity- and diesel-based pumping systems (Chandel *et al.*, 2015). Studies suggest that the payback period of an efficient water supply system based on PV power is about five years (Dursun and Özden, 2012).

However, a solar irrigation pump system method has to take into account the fact that the demand for irrigation water will change throughout the year. During the irrigation season, the peak demand is often more than twice the average demand. This means that solar pumps for irrigation are underutilized for much of the year. Attention should be paid to the system of irrigation water distribution and application to the crops (Shinde and Wandre, 2015).

Overall, it has been demonstrated that solar PV water supply systems offer good prospects and may help many regions of the world to solve their socioeconomic problems, in particular those relating to food production.

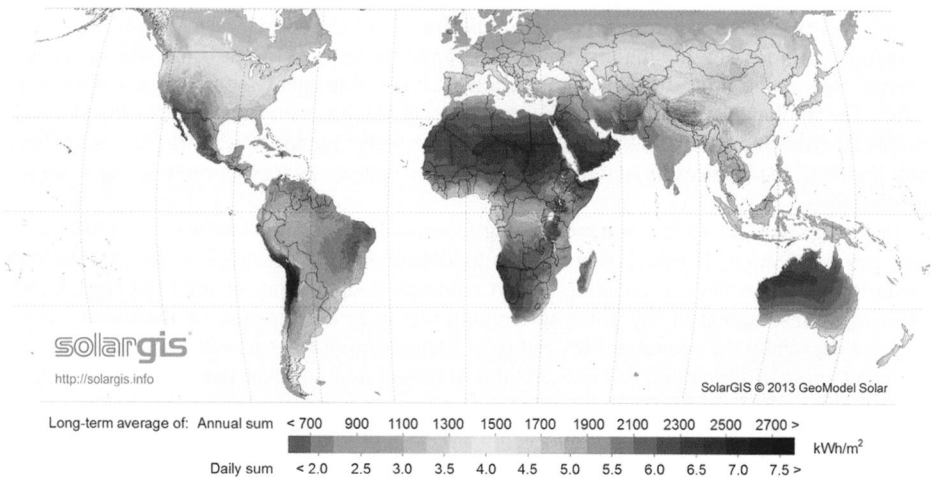

Figure 10.1.   Annual and daily irradiation sums on the surface of the earth.

10.2   ENERGY DEMAND OF WATER PUMPING

Pumping of water means an increase in the hydraulic energy level of that water. The water-pumping energy requirement is determined by the relationship between the water volume requirements and the energy needed. The formula connecting these two parameters is the following:

$$E_h = \rho g h V \quad [\text{J}] \tag{10.1}$$

where:

E_h: hydraulic energy [J]
ρ: water density standard [10^3 kg m^{-3}]
g: acceleration due to gravity [9.81 m s^{-2}]
h: total head of water [m]
V: required volume of water [m^3]

The hydraulic power (P_h) required for lifting depends on the flow rate Q [m^3s^{-1}]. The formula for hydraulic power is simply obtained from the formula for energy, Equation (10.1), by replacing V with flow rate:

$$P_h = \rho g h Q \quad [\text{W}] \tag{10.2}$$

The total power demand of any pumping system is higher than the hydraulic power required because of energy losses. Figure 10.2 shows the elements that make up the total head (h):

$$h = h_s + h_p + h_{out} + h_l \quad [\text{m}] \tag{10.3}$$

where:

h_s: suction head (if the pump is located above the water level, see Figure 10.2a; for submersible pumps – see Figure 10.2b – h_s is negative)

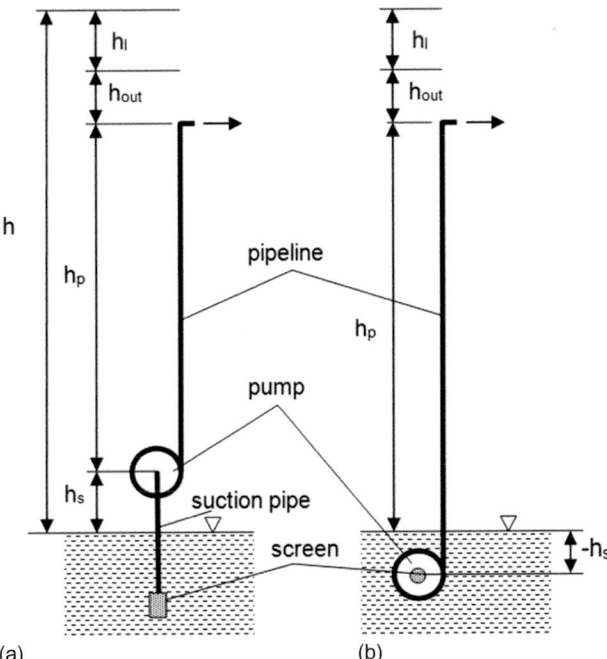

Figure 10.2.   Elements of total head in pumps that are: (a) above the surface; (b) submersible.

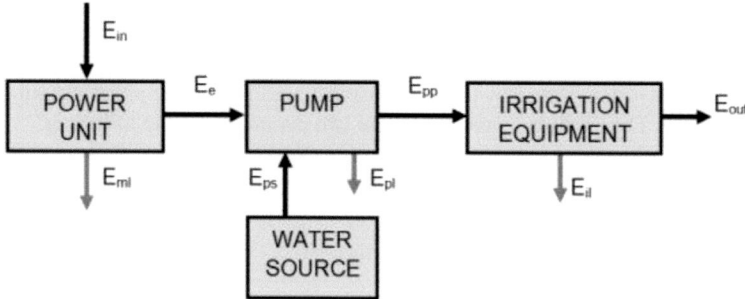

Figure 10.3.   The energetic process of irrigation.

h_p: head of the pressure side of the pump
h_{out}: head in outlet
h_l: the total head of losses.

Instead of deriving the different heads in meters, it is often practical to use the pressure (p) in Pascal [Pa] or bar units. Any type of head has an equivalent pressure that can be calculated as:

$$p = \rho g h \quad [\text{Pa}] \tag{10.4}$$

or:

$$p = 10^{-5} \rho g h \quad [\text{bar}] \tag{10.5}$$

where the parameters are as previously defined.

10.2.1   *The energetic process of irrigation*

Irrigation is an energy-transforming process; some energy transformation appears in the elements of the system during irrigation periods. The irrigation water needs defined energy for transportation and distribution, depending on the system and the specification of elements. The general scheme of this energetic system can be seen in Figure 10.3.

The *input energy* (E_{in}) is transformed to effective power energy (E_e) by the power unit. The motoric losses of the power unit (E_{ml}) depend on the nature of the unit: the maximum efficiency is 40–45% for engines, and 80–85% for motors. However, the actual efficiency is influenced strongly by the energetical adaptation of the power unit and the pump, and also determined by the operational conditions. The advantage of engine-driven pumps is the variable revolution of the engine. In this case, a relatively high efficiency can be obtained. For motoric driving, the same possibility can be realized if the motor is also variable revolution only. Such installations are expensive but in a long-term operation the extra investment provides good returns.

The *effective energy* (E_e) for the pumping covers the energy demand of lifting the water from the water source (if there is any), the increase in the water pressure energy after passing through the pump (E_{pp}), and all the energy losses of the pump (E_{pl}). From the aspect of the irrigation equipment, the E_{pp} is the most important, because its value determines the quality of irrigation and the correct operational conditions of the equipment.

The *hydraulic power demand* is determined by the irrigation equipment:

$$P_{hp} = 10^2 \, Q \, p_{pp} \quad [\text{kW}] \tag{10.6}$$

where:

Q: the rate of flow needed [m³ s⁻¹]
p_{pp}: the pressure of the pump at the outlet [bar]

The *pressure demand* of the irrigation equipment consists of two parts:

$$p_{pp} = p_t + p_{out} \quad [bar] \tag{10.7}$$

where:

p_t: the pressure demand of the transport [bar]
p_{out}: the pressure demand of the outlet (the pressure demand of the outlet elements) [bar]

The pressure demand of the transport (the water streaming in the tubes of equipment) is calculated by a well-established analytical method. In this way, the *total pressure losses* of the equipment can be determined as follows:

$$p_t = \sum_{i=1}^{n} \Delta p_i = \sum_{i=1}^{n} \lambda_i \frac{l_i}{d_i} \frac{\rho}{2} v_i^2 + \sum_{j=1}^{m} \xi_j \frac{\rho}{2} v_j^2 \quad [Pa] \tag{10.8}$$

where:

λ_i: frictional coefficient of the tube section; its value is usually 0.01–0.02
l_i: length of the pipe section [m]
d_i: inner diameter of the pipe section [m]
v_i: velocity of the water in the pipe section [m s^{-1}]
ξ_j: local loss coefficient
v_j: velocity of the water at the place of the local loss [m s^{-1}]
ρ: density of the water [kg m^{-3}]
(If the irrigation equipment is very complicated, the use of computer software is suggested.)

Equation (10.8) shows that the pressure losses (the energy losses of the transition) can be reduced by increasing the diameter of the tube sections. But, of course, if the diameter of the pipelines is increased, the investment costs will also grow. It is an important task to find the optimum diameter of the pipe sections when planning the irrigation equipment. It follows that if the irrigation equipment has more distributor elements, the outlet pressure changes at each element. During the sizing process it is important to check that the outlet pressure at the very last element (along the pipelines) is satisfactory.

10.2.2   *Energetics of the irrigation equipment*

The demand of the transition and outlet pressure (namely the hydraulic power demand) is determined by the irrigation system and the construction of the actual equipment. But the general rule is that the energy demand is influenced by the outlet pressure if the distributed flow rate is constant. Based on Equation (10.6), the *energy demand* of the irrigation equipment is:

$$E = 10^2 \, Q \, p_t \, t_i \quad [kJ] \tag{10.9}$$

where:

Q: distributed flow rate [m^3 h^{-1}]
p_t: total pressure (pressure demand of the suction + transition + outlet) [bar]
t_i: operation time of the irrigation [h]

The *irrigation time* (t_i) can be calculated from the intensity of irrigation (i) [mm h^{-1}] and the dose of irrigated water, h [mm]:

$$t_i = \frac{h}{i} \quad [h] \tag{10.10}$$

The effective *average intensity* of the irrigation can be calculated as:

$$i = 10^3 \frac{Q}{A} \quad [mm \, h^{-1}] \tag{10.11}$$

where:

A: area irrigated at same time [m^2]

The energy demand of irrigation is based on Equations (10.10) and (10.11):

$$E = 10^2 A \, i p_t \frac{h}{i} = 10^2 A \, p_t h \quad [\text{J}] \tag{10.12}$$

The basis of the evaluation or comparison of different irrigation methods or equipment may also be other specific parameters. Three specific values can be compared because in Equation (10.12) there are three variable parameters:

- The specific energy of the water dose unit:

$$E_h = \frac{E}{h} = 10^2 A p_t \quad [\text{J mm}^{-1}] \tag{10.13}$$

- The specific energy of the irrigated area unit:

$$E_A = \frac{E}{A} = 10^2 p_t h \quad [\text{J m}^{-2}] \tag{10.14}$$

- The specific energy demand of the flow rate unit of the irrigated water:

$$E_Q = \frac{E}{Q} = 10^2 \frac{h}{i} p_t = 10^2 t_i p_t \quad [\text{kJ m}^{-3} \text{h}^{-1}] \tag{10.15}$$

From an energetic aspect, Equation (10.10) may be suggested as the basis for comparison of different irrigation techniques. However, in the course of the evaluation process there can be problems with the establishment of the intensity value. Equation (10.15) illustrates the inverse proportionality of the energy demand and the intensity (Fig. 10.4). Because of the many different water distribution methods, the intensity of irrigation is specific to the irrigation system and the actual equipment.

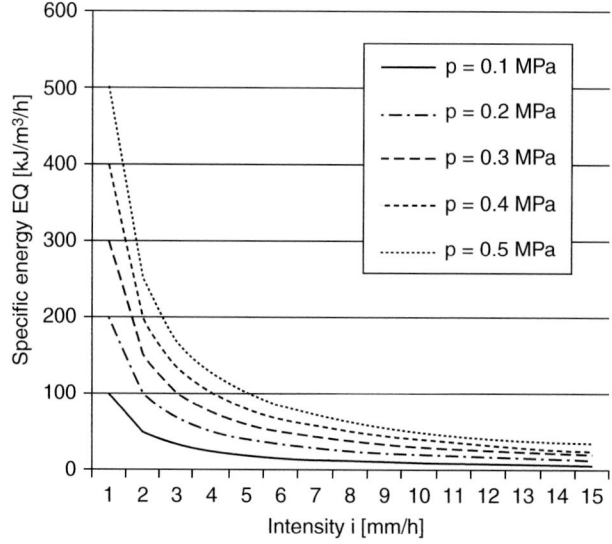

Figure 10.4.   Specific energy demand (E_Q) as a function of intensity (i).

10.3   ENERGY DEMAND OF IRRIGATION METHODS

The power source is determined by the pumping conditions needed, namely the flow rate (Q) and the total pressure required (p_t). A solar PV power source can be applied economically if a water- and energy-saving irrigation method is used.

10.3.1   *Sprinkler irrigation*

As with rainfall, the total soil surface is usually wetted with sprinkler irrigation systems. In this method, the outlet elements are different kinds of sprinklers in either a static or moving operational mode. As the intensity of the water varies along the radius of the spread, an average intensity can be defined at a single sprinkler, and can be calculated from the actual operational data.

When using sprinkler irrigation for the supplementation of rainfall, the main objective is to distribute the water uniformly on the surface of the soil. In this case, every single plant gets almost the same volume of water. In sprinkler irrigation, the total surface is wetted by the irrigated water and the cultivation area takes in water approximately evenly. In this way the plants can maximize their use of the nutrients of the soil (all of the wetted cubic capacity of the soil is available to the plants).

In sprinkler irrigation, the average intensity of water and the total pressure can vary across a relatively wide range of values, depending on the construction of the equipment. In general, higher intensities require greater pressures.

In order to compare the different sprinkler irrigation techniques from the point of view of energy demand, a new specific parameter is needed. Based on Equation (10.15), the following equation can be derived:

$$E_{Qh} = \frac{E_Q}{h} = \frac{10^2 p_t}{i} \quad [\text{kJ m}^{-3}\,\text{h}^{-1}\,\text{mm}^{-1}] \tag{10.16}$$

E_{Qh} represents the specific energy demand, which correlates with the flow rate unit and the dose unit of water. Equation (10.16) is used to calculate the E_{Qh} values of the practical ranges of total pressure and intensity. The results can be seen in Table 10.1.

As can be observed in Table 10.1, there is no significant difference among the specific energy demands of different sprinkler irrigation systems, with the exception of micro-sprinkler irrigation, where the value of the specific energy can be as high as $200\,\text{kJ m}^{-3}\,\text{h}^{-1}\,\text{mm}^{-1}$. Therefore, in terms of energy, micro-sprinkler irrigation is relatively expensive for rainfall-replacement irrigation. But the water losses – and hence the energy losses – should be taken into account in sprinkler irrigation. Evaporation in the atmosphere, the transition effect of the wind, and evaporation at the soil surface can all be reasons for water losses. In unfavorable conditions, the water and energy losses can reach values as high as 30–50%.

Table 10.1.   Specific energy demand of sprinkler irrigation (Patay *et al.*, 2012).

Irrigation equipment	Pressure p_t [MPa]	Average intensity i [mm h^{-1}]	Specific energy demand E_{Qh} [kJ m^{-3} h^{-1} mm^{-1}]
Micro sprinklers	0.2–0.4	1–5	40–200
Moving systems	0.3–0.4	5–10	60–80
Reel system with console	0.4–0.5	5–10	40–50
Reel system with single sprinkler	0.5–1	10–20	20–100

10.3.2   *Drip irrigation*

Drip or trickle irrigation is usually applied in plantations, horticulture and greenhouses. The common characteristic of dripping systems is spot water distribution along the pipelines. If the pipeline is made from porous material, the water can percolate through the wall of the pipeline. In this case, the wetted area is linear. In drip irrigation systems, the intensity is not an evident parameter.

If the average intensity is calculated as the irrigated water devided by the wetted area (or by the wetted cross section in the root zone), the intensity reaches a relatively high value. It is also worth mentioning that the cultivation surface is only partly irrigated. However, if the referred surface is the total surface (or the cultivation surface of the plants), the intensity is much lower. For a correct comparison of drip irrigation to sprinkler systems, the average intensity must be calculated for the total surface.

The cultivation surface depends on the row spacing and the plant distance. The cultivation area is precisely determined by the needs of the plant and influenced by the optimum utilization of the land. If the cultivation surface is only partially irrigated, the roots of the plants are concentrated in the wetted area or cubic volume. Other parts of the cultivation surface are largely unused.

The definition of intensity in drip irrigation systems is only really correct if the irrigated water volume refers to the total irrigated area:

$$i = \frac{q_A}{A_b} = \frac{n\,q_1}{a\,b} \quad [\text{mm h}^{-1}] \tag{10.17}$$

where:
  q_A: irrigated flow rate of the cultivation area $[\text{dm}^3\,\text{h}^{-1}]$
  q_1: flow rate of the single dripper element $[\text{dm}^3\,\text{h}^{-1}]$
  n: number of drippers per area of cultivation
  A_b: cultivation area per plant $[\text{m}^2]$
  a and b: distance of rows and plants [m]

If one dripper pipeline is installed to irrigate a row, the number of drippers per cultivation area is $n = b/t$, where t is the distance [m] of the dripper elements along the pipeline. The intensity in this case is:

$$i = \frac{q_1}{ta} \quad [\text{mm h}^{-1}] \tag{10.18}$$

The specific energy demand of a drip irrigation system is given in Equations (10.11) and (10.13):

$$E_{Qh} = \frac{102\,p_t\,t\,a}{q_1} \quad [\text{kJ m}^{-3}\,\text{h}^{-1}\,\text{mm}^{-1}] \tag{10.19}$$

Equation (10.19) shows clearly how the specific energy demand is influenced by the operational parameters. In Table 10.2, data derived from the equation for some typical operational conditions can be seen.

Table 10.2 clearly shows that the specific energy demand changes by much greater intervals in drip irrigation than with the sprinklers shown in Table 10.1. The conclusion is that drip irrigation may be energy-saving or energy-intensive depending on the operational characteristics, namely on the scheme of the pipelines (a), the specific number of drippers along the pipeline (t), the pressure, and the flow rate of the drippers. It is also an important conclusion of the theoretical test that drip irrigation may be a water- and energy-saving irrigation system but the price of this is that the cultivation area of the plants is irrigated unevenly. At the same time, the effective energy demand of drip irrigation is reduced by the minimal water losses, and the distributed water can be utilized effectively by the plants.

Table 10.2. Specific energy demand of trickle irrigation (Patay *et al.*, 2012).

Pressure p_t [MPa]	Distance of elements t [m]	Distance of rows a [m]	Flow rate of drippers q_1 [dm^3 h^{-1}]	Specific energy E_{Qh} [kJ m^{-3} h^{-1} mm^{-1}]
0.1	0.2	0.5	1.5	7
	0.4	1.0	2.0	20
	0.6	1.5	2.5	36
	0.8	3.0	3.0	60
0.2	0.2	0.5	1.5	13
	0.4	1.0	2.0	40
	0.6	1.5	2.5	72
	0.8	3.0	3.0	160
0.3	0.2	0.5	1.5	20
	0.4	1.0	2.0	60
	0.6	1.5	2.5	108
	0.8	3.0	3.0	240

10.4 ELECTRIC WATER PUMPS FOR WATER SUPPLY AND IRRIGATION

10.4.1 *Types of solar pumps*

There are various types of electric water pumps for different operational conditions. These include (Goswami, 1986; Stetson and Mecham, 2012):

- *Jack pump*: experience has shown that the currents to the motor vary by a factor of five during operation. Unless batteries are used in the system, this promotes extremely inefficient use of the PV energy source.
- *Jet pump*: the efficiency of this system is between 5–10% and it is not suitable for use with photovoltaics.
- *Centrifugal pump* with DC motor on surface: the main advantage of this pump is that the surface-mounted DC motor is easily accessible for periodic replacement of brushes.
- *Submersible centrifugal pump* with AC motor: a major problem with AC motors is the high starting current required. If the AC motor is connected to a pump the starting current is of the order of five to eight times the running current. It needs suitably high-capacity batteries in a solar PV system.
- *Centrifugal pump with brushless DC motor*: this motor has an electronic controller instead of brushes. The main advantages of this system are high efficiency (over 90%), long life expectancy and low maintenance costs. Some examples of this pump type can also be operated in AC mode directly, without any reinstallation.

Nowadays, solar pumps fall into two major categories, as follows:

- *Surface pump* – including pressure, delivery, and booster pumps.
- *Submersible pump* – primarily submersible well pumps.

Nearly all solar pumps are the centrifugal type but there are other types too (piston, helical).

Delivery surface pumps are used to move water from one place to another. Some are capable of creating high pressure while others are intended mainly for moving large volumes at low pressure (such as moving water from a cistern to a stock-watering tank). Flows can range from small (e.g. 2–3 L min^{-1}) to substantial (e.g. 110–150 L min^{-1}). It is worth mentioning that surface pumps of all types get water from a tank, spring, cistern, or similar. The minimum suction lift is just a few meters. The maximum possible suction limit for even a perfect surface pump is just under

Table 10.3.   Main technical data of solar pumps.

Type of solar pump	Total dynamic head [m]	Flow rate [$m^3 h^{-1}$]	DC voltage [V]
Piston pumps	10–60	1–2	12–48
Centrifugal pumps	10–40	2–10	12–36
Booster pumps	10–30	0.5–1	12–24
Centric submersible pumps	20–40	1–5	12–24
Brushless submersible pumps	20–50	0.5–30	30–180
Helical submersible pumps	max. 50	1–2	12–50

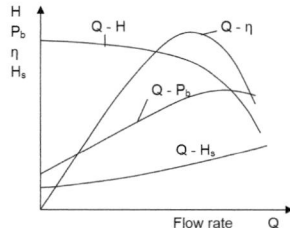

Figure 10.5.   Characteristic curves for a single-stage centrifugal pump.

ten vertical meters at sea level (consequent on atmospheric pressure of 0.1 MPa). However, there are no perfect pumps – a suction lift of 2 to 7.5 m is generally the maximum achievable.

Table 10.3 shows the technical data of typical solar pumps to demonstrate the possibilities of solar water pumping and irrigation.

10.4.2   *Characteristics of pumps*

Flow rate (or capacity), pressure head, suction head, power demand and efficiency are the parameters that describe a pump's performance. These parameters may change under different operational conditions. The relationships between them are utilized to describe the operating properties of a pump, and the resulting set of four curves is known as the pump's characteristic curves. Pump manufacturers usually publish a set of characteristic curves for each pump model they make. These curves provide the basis for planning of different pumping systems.

In the simplest case the pump operates at constant speed, that is, the revolution of the pump shaft is nearly constant and does not depend on the operational conditions. The characteristic curves of a typical single-stage pump are illustrated in Figure 10.5. The head versus pump capacity (Q–H) curve relates the head produced by a pump to the flow rate being pumped. Generally, the head produced steadily decreases as the flow rate increases. In centrifugal pumps, the Q–H curve is determined mainly by the type and measure of the impeller.

The shape of the Q–P_b curve depends also on the pump's specific speed and impeller design. For radial flow irrigation pumps, P_b generally increases from a value above zero to a peak and then declines slightly as Q increases from zero.

The efficiency (η) generally shows the quality of the process in which the energy transformation happens. The Q–η for a pump steadily increases to a peak as Q increases from zero, and then declines. The actual efficiency value can be calculated from the actual value of break power and the hydraulic power (P_h):

$$\eta = \frac{P_h}{P_b}100 = \frac{QH}{P_b}100 \quad [\%] \tag{10.20}$$

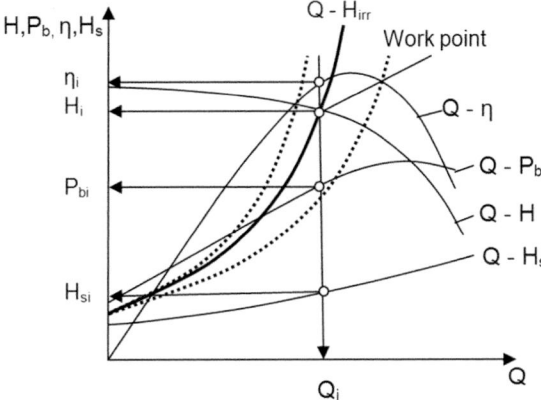

Figure 10.6. Work points of the pump and the operational parameters.

From Equation (10.20), the required break power can be calculated as:

$$P_b = \frac{Q \rho g H}{\eta} \quad [W] \tag{10.21}$$

In general, there is only one peak efficiency value for a specified impeller, which is in practice a constant-speed pump. Efficiency η is also related to the specific speed and impeller design and is influenced by the type of materials used in the construction, the finish on castings, and the quality of bearings used.

The fourth characteristic curve, the $Q–H_s$ curve, shows that the net positive suction head steadily increases in a typical radial flow pump.

The actual operational parameters of a pump are determined by the hydraulic system (irrigation equipment) connected to the pump. The irrigation equipment usually consists of different pipelines, fittings and emitters. Friction losses occur along the water path in the system, as per Equation (10.8). The Δp pressure or ΔH head losses depend on the Q^2 so the $Q–H$ curve of the irrigation equipment is an exponential one (Fig. 10.6). The points of intersection of the system and pump $Q–H$ curves show the actual operating data, the so-called work points.

It is an economically important condition that the efficiency of the system should be as high as possible. This means that the work point should be aligned closely to the maximum efficiency field of Q. A simple method of control is regulation by valves. A partly opened valve means a local pressure loss in the system, so by changing the setting of the valve the $Q–H_{irr}$ curve can be modified (see the dotted curves in Figure 10.6). However, this regulation is not economical because of the direct energy losses at the valve. It is better if the regulation occurs through some other hydraulic parameters, for example by the characteristics or number of emitters of a single-stage pump.

From a regulation perspective, it is best to apply a multistage (multi-speed) motor for pump driving. In this case the work point is moving along the static $Q–H_{irr}$ curve depending on the actual motor speed, and there is no hydraulic loss in the irrigation system (Fig. 10.7).

10.4.3 *Solar PV pumping stations*

The establishment of a solar PV-powered pumping station basically depends on the available water resources. The main parts of a solar PV pumping station are:

- Solar PV panels
- Electric motor-powered pump

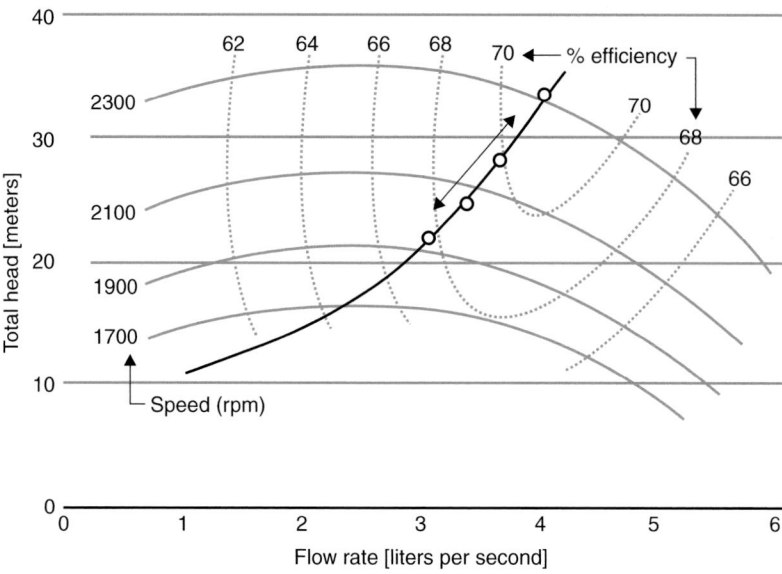

Figure 10.7.   System regulation by a multi-speed motor pump (Yiasoumi, 2003)

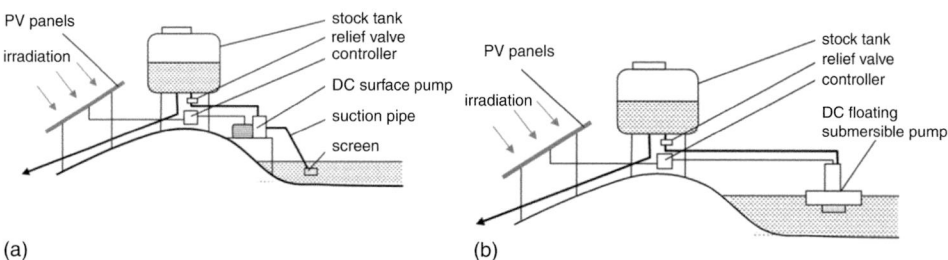

Figure 10.8.   Solar PV pumping stations for low-pressure irrigation.

- Controller
- Water storage tank
- Batteries (instead of tank).

One type of pumping station is based on water collecting in an elevated (stock) tank between the irrigation periods. This system can be realized both by surface pumps (Fig. 10.8a) and floating submersible pumps (Fig. 10.8b) if there is a surface water source or a well. There are no batteries required, but the static pressure for the irrigation equipment is strongly limited. These arrangements can only be used for low-energy-demand surface and drip irrigation. If the pressure required is higher (for example in micro-sprinkler irrigation) a battery system can be used.

If the irrigation is based on a well, the pumping station can also be a low-pressure construct with a stock tank (Fig. 10.9a) or a high-pressure construct with direct pumping of water (Fig. 10.9b). Between the irrigation periods, the energy collected by the PV panels is stored in the batteries. The process of charging is regulated by the controller unit, which is also usually able to control the irrigation process.

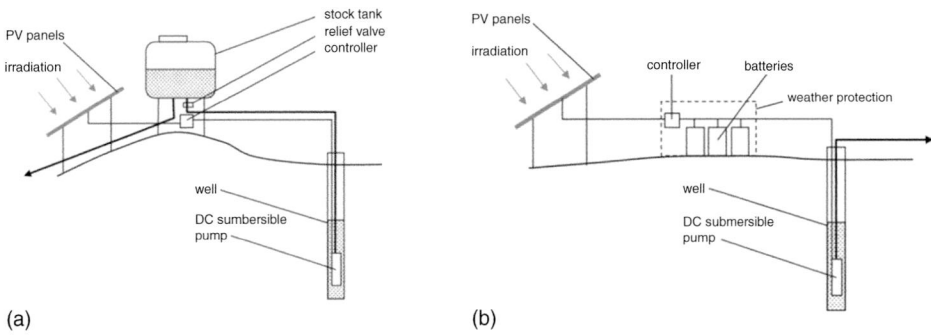

Figure 10.9.   Solar PV pumping stations based on a well.

10.5   SOLAR PV PANELS FOR PUMPING

Photovoltaic cells are able to turn the energy of solar radiation into electricity due to an energy transfer that occurs at the subatomic level. The intensity of the electrical energy production of PV cells or panels depends on the intensity of the solar radiation. The quality of the energy transformation or the efficiency of the process is determined by the type of cells. All commercial PV panels consist of silicon cells because of their relatively low price. Dependent on the manufacturing process, the four types of cells that are in general use in support of the pumping process are:

- Monocrystalline cells
- Polycrystalline cells
- Amorphous silicon cells
- Hybrid PV cells.

For solar PV pumping, the monocrystalline and polycrystalline panels are usually recommended because of their favorable price/efficiency relationship (Table 10.4). Monocrystalline PV cells have efficiencies of 13–17% and are the most efficient of the three basic types of silicon PV cells. However, they require more time and manufacturing energy than the polycrystalline silicon PV cells, and therefore they are slightly more expensive. Polycrystalline silicon is also produced from molten and highly pure molten silicon, as is the monocrystalline type, but uses a casting process. Mass-produced polycrystalline PV cell modules have an efficiency of 11–15%.

Amorphous silicon PV cells are made from non-crystalline semiconductor material with a thin-film technology. The layer of semiconductor material is only 0.5–2.0 μm thick. These PV cells have an efficiency of between 6 and 8%. Hybrid PV cells are classified as PV cells that use two different types of PV technology. The hybrid PV cell shown in Table 10.4 is made by Sanyo and comprises a monocrystalline PV cell covered by an ultra-thin amorphous silicon PV layer. The advantages of these types of cells are that they perform well at high temperatures and maintain higher efficiencies (over 18%) than conventional silicon PV cells. However, these cells come at a cost premium.

The solar PV panel is a unit that consists of cells in crystalline-type systems. The panel is a DC electrical source with a voltage of 12 or 24 V at the outlet. The simplest application is when the solar panel or the connected panels operate the DC pump directly. The theoretical effective power of the motor is calculated as:

$$P_e = UI \quad [W] \tag{10.22}$$

where U is the voltage [volt, V] and I is the current [ampere, A].

While the voltage of the PV panel is given and is approximately constant, the effective motor power depends on the actual current. Otherwise, the current depends on the actual irradiation

Table 10.4.   Main characteristics of PV cells (EvoEnergy, 2012).

Type of cell	Surface	Efficiency [%]	Relative price
Monocrystalline		13–17	1
Polycrystalline		11–15	0.6
Amorphous		6–8	0.2–0.7
Hybrid PV		>18	1.5–3

of the solar panels, as shown in Figure 10.10. After the initial start-up voltage, the current and voltage components of a DC electrical load (pump) form a straight line on an *I-U* graph, which increases at a constant current-to-voltage ratio.

If a PV array is designed to produce 24 V, but the pump requires only 12 V, the load will only be drawn from the PV array, the power of which corresponds to 12 V on the *I-U* curve, even though the PV array is able to produce more power. Hence, when selecting the appropriate PV array and DC pump it is very important to consider the nominal voltage levels. The best practice is that the nominal P_e of the motor is equivalent to the peak power of the PV array. (Peak power means the maximum power output of the PV system in standard operational conditions.) In this case, the work point of the PV array and the DC motor are optimal if the irradiation is $1 \, \mathrm{kW \, m^{-2}}$. However, if the irradiation decreases, the effective power of the motor also decreases and will be much less than at the optimum work point associated with the actual irradiation.

For this reason, the direct operation of solar DC pumps by solar PV panels is usually not advantageous. It is acceptable only in stock tank or reservoir systems because of the relatively low costs.

Pressurized precision irrigation systems need static pump operation conditions that do not depend on the actual irradiation. This can be ensured by the use of solar batteries for the PV systems, where both the charging process and the irrigation can be controlled precisely with good system efficiency.

10.6   DESIGN OF SOLAR PV-BASED IRRIGATION

10.6.1   *Theory*

For the design of a solar PV-based irrigation system some basic data and information are needed. The most important data are:

- *Farm data*: the size of the land to be irrigated, soil type and its water management, land topography, the crop or crops produced.

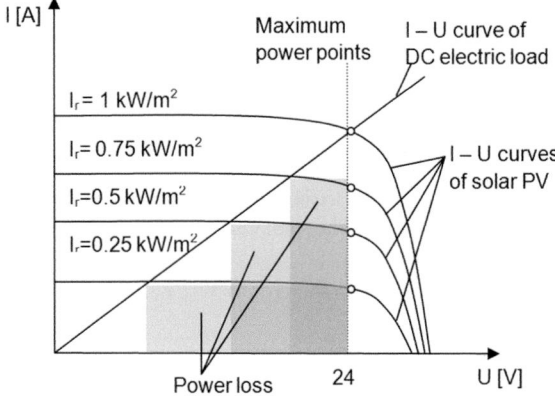

Figure 10.10. Graph showing *I-U* curves of solar PV panel at different irradiations (I_r) (Helikson *et al.*, 1991).

- *Weather data*: average annual rainfall, average rainfall in the growing period, solar irradiation in the growing period (the "average" means the average values over many years, based on local meteorological data).
- *Water source data*: type of water source available, quantity and quality of the water, location of water source and irrigated land.

As a first step, the calculation of the annual irrigation water demand is needed. If the average annual rainfall is h_a [mm] and the water demand of the crop during the growing period is h_d [mm], the deficit in water is expressed as:

$$h_i = h_d - h_a \text{ [mm]} \tag{10.23}$$

where h_i is the expected water demand to be ensured by irrigation. This value can be modified if the average rainfall in the growing/vegetation period (h_v) is very low and the crop is a shallow-rooted type. In this case, the water deficit can be estimated as:

$$h_i = h_d - h_v \text{ [mm]} \tag{10.24}$$

The annual water volume (V_a) required for irrigation can be calculated from the area of the irrigated land (A [m^2]) and (h_i):

$$V_a = 10^{-3} A h_i \text{ [m}^3 \text{ year}^{-1}] \tag{10.25}$$

The most suitable irrigation method can be selected according to the irrigation water demand, the characteristics of the crop (spacing, rooting, and microclimate demands) and the soil conditions. When selecting the irrigation method it is necessary to consider that the planned crop yield should be obtained with as little water and energy as possible. The irrigation equipment (pipeline tracks, emitters, emitter spacing, etc.) is determined by the irrigation method selected, and can be designed in detail.

On the basis of the irrigation system plan, the total pressure demand (p_t) can be determined. The energy demand of irrigation can be calculated from pressure demand p_t [bar] and annual water volume V_a [m^3]:

$$E_a = \frac{p_t - V_a}{36} \text{ [kWh]} \tag{10.26}$$

The value of E_a indicates the net energy demand expected from the solar PV array during the irrigation season. The next step is to determine the capacity of the PV panels. The input energy is the irradiation energy sum (ΣI_r) collected during the irrigation period. It can be calculated from the daily or monthly irradiation data. Another way of calculating ΣI_r is based on the calculation of the average irradiation (I_{ra} [kW m^{-2}]) and the number of hours of sunlight (n_{hs}) during the irrigation season:

$$\sum I_r = I_{ra} \cdot n_{hs} \quad [\text{kW m}^{-2}] \tag{10.27}$$

The effective energy demand E_a refers to the hydraulic energy of the water. Due to the different energy losses of elements in the system, the total energy requirement is:

$$E_{at} = \frac{E_a}{\eta_s} \quad [\text{kWh}] \tag{10.28}$$

where η_s is the system efficiency.

The system efficiency depends on the individual efficiency of system elements. The main components of η_s are the:

- Efficiency of solar PV panels, η_{PV}
- Efficiency of electric circuits and devices, η_c
- Efficiency of storage (batteries), η_{st}
- Efficiency of pump (including the motor efficiency), η_p

Thus, the system efficiency is:

$$\eta_s = \eta_{PV} \, \eta_c \, \eta_{st} \, \eta_p \tag{10.29}$$

The required area of PV panels, A_{PV}, is:

$$A_{PV} = \frac{E_{at}}{\sum I_r} \quad [\text{m}^2] \tag{10.30}$$

The capacity of batteries depends on the frequency of irrigation and the irrigated water h_i. The battery capacity required can be decreased if the water demand, h_i, is met with frequent irrigation and small doses of water. If f represents the irrigation frequency, the energy storage capacity (E_{st}) is:

$$E_{st} = \frac{E_{at}}{f} \quad [\text{kWh}] \tag{10.31}$$

The definition of f is one of the main elements of irrigation strategy. The frequency of irrigation depends on the weather conditions (rainfall, temperature, irradiation, relative humidity of air etc.) and the water demand of the crop over time. However, an average f can be forecast.

If the voltage of batteries is U (12 or 24 V), the battery capacity (C_b) required is:

$$C_b = \frac{10^3 E_{st}}{U} \quad [\text{Ah}] \tag{10.32}$$

10.6.2 *Application*

Consider the example of a plantation of area, $A = 1$ ha. The average annual rainfall is $h_a = 500$ mm per year. For reliable and high quality production, the necessary annual water volume, $h_d = 620$ mm. The extra water demand is:

$$h_i = h_d - h_a = 620 - 500 = 120 \, \text{mm year}^{-1} \tag{10.33}$$

The water volume needed for irrigation is:

$$V = 10^{-3} A \, h_i = 10^{-3} \times 10^4 \times 120 = 1200 \, \text{m}^3 \, \text{year}^{-1} \tag{10.34}$$

If the selected irrigation system is trickle irrigation, the total pressure required is $p_t = 2$ bar. The irrigation is based on a solar PV power source with battery energy storage. The energy demand of the irrigation is:

$$E_a = \frac{p_t V}{36} = \frac{2 \times 1200}{36} = 66.7 \, \text{kWh year}^{-1} \tag{10.35}$$

Based on the meteorological data, the sum of irradiation during the days of the irrigation season, $n_{id} = 90$, is $\Sigma I_r = 270 \, \text{kWh m}^{-2}$. The total annual input energy can be calculated if the system efficiency is known, the elements of η_s being the:

- Efficiency of polycrystalline PV panels, $\eta_{PV} = 0.1$
- Efficiency of circuit and controller, $\eta_c = 0.95$
- Efficiency of battery storage, $\eta_{st} = 0.85$
- Efficiency of brushless DC motor pump, $\eta_p = 0.9$

Using these data, the system efficiency is:

$$\eta_s = \eta_{PV} \, \eta_c \, \eta_{st} \, \eta_p = 0.1 \times 0.95 \times 0.85 \times 0.9 = 0.073 \, (7.3\%) \tag{10.36}$$

The total input energy demand is:

$$E_{at} = \frac{E_a}{\eta_s} = \frac{66.7}{0.073} = 914 \, \text{kWh year}^{-1} \tag{10.37}$$

The surface area of the PV array is:

$$A_{PV} = \frac{E_{at}}{\Sigma I_r} = \frac{914}{270} = 3.38 \, \text{m}^2 \tag{10.38}$$

The capacity of the batteries depends on the frequency of irrigation. If the frequency of irrigation, $f_i = 6$ days, the number of irrigation processes during the n_{id} irrigation season is:

$$n_i = \frac{n_{id}}{f} = \frac{90}{6} = 15 \tag{10.39}$$

The energy stored between two irrigation processes is approximately:

$$E_{st} = \frac{E_{at}}{n_i} = \frac{914}{15} = 61 \, \text{kWh} \tag{10.40}$$

If the voltage of the DC electric system, $U = 24$ V, the capacity of battery required is:

$$C_b = \frac{10^3 \cdot E_{st}}{U} = \frac{61000}{24} = 2542 \, \text{Ah} \tag{10.41}$$

Selecting a type of battery with capacity, $C_{b1} = 230$ Ah, the number of batteries needed is:

$$n_b = \frac{C_b}{C_{b1}} = \frac{2542}{230} = 11 \, \text{units} \tag{10.42}$$

These are only the most universal parameters of solar PV-powered pumping stations. There is not space here to discuss many important local details.

10.7 SOLAR WATER PUMPING IN ECONOMIC COMPARISON WITH DIESEL- AND GRID ELECTRICITY-POWERED PUMPING

Solar water pumping is no novelty. It has existed since the early 1980s, and has since been continuously improved so that, nowadays, it constitutes a well-established, globally used technology.

There are good reasons why solar water pumping is being used more and more; however, it has still not made a major breakthrough (Varadi, 2014).

Solar pumping for agricultural purposes is an important application of PV. Solar water pumps are available from different providers. They come in a large range of sizes. One of the largest providers offers them in a pump power range of between 0.15 and 21 kW, with lifts of up to 350 m and flow rates of up 130 m^3 h^{-1} (Varadi, 2014).

Today, solar water-pumping systems (using fixed or tracking arrays) are easy to install, operate and maintain. When compared with grid-powered or diesel-driven pumps, solar water-pumping systems have higher initial capital costs. However, the running and maintenance costs are much lower because no diesel or electricity needs to be purchased, meaning that the life-cycle cost is lower, given that a typical system lasts 20–30 years. According to Varadi (2014), the current investment cost of a solar-powered pump can be recovered within 3–4 years and, from then on, the money that would otherwise be needed for diesel or electricity purchase constitutes economic savings. However, a detailed study is required for each particular case because sunshine conditions vary from site to site. Fortunately, the greatest water needs arise in the sunniest seasons (spring, summer, autumn). In addition, as the pumping requirements vary from case to case, the respective system needs to be correspondingly designed and installed. In some cases (e.g. irrigation, dairy industry), pressurized water systems are required. In such cases, a solar pump works as a standard AC pump but its size must be carefully determined (depending on the required pressure) and it must be connected to a closed pressure tank. If a permanent water supply is required, the installation of a water reservoir (with a capacity large enough to sustain 2–5 days' water consumption – again depending on the specific case) is the cheapest version (based on intermittent water pumping when sunshine is available to power the system) (Bacusmo, 2013). In contrast, the installation of batteries for energy storage in support of continuous pumping can significantly increase overall installation costs and require specific maintenance; it is not recommended by several providers (Solar Pumping Solutions, 2016).

When compared with grid-connected pumps, another advantage of small- to medium-sized solar-powered pumps is that they can be easily transported, for example on a trailer, which allows seasons to be lengthened or facilitates the rotation of grazing areas and so increases the economic benefit.

In addition, in many cases, the permanent on-farm availability of diesel/petrol may require the installation of large storage tanks, the investment costs of which can be of the order of the price of a solar pumping system, which has continually decreased in recent years and is expected to decrease further still in the future (Varadi, 2014).

When comparing the solar option with fossil fuel generators or grid-powered pumps, the highest cost reductions can be expected in cases where rivers are remote or cannot cover the water demand, and where access to the electricity grid is absent or distant. In Australia, farm water sources are often distributed over large areas, and there are few power lines; it would therefore be costly to install and maintain electrical connections to the pumps and/or it would require time-consuming, labor-intensive and costly refueling of diesel generators (a further consideration is the continually increasing price of diesel). Moreover, diesel generators require significant maintenance and the continuous supply of oil could prove to be a problem, further increasing the cost. A study of Australian conditions has shown that, if the electricity grid is more than one kilometer away, solar-driven pumps are the most economical option (Solar Pumping Solutions, 2016).

A solar-versus-diesel calculator can be found at http://www.lorentz.de/svd. This calculator estimates lifetime diesel costs and compares them with those of the solar option by using four parameters: (i) location based on Google Maps; (ii) water depth; (iii) water demand per day; (iv) current diesel fuel costs. The calculator further allows the input of numerous assumptions such as those for the present diesel option:

- annual fuel price increase
- fuel delivery costs
- annual service cost

- replacement pump cost
- replacement generator cost
- life of diesel pump
- life of generator;

and also those for the solar option:

- cost of solar pump system
- cost of solar generator
- life of solar pump system
- typical service cost.

An additional general parameter, the general price inflation rate, is also included in the calculator.

10.8   CASE STUDIES

Several studies have been performed to compare the economics of solar-powered versus petrol/diesel-generator-powered water-pumping systems.

10.8.1   *USA*

The Bureau of Land Management at Battle Mountain, Nevada, found that for a 228 gph system [863 L] (275 foot design head [83.8 m], i.e. head difference between water table and storage tank), a PV system cost only 64% as much over 20 years as a generator-powered system cost over just ten years (Sinton *et al.*, 2005).

10.8.2   *India*

Agriculture accounts for nearly 80% of India's entire freshwater use (IRENA, 2015). Pumping is essential for India's agriculture as otherwise it would depend on seasonal rainfall, which would reduce agricultural production. At present, about 35 million hectares are under irrigation, using 212 billion (10^9) m^3 of groundwater annually, making irrigation unsustainable (Tweed, 2014). However, switching from "expensive" irrigation water pumped by diesel generator or grid electricity-driven pumps to the "cheaper" solar pumping option (the operational cost of PV pumps is negligible) could increase the water volumes pumped by farmers, which would further increase the unsustainability of groundwater exploitation in India. This risk might be overcome by combining governmental or state subsidies to purchase solar water pumps with the obligation to switch to drip irrigation (Tweed, 2014).

Nationwide, there are estimated to be roughly 26 million agricultural pumps. There are at least 12 million electric (grid-based) and 9 million diesel-powered irrigation pump sets (C-STEP, 2010). Both grid-based electricity and diesel are highly subsidized (IGEP, 2013; C-STEP, 2010; Tweed, 2014). Farmers pay only about 13% of the true cost of electricity. Therefore, these subsidies represent a high economic burden on the government: India pays over US$ 6 billion annually on energy subsidies (Casey, 2013). In addition, the cheap subsidized energy leads to increased groundwater pumping and unsustainable groundwater extraction (IRENA, 2015).

Recognizing the potential benefits, and to remove the need to pay energy subsidies to farmers, India plans to replace many of its 26 million groundwater pumps for irrigation with solar pumps (Tweed, 2014). In a first phase, it is planned to switch the first 200,000 most easily replaceable pumps to solar, which will require about US$ 1.6 billion in investment during the next five years (Bloomberg, 2014). Furthermore, India has announced the deployment of 20,000 solar pumps in selected rural areas to expand access to piped water (Indian Ministry of New and Renewable Energy, 2014).

From an environmental point of view, the significance of emission reduction is evident if we consider the following numbers. The replacement of five million diesel pump sets with solar pumping systems would save about 18.7 GW of installed capacity and 23.4 billion MWh (or 23.4×10^{15} MWh) of consumed grid-electricity produced from 10 billion liters, resulting in a CO_2 emission reduction of 26 million tons per year (this calculation assumes that an average pump has a capacity of 3.73 kW and operates during 1250 h annually (CEEW, 2013). As part of the Indo-German Energy Program, the German Agency for International Technical Cooperation (GTZ, now GIZ), researched the feasibility of solar pumping in the Indian state of Bihar (GIZ, 2013). It compared the ten-year life-cycle cost of diesel with a solar-powered pump for the case of a 1-hp pump powered by a 2 kW diesel generator and a solar PV water-pumping system of equal capacity. The investment cost of the solar-powered pump was eight times higher than the diesel pump with generator (200,000 vs 25,000 Indian Rupees, IR), but the net present maintenance cost of the solar option was about four times lower than the diesel option (3072 vs 12,298 IR). However, it was the net fuel cost of nil in the case of solar versus 278,993 IR in the case of the diesel option that made the solar option the more economic one (life-cycle cost: 203,072 vs 316,282 IR). In situations where labor costs are high, the difference would be even more significant.

10.8.3 *Morocco*

In the case study of Ain Sfa, Oujda, Morocco, solar-powered pumping from a depth of 40 m produces 120 m^3 water daily (one well pump, one reservoir pump, each 120 m^3 day^{-1}, powered by 10 kWp PV modules) for irrigation of a large area of 6 ha, which was formerly irrigated by fossil-fuel-powered pumps. The justification for the technology shift was a usage and cost analysis, which indicated that the five-year cost per m^3 of water was US\$ 0.12 for the diesel option and only US\$ 0.08 for the solar option (Lorentz, 2013).

10.9 CONCLUSIONS

Solar PV pumping techniques have been developed recently as a result of a reduction in the price of PV panels as well as an increase in efficiency. Today, an efficient solar pumping system can transform average daily solar energy power to hydraulic power with an efficiency of more than 4%. The benefits of a solar pumping system are highly dependent on good irrigation system design, derived from accurate site and demand data. An important aspect of solar PV pumping stations is the possibility for on-grid operation because solar PV plants for irrigation are underutilized for most of the year. A new momentum can be expected with the new types of PV cells. The multilayer technologies and the concentrator-type cells may bring a strong increase in panel efficiency. The subsystem elements (motor, pump, power conditioning) are also under active development and higher efficiencies are likely in the future.

Summarizing the experiences of solar water pumping and irrigation, it can be stated that there are important advantages but also limits to its application. The advantages are:

- The absence of traditional fuel (electricity, diesel oil) to operate the pumping station means the operating costs are minimal;
- Well-designed and located PV panels and pumps do not involve significant maintenance costs;
- There is no environmental pollution or pressure associated with their operation;
- The pumping station design can be flexible because the PV panels do not need to be next to the water source.

The main limitations are:

- Solar pumping is not suitable when the required flow rate is high;
- The energy yield is cyclic as a result of its dependence on sunlight;
- In some areas, effective precautions may be needed to protect against theft.

REFERENCES

Bacusmo, J. (2013) Solar pumping systems. Groundwater Engineering (GWE). Available from: http://www.groundwaterinternational.com/uploads/groundwater_engineering/files/Solar_Water_Pumping_Systems.pdf [accessed December 2016].

Bloomberg (2014) Solar water pumps wean farmers from India's archaic grid. Available from: http://www. bloomberg.com/news/articles/2014-02-07/solar-water-pumps-wean-farmers-from-india-s-archaic-grid [accessed December 2016].

Casey, A. (2013) Reforming energy subsidies could curb India's water stress. Worldwatch Institute, Washington, DC. Available from: http://www.worldwatch.org/reforming-energy-subsidies-could-curb-india's-water-stress-0 [accessed October 2016].

CEEW (2013) Renewables beyond electricity. Council on Energy, Environment and Water, New Delhi, India. Available from: http://ceew.in/pdf/CEEW-WWF-RE+Renewables-Beyond-Electricity-Dec13.pdf [accessed December 2016].

Chandel, S.S., Nagaraju Naik, M. & Chandel, R. (2015) Review of solar photovoltaic water pumping system technology for irrigation and drinking water supplies. *Renewable and Sustainable Energy Reviews*, 49, 1084–1099.

C-STEP (2010) Harnessing solar energy: options for India. Centre for Study of Science, Technology and Policy (C-STEP), Bangalore, India.

Dursun, M. & Özden, S. (2012) Design and implementation of drip irrigation system with solar energy. *International Review of Automatic Control*, 5(2), 156–165.

EvoEnergy. (2012) Solar PV Cell Comparison. EvoEnergy Ltd., Nottingham, UK. Available from: http://www.evoenergy.co.uk/services/our-technology/pv-cell-comparison/ [accessed December 2016].

GIZ (2013) Solar water pumping for irrigation: potential and barriers in Bihar, India. Deutsche Gesellschaft für Internationale Zusammenarbeit (GIZ) GmbH, New Delhi, India; Ministry of New and Renewable Energy (MNRE), New Delhi, India. Available from: http://www.igen-re.in/files/giz__2013__factsheet_solar_water_pumping_for_irrigation_in_bihar.pdf [accessed December 2016].

Goswami, D.Y. (ed.) (1986) *Alternative Energy in Agriculture*, Vol. II. CRC Press, Boca Raton, FL.

Helikson, H.J., Haman, D.Z. & Baird, C.D. (1991) Pumping water for irrigation using solar energy, Fact Sheet EES-63. University of Florida, Gainesville.

Indian Ministry of New and Renewable Energy (MNRE) (2014) Solar energy based dual pump piped water supply scheme. New Delhi, India. Available from: http://www.mdws.gov.in/solar-energy-based-dual-pump-piped-water-supply-scheme [accessed December 2016].

IGEP (2013). Solar water pumping for Irrigation. Deutsche Gesellschaft für Internationale Zusammenarbeit (GIZ) GmbH, New Delhi, India; Ministry of New and Renewable Energy (MNRE), New Delhi, India. Available from: http://igen-re.in/files/giz__2013__report_solar_water_pumping_for_irrigation_in_bihar.pdf [accessed December 2016].

IRENA (2015) Renewable energy in the water, energy & food nexus. International Renewable Energy Agency, Masdar City, UAE.

Johanson, T.B., Kelly, H., Reddy, A.K.N. & Williams, R.H. (eds.) (1993) *Renewable Energy. Sources for Fuels and Electricity*. Island Press, Washington, DC.

Lorentz. (2013) Solar water pumping for irrigation in Oujda, Morocco. Bernt Lorentz GmbH & Co. KG, Henstedt-Ulzburg, Germany. Available from: http://www.lorentz.de/pdf/lorentz_casestudy_irrigation_morocco_en-en.pdf [accessed December 2016].

Patay, I., Montvajszki, M. & Gergely, Z. (2012) Energy demand of irrigation. Energy and carbon saving in the practice. *Hungarian Agricultural Engineering*, 24, 26–29.

Shinde, V.B. & Wandre, S.S. (2015) Solar photovoltaic water pumping system for irrigation: a review. *African Journal of Agricultural Research*, 10(22), 2267–2273.

Sinton, C.W., Butler, R. & Winnett, R. (2005) Guide to solar-powered water pumping systems in New York State. New York State Energy Research and Development Authority (NYSERDA), Albany. Available from: https://www.nyserda.ny.gov/-/media/Files/EERP/Renewables/Guide-solar-powered-pumping-systems-NYS.pdf [accessed December 2016].

Solar Pumping Solutions. (2016) Solar Pumps – Solar Frequently Asked Questions – Why we don't recommend batteries in water pumping systems. SPS Solar, Mudgee, Australia. Available from: http://www.solarpumping.com.au/pumps_solar_faq.htm [accessed December 2016].

Stetson, V.E. & Mecham, B.Q. (eds.) (2012) *Irrigation*, 6th edn. Irrigation Association, Falls Church, VA.

Tweed, K. (2014) India plans to install 26 million solar-powered water pumps. Available from: http://spectrum.ieee.org/energywise/green-tech/solar/india-plans-for-26-million-solar-water-pumps [accessed December 2016].

Varadi, P.F. (2014) *Sun Above the Horizon: Meteoric Rise of the Solar Industry*. Pan Stanford, Singapore.

Yiasoumi, B. (2003) Selecting an irrigation pump. Agfact E5.8, 3rd edition. Available from: http://www.dpi.nsw.gov.au/agriculture/resources/water/irrigation/systems/pumps/selecting [accessed December 2016].

CHAPTER 11

Solar drying

Om Prakash, Anil Kumar & Atul Sharma

11.1 INTRODUCTION

It is established fact that food is a prime necessity for human beings. The Food and Agriculture Organization (FAO) reported in 2015 that 794.6 million people were undernourished in the world in 2014–2016, which is 10.9% of the world population (www.fao.org). Hence, agricultural production is required to increase to provide food security to people. It is predicted that, in the next 25 years, about 50% increase in food production will be needed in developing countries to meet population growth. An alternative solution to this problem is to minimize food loss (Prakash, 2015). Food production losses happen mainly at three different stages, namely, harvest, post-harvest and marketing.

It is calculated that 80–85 × 10⁶ tons of food loss occurs in India due to lack of suitable post-harvest technologies (Tiwari and Barnwal, 2011). Table 11.1 shows various post-harvest losses for selected food products and the estimated loss of money.

Agricultural produce with a moisture content above 80% falls into the category of high-moisture content crops or highly perishable commodities (Prakash et al., 2016a). In order to keep crops fresh, low-temperature storage technology is applied; however, this approach is found to be very costly.

Drying for preservation is one of the most important post-harvest methods to minimize food losses. It is an ancient preservation technique. The dried product can be stored for an extended period. Experts in drying techniques identify drying temperature and safe moisture content as the most important parameters, and Table 11.2 presents values of these for some selected crops.

Drying is a complicated heat and mass transfer process in which the heat surrounding the product transfers to the external surface of the product. Some part of this external heat is transferred to the inside of the product (Prakash and Kumar, 2014g; Prakash et al., 2015). This leads to a rise in temperature and the formation of water vapor. The remaining heat evaporates the moisture from the surface of the product. The moisture and water vapor from the internal region of the product come to the surface and replace the loss of moisture. In the initial stage, moisture removal is rapid due to excess moisture. However, later, it depends on the material and the moisture content of the product.

Table 11.1. Post-harvest loss for food products (Tiwari and Barnwal, 2011).

Product	Loss [%]	Value [billion US$]
Grains	10	26.32
Pulses	15	3.19
Fruits	30	21.69
Vegetables	30	22.49
Floriculture	40	0.64
Dairy (milk)*	1	1.44

*Handling losses based on limited study in Karnal, Haryana, India.

235

Table 11.2.   Maximum allowable temperature and moisture content on wet basis (wb) for drying of selected agricultural produce (Prakash and Kumar, 2013a, Prakash *et al.*, 2016b).

Type of crop	Crop	Initial moisture content [% wb]	Final moisture content [% wb]	Maximum allowable temp. [°C]
Grains	Paddy, raw	22–24	11	50
	Paddy, parboiled	30–35	13	50
	Maize	35	15	60
	Wheat	20	16	45
	Corn	24	14	50
	Rice	24	11	50
	Pulses	20–22	9–10	40–60
	Oil seed	20–25	7–9	40–60
Vegetables	Green peas	80	5	65
	Cauliflower	80	6	65
	Carrot	70	5	75
	Green bean	70	5	75
	Onion	80	4	55
	Garlic	80	4	55
	Cabbage	80	4	55
	Sweet potato	75	7	75
	Potato	75	7	75
	Chili	80	5	65
	Tomato	96	10	60
	Eggplant	95	6	60
Fruits	Apricot	85	18	65
	Apple	80	24	70
	Grape	80	15–20	70
	Banana	80	15	70
	Guava	80	7	65
	Okra	80	20	65
	Pineapple	80	10	65

Drying of agricultural produce is highly energy-intensive. In developed countries, 8–12% of primary energy is consumed by the drying operation (Imre, 1993). Due to the oil crisis in the 1970s, people started to think about using renewable energy to fulfill their energy demands, including the drying of agricultural produce. Solar energy is one of the most readily available renewable energy sources and can be used in many ways to preserve precious conventional fuels.

11.1.1   *Importance of renewable energy*

The demand for fossil fuels is increasing day by day, but the supply is decreasing and as a result the cost of fossil fuels is increasing. Hence, it is necessary to look for a different source of energy to meet energy requirements. Such sources should be abundant in nature, economical, have low maintenance costs and be pollution-free (Raman and Tiwari, 2008). At the global level, the generation of electrical energy mostly depends on fossil fuels. Fossil fuels are, however, depleting due to extensive and continuous use prompted by rising levels of lifestyle. Moreover, burning of fossil fuels is the main cause of CO_2 emissions, leading to air pollution, acid rain, the greenhouse effect and environmental degradation. To overcome the problem, several developed and developing countries are working to harness renewable energy resources because of their free availability and non-polluting nature, and to reduce the fast depletion of conventional energy resources. Hence, there is a need to conserve fossil fuels and to explore possible alternatives. From this perspective, awareness of the utilization of renewable energy sources has gained rapid acceptance at a universal level. Solar energy is one type of renewable energy resource that is generally available throughout the world. Solar energy has large potential to fulfill our energy needs and it avoids most of the harmful impacts of the use of fossil fuels. Most of the Indian subcontinent has high potential for solar radiation, which is suitable for the development of a solar photovoltaic (PV) system for power generation and other solar thermal applications. Other options include nuclear energy, wind energy, biomass, fuel cells and ocean/wave energy (Kumar and Tiwari, 2009; Nawaz and Tiwari, 2006).

11.2   BASICS OF DRYING

11.2.1   *Moisture content*

Moisture content is the most important parameter in the storage of agricultural produce. It indicates the amount of water present in the product. It can be represented as percentage or in decimal fraction. There are two types of moisture content, namely, wet basis and dry basis.

Moisture content on wet basis (MC_{wb}) is the ratio of actual weight of moisture present in the product to the total weight of the product (both moisture content and dry matter):

$$\text{Decimal} \quad MC_{wb} = \frac{\text{Wt. of } H_2O}{\text{Wt. of } H_2O + \text{dry matter}} \quad (11.1)$$

$$\text{Percentage} \quad MC_{wb} = \frac{\text{Wt. of } H_2O}{\text{Wt. of } H_2O + \text{dry matter}} \times 100 \quad (11.2)$$

The moisture content on a dry basis (MC_{db}) is the ratio of actual weight of moisture present in the product to the weight of dry matter in the product:

$$\text{Decimal} \quad MC_{db} = \frac{\text{Wt. of } H_2O}{\text{Wt. of dry matter}} \quad (11.3)$$

$$\text{Percentage} \quad MC_{db} = \frac{\text{Wt. of } H_2O}{\text{Wt. of dry matter}} \times 100 \quad (11.4)$$

Ekechukwu (1999) developed the relationship between wet basis moisture content [%] and dry basis moisture content [%], as shown in Figure 11.1.

11.2.2   *Moisture content evaluation*

It is very important to know the safe moisture content of a product for effective drying. Neither overdrying nor underdrying is acceptable. Overdrying causes the biochemical properties of the product to deteriorate, as well as causing an unnecessary reduction in weight. Underdrying leads to spoilage of the product because of excess moisture. Table 11.3 shows the methods used to evaluate safe moisture content (Tiwari and Barnwal, 2011).

11.2.3   *Theory of drying*

There are two types of drying, namely, thin-layer and thick-bed drying. In a solar drying process, thin-layer drying is generally used.

11.2.3.1   *Thin-layer drying*
In thin-layer drying, there are two basic assumptions, namely, that the proportion of air per crop is infinitely large, and that the whole product is kept under almost constant drying air and relative humidity. Figure 11.2 shows an example of thin-layer drying under natural or open sun drying.

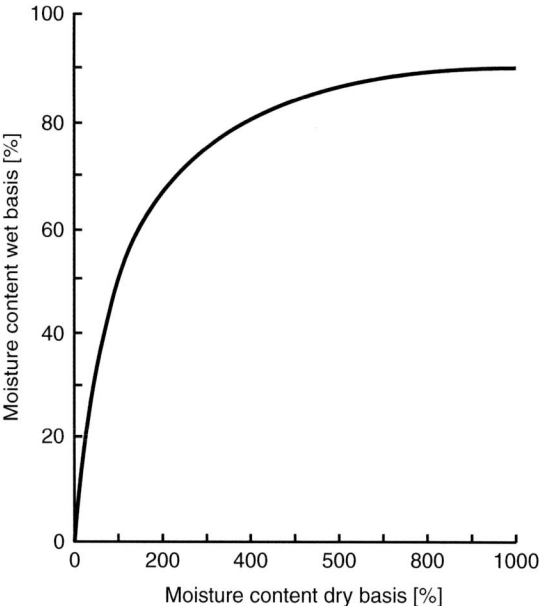

Figure 11.1.   Variation of moisture content on wet basis and dry basis (Ekechukwu, 1999).

Table 11.3.   Methods for moisture evaluation.

Direct methods	Oven	After measurement of the weight, the sample is placed inside an oven for a period of 0.75–24 hours, based on crop sample.
	Distillation	This technique is based on the concept of co-distillation of the moisture of the crop with an immiscible solvent with a high boiling point. Then the volume of water in the collected mixture, which is distilled off, is measured.
	Infrared lamp	In this method, the weighted sample is placed under an infrared (IR) lamp for a specified time span based on crop sample.
	Microwave radiation	In this methodology, the measured sample is placed inside a microwave oven for a certain span of time, with temperature based on the sample requirement.
	Chemical reaction	This method is based on volumetric titration of the sample crop and is applied to crops that show changeable results when heat is applied. This is mainly useful for low moisture content crops. It is based on the concept of Karl Fischer titration.
Indirect methods	Electrical resistance	In this approach, the bean, for example, is placed in between two electrodes and a current applied between them.
	Capacitance	In this methodology, a current is applied to the two plates of a condenser and the crop is placed in between them.

Several researchers have proposed various drying kinetics models (semi-empirical/fully empirical) for thin-layer drying. A consolidated list is presented in Table 11.4.

11.2.3.2   *Thick-bed drying*

In thick-bed drying, drying starts from the bottom of the bin chamber, with which the hot air first comes into contact. The hot air is supplied through a fan. In general, this mode of drying

Figure 11.2.   Photograph of thin-layer drying (Prakash and Kumar, 2014b).

Table 11.4.   Different mathematical models for thin-layer drying curves (Prakash and Kumar, 2014a).

No.	Model name	Model equation	Reference
1	Lewis	$M_r = \exp(-kt)$	Prakash and Kumar (2014b)
2	Page	$M_r = \exp(-kt^n)$	Prakash and Kumar (2014b)
3	Modified Page	$M_r = \exp[-(kt)^n]$	Tiwari and Barnwal (2011)
4	Henderson and Pabis	$M_r = a\exp(-kt)$	Prakash and Kumar (2014b)
5	Modified Henderson and Pabis	$M_r = a\exp(-kt) + b\exp(-gt) + c\exp(-ht)$	Prakash and Kumar (2014b)
6	Logarithmic	$M_r = a\exp(-kt) + c$	Prakash and Kumar (2014b)
7	Two-term	$M_r = a\exp(-k_0 t) + b\exp(-k_1 t)$	Prakash and Kumar (2014b)
8	Two-term exponential	$M_r = a\exp(-kt) + (1-a)\exp(-kat)$	Tiwari and Barnwal (2011)
9	Wang and Singh	$M_r = 1 + at + bt^2$	Prakash and Kumar (2014b)
10	Thompson	$t = a\ln(M_r) + b[\ln(M_r)]^2$	Tiwari and Barnwal (2011)
11	Approximation of diffusion	$M_r = a\exp(-kt) + (1-a)\exp(-gt)$	Tiwari and Barnwal (2011)
12	Verma *et al.*	$M_r = a\exp(-kt) + b\exp(-gt) + c\exp(-ht)$	Tiwari and Barnwal (2011)
13	Simplified Fick's diffusion equation	$M_r = a\exp[-c(t/L^n)]$	Tiwari and Barnwal (2011)
14	Modified Page equation-II	$M_r = \exp[-k(t/L^2)^n]$	Tiwari and Barnwal (2011)
15	Midilli and Kucuk	$M_r = \exp(-kt^n) + bt$	Tiwari and Barnwal (2011)
16	Prakash and Kumar	$M_r = at^3 + bt^2 + ct + d$	Prakash and Kumar (2014b)

a, b, c, d, k, k_0, L, g and h are constants and t is the drying time.

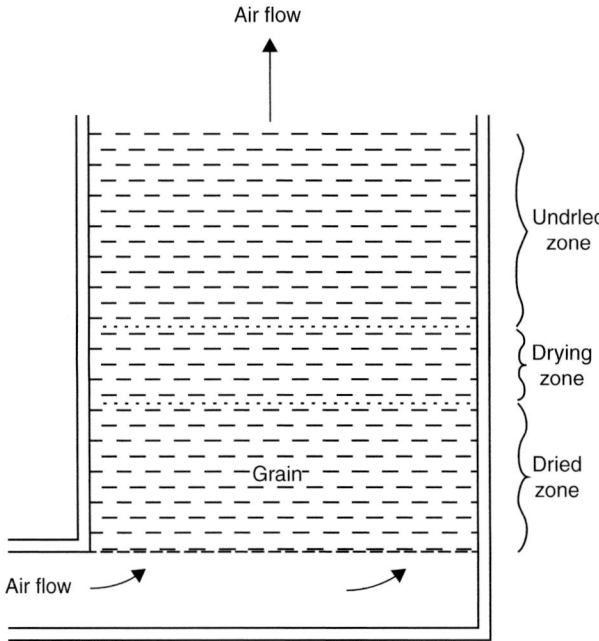

Figure 11.3.   Schematic diagram of thick-bed drying (Proctor, 1994).

is applied to grains. The drying bed is usually 0.30–0.60 m thick. In this mode of drying, the overall drying rate remains constant. The supplied hot air absorbs the moisture of the crop until it becomes saturated. The modes of heat transfer are mainly conduction and convection.

Figure 11.3 illustrates the actual mechanism of thick-bed drying. Here hot air moves from the bottom to the top. In this way the crop is heated, the moisture evaporates and the crop dries out.

11.3   SELECTION OF SOLAR DRYERS FOR THE PRODUCT

It is very difficult to select a suitable solar dryer due to differences in moisture content and other biochemical properties of the agricultural produce. Table 11.5 presents a sample checklist for the evaluation and selection of a solar dryer based on the product to be dried (Visavale, 2012).

11.4   TYPES OF SOLAR DRYERS

Solar dryers are mainly classified into two groups: natural solar dryers and artificial solar dryers. Artificial solar dryers are also called controlled solar dryers. Artificial solar dryers can be operated in two different modes of heat transfer, namely, natural convection and forced convection. Hence, artificial solar dryers can be classified into two groups: active dryers and passive dryers. Active dryers operate under forced convection of heat transfer and passive dryers operate under natural convection of heat transfer. Both active and passive dryers can be operated in three modes, namely, direct mode, indirect mode and mixed mode. A detailed classification of solar dryers is shown in Figure 11.4.

Table 11.5. Checklist for preliminary evaluation and selection of solar dryers (Visavale, 2012).

No.	Parameter	Features
1	Physical features of dryer	Type, size and shape
		Collector area
		Drying capacity/loading density (kg/unit tray area)
		Tray area and number of trays
		Loading/unloading convenience
2	Thermal performance	Solar insulation
		Drying time/drying rate
		Dryer/drying efficiency
		Drying air temperature and relative humidity
		Airflow rate
3	Properties of the material being handled	Physical characteristics (wet/dry)
		Acidity
		Corrosiveness
		Toxicity
		Flammability
		Particle size
		Abrasiveness
4	Drying characteristics of the material	Type of moisture (bound, unbound, or both)
		Initial moisture content
		Final moisture content (maximum)
		Permissible drying temperature
		Probable drying time for different dryers
5	Flow of material to and from the dryer	Quantity to be handled per hour
		Continuous or batch operation
		Process prior to drying
		Process subsequent to drying
6	Product qualities	Shrinkage
		Contamination
		Uniformity of final moisture content
		Decomposition of product
		Overdrying
		State of subdivision
		Appearance
		Flavor
		Bulk density
7	Facilities available at site of proposed installation	Space
		Temperature, humidity, and cleanliness of air
		Available fuels
		Available electric power
		Permissible noise, vibration, dust, or heat losses
		Source of wet feed
		Exhaust-gas outlets
8	Economics	Cost of dryer
		Cost of drying
		Payback

11.4.1 *Natural solar dryers*

The natural solar dryer is also commonly known as the open sun dryer. This is an ancient method of drying and people have been using this method for generations. Figure 11.5 shows the actual methodology of natural solar drying. In this drying process, the agricultural produce is kept in the

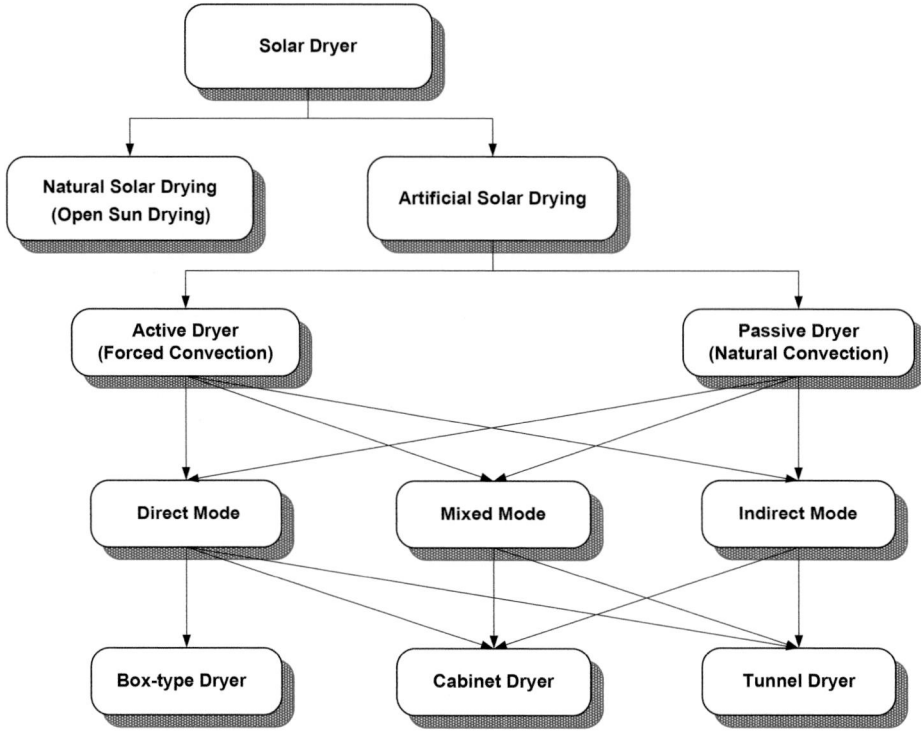

Figure 11.4.   Classification of solar dryers.

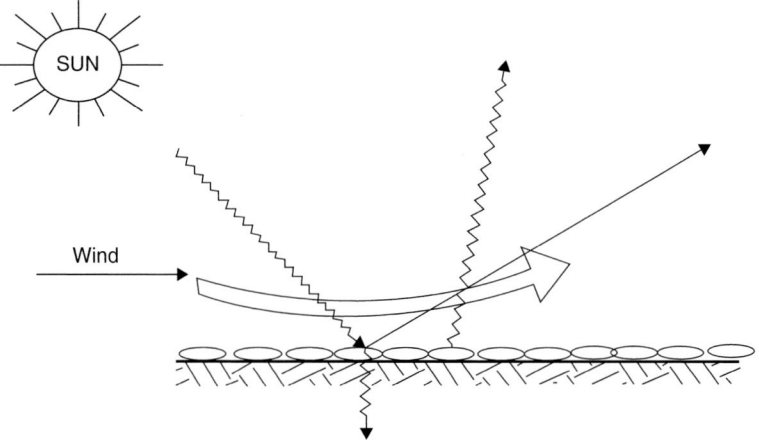

Figure 11.5.   Schematic diagram of natural solar dryer (Prakash, 2015).

sunlight over a horizontal surface. The surface may be either concrete or bare ground. Figure 11.6 illustrates the drying of paddy or rice on a concrete floor.

The shorter wavelengths of the solar radiation fall on the crop surface. Certain parts of this solar radiation are absorbed in the crop and the remaining parts are reflected back to the atmosphere. The absorbed radiation increases the temperature of the crop and thus moisture evaporates from the crop (Sodha *et al.*, 1985).

Figure 11.6. Drying of paddy on concrete floor (Prakash and Kumar, 2014c).

Table 11.6. Comparison between open sun drying and solar dryers (Aware and Thorat, 2012).

Open sun drying	Solar dryer
Classical method	Scientific method
Long-span drying time	Short-span drying time
High chance of serious contamination by birds, insects, etc.	No chance of any contamination
Low hygienic standard	High hygienic standard
Low-quality products	High-quality products
May not meet Good Manufacturing Practice (GMP) standard	Meets GMP standard
Possible only on full sunny days	Possible all year round
High chance of over-/underdrying	Even drying
Low profit and more space is required	More profit and less space is required

Although this method of drying is simple and free of any pollution or cost, it nevertheless has certain limitations. The major shortcomings of natural drying are: contamination of the product due to dirt and insects; wastage by birds/mice; spoilage due to sudden and unpredicted rain; no control of temperature, leading to potential overdrying. Overdrying may cause serious loss of germination power, nutritional changes and sometimes total damage.

A comparison between open sun drying and a solar dryer is presented in Table 11.6.

11.4.2 *Direct solar dryers*

In direct solar dryer, the solar radiation is being directly used for the purpose of the drying. There are many solar dryers which comes in the category of the direct solar dryer namely greenhouse dryer, cabinet dryer, tent dryer etc. The greenhouse dryer is a multipurpose structure which is used in various applications, such as cultivation, drying, aquaculture, and solariums (Kumar *et al.*, 2006). The structure of the greenhouse is made of a transparent wall using suitable polymer materials. It operates on the basis of the greenhouse effect and is effective for medium thermal drying.

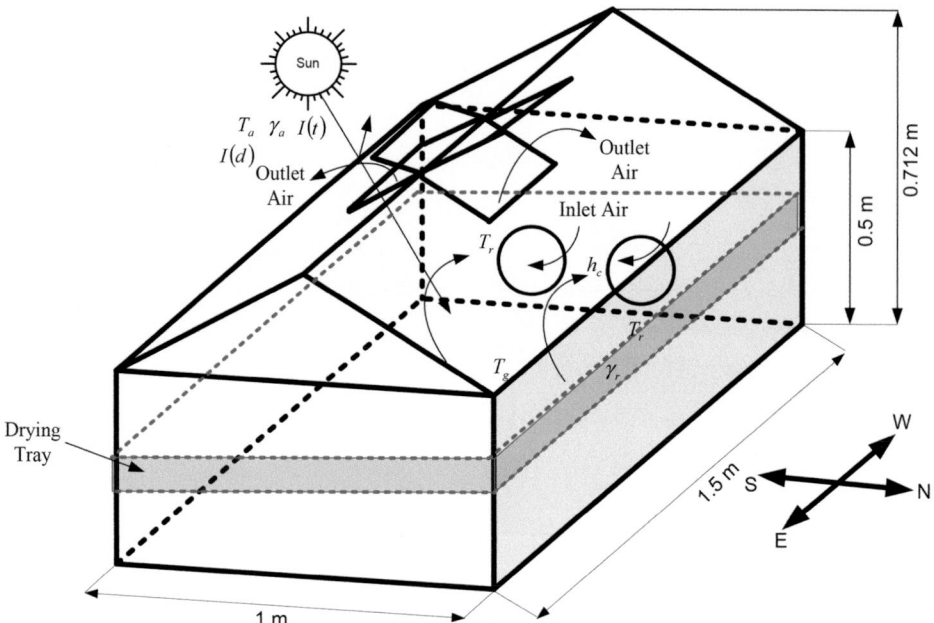

Figure 11.7.   Schematic diagram of modified greenhouse dryer in passive mode (Prakash and Kumar, 2014d).

Prakash and Kumar (2014c) have reviewed all the research works in the field of greenhouse drying. They have broadly classified the greenhouse dryer into two types, namely, active mode greenhouse dryers and passive mode greenhouse dryers. The active mode greenhouse dryer is based on heat transfer by forced convection, while the passive mode greenhouse dryer is based on natural convection. Various researchers have studied the conventional greenhouse dryer for drying various agricultural produce (Jain and Tiwari, 2004; Kumar and Tiwari, 2006a, 2006b, 2006c, 2007). However, due to various thermal losses, its applications are restricted. Two main thermal losses occur, namely, the loss of incident solar radiation through the north wall and loss of thermal energy from the ground or floor. Prakash and Kumar (2014d, 2015) incorporated a solution to these problems: they made the north wall opaque with a mirror, which restricts the loss of radiation through the north wall, and the concrete floor of the greenhouse was covered with a double-layer black PVC sheet with holes in the upper layer. In order to enhance the drying efficiency, a further modification in the structure was made by using extended steel wire mesh, which maximizes the inner area of the dryer. Figure 11.7 shows a schematic diagram of the modified greenhouse in passive mode, and Figure 11.8 shows the modified greenhouse dryer in active mode.

In order to validate the modification, a study of the modified greenhouse dryer was conducted in the no-load condition in both passive mode and active mode (Prakash and Kumar, 2013b, 2014f). The modified greenhouse dryer under active mode is operated in the load condition for drying of tomato flakes (Prakash and Kumar, 2014b). Crops dried in the dryer were found to be superior in quality and with a higher drying rate and reduced drying time by comparison with natural sun drying. The authors also proposed a mathematical drying kinetics model (equation), which shows superior performance when compared with all existing models.

Aware and Thorat (2012) have used the solar cabinet dryer for the drying of grapes. A photograph of this is shown in Figure 11.9. This dryer has 32 aluminum trays, each with an area of $0.46\,\mathrm{m}^2$. The air was heated in the solar collector and distributed in the crop-drying chamber by a fan. The internal parameters of temperature, relative humidity and air velocity vary from 53 to 57°C, 35 to 45% and 0.9 to $1.0\,\mathrm{m\,s}^{-1}$, respectively.

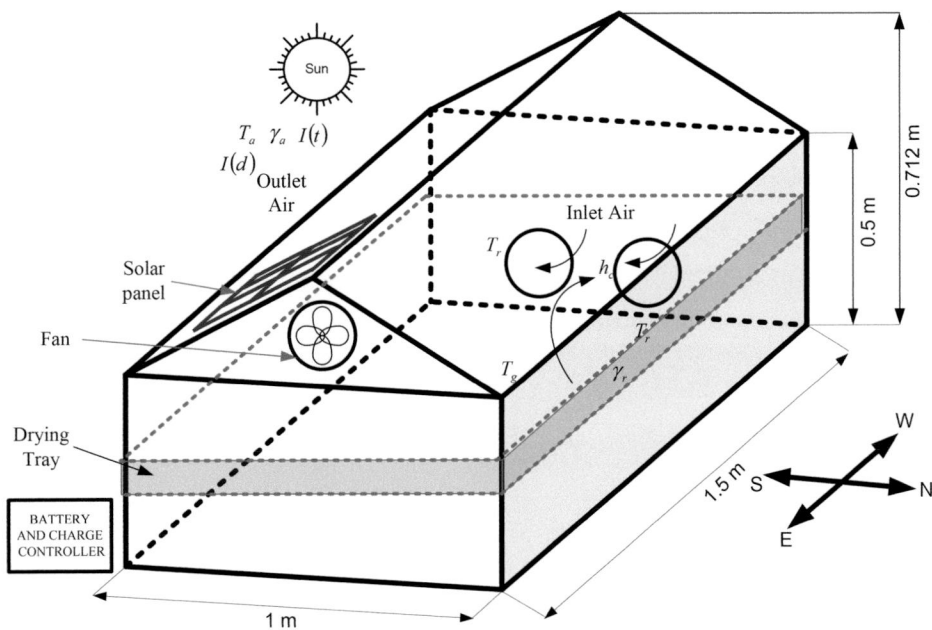

Figure 11.8.   Schematic diagram of modified greenhouse dryer in active mode (Prakash and Kumar, 2014e).

Figure 11.9.   Photograph of solar cabinet dryer (Aware and Thorat, 2012).

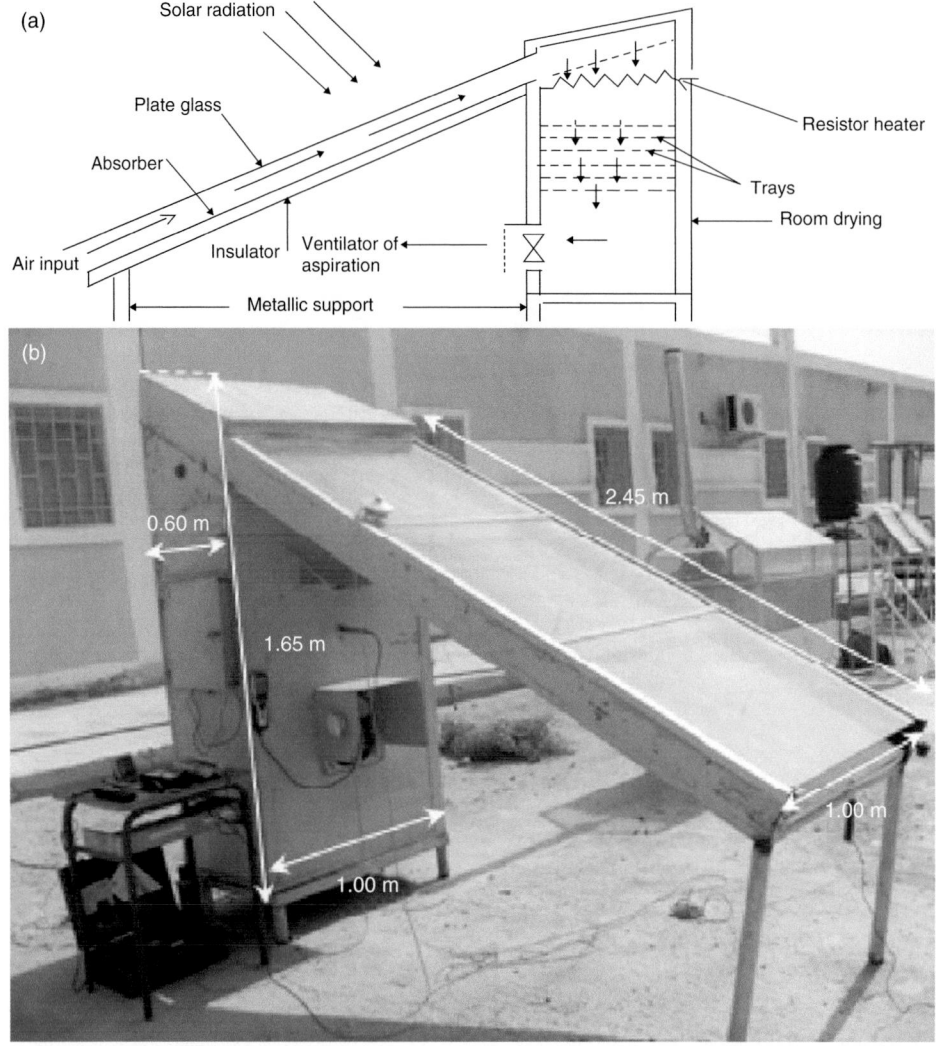

Figure 11.10.   Indirect solar active electric dryer: (a) schematic; (b) photograph (Boughali *et al.*, 2009).

11.4.3   *Indirect solar dryer*

The direct solar dryer has a number of limitations, namely, overheating, overdrying, decoloriza-tion, and degradation of the biochemical properties. These are mainly due to direct exposure of the crop to sunlight.

In the indirect solar dryer, the crop is exposed to sunshine in an indirect manner. With the help of solar radiation, flowing air is heated up and is passed through the drying chamber where the crop is kept. Due to the presence of hot air, the temperature of the crop increases and evaporation of moisture takes place. An arrangement within the dryer ensures the removal of the humid air.

Indirect solar dryers are classified into two groups, namely, indirect passive dryers and indirect active dryers, as presented in Figure 11.10 (Boughali *et al.*, 2009). The indirect passive dryer is not commonly used due to its low drying rate. The indirect active solar dryer is more commonly used. In this dryer, the flowing air is heated up by an electric heater/solar air heater prior to reaching the crop in the drying chamber.

Tomato slices were dried in the indirect solar active electric dryer and various drying kinetics models were applied (see Table 11.4). Midilli's model showed superior prediction ability when compared with other models. The payback period of this dryer was found to be very low, at only 1.27 years (Boughali *et al.*, 2009).

Mint leaf was dried in two different drying modes, namely, open sun or natural drying and an indirect solar dryer with a solar air heater. The solar dryer showed the superior drying performance. Ten different mathematical models from the literature for thin-layer drying were applied for both modes. Of all models, the Wang and Singh model showed superior performance when compared with the other models (Akpinar, 2010).

11.4.4 *Mixed solar dryer*

The mixed-mode solar dryer is a combination of both direct and indirect solar dryer and incorporates the advantages of both types. It can be operated either by forced or natural convection modes. Forson *et al.* (2007) have studied the mixed-mode solar dryer using natural convection for the drying of cassava and other crops. Singh and Kumar (2012) have developed a thermal test procedure (no-load performance index, NLPI) for testing the direct, indirect and mixed modes of solar dryers in both natural and forced convection modes. The experimental setup for the forced mode of the mixed-mode solar dryer is presented in Figure 11.10. The study shows that the mixed-mode solar dryer has the highest NLPI when compared with the other dryers in both natural and forced convection modes.

11.5 TESTING OF SOLAR DRYER

11.5.1 *No-load testing of the solar dryer*

A generalized methodology has been developed to evaluate the thermal performance of solar dryers under passive and active modes. Singh and Kumar (2012) have proposed a dimensionless mathematical model (NLPI) to test all types of solar dryer. This model has been developed from Duffie and Beckman (1991) for the basic, traditional solar air heater-dryer on the basis of basic heat transfer equations in the steady-state no-load condition. Here, air flows between the absorber plate and transparent glass.

For the absorber plate, the heat balance equation will be:

$$S' = U_b'(T_{ap} - T_{am}) + h_{cpf}'(T_{ap} - T_{af}) + h_{rpf}'(T_{pa} - T_{ag}) \tag{11.5}$$

For the glass cover, the heat balance equation will be:

$$U_t'(T_{ag} - T_{am}) = h_{cfg}'(T_{af} - T_{ag}) + h_{rpf}'(T_{ap} - T_{ag}) \tag{11.6}$$

For the inside moving air, the heat balance equation will be:

$$\dot{m}C_p(T_{afo} - T_{afi}) = A[h_{cpf}'(T_{ap} - T_{af}) - h_{cfg}'(T_{af} - T_{ag}) \tag{11.7}$$

By solving Equations (11.5) and (11.7):

$$h_{cpf}' = \frac{S' - U_b'(T_{ap} - T_{am}) - h_{rpf}'(T_{pa} - T_{ag})}{T_{ap} - T_{af}} \tag{11.8}$$

$$h_{cfg}' = \frac{U_t'(T_{ag} - T_{am}) - h_{rpf}'(T_{ap} - T_{ag})}{T_{af} - T_{ag}} \tag{11.9}$$

$$\dot{m}C_p(T_{afo} - T_{afi}) = Ah_{cpf}'[(T_{ap} - T_{af}) - \frac{h_{cfg}'}{h_{cpf}'}(T_{af} - T_{ag})] \tag{11.10}$$

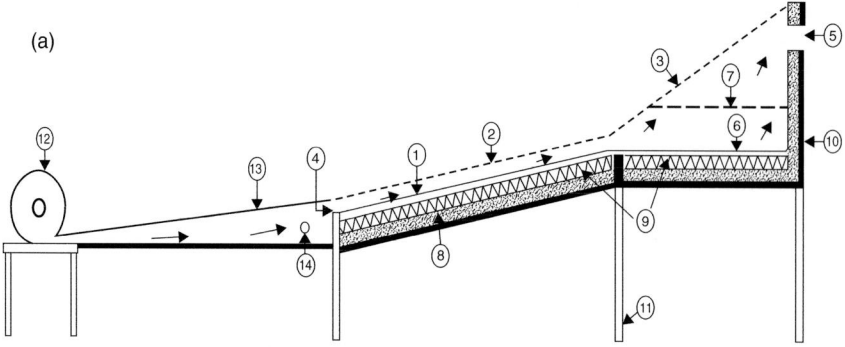

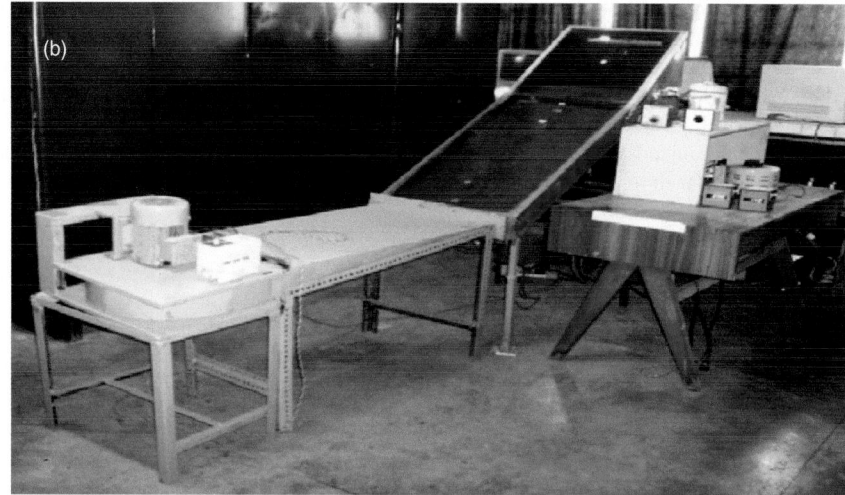

Figure 11.11.   Mixed-mode solar dryer: (a) schematic (key below); (b) photo (Singh and Kumar, 2012).
(1: collector absorber plate, 2: collector glass cover, 3: drying chamber glass cover, 4: inlet
vent, 5: outlet vent, 6: drying chamber absorber plate, 7: wire mesh, 8: insulation, 9: electric
heating plates, 10: wooden case, 11: angle iron stand, 12: blower, 13: divergent duct, 14: small
hole for measuring air flow rate.)

By solving Equations (11.8) and (11.9):

$$\frac{h'_{\text{cfg}}}{h'_{\text{cpf}}} = \left[\frac{U'_{\text{t}}(T_{\text{ag}} - T_{\text{am}}) - h'_{\text{rpf}}(T_{\text{ap}} - T_{\text{ag}})}{S' - U'_{\text{b}}(T_{\text{ap}} - T_{\text{am}}) - h'_{\text{rpf}}(T_{\text{pa}} - T_{\text{ag}})} \right] \left[\frac{T_{\text{ap}} - T_{\text{af}}}{T_{\text{af}} - T_{\text{ag}}} \right] \tag{11.11}$$

By solving Equations (11.10) and (11.11):

$$\dot{m}C_{\text{p}}(T_{\text{afo}} - T_{\text{afi}}) = Ah'_{\text{cpf}} \left[(T_{\text{ap}} - T_{\text{af}}) - \frac{U'_{\text{t}}(T_{\text{ag}} - T_{\text{am}}) - h'_{\text{rpf}}(T_{\text{ap}} - T_{\text{ag}})}{S' - U'_{\text{b}}(T_{\text{ap}} - T_{\text{am}}) - h'_{\text{rpf}}(T_{\text{pa}} - T_{\text{ag}})} \left(T_{\text{ap}} - T_{\text{af}} \right) \right]$$

$$\tag{11.12}$$

After re-arrangement of this equation:

$$\frac{\dot{m}C_{\text{p}}}{Ah'_{\text{cpf}}} = \frac{(T_{\text{ap}} - T_{\text{af}})}{(T_{\text{afo}} - T_{\text{afi}})} \left[\frac{S' - U'_{\text{b}}(T_{\text{ap}} - T_{\text{am}}) - U'_{\text{t}}(T_{\text{ag}} - T_{\text{am}})}{S' - U'_{\text{b}}(T_{\text{ap}} - T_{\text{am}}) - h'_{\text{rpf}}(T_{\text{pa}} - T_{\text{ag}})} \right] = NLPI \tag{11.13}$$

Computation of various heat transfer parameters used in the above equations are given as (Duffie and Beckman, 1991):

$$h'_{rpf} = \frac{\sigma[(T_{ap} + 273)^2 + (T_{ag} + 273)^2][(T_{ap} + 273) + (T_{ag} + 273)]}{\left[\dfrac{1}{\varepsilon_p} + \dfrac{1}{\varepsilon_g} - 1\right]} \tag{11.14}$$

$$U'_b = \frac{K'_b}{L'_b} \tag{11.15}$$

$$U'_t = h'_{rga} + h'_{cga} \tag{11.16}$$

$$h'_{rga} = \frac{\sigma\varepsilon_g[(T_{ag} + 273)^4 + (T_{am} + 273)^4]}{(T_{ag} - T_{am})} \tag{11.17}$$

$$Nu = h'_{cga} \times \frac{L}{K} = 0.54(Ra)^{0.25} \tag{11.18}$$

$$h'_{cga} = 0.54(Ra)^{0.25} \times \frac{K}{L} \tag{11.19}$$

11.5.2   *Testing the solar dryer under load conditions*

When the solar dryer is operated under load conditions, the following parameters are calculated.

11.5.2.1   *Heat collection efficiency*

The heat collection efficiency evaluates the performance of the solar dryer collector. It is calculated as (Leon *et al.*, 2002):

$$\eta_c = \frac{E_a}{E_c} \tag{11.20}$$

where E_a is the energy absorbed by the flowing air and E_c is the global solar radiation on the collection area in J.

11.5.2.2   *Pick-up efficiency*

The pick-up efficiency evaluates the evaporation of moisture from the flowing hot air. It decreases as the moisture content decreases.

$$\eta_p = \frac{h_o - h_i}{h_{as} - h_i} \tag{11.21}$$

where h_i is the absolute humidity of the inlet air [%], h_o is the absolute humidity of the outlet air [%] and h_{as} is the absolute humidity of the inlet air at the point of adiabatic saturation [%].

11.5.2.3   *Drying efficiency*

The drying efficiency illustrates how well solar radiation is being used for the purpose of drying in the system. The generalized drying efficiency (η_s) is given as:

$$\eta_s = \frac{W_m L_w}{I_g A_d} \tag{11.22}$$

where, W_m is weight of water evaporated from the product [kg], L_w is latent heat of vaporization of water at exit air temperature [J kg^{-1}], I_g is hourly average solar radiation on the dryer [kW h^{-1}] and A_d is aperture area of the dryer [m^2].

For the active mode dryer (Leon *et al.*, 2002):

$$\eta_s = \frac{W_m L_w}{I_g A_d + E_c} \tag{11.23}$$

For the hybrid mode (Leon *et al.*, 2002):

$$\eta_s = \frac{W_m L_w}{I_g A_d + E_c + (m_c \times LCV)} \tag{11.24}$$

where, m_c is mass of fuel consumed [kg] and LCV is lower calorific value of fuel [J kg^{-1}].

11.5.2.4 *Drying rate*
The drying rate is the rate of moisture evaporation per unit of time during the product-drying process. It is calculated as (Prakash and Kumar, 2014b):

$$DR = \frac{ME}{t \times DM} \tag{11.25}$$

11.6   ENERGY ANALYSIS

11.6.1 *Embodied energy*

Embodied energy is an accounting methodology that is used to find the sum of the energy required for a product life cycle or service (Baird *et al.*, 1997). This concept is useful in evaluating the effectiveness of energy-saving devices/services. It also serves as a measure of the extent to which a product or service will contribute to or mitigate global warming. It can be classified into two groups on the basis of the methodology used: the first is life-cycle analysis and the second is dependent on input-output analysis. Life-cycle analysis is a method of estimation of direct and indirect input energy, materials, labor services and process energy (Li *et al.*, 2014).

11.6.2 *Embodied energy for materials*

The embodied energy of selected materials is given in Table 11.7.

Table 11.7.   Embodied energy of different materials (Prakash and Kumar, 2014a).

Material	Embodied energy coefficient [kWh kg^{-1}]	Reference
Aluminum	55.28	(Baird *et al.*, 1997)
Galvanized steel	9.67	(Baird *et al.*, 1997)
Copper	19.61	(Baird *et al.*, 1997)
Glass	7.28	(Baird *et al.*, 1997)
Low-density polyethylene	28.61	(Baird *et al.*, 1997)
Polyvinyl chloride (PVC)	19.44	(Baird *et al.*, 1997)
Wood	0.31	(Baird *et al.*, 1997)
Wood board	2.89	(Baird *et al.*, 1997)
Paint	25.11	(Baird *et al.*, 1997)
Toughened glass	27	(www.level.org.nz)

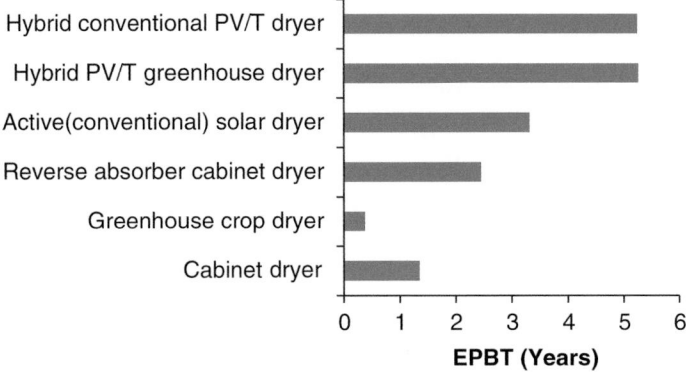

Figure 11.12. Energy payback time for selected solar dryers (Tiwari and Barnwal, 2011).

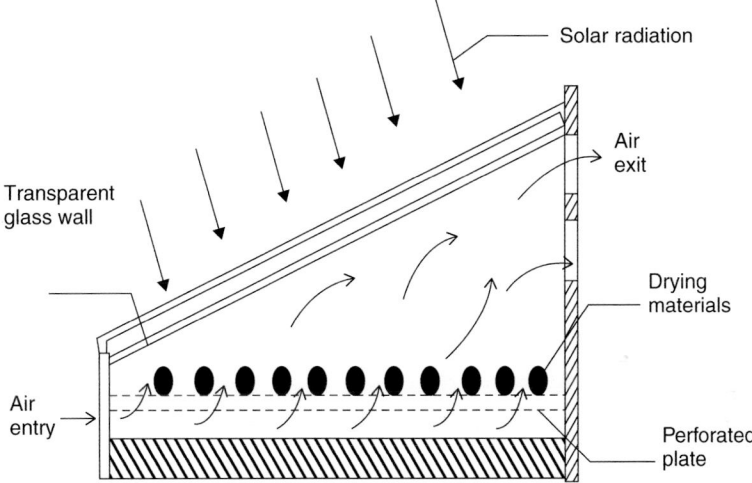

Figure 11.13. Schematic diagram of solar cabinet dryer (Prakash and Kumar, 2013a).

11.6.3 *Energy payback time*

Energy payback time (EPBT) is the time taken for a product to pay back its embodied energy. It is calculated as (Prakash and Kumar, 2016c):

$$\text{Energy payback time (EPBT)} = \frac{\text{embodied energy}}{\text{annual energy output}} \qquad (11.26)$$

The payback times of selected solar dryers are presented in Figure 11.12.

11.6.4 *Analysis of EPBT of solar dryers*

The energy payback times for a cabinet dryer and a modified greenhouse dryer have been calculated.

11.6.4.1 *Cabinet dryer*

A schematic diagram of a traditional cabinet dryer is shown in Figure 11.13. The various materials involved in the fabrication of this dryer are presented in Table 11.8. Figures 11.14 and 11.15 show the breakdown by mass and embodied energy of the cabinet dryer.

Table 11.8.   Total embodied energy of the solar cabinet dryer (Tiwari, 2006).

Component	Mass [kg]	Energy density [kWh kg^{-1}]	Total energy [kWh]
Glass cover	4.00	7.28	29.12
Wood material	10.00	2.89	28.9
Galvanized iron	1.00	9.67	9.67
Total embodied energy of the cabinet dryer			67.69

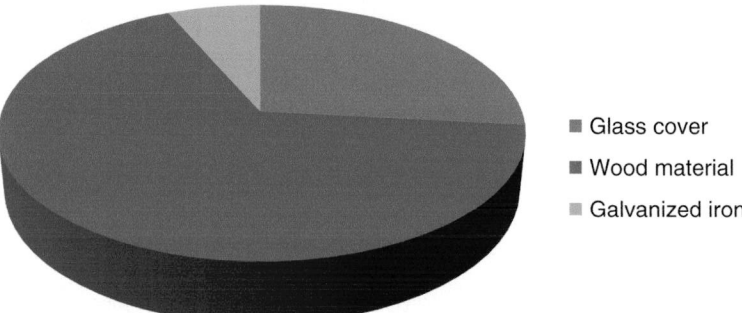

Figure 11.14.   Breakdown of the materials used in the fabrication of the cabinet dryer.

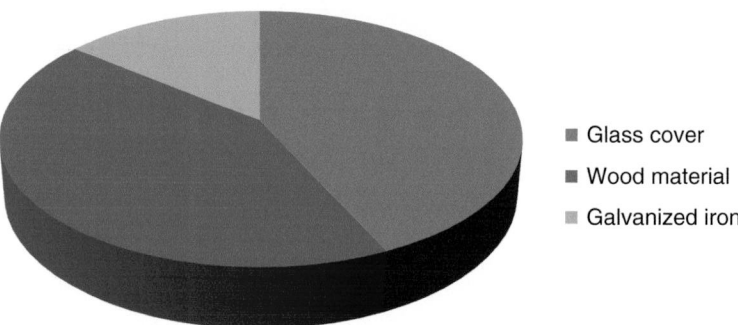

Figure 11.15.   Breakdown of the embodied energy of the cabinet dryer.

The annual useful energy for the 0.5 m^2 solar cabinet dryer is 50 kWh (Tiwari, 2006), which gives an energy payback time:

$$\text{EPBT} = \frac{67.69}{50} = 1.35 \text{ years}$$

11.6.4.2   Modified greenhouse dryer
Figure 11.16 shows the modified greenhouse dryer in active mode. The details of the materials used in the fabrication of the dryer, along with their energy density, are presented in Table 11.9. Figures 11.17 and 11.18, respectively, show the breakdown of the materials in terms of their energy density and mass.

Figure 11.16. Photograph of modified greenhouse dryer in active mode (Prakash and Kumar, 2014a).

Table 11.9. Embodied energy for the fabrication of modified greenhouse dryer in active mode (Prakash *et al.*, 2016c).

Material	Quantity	Embodied energy coefficient [kWh]	Total [kWh]
Polycarbonate sheet	15.60 kg	10.20 kg^{-1}	159.08
Glass	5.40 kg	7.28 kg^{-1}	39.31
Silver coating	0.75 m^2	0.28 m^{-2}	0.21
Black PVC sheet	0.32 kg	19.44 kg^{-1}	6.32
Wire mesh steel tray	0.70 kg	9.67 kg^{-1}	6.77
Aluminum sections			
(i) 1″ × 1 mm section	3.59 kg	55.28 kg^{-1}	198.46
(ii) 4″ × 1 mm section	0.82 kg	55.28 kg^{-1}	45.33
(iii) 1″ × 3 mm angle	0.08 kg	55.28 kg^{-1}	4.42
Fittings			
(i) Hinges	0.20 kg	55.28 kg^{-1}	11.06
(ii) Door lock	0.03 kg	55.28 kg^{-1}	1.38
(iii) Handle	0.10 kg	55.28 kg^{-1}	5.53
(iv) Steel screw	0.25 kg	9.67 kg^{-1}	2.42
DC fan			
(i) Plastic	0.12 kg	19.44 kg^{-1}	2.33
(ii) Copper wire	0.05 kg	19.61 kg^{-1}	0.98
Polycrystalline solar cell	0.06 m^2	1130.60 m^{-2}	66.14
Battery			46.00
Solar charge controller			33.00
Grand total [kWh]			628.73

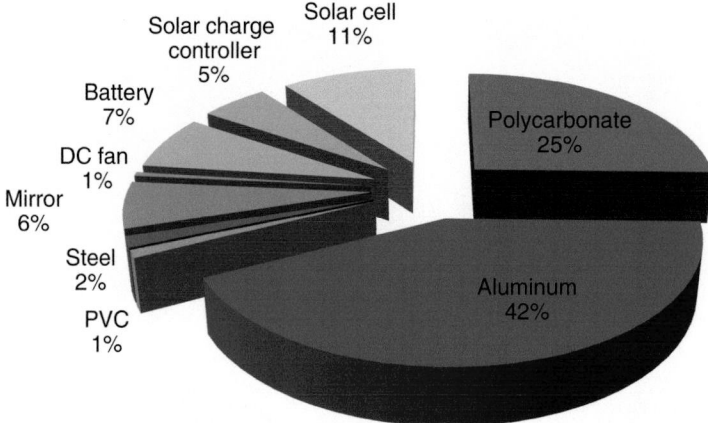

Figure 11.17.   Breakdown of embodied energy [kWh] of different materials used in modified greenhouse dryer in active mode (Prakash and Kumar, 2014b).

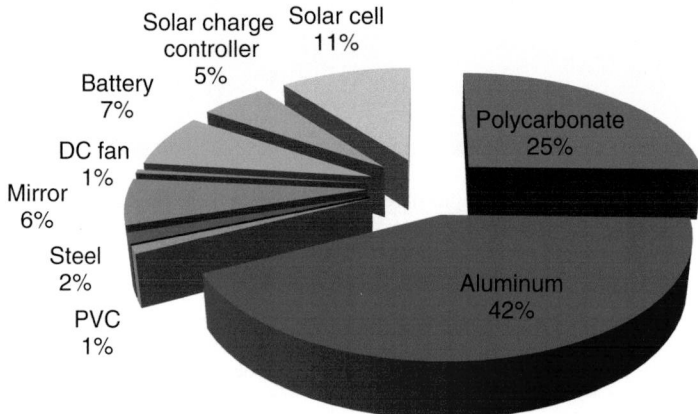

Figure 11.18.   Breakdown of items used in the fabrication of modified greenhouse dryer in active mode (Prakash and Kumar, 2014b).

The annual useful energy for the modified greenhouse dryer under active mode is 551.52 kWh (Prakash *et al.*, 2016c), making the energy payback time:

$$\text{EPBT} = \frac{628.7287}{551.5164} = 1.14 \, \text{years}$$

11.6.5   *CO_2 emissions*

According to Watt *et al.* (1998), the average CO_2 emissions from coal for generating electric power are 0.98 kg kWh^{-1}.

The CO_2 emissions per year for each component is calculated as follows (Prakash and Kumar, 2014b):

$$\text{CO}_2 \text{ emissions per year} = \frac{\text{Embodied energy} \times 0.98}{\text{Lifetime}} \qquad (11.27)$$

The CO_2 emissions per year for selected solar dryers are presented in Figure 11.19.

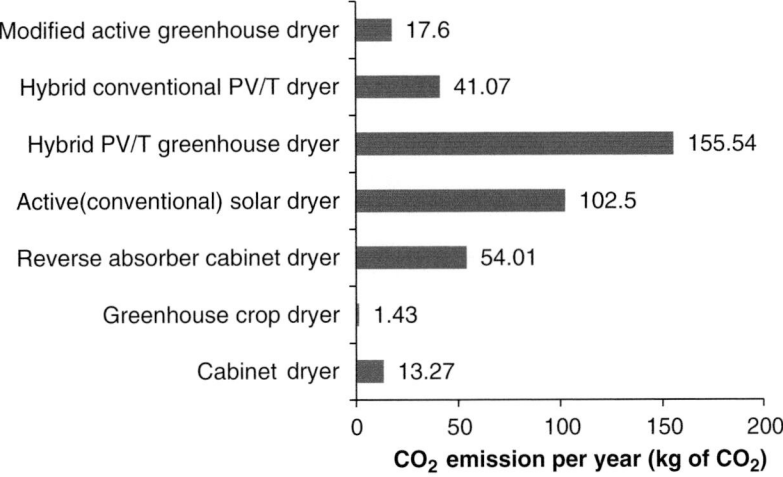

Figure 11.19. CO_2 emissions of selected solar dryers (Prakash and Kumar, 2014b).

Table 11.10.   Properties of dried tomato in both drying modes (Prakash and Kumar, 2014b).

Tomato property	Drying mode	
	Open	Active
Moisture [wb %]	13.04	9.09
Ascorbic acid [mg/100 g]	14.23	29.20
Lycopene [mg/100 g]	24.02	34.63
β-carotenoid [mg/100 g]	2.02	2.64
Total carotenoid [mg/100 g]	4.10	3.14
Nitrogen [mg/100 g]	10.10	13.10
Reducing sugars [μg/100 mg]	78.68	80.22
Polyphenols [mg/100 mg]	78.20	69.14

11.7   IMPACT OF SOLAR DRYING ON PRODUCT QUALITY

Prakash and Kumar (2014b) dried tomato flakes in the two different modes, namely, under natural drying and in the modified solar greenhouse dryer in active mode. The biochemical properties of the dried tomato flakes from the different drying modes were tested at the Indian Agricultural Research Institute, New Delhi. The test report is presented in Table 11.10. The results reveal that the crop dried in the solar dryer was more nutrient-rich than the crop dried naturally.

11.8   CONCLUDING REMARKS

At present, food security is one of the major challenges in the world. One of the solutions is to minimize the post-harvest loss of agricultural produce. Drying is one of the principal methods of reducing post-harvest loss. Apart from the solar dryer, all prominent drying methodologies utilize fossil fuels, either directly or indirectly. Due to the many limitations of such fossil-fuel-based dryers, solar dryers are proven to be more promising, reliable, cost-effective and eco-friendly. Although natural drying or open sun drying has been used for many generations, this process of drying has had limited use due to the disadvantages associated with traditional methods.

An attempt has been made to provide a general consolidated overview of solar drying. A detailed classification of the solar dryer has been presented. All major types of solar dryer have been discussed. A complete performance analysis of the solar dryer has been presented.

NOMENCLATURE

A	cross-sectional heat absorber area [m²]
A_d	aperture area of the dryer [m²]
C_p	specific heat of air at constant pressure [J kg⁻¹ K⁻¹]
E_c	energy consumed by the exhaust fan [J]
h'_{cfg}	convective heat transfer coefficient from hot air to transparent polymer cover of the system [W m⁻² K⁻¹]
h'_{cga}	convective heat transfer coefficient from transparent glass/polymer cover to ambient air of the system [W m⁻² K⁻¹]
h'_{cpg}	convective heat transfer coefficient from absorber plate to hot air of the system [W m⁻² K⁻¹]
h'_{rga}	radiative heat transfer coefficient from transparent glass/polymer cover to ambient air of the system [W m⁻² K⁻¹]
h'_{rpg}	radiative heat transfer coefficient from absorber plate to transparent glass/polymer cover of the system [W m⁻² K⁻¹]
I_g	global solar radiation on mean hourly basis [kWh] on the aperture of the system
K'_b	insulation conductivity [W m⁻¹ K⁻¹]
L	characteristic length [m]
L_w	latent heat of vaporization of water at the outlet air temperature
L_b	length of insulation [m]
LCV	lower calorific value of the fuel [J kg⁻¹]
$\dot{m}$	mass flow rate of the air [kg s⁻¹]
m_f	mass of fuel consumed [kg]
PV/T	photovoltaic thermal
S	flux absorbed in the absorber plate [W m⁻²]
T_{ap}	mean temperature of absorber plate [°C]
T_{ag}	mean temperature of transparent glass/polymer cover [°C]
T_{af}	mean temperature of hot air [°C]
T_{afo}	mean temperature of outlet hot air [°C]
T_{afi}	mean temperature of inlet air [°C]
T	drying time [s]
T_{am}	ambient temperature [°C]
W_m	weight of moisture evaporated from the crop [kg]

Greek symbols

ε_p	emissivity of the absorber plate
ε_g	emissivity of the transparent glass/polymer cover
σ	Stefan–Boltzmann constant [5.67×10^{-8} W m⁻² K⁻⁴]

REFERENCES

Akpinar, E.K. (2010) Drying of mint leaves in a solar dryer and under open sun: modelling, performance analyses. *Energy Conversion and Management*, 51(12), 2407–2418.

Aware, R. & Thorat, B.N. (2012) Solar drying of fruits and vegetables. *Solar Drying: Fundamentals, Applications and Innovations*, 51.

Baird, A.G., Alcorn, A. & Haslam, P. (1997) The energy embodied in building materials – updated New Zealand coefficients and their significance. *IPENZ Transactions*, 24, 46–54.

Boughali, S., Benmoussa, H., Bouchekima, B., Mennouche, D., Bouguettaia, H. & Bechki, D. (2009) Crop drying by indirect active hybrid solar-electrical dryer in the eastern Algerian septentrional Sahara. *Solar Energy*, 83, 2223–2232.

Duffie, J.A. & Beckman, W.A. (1991) *Solar Engineering of Thermal Processes*. John Wiley and Sons, New York.

Esper, A. & Muhlbauer, W. (1998) Solar drying – an effective means of food preservation. In: Sayigh, A.A.M. (ed.) *Proceedings World Renewable Energy Congress V, 20–25 September, 1998, Florence, Italy, Renewable Energy*, 15, 95–100.

Ekechukwu, O.V. (1999) Review of solar-energy drying systems. I. An overview of drying principles and theory. *Energy Conversion and Management*, 40, 593–613.

Forson, F.K., Nazha, M.A.A. & Rajakaruna, H. (2007) Modelling and experimental studies on a mixed-mode natural convection solar crop dryer. *Solar Energy*, 81(3), 346–357.

Imre, L. (1993) Energy aspects of drying. *Proceedings of the First International Workshop Energy Perspective in Plantation Industry, 10–12 February 1993, Coonoor, India*. pp. 122–133.

Jain, D. & Tiwari, G.N. (2004) Effect of greenhouse on crop drying under natural and forced convection I: Evaluation of convective mass transfer coefficient. *Energy Conversion and Management*, 45, 765–783.

Kumar, A. & Tiwari, G.N. (2006a) Thermal modelling of a natural convection greenhouse drying system for jaggery: an experimental validation. *Solar Energy*, 80, 1135–1144.

Kumar, A. & Tiwari, G.N. (2006b) Effect of shape and size on convective mass transfer coefficient during greenhouse drying (GHD) of jaggery. *Journal of Food Engineering*, 73, 121–134.

Kumar, A. & Tiwari, G.N. (2006c) Thermal modelling and parametric study of a forced convection greenhouse drying system for jaggery: an experimental validation. *International Journal of Agricultural Research*, 1, 265–279.

Kumar, A. & Tiwari, G.N. (2007) Effect of mass on convective mass transfer coefficient during open sun and greenhouse drying of onion flakes. *Journal of Food Engineering*, 79, 1337–1350.

Kumar, S. & Tiwari, G.N. (2009) Solar life cycle cost analysis of single slope hybrid (PV/T) active solar still. *Applied Energy*, 86, 1995–2004.

Kumar, A., Tiwari, G.N., Kumar, S. & Pandey, M. (2006) Role of greenhouse technology in agriculture engineering. *International Journal of Agricultural Research*, 1(4), 364–372.

Leon, M.A., Kumar, S. & Bhattacharya, S.C. (2002) A comprehensive procedure for performance evaluation of solar food dryers. *Renewable and Sustainable Energy Reviews*, 6, 367–393.

Li, J.S., Chen, G.Q., Wu, X.F., Hayatb, T., Alsaedi, A. & Ahmad, B. (2014) Embodied energy assessment for Macao's external trade. *Renewable and Sustainable Energy Reviews*, 34, 642–653.

Nawaz, I. & Tiwari, G.N. (2006) Embodied energy analysis of photovoltaic (PV) system based on macro- and micro-level. *Energy Policy*, 34, 3144–3152.

Neria, E., Rugani, B., Benetto, E. & Bastianoni, S. (2014) Energy evaluation vs. life cycle-based embodied energy (solar, tidal and geothermal) of wood biomass resources. *Ecological Indicators*, 36, 419–430.

Prakash, O. (2015) *Design and Performance Analysis of Modified Greenhouse Dryer*. PhD Thesis in Department of Energy, Maulana Azad National Institute of Technology, Bhopal, India.

Prakash, O. & Kumar, A. (2013a) Historical review and recent trends in solar drying systems. *International Journal of Green Energy*, 10, 690–738.

Prakash, O. & Kumar, A. (2013b) ANFIS prediction model of a modified active greenhouse dryer in no-load conditions in the month of January. *International Journal of Advanced Computer Research*, 3(1), 220–223.

Prakash, O. & Kumar, A. (2014a) Application of artificial neural network for prediction of jaggery mass during drying inside natural convection greenhouse dryer. *International Journal of Ambient Energy*, 35(4), 186–192.

Prakash, O. & Kumar, A. (2014b) Environomical analysis and mathematical modelling for tomato flakes drying in a modified greenhouse dryer under active mode. *International Journal of Food Engineering*, 10(4), 669–681.

Prakash, O. & Kumar, A. (2014c) ANFIS modelling of a natural convection greenhouse drying system for jaggery: An experimental validation. *International Journal of Sustainable Energy*, 33(2), 316–335.

Prakash, O. & Kumar, A. (2014d) Design, development and testing of modified greenhouse dryer under natural convection. *Heat Transfer Research*, 45(5), 433–451.

Prakash, O. & Kumar, A. (2014e) Thermal performance evaluation of modified active greenhouse dryer. *Journal of Building Physics*, 37(4), 395–402.

Prakash, O. & Kumar, A. (2014f) Performance evaluation of greenhouse dryer with opaque north wall. *Heat and Mass Transfer*, 50(12), 493–500.

Prakash, O. & Kumar, A. (2014g) Solar greenhouse drying: a review. *Renewable and Sustainable Energy Reviews*, 29, 905–910.

Prakash, O. & Kumar, A. (2015) Annual performance of modified greenhouse dryer under passive mode in no-load conditions. *International Journal of Green Energy*, 12, 1091–1099.

Prakash, O., Kumar, A., Kaviti, A. & Vishwanatha, P. (2015) Prediction of moisture evaporation rate from jaggery in green house drying using Fuzzy Logic. *Heat Transfer Research*, 46(10), 923–935.

Prakash, O., Kumar, A. & Sharaf-Eldeen, Y. (2016a) Review of Indian solar drying status. *Current Sustainable/Renewable Energy Reports* 3(3), 113–120.

Prakash, O., Laguri, V., Pandey, A., Kumar, A. & Kumar, A. (2016b) Review on various modelling techniques for the solar dryers. *Renewable and Sustainable Energy Reviews*, 62, 396–417.

Prakash, O., Kumar, A. & Laguri, V. (2016c) Performance of modified greenhouse dryer with thermal energy storage. *Energy Report*, 2, 155–162.

Proctor, D.L. (ed.) (1994) Grain storage techniques: Evolution and trends in developing countries. Food and Agriculture Organization of the United Nations (FAO), Rome.

Raman, V. & Tiwari, G.N. (2008) Life cycle cost analysis of HPVT air collector under different Indian climatic conditions. *Energy Policy*, 36, 603–611.

Sharma, A., Chen, C.R. & Lan, N.V. (2009) Solar-energy drying systems: a review. *Renewable and Sustainable Energy Reviews*, 13, 1185–1210.

Singh, S. & Kumar, S. (2012) Testing method for thermal performance based rating of various solar dryer designs. *Solar Energy*, 86, 87–98.

Sodha, M.S., Dang, A., Bansal, P.K. & Sharma, S.B. (1985) An analytical and experimental study of open sun drying and a cabinet type drier. *Energy Conversion and Management*, 25(3), 263–271.

Tiwari, G.N. (2006) *Solar Energy Technology Advances*. Nova Science, New York.

Tiwari, G.N. & Barnwal, P. (2011) *Fundamentals of Solar Dryers*. Anamaya, New Delhi, India.

Visavale, G.L. (2012) Principles, classification and selection of solar dryers. In: Hii, C.L., Ong, S.P., Jangam, S.V. & Mujumdar, A.S. (eds) *Solar Drying: Fundamentals, Applications and Innovations*. University of Nottingham (Malaysia Campus)/National University of Singapore. pp. 1–50.

Watt, M., Johnson, A., Ellis, M. & Quthred, N. (1998) Life cycle air emission from PV power system. *Progress in Photovoltaic Research Application*, 6(2), 127–136.

CHAPTER 12

Small-scale wind power energy systems for use in agriculture and similar applications

Wojciech Miąskowski, Krzysztof Nalepa, Paweł Pietkiewicz & Janusz Piechocki

12.1 INTRODUCTION

Action to find alternative sources of energy, including renewable energy sources, has been forced not just by the exhaustion of primary energy sources but also and, in fact, primarily, by the increase of emissions of carbon dioxide and other greenhouse gases to the atmosphere.

Wind has been used as an energy carrier since ancient times. Wind energy has long been used to generate a driving force for sailing ships. The first wind turbines appeared in China and Mediterranean countries about 1800 years ago. They were used to pump water or grind grain. As early as the 8th century, large four-wing windmills appeared in Europe, and the greatest role played by wind energy was in the 16th century (Hau and von Renouard, 2013; Hill, 1991; Lucas, 2006). By the end of the 19th century, interest in wind power had declined, but interest in this energy carrier was renewed following the energy crisis of 1973 (Hau and von Renouard, 2013). Wind energy is a converted form of solar energy, caused by the movement of air masses induced by irregular heating of the earth's surface. Approximately a quarter of this energy involves the movement of air masses directly adjacent to the surface of the ground. However, given the potential deployment of devices to convert wind energy, only a small proportion of such resources can be used. Technologically usable wind resources are estimated at about 40 TW (Lewandowski, 2007).

Currently, wind turbines are used to drive generators, water pumps, and air compressors, among other devices. The span of available wind turbines is increasing, ranging from small domestic wind turbines up to very large constructions incorporated into the main energy grids. The electrical power of a wind turbine approaches more than 8 MW with a rotor diameter of $\sim$164 m (Windpowermonthly, n.d.). In operation, wind turbines are an energy source that produces no pollution whatsoever. However, when operating, noise is generated from the rotor, as blades cut through the air, and from the gearbox. Ongoing development of wind turbines tends to increase efficiency and reduce the level of noise generated during their operation (Guidati et al., 2000).

12.2 ASSESSMENT OF WIND PARAMETERS FOR THE PURPOSE OF WIND TURBINE LOCATION

12.2.1 Key parameters of wind as a source of energy

Wind is the movement of air masses from areas of high pressure to areas of low pressure. Solar power is the primary cause of air flow on earth. Solar radiation heats the surface of the earth, and increases the temperature of air. Heat energy is not evenly distributed. Hot air rises, and it is replaced by masses of cold air (Fig. 12.1). The earth rotates around its own axis, and the sun heats only a part of the earth's surface in a given period of time. Air circulation is also inhibited by land relief, the uneven distribution of land and water on the earth's surface, uneven exposure to light in different regions, and other factors which lead to variations in wind speed and direction. Wind parameters also change over time.

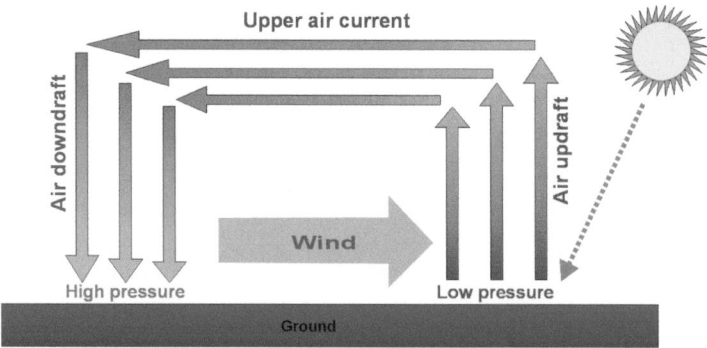

Figure 12.1.   Wind-forming mechanism.

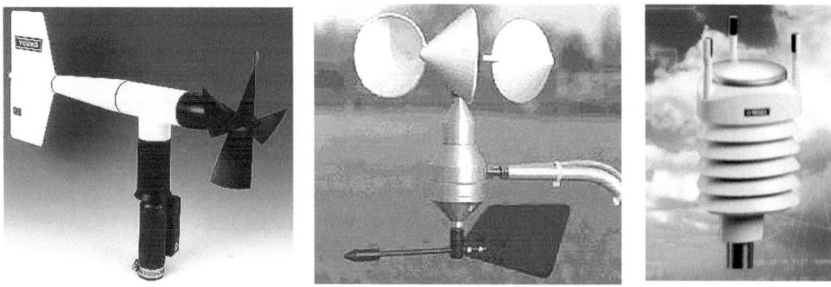

Figure 12.2.   LB-746 and LB-747 anemometers and WXT520 ultrasonic wind sensor (LAB-EL, n.d.).

Unfortunately, local wind conditions do not allow the building of wind farms in all regions of the world. Weather conditions may also differ dramatically within the same region.

Wind turbines are sited on the basis of an assessment of local wind conditions. A preliminary analysis can be performed with the use of wind maps and atlases (for large-scale wind turbines). The success of a planned wind turbine is determined by the reliability of data about wind speed, wind direction and the frequency and duration of various wind speeds in the proposed location. Long-term reference data is required for a reliable assessment of wind power potential. According to general standards, wind parameters should be measured continuously over a period of at least one year. Wind speed and direction are measured with windmill anemometers, ultrasonic anemometers (see examples in Fig. 12.2) and laser Doppler sensors (Cuerva and Sanz-Andres, 2000).

Measurement results are stored in the device's memory, and can be transmitted to computers for processing and analysis. Anemometer data is used to analyze wind speed and the frequency of different wind speeds, and to assess wind power potential in a given location (Abiven et al., 2011; Palma et al., 2008).

The key parameter of wind energy is wind speed. In systems that rely on a yawing mechanism, wind direction is also an important factor that influences control processes in the wind turbine. Wind speed varies on a daily, monthly and seasonal basis. Because wind is caused by uneven heating of the earth's surface by the sun, wind speed is a function of location and time:

$$v = f(x, y, z, t) \tag{12.1}$$

where:
   x, y, z: coordinates in space
   t: time

Wind energy is the most important parameter for energy production (Betz, 1966):

$$E = \int \frac{1}{2} \rho v^3 dt \tag{12.2}$$

where:

ρ: instantaneous density of air
v: instantaneous airstream velocity
dt: flow time of air with parameters ρ and v

Equation (12.2) indicates that, in addition to speed, wind energy is also determined by air density. Air density is influenced by humidity and static pressure, which is caused by atmospheric pressure. Air can be regarded as a fluid characterized by high compressibility. Air density varies across different locations. Therefore, similarly to wind speed, air density is also a function of location and time:

$$\rho = f(x, y, z, t) \tag{12.3}$$

Wind parameters, in particular wind energy, are generally evaluated in specific locations. All measurements and calculations are performed at constant values of coordinates x, y and z. In this case, Equation (12.1) can be simplified to:

$$v = f(t) \tag{12.4}$$

The pressure exerted by wind on structural components, such as billboards or power posts, as well as turbine rotor blades, is calculated on the basis of constant air density. In this case, Equation (12.2) then takes the following form:

$$E = \overline{\rho} \int \frac{v^3}{2} dt \tag{12.5}$$

where:

$\overline{\rho}$: average air density applied in calculations

Equation (12.5) indicates that assessments of wind energy that can be extracted from moving air masses are limited to measurements and calculations of wind speed. In practice, wind energy is calculated as the sum of products (Nałęcz, 2007):

$$E = \sum_{n} \overline{\rho} \frac{v^3}{2} \Delta t \tag{12.6}$$

where:

Δt: unit of time during which wind speed is regarded as constant (time interval between measurements of average wind speed)
v: average wind speed observed in the time unit Δt
n: number of time units during which wind speed was measured

The accuracy of the calculations performed with the use of Equation (12.6) increases with the frequency of measurements of average wind speed.

Wind assessment reports developed for turbine feasibility studies evaluate the average wind speed in a given area. The use of average wind speed values significantly simplifies the calculations and the algorithm for selecting the optimal wind turbine.

The diagram in Figure 12.3 presents a theoretical example of variations in wind speed over time (red and blue lines) and the average value calculated for every measurement (green line). Although wind speed in the third hour of measurement reached $16 \, \text{m s}^{-1}$ during P2 and only $6 \, \text{m s}^{-1}$ during P1, the average speed throughout the entire time of measurement (P1 plus P2) was determined at $4.6 \, \text{m s}^{-1}$. An analysis of individual values in each set of measurements indicates that P2 offers

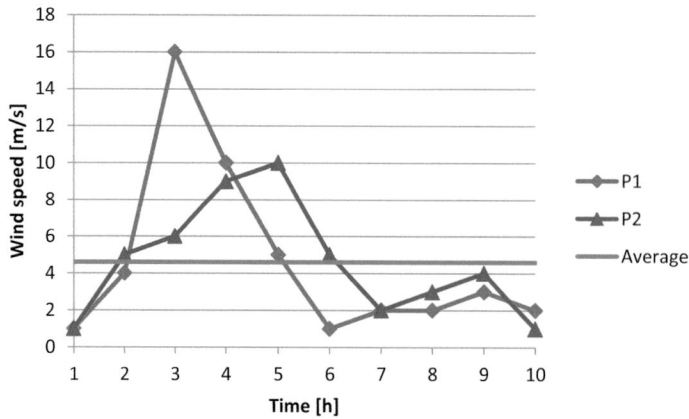

Figure 12.3.   Measurements of wind speed with constant average value.

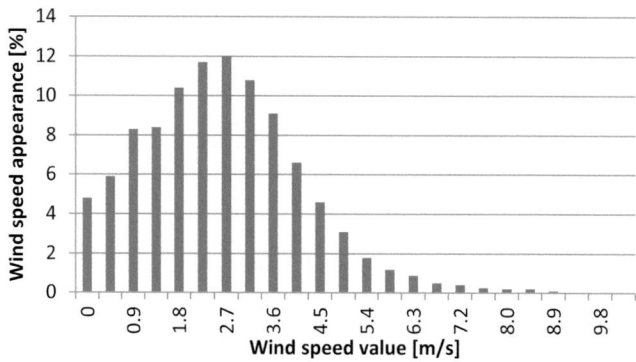

Figure 12.4.   Frequency of average wind speeds during a study at UWM, Olsztyn.

a more efficient scenario. An examination of variations in wind speed (Fig. 12.3) suggests that despite significant analytical simplifications, the use of average wind speeds produces divergence between the results of calculation and reality.

Maximum speeds are observed in every region under particular weather conditions, such as wind gusts and rain storms. However, maximum wind speeds are momentary or are observed only over very short periods of time. Maximum speed is not the decisive parameter in projects aiming to convert wind energy into a useful form of energy. A wind turbine designed to be powered by wind with maximum observable speed would operate only for the short periods of time during which maximum wind speeds were achieved. Nevertheless, this parameter of the wind is very important at the stage of designing wind turbines and other outdoor sites. Structures that are exposed to wind have to be able to resist the aerodynamic forces resulting from maximum wind speeds in a given location.

Wind resource assessments for turbine feasibility studies should analyze the distribution of wind speeds over time (Kose *et al.*, 2004).

For the purpose of this publication, studies were carried out at the premises of the University of Warmia and Mazury (UWM) in Olsztyn, Poland, to assess the placement of a small wind turbine on the roof of one of the campus buildings. The distribution of wind speeds measured during a long-term study is presented graphically in Figure 12.4. The observed wind speeds are average values determined over short periods of time, Δt. The data presented indicates that the most prevalent wind speed during the study was $2.7\,\mathrm{m\,s^{-1}}$ (accounting for 12% of recorded measurements).

Wind speeds above $6\,\mathrm{m\,s^{-1}}$ were observed, but they accounted for a minor proportion of the results. The distribution of wind speeds over time can be used to determine wind energy (Nałęcz, 2007):

$$E = \sum_n \bar{\rho}\frac{v^3}{2}\,\Delta t\,u_\%$$ (12.7)

where:

Δt: unit of time during which wind speed is regarded as constant (time interval between measurements of average wind speed)

v: average wind speed observed in a time unit Δt

$u_\%$: percentage share of time units during which speed v was observed

n: number of wind speed values recorded during the study

In the results presented in Figure 12.4, a wind speed between 1.8 and $3.6\,\mathrm{m\,s^{-1}}$ is the most frequent and this is a typical distribution of wind speed, described by the Weibull distribution (Carta *et al.*, 2009; Morgan *et al.*, 2011).

The results of the study carried out on the UWM campus confirmed that, in the place of measurement, wind speeds of up to $4.5\,\mathrm{m\,s^{-1}}$ were prevalent for around 80% of the time of the occurrence of wind (observed on campus). Higher speeds could be expected only during the remaining 20% of the time. The cut-in speed of contemporary wind micro-turbines is as low as 2–$3.5\,\mathrm{m\,s^{-1}}$, but the nominal power is obtained at a wind speed of about $10\,\mathrm{m\,s^{-1}}$. The results of the experiment indicated that, despite occasional wind speeds above $8.5\,\mathrm{m\,s^{-1}}$, these occurred so rarely that the place assessed was not a suitable site for the location of a wind turbine (a wind turbine sited at this location would produce only a small amount of energy). This shows how important it is to measure wind parameters at a site before making the decision to install a wind turbine there.

In estimating the efficiency of wind power, one should take into account the distribution of the wind speed at the place of planned wind turbine location. Figure 12.5 shows the power generated in two types of wind turbine and the annual quantity of energy production. The comparison involves two kinds of 3 kW wind turbines (a three-bladed horizontal-axis wind turbine and a twin-bladed H-rotor) working in various wind strengths (average wind speeds of 3 and $6\,\mathrm{m\,s^{-1}}$). The calculated values for annual energy production show that correct selection of the wind turbine for the local wind conditions determines the effects obtained.

Global wind energy resources are estimated at about 400 TW (Adams and Keith, 2013). However, of this, the utilization for which there is technical capability is estimated at about 40 TW (Lewandowski, 2007), of which about 486 MW (GWEC, 2017) is currently used. Not all wind energy can be converted to useful energy. The maximum efficiency of wind conversion arising from wind turbine aerodynamics determined by Betz (Schaffarczyk, 2014; Hansen, 2015). The effectiveness of each wind turbine type is shown in Figure 12.6.

The power with which the wind acts on the rotor surface of a wind turbine is described (Schaffarczyk, 2014) by:

$$P_\mathrm{w} = \frac{\rho}{2}\,A_\mathrm{r}\,v^3$$ (12.8)

where:

P_w: wind power

A_r: active face of wind turbine's rotor

ρ: air density

v: wind speed

The power coefficient c_p determines what proportion of the wind power is converted into the mechanical power of the turbine, P_t (Schaffarczyk, 2014):

$$P_\mathrm{t} = c_p\,\frac{\rho}{2}\,A_r\,v^3$$ (12.9)

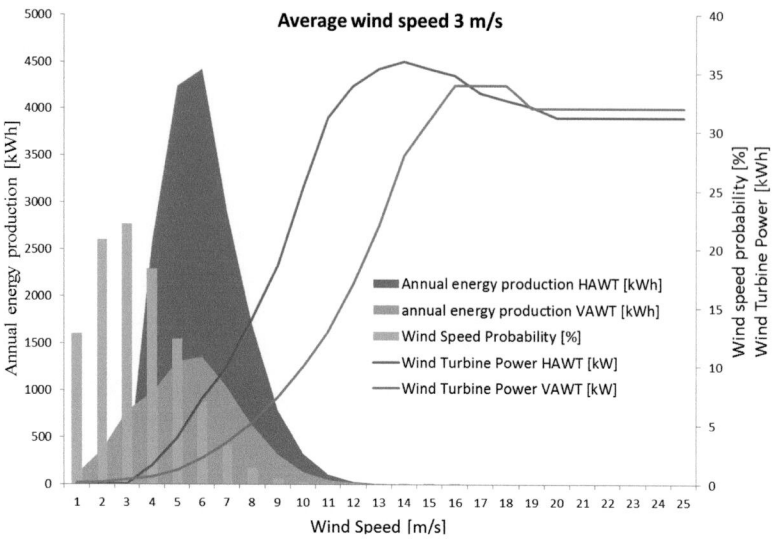

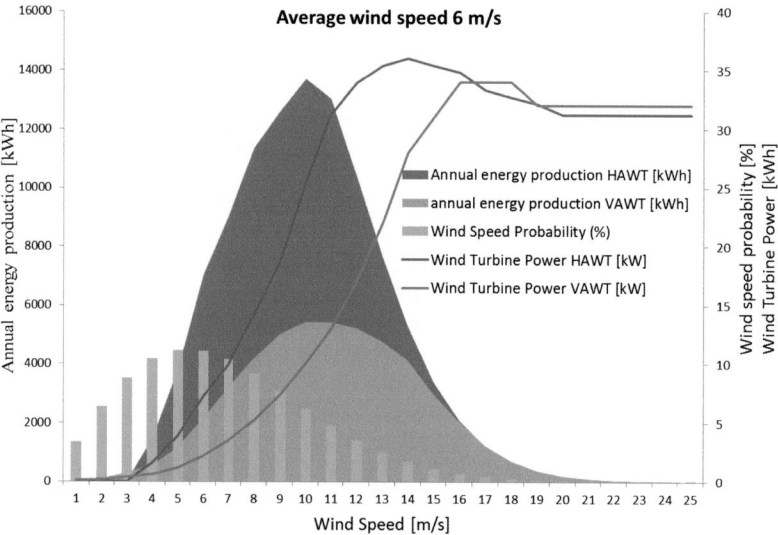

Figure 12.5.   Comparison of power generated in two kinds of 3 kW wind turbines and their annual quantities of energy production.

The specific speed coefficient λ is the ratio of the velocity of the blade's end v_{tip} to the wind speed v_{wind}:

$$\lambda = \frac{v_{tip}}{v_{wind}} = \frac{\omega \, R_{tip}}{v_{wind}} \tag{12.10}$$

where:

ω: angular velocity of rotor blades

R_{tip}: blade radius – length from the rotation axis of the rotor to the end of the blade

The wind turbine torque is:

$$M = \frac{P}{\omega} \tag{12.11}$$

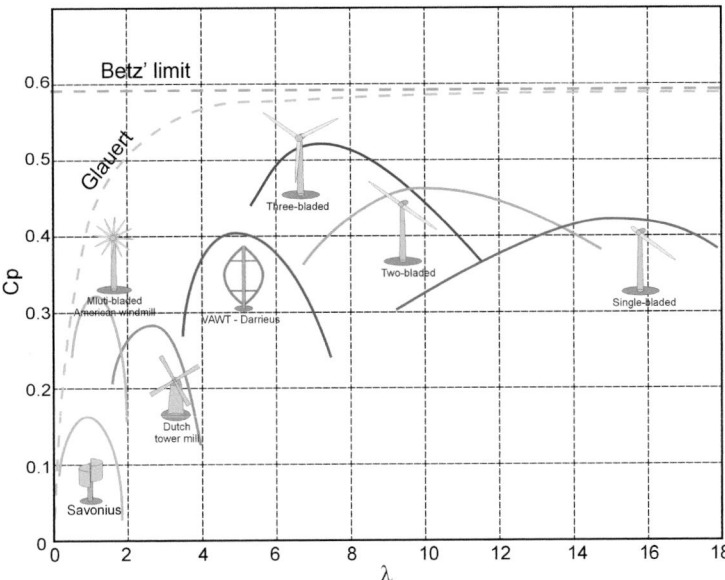

Figure 12.6.   Power coefficient c_p versus specific speed coefficient λ for different types of wind turbines.

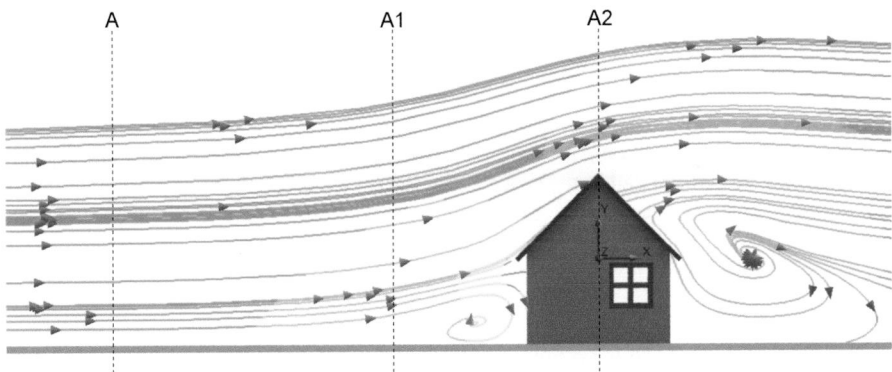

Figure 12.7.   Air compression on the windward side of an artificial obstacle.

12.2.2   *Maximizing the potential of wind power on local ground*

However, numerous research studies have indicated that small-scale wind turbines can be built under supportive local conditions to generate substantial benefits, including in regions that rank low in wind resource assessments.

At higher altitudes above the ground, the speed of wind blowing over a large area is relatively consistent over a given period of time. In an attempt to increase the efficiency of wind energy, airstream velocity has to be maximized locally (van Bussel and Mertens, 2005). Methods for increasing local airstream velocity rely on natural relief features and man-made structures to compress air on the windward side of an obstacle. Compressed wind is a fragment of space with increased airstream velocity. This concept is illustrated in Figure 12.7.

An ideal wind speed profile in an area that is free of obstacles is shown on the left side of Figure 12.7. If we disregard the effect of air flow on the ground, it can be assumed that wind speed is constant along section *A* (Mavroidis *et al.*, 2003). If wind conditions remain constant

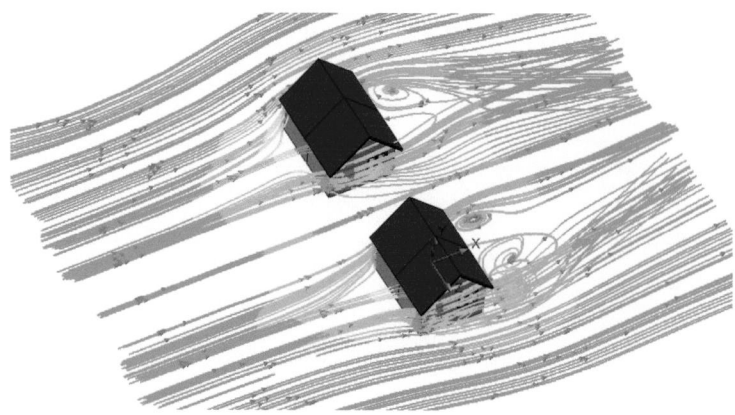

Figure 12.8.   Compressed air on the windy side of an artificial obstacle.

over time, Equation (12.12) can be used to calculate the achievable energy extraction of wind power:

$$E = \frac{\rho\, c^3}{2}t \tag{12.12}$$

where:

ρ: air density

c: average wind speed along section A

t: time of measurement or time interval used in calculations

An obstacle to air flow, such as a residential building, is shown on the right side of Figure 12.7. When the obstacle is suitably positioned, the airstream will become deformed. The streams of flowing air will be compressed in the space above the building's roof. If a specific section above the building is analyzed, we can use the continuity Equation (12.13) to calculate the speed of wind near the obstacle:

$$\rho\, c_1\, A_1 = \rho\, c_2\, A_2 \tag{12.13}$$

where:

c_1, c_2: average wind speed in sections A_1 and A_2

Since section A_2 is reduced in size by the obstacle, the flowing air mass has to find a different travel path and, therefore, the same mass of air flows through section A_2, which is smaller than section A_1, over the same time interval. Thus, the wind speed in section A_2 must increase relative to section A_1 (according to Equation 12.13). This mechanism is responsible for increasing local airstream velocity.

A similar effect is observed in the horizontal plane when natural or artificial obstacles narrow the path traveled by a stream of air (Fig. 12.8).

By increasing local airstream velocity, the compressed air effect supports the use of small wind turbines that can be operated at wind speeds as low as $3\,\mathrm{m\,s^{-1}}$. At higher wind speeds, the compression effect increases the airstream velocity near the turbine to 10–$12\,\mathrm{m\,s^{-1}}$, which is sufficient to power several types of wind turbine.

12.3   TYPES OF WIND TURBINES

Wind turbines are usually used to produce electricity, but they can also be used to pump water, and to drain and irrigate fields. Wind turbines can be connected to the power grid or act as autonomous power systems and offer the possibility of using wind energy to produce energy forms (electricity) that are ready for use by consumers.

Wind turbines convert wind energy into electricity, mechanical energy and other types of energy. The most important part of a wind turbine is the rotor, which converts the kinetic energy of moving air particles into mechanical energy in the form of rotational motion, which is then transferred to actuating devices (generator, pump, etc.) (Gipe, 2004).

12.3.1 *Classification of wind turbines*

Wind turbines can be classified according to various criteria. Some classification methods are presented below.

12.3.1.1 *Classification based on capacity*
Wind turbines can be divided into micro-turbines, small-scale turbines and large-scale turbines, according to their power output. The first two types of turbines are most often used to provide power to homes and small farms.

- Micro-turbines have an output of up to 100 W. They are used to charge batteries that feed autonomous power systems where a power grid is not available or is not used. The energy produced is sufficient to light individual lamps, entire rooms or even power small devices.
- Small-scale wind turbines have an output of 100 W to 50 kW. Small-scale wind turbines can power individual households and small businesses. The most popular household turbines have an output of 3–5 kW. With battery backup, the turbine can power lighting systems, pumps and household equipment.
- Large-scale wind turbines (with an output in excess of 100 kW) can power households, but they are mainly used to generate electricity that is sold to the power grid.

12.3.1.2 *Classification based on size*
Wind speed is influenced by many factors, including weather conditions and site altitude (the higher a wind turbine is sited, the greater its output). Large-scale wind turbines are mounted on towers with a height of 70 to 120 m. Small-scale turbines are usually positioned on masts with a height of between 1.5 m (if sited on the roof of a building), and 15–20 m (if sited on the ground).

12.3.1.3 *Classification based on axis of rotation*
The axis of rotation is another criterion for classifying wind turbines. Turbines typically fall into two categories:

- Horizontal-axis wind turbines (HAWT).
- Vertical-axis wind turbines (VAWT).

12.3.1.4 *Classification based on other criteria*
Other criteria that can be used to classify wind turbines include:

- Types of generated energy: electricity or mechanical energy (e.g. wind pumps).
- Number of rotor blades: single-, twin-, three-, four- or multi-bladed turbines.
- Rotor orientation relative to wind direction (in HAWTs): upwind or downwind turbines.
- Rotational speed: low-, medium- or high-speed wind turbines.

12.4 HORIZONTAL-AXIS WIND TURBINES (HAWTs)

The majority of HAWTs have rotors with one, two or three conventional blades. Turbines with a higher number of blades are used mainly for pumping water, powering mills and other applications

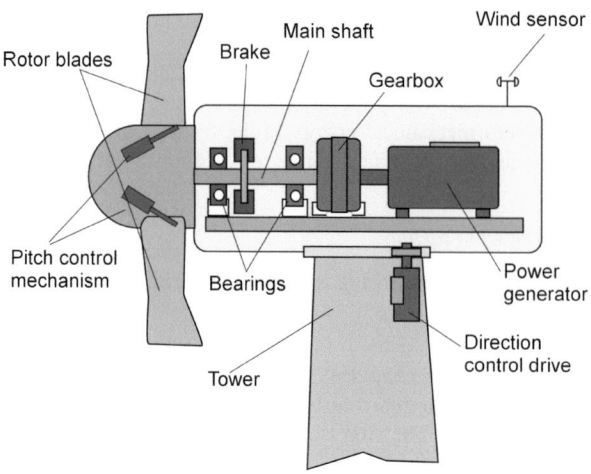

Figure 12.9.   Simplified diagram of a wind turbine.

that require mechanical energy. At present, HAWTs account for around 95–98% of the solutions installed worldwide (IRENA, 2012).

A typical wind turbine is shown in Figure 12.9. The diagram presents a large-scale turbine, but the majority of structural sub-assemblies shown are applied in all types of wind turbine. The key element of a wind turbine is the rotor, which converts wind energy to the mechanical energy that is transmitted to the generator. In most HAWTs, rotors have one or more blades. Blades, depending on the wind turbine type, are made of wood, aluminum and its alloys, steel, reinforced glass or carbon fiber. Each blade consists of two halves mounted on a shaft (Golfman, 2012; Hansen *et al.*, 2006).

In some solutions, the blade pitch angle can be modified with the use of a hydraulic actuator. The rotor is mounted on a low-speed shaft which transmits power to a high-speed shaft via a gearbox. The high-speed shaft is coupled with the generator shaft. Some wind towers operate without a gearbox. The standard rotor speed is 10–50 revolutions per minute (rpm), and the gearbox increases rotor speed to 1500–3000 rpm (Hansen, 2015).

During start-up, generators are connected to the grid by thyristors with shunt contactors. A microprocessor controller monitors turbine operation and carries out precision diagnostics on the principal turbine components. The nacelle also houses lubricating systems, cooling systems and a brake disk. The nacelle and the rotor are moved in the direction of the wind by the yaw drive and yaw gear, which are mounted on top of the turbine tower. The tower is made of steel or reinforced concrete. Most towers are built from large-diameter pipes; lattice towers are less frequently encountered.

The turbine is also a key structural element, which transforms the wind energy into the mechanical energy of rotational motion, which is then converted into electrical energy in the generator. Turbine parameters determine the power output and the rotational speed of the entire wind-powered generator. The size of a turbine is selected to accommodate the rotor. The remaining elements, including the generator, gearbox, tower and nacelle, are also selected according to parameters of the wind turbine (mechanical power converted from wind by turbine) (Adaramola, 2014).

As previously indicated, wind turbines can also be classified according to the rotor's position in the unit relative to wind direction as shown in Figure 12.10. Thus, in a downwind turbine, the rotor is effectively at the back of the unit, and in an upwind turbine, the rotor is positioned at the front of the tower, relative to the wind. Downwind turbines are less popular due to losses caused by the wind shadow of the tower (Wang and Coton, 2001).

A schematic visualization of various types of HAWT is presented in Figure 12.11.

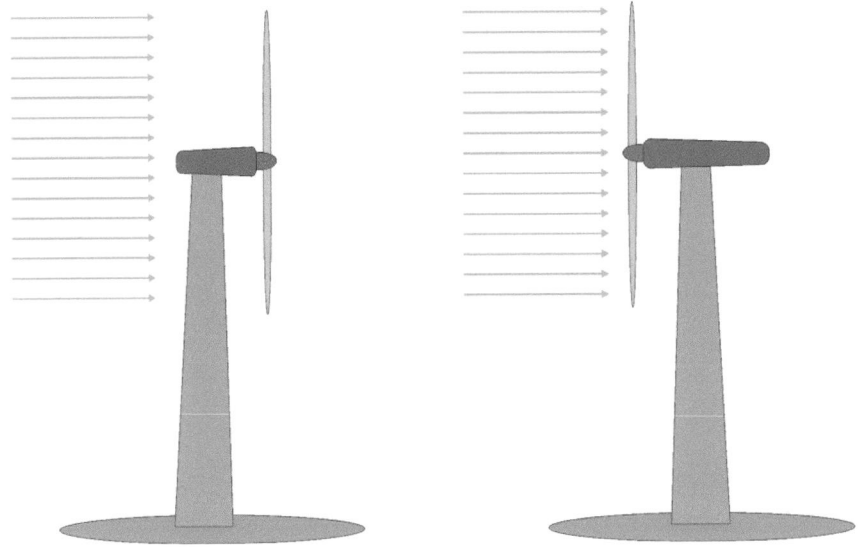

Figure 12.10. Downwind (left) and upwind (right) horizontal-axis wind turbines.

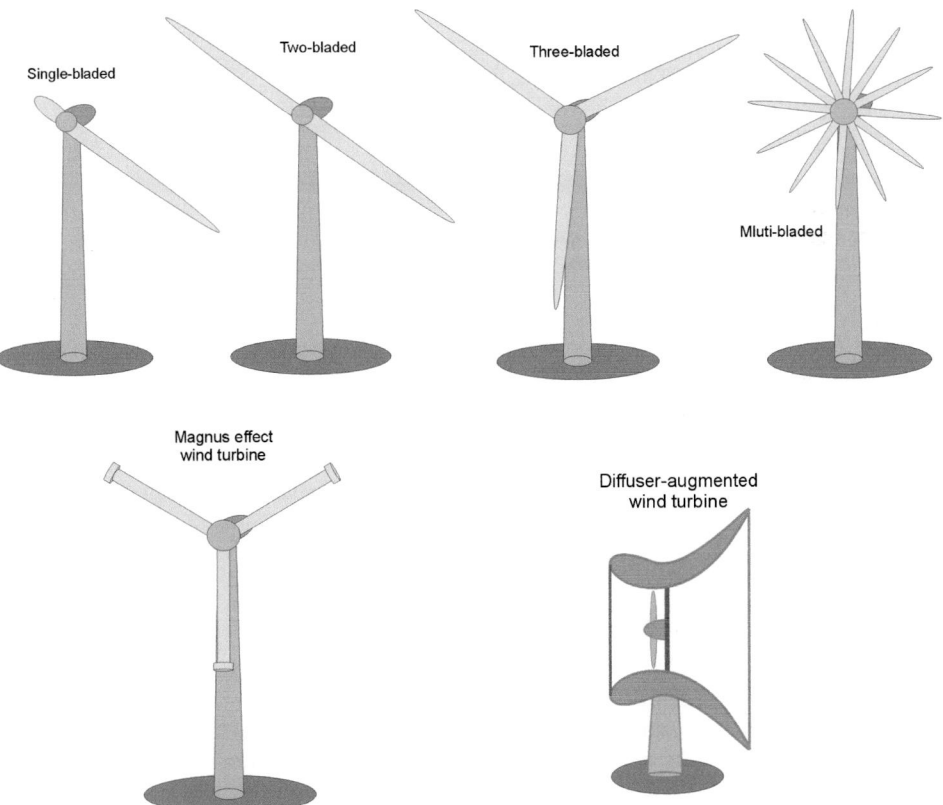

Figure 12.11. Types of horizontal-axis wind turbine.

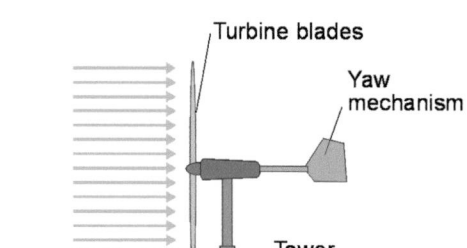

Figure 12.12.　Working principle of a horizontal-axis wind turbine.

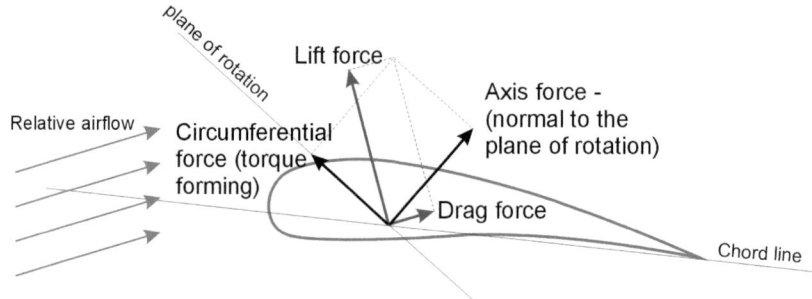

Figure 12.13.　Lift force generation on a rotor blade.

12.4.1　*Basic working principle of HAWTs*

A stream of air acts on a turbine's blades (Fig. 12.12), adjusted to an optimal angle, and sets the rotor in motion (Nagai *et al.*, 2009). The energy of a moving rotor can be used directly to power mechanical devices (such as a pump), or it can be directed to a generator to produce electricity. The yaw mechanism orientates the rotor into the wind to optimize the turbine's performance.

A wind turbine's output is determined by several factors, including wind speed and system efficiency (Kishinamia *et al.*, 2005). The formation of the aerodynamic lift on a rotor blade is shown in Figure 12.13. The circumferential component of this force sets the rotor in rotational motion and is a source of the rotor's output torque (Manwell *et al.*, 2002).

12.4.2　*Examples of structural solutions*

12.4.2.1　*Structural solutions with three-bladed rotors*

Structural solutions incorporating three-bladed rotors are presented in Figures 12.14 and 12.15. Three-bladed rotors are the most commonly applied in large-, medium- and small-scale wind turbines. For example, SWIND turbine rotors (SWIND Elektrownie Wiatrowe, Jaworowa, Poland) are made of composite materials used in aircraft.

Composite materials are highly resilient and light, which is of particular importance in wind turbines. Blades are suitably profiled to minimize noise emissions. SWIND 3200 and 6000 wind

Figure 12.14.   SWIND 3200 three-bladed wind turbine (SWIND, n.d.).

Figure 12.15.   SWIND 6000 three-bladed wind turbine (SWIND, n.d.).

Table 12.1.   The parameters of SWIND wind turbines (SWIND, n.d.).

	SWIND 3200	SWIND 6000
Rated power	3.2 kW	6 kW
Rotor diameter	3.5 m	4.8 m
Rotor	3-bladed	3-bladed
Generator	3-phase AC synchronous generator	3-phase AC synchronous generator
Nominal wind speed	$12\,\mathrm{m\,s^{-1}}$	$12.5\,\mathrm{m\,s^{-1}}$
Weight of nacelle with rotor	$\sim 80\,\mathrm{kg}$	$\sim 210\,\mathrm{kg}$
Yaw	Yaw vane	Yaw vane

turbines (see Table 12.1) are used mainly for:

- Generating electricity for household needs
- Heating water
- Supporting central heating systems
- Supplying heat.

KOMEL wind turbines (KOMEL Institute, Katowice, Poland) (Fig. 12.16 and Table 12.2) have aluminum blades on a support frame mounted on a steel pipe. The design of the support frame

(a)   (b)

Figure 12.16.   Three-bladed turbines: (a) household wind turbine; (b) JSW 750-12 wind micro-turbine
(KOMEL, n.d.).

Table 12.2.   The parameters of KOMEL wind turbines (KOMEL, n.d.).

	KOMEL	JSW 750-12
Rated power	4.0 kW	120 W
Rotor diameter	5.8 m	0.75 m
Rotor	3-bladed	3-bladed
Generator	PMGg 180L16 spec ($3\,kVA$, 180 rpm, $U_N = 3 \times 170\,V \to$ $230\,V\ U_{DC}$, $f_N = 24\,Hz$),	3-phase AC synchronous generator
Nominal wind speed	$10\,m\,s^{-1}$	$12\,m\,s^{-1}$
Yaw	Yaw vane	Yaw vane

facilitates blade assembly. Angles of incidence and attack are set by placing the support frame on even ground during assembly. The angle of incidence has to be correctly set to optimize the power output of the wind turbine (KOMEL, n.d.).

A wind micro-turbine (Fig. 12.16b) is a cheap, quiet and pollution-free source of electricity. This type of wind turbine is used to charge batteries that supply various off-grid systems, including:

- Lighting
- Irrigators
- Remote telemetry in precision agriculture, data acquisition systems
- Electric pumps
- Electric air pumps for pond aeration.

Three-bladed wind turbines with similar structural designs are supplied by many other manufacturers, too. The energy they produce, depending on the size of the wind turbine, can be used in agriculture for local needs (micro-turbines and small wind turbines), or sold in to the electrical power system (medium- and large-scale turbines).

12.4.2.2 *Single- and twin-bladed wind turbines*
Twin blades lower investment costs and reduce rotor weight. However, greater noise and a less harmonious appearance detract from the popularity of twin-bladed turbines. Such structures require a tilt rotor hub to attenuate the load imposed by blades passing the tower.

Single-bladed rotors are very rarely used. They share the disadvantages of twin-bladed rotors, requiring even higher rotational speeds and generating even more noise. Their only advantage is lower installation costs.

12.4.2.3 *Multi-bladed wind towers*
Low-speed, multi-bladed wind turbines (Fig. 12.16) are well suited for operation in closed loop systems, and they are very popular in weakly populated regions of the US. Multi-bladed turbines are not used industrially, but they are efficient sources of additional electricity or mechanical power (e.g. water pumps) for households and farms. Their greatest advantages include low cut-in speeds, high torque, simple structure (specially profiled airfoils are not required), and low price in comparison with two- and three-bladed wind towers (Valdès and Ramamonjisoa, 2006).

To illustrate, a 5 kW multi-bladed turbine with a rotor diameter of 5.5 m produces an energy-to-swept-area ratio of 0.21 kW m^{-2}. A better result can be achieved with a standard three-bladed wind turbine with the same rating (0.26–0.36 kW m^{-2} for small-scale generators).

It should be noted, however, that the cut-in speed of a multi-bladed rotor is just 2.1 m s^{-1}, whereas three-bladed turbines can only begin to operate at wind speeds of 3–4 m s^{-1}.

The yaw mechanism is a tail vane made of steel sheet and angled sections. The turbine's rotational speed is usually controlled by directing the rotor away from the direction of the wind with a guide vane mounted on the side. The rotor is shut down by pulling down the tail vane and the side guide with a string wound on a reel.

If rotor torque increases due to higher wind speed, this increase will be manifested mainly by higher rotational speed. Since torque decreases with a rise in rotational speed, the rotational speed will increase until the rotor torque becomes equal to the load torque. If load torque increases or rotor torque decreases for any reason, the turbine's rotational speed will be lowered. Since torque increases with a drop in rotational speed, the rotational speed will increase until torque becomes equal to load torque. For this reason, a multi-bladed turbine with a constant torque engine is characterized by stable operation. A multi-bladed turbine will continue to operate when overloaded, but its rotational speed and power coefficient will decrease. In a multi-bladed turbine, axial force from wind is greatest during start-up, and it quickly drops with an increase in rotational speed (KOMEL, n.d.).

12.4.2.4 *Diffuser-augmented wind turbines*
The use of a diffuser marked the next stage of the development of HAWTs. The placement of a rotor inside a diffuser theoretically increases a turbine's efficiency. According to Bernoulli's principle explaining the behavior of a flow medium (e.g. gas) in a pipeline, changes in pipeline diameter will affect flow speed. If a conventional rotor is mounted inside a diffuser (in its narrowed passage), it will move faster in the stream of air that flows at a higher speed than the air (wind) outside the diffuser. This approach increases the turbine's performance (Matsushima *et al.*, 2006; Ohyaa *et al.*, 2008).

The rated capacity of wind turbines using a diffuser is three times higher in comparison with turbines of a similar size where the blades are not surrounded by a diffuser. Fast blade movement requires a lower gear ratio, and a smaller size increases structural rigidity. The diffuser improves aerodynamic efficiency, and it safeguards the contents of the rotor during threats posed by strong winds.

The parameters of a diffuser-augmented wind turbine support operation at wind speeds lower than $4\,\mathrm{m\,s^{-1}}$. This implies that such a solution could, for example, be applied in practically all locations throughout Poland. Turbines with a smaller diameter can also be mounted on lower towers to significantly lower installation costs.

On an annual basis, diffuser-augmented wind turbines are designed to operate for 3200 hours longer than conventional devices. The annual output of a diffuser-augmented wind turbine with a capacity of $Ps = 660\,\mathrm{kW}$ reaches 4.3 GWh, which is approximately 3 GWh higher than an equivalent conventional turbine, which generates 1.3 GWh energy per year. For this reason, diffuser-augmented turbines can be operated in regions with less than optimal wind conditions (Franković and Vrsalović, 2001).

12.4.2.5   *Energy Ball wind turbines*

The Energy Ball wind turbine (Home Energy International BV, Hoofddorp, The Netherlands) is a novel solution that breaks with conventional turbine design by using a spherical structure. The Energy Ball rotates parallel to the earth's axis (horizontally).

The Energy Ball has been designed in the shape of a sphere for more efficient energy generation, lower noise emissions and greater operating safety. The turbine's spherical structure contributes to higher aerodynamic efficiency and soundless operation. It can be used to power households with low energy requirements (HomeEnergy, n.d.).

Detailed structural data has not been revealed, but the smallest Energy Ball can meet around 15% of the energy requirements of an average household. As an additional advantage, the device can generate energy at very low wind speeds by relying on the Venturi effect. The Energy Ball uses a nozzle, which was developed by Giovanni Battista Venturi to measure the flow rate of fluids and gases. The wind turbine operates according to Bernoulli's principle: the flow rate of gas or fluid moving along a pipe increases when the medium enters a constricting throat. In mathematical terms, the squared speed of a flowing medium is directly proportional to the difference in pressure ahead of and behind the narrowed section of the pipe. A nozzle thus creates negative pressure.

12.5   VERTICAL-AXIS WIND TURBINES (VAWTs)

Progress in the development of VAWTs was considerably slower than the advances made in the HAWT sector. In comparison with HAWTs, VAWTs have a small share of the installations that have been put into service. VAWTs rely on three basic structural solutions: the Savonius turbine, the Darrieus turbine and the H-rotor turbine (Bhutta *et al.*, 2012).

In 1922, Finnish engineer Sigurd J. Savonius developed a VAWT. The turbine's working principle is shown in Figure 12.17. The Savonius turbine is one of the simplest turbines. It is used by amateurs to develop wind turbines for residential purposes, as well as by professional wind

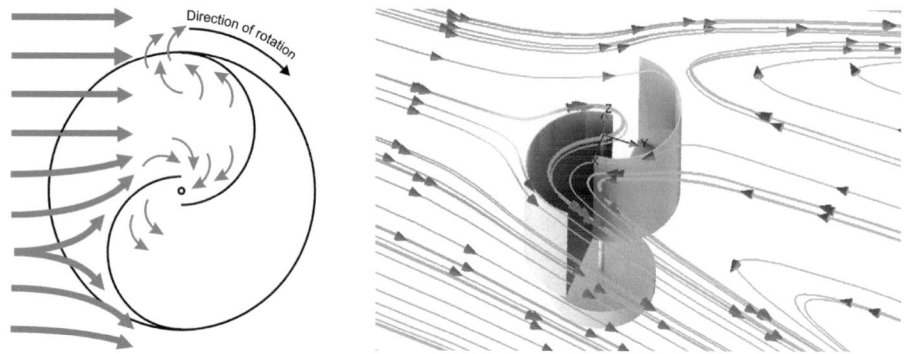

Figure 12.17.   Working principle of a Savonius turbine and its computer simulation results.

turbine suppliers. Savonius turbines extract much less of the wind's power than lift-type turbines. Despite this, Savonius turbines continue to be popular due to their reliability, simple structure and low cost (Menet, 2004).

The Darrieus turbine was patented by its inventor, Georges Jean Marie Darrieus, in 1931. Turbines of this type have zero starting torque, and an external energy source is required to set the device in motion. Most Darrieus turbines are started by an electric motor, but alternative solutions are also applied (Islam *et al.*, 2008). The structure of a typical Darrieus turbine is presented in Figure 12.18. A Darrieus turbine equipped with two Savonius turbines is shown in Figure 12.19. This unconventional solution produces the required torque for start-up.

The Darrieus turbine uses the lift force created by air flowing around the airfoils (Deglaire *et al.*, 2009). The principle responsible for the generation of lift force in Darrieus and H-rotor turbines is illustrated in Figure 12.20.

Figure 12.18.   Darrieus wind turbine (left) and modified Darrieus-type wind turbine (right), named H-rotor wind turbine.

Figure 12.19.   Off-grid 3 kW magnetic levitation wind turbine (TYPMAR, n.d.).

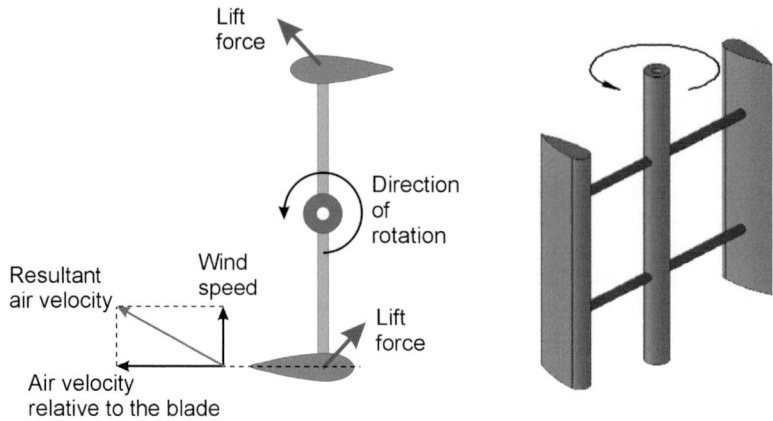

Figure 12.20.   Generation of lift force by air flowing around airfoils.

12.5.1   *Structural solutions*

12.5.1.1   *Wind-powered generators that rely on the Savonius turbine*

A typical Savonius turbine is characterized by low efficiency and high torque variability during operation (Altan and Atılgan, 2008; Irabu and Roy, 2007). Various remedies to these problems have been proposed by scientists and wind turbine suppliers. The most successful modifications include a turbine that relies on wind tunnels to increase the speed of air flowing through the turbine (Altan *et al.*, 2008), and spiral turbines. A spiral turbine is an evolution of the Savonius turbine in which rotor blades have been replaced with spiral sails parallel to the rotational axis (Windside, n.d.). This modification reduces torque variability during rotor operation (Kamoji *et al.*, 2009).

Similarly to spiral turbines, the Helix Wind turbine featured "wind pockets" which were positioned parallel to the rotational axis. These wind pockets were separate elements that were joined to form a system of airfoils. The helical turbine was characterized by comparatively low torque variability and high energy efficiency for a modified Savonius turbine. It had a modular structure and was delivered to the user as a self-assembly kit. The turbine had a unique design which contributed to its aesthetic appeal (Inhabitat, n.d.). Although the company responsible for the Helix Wind turbine has ceased trading, its segmented wind turbine solution is an interesting proposition (visually and construction-wise) in the approach to small wind turbines.

12.5.1.2   *H-rotor turbines*

Modified H-rotor turbines account for a large number of the VAWTs currently manufactured in the world. In this type of turbine, lift force generated on airfoils produces torque. An H-rotor turbine is a variant of a Darrieus turbine. The available solutions have two to five airfoils. H-rotors are applied in micro-turbines (less than 100 W) as well as in large-scale turbines that generate several dozen kilowatts. An example is the Ropatec generators (Ropatec, Bolzano, Italy), which are VAWTs, the rotors of which have been optimized for operation in low-speed winds (Ropatec, n.d.).

The qr5 wind turbine (Quietrevolution, St. Ives, UK) is another brand of H-rotor wind-powered generator. Its structure has been optimized to allow operation in sites where wind direction is highly variable. The turbine is powered not only by winds that exert a perpendicular force on the rotational axis, but also utilizes upward wind flows. The purpose of this modification was to adapt a VAWT for operation in densely developed urban areas with tall buildings. In urban environments, powerful lift forces are generated by fast moving streams of air (Quietrevolution, n.d.).

Figure 12.21.   PacWind (USA) turbines: (a) 500 W Sea Hawk; (b) 5 kW Aeolian (Desert Power, n.d.).

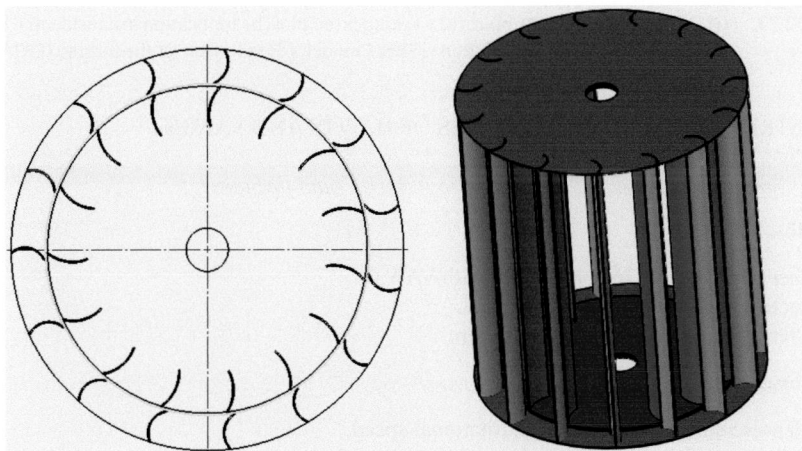

Figure 12.22.   Dual-stage vertical-axis wind turbine patented by Z. Pawlak.

12.5.1.3 *Drum-rotor turbines*

Wind-powered generators with drum-shaped rotors are yet another modification of VAWTs. Variants with and without wind tunnels are available, as shown in Figure 12.21 (Desert Power, n.d.).

In 1985, Polish engineer Z. Pawlak patented a dual-stage VAWT characterized by high starting torque, low tip-speed ratio and low noise emission (Fig. 12.22). The patented design can be used to capture wind energy or convert the energy of sea waves and currents (Pawlak, n.d.).

Turbines with a similar structure are manufactured by GUAL Industrie of France (GUAL, n.d.). Diagrams of the turbine, its working principle, possible applications and parameters are shown in Figure 12.23.

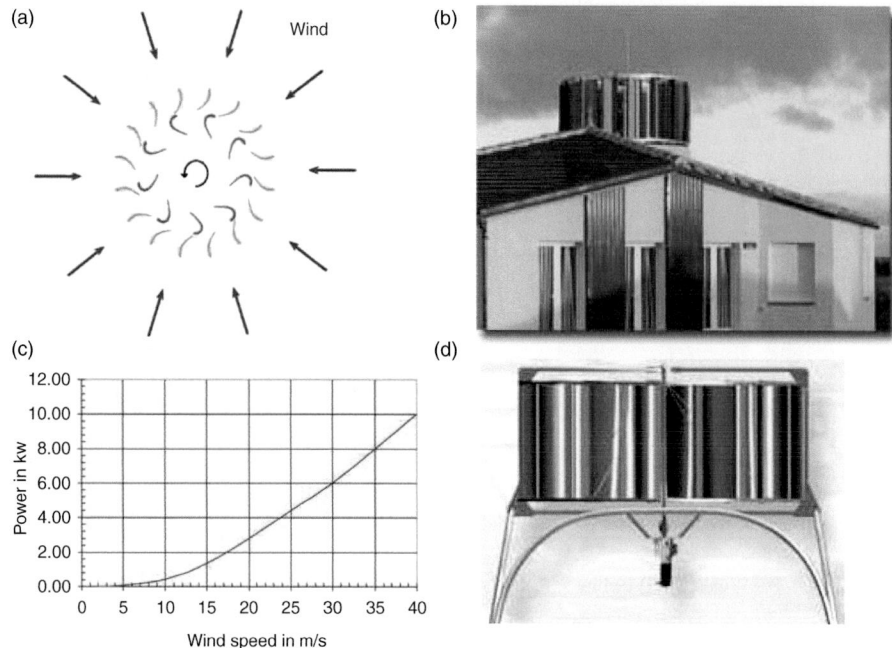

Figure 12.23.   GUAL Industrie wind turbine: (a) working principle; (b) application in a residential building; (c) parameters of the StatoEolien GSE8/3 model; (d) side view of the turbine (GUAL, n.d.).

12.6   STRENGTHS AND WEAKNESSES OF HAWTS AND VAWTS

12.6.1   *Strengths and weaknesses of horizontal-axis wind turbines*

Strengths:

- Higher power output in comparison with VAWTs.
- Attractive and harmonious appearance.
- Higher efficiency and power coefficient.

Weaknesses:

- High noise emissions due to high rotational speed.
- A furling system is required to protect the turbine from strong winds.
- A yaw mechanism is required to turn the rotor into the wind.
- Slip rings are required if the generator is installed inside a nacelle.

12.6.2   *Strengths and weaknesses of vertical-axis wind turbines*

Strengths:

- Stable operation regardless of wind direction; simple structure and control; a yaw mechanism is not required.
- Easy assembly on existing structures; tall towers are not required.
- Can be installed on roofs, posts, existing mast structures, etc.
- Low noise emissions even at maximum rotational speed.
- Resistant to strong winds; the shape of the rotor limits rotational speed aerodynamically; do not have to be shut down, even at wind speeds of $40\,\mathrm{m\,s^{-1}}$.

- Resistant to hard rime, frost and wet snow; due to small diameter and low rotational speed, the rotor's static imbalance caused by unfavorable weather conditions does not produce dangerous vibrations.
- Service-free operation of the generator; slip rings are not required.
- Portable structure, which is easy to assemble and disassemble.
- Relatively low cost in comparison with conventional HAWTs.
- Attractive appearance; can be used as a medium for outdoor advertising or to enhance the landscape.

Weaknesses:

- Low efficiency; much larger turbines are required to generate the same amount of energy as HAWTs.
- Due to low rotational speed, a low-speed generator or a gearbox are required, which further decreases efficiency and increases noise emissions.

12.7   APPLICATIONS OF WIND TURBINES

12.7.1   *Possible applications of wind micro-turbines*

- Generation of electricity on a local or national scale.
- Source of light for billboards and noticeboards – VAWTs do not require tall towers.
- VAWTs can be operated to power holiday cottages and hotels in coastal areas and mountains where strong winds are prevalent. The efficiency of VAWTs increases under extreme weather conditions.
- Micro-turbines can be installed on building roofs. According to computer simulations, wind speed increases by 30% several meters above a roof, in comparison with undeveloped locations. Air is compressed on the windward side of an obstacle such as a building, which more than doubles the turbine's power coefficient.
- Balconies and terraces.
- Single-family houses, holiday cottages, gardens, gazebos.
- Households situated remotely from power grids.
- Water heating in public utility buildings, kindergartens, bars, etc.
- Camping sites, camp trailers.
- Lighting for streets, pavements, piers, etc.
- Waterway signs, yachts.
- Agriculture – for powering electrical devices and household equipment.
- Powering drainage pumps.
- Fishponds, water features, fish tanks, fish farms.
- Aeration devices, water treatment and recycling systems, water-heating systems.
- Water storage and pumping.

12.7.2   *The use of small-scale wind turbines in rural areas*

Wind farms comprising more than one turbine are most popular in rural areas. Their combined output scales to several hundred megawatts. Small-scale turbines with output below 100 kW are applied in individual households. Houses and farms usually use wind turbines with a generation capacity of 1 to 20 kW.

Wind turbines generate electricity that is either used locally (for lighting, heating and producing hot water in households) or is supplied to the power grid. The farming sector and the food processing industry require various types of energy, and wind turbines are applied to:

- Generate electricity.
- Heat households and farm buildings.

- Heat household water and water for agricultural and feedstuff processes.
- Power processing devices such as fans for drying agricultural produce, feedstuff processing devices, compressors, etc.
- Power water pumps to collect and store water, irrigate fields, pump water from basins, fill fishponds, etc.
- Power water aerators in fishponds and lakes.

Wind turbines can heat households and farm buildings and produce hot water (process and household water). Such systems rely on a heater (heating assembly) placed in the medium to be heated (e.g. water). The advantage of this solution is that practically all of the electricity generated from wind energy (even low-quality electrical power generated at low wind speed) is converted to heat. This increases the utility of the wind power system.

Wind turbines can be equipped with pump units for pumping, recycling and aerating water reservoirs, in particular in lakes, fishponds and settlement tanks in sewage treatment plants. A wind-powered water pump is presented in the following section (see 12.8.5).

Deserts, semi-arid regions and steppes are deficient in water and have weakly developed power grids. They are characterized by strong winds; therefore, pumps driven by wind energy are an ideal solution for pumping groundwater in these areas. The inhabitants of Karoo, South Africa, have relied on this solution for generations. Wind pumps are used in farms around the world to collect and store water (Valdès and Raniriharinosy, 2001). They are usually equipped with a multi-bladed wind turbine, and they are popularly referred to as windmill pumps (Harries, 2002; La Rotta and Pinilla, 2007).

Wind turbines equipped with a water aeration system are used to treat and recycle water reservoirs and supply air to fishponds. The use of such devices is required in water reservoirs characterized by low concentrations or an absence of oxygen. Oxygen deficiencies can result from impaired biological balance due to organic or chemical contamination, or excessive growth of algae and aquatic vegetation. This intensifies the decomposition of organic matter in water, depletes oxygen resources and leads to the death of aquatic plants and animals.

Wind-powered aerators can also be used at sites with a high oxygen demand, such as fishponds. Wind-powered devices can cater effectively to the high demand for various types of energy in agriculture as well. Modular systems should be developed in locations where the parameters of the wind turbine and actuating devices have to be individually selected and tailored to meet the users' specific needs.

12.8   WIND ENERGY SOLUTIONS FOR HOUSEHOLDS

An analysis of a wind turbine's structural elements supports the identification of the following basic functional units (Fig. 12.24):

- Conversion of the kinetic energy of air to mechanical energy in rotational motion.
- Processing of mechanical energy in rotational motion to generate other types of energy (electricity, heat).
- Energy storage (mechanical energy – compressed air, kinetic and potential energy batteries; electrical energy – electrochemical batteries, electric double-layer capacitors, power grid; heat energy – direct heating of working medium, heat from phase transition of the working medium; chemical energy – storage of energy in the form of hydrogen for powering fuel cells and combustion engines).

In the example solutions that follow, modular wind turbine systems have been developed to convert wind energy into useful forms of energy. In each solution described, wind turbine systems are applied to generate one type of useful energy, but technologies that produce other types of energy are also available.

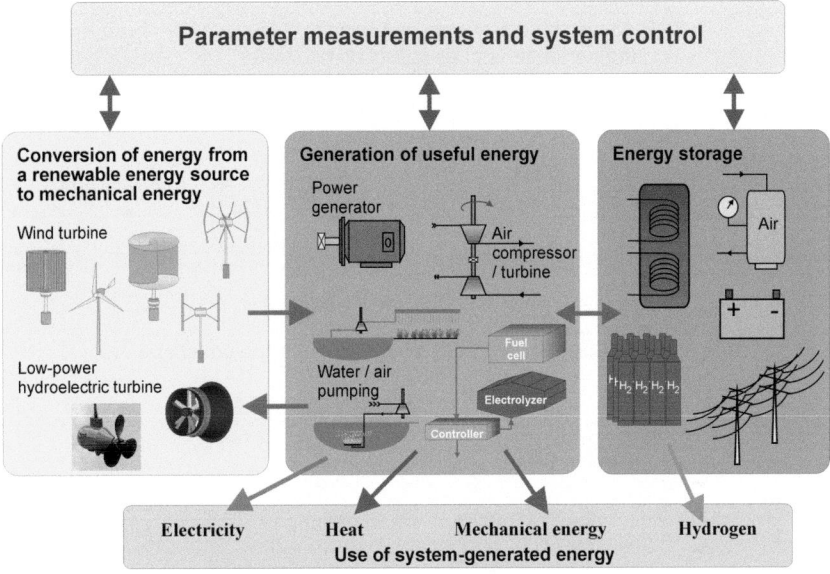

Figure 12.24.   Functional units in wind turbine systems (Nalepa *et al.*, 2011).

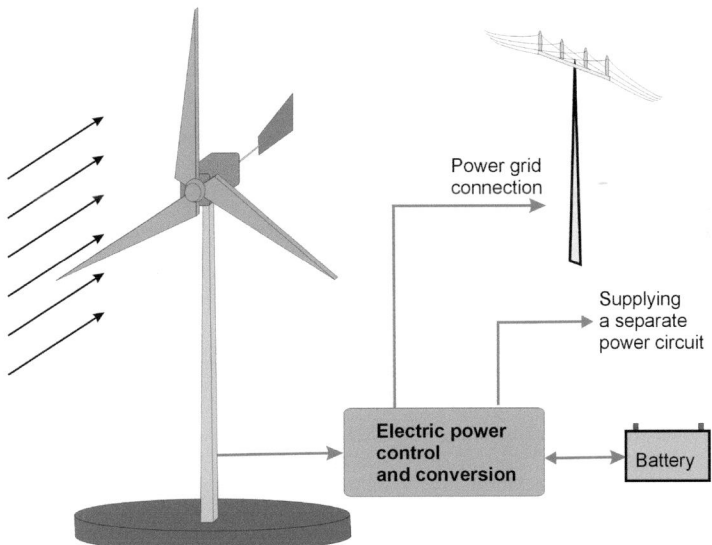

Figure 12.25.   A wind turbine that converts wind energy to electricity can be connected to an energy storage
system or a local distribution system.

12.8.1   *Solution 1 – Conversion of wind energy to electricity*

Electricity generated by a wind turbine can be used to cater to local household needs or can be
sold to the power grid (Fig. 12.25).

In order to sell electricity to a power grid, the wind turbine has to generate electrical power of a
high quality. The parameters (frequency and voltage) of the electricity produced by a wind turbine
should be identical to the parameters of the grid being supplied. Automatic control systems are
applied to synchronize the wind turbine's output with the power grid (Fig. 12.26).

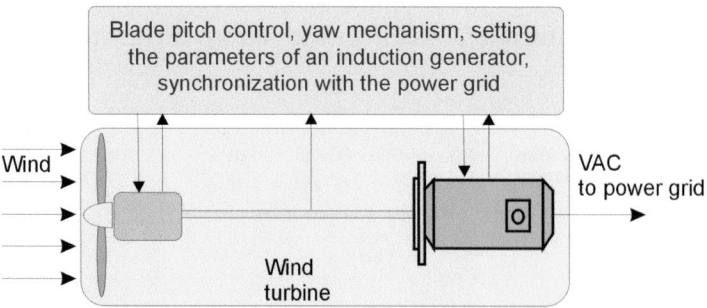

Figure 12.26.   Diagram of a system synchronizing a wind turbine with a power grid.

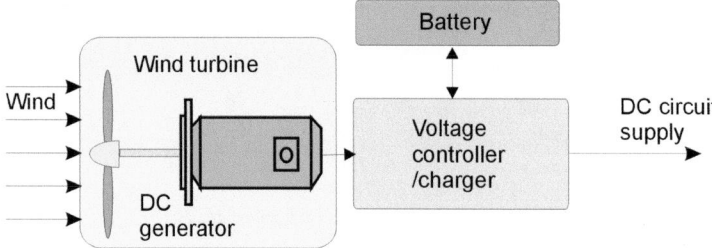

Figure 12.27.   Diagram of an autonomous wind turbine system with a DC generator and a DC branch circuit.

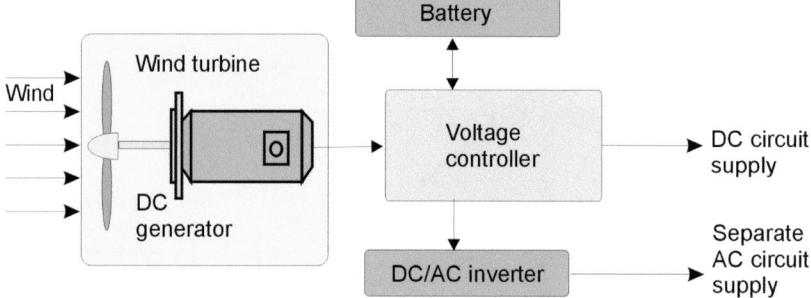

Figure 12.28.   Diagram of an autonomous wind turbine system with a DC generator and a DC/AC branch circuit.

A wind turbine that generates energy for local needs is an autonomous power source. In such systems, electricity is produced by direct current (DC) generators or small alternating current (AC) generators, which are often equipped with permanent magnets that do not require additional excitation systems. Because the quality and quantity of generated energy is determined by wind speed, such wind turbines are often connected to batteries. A diagram of such a system is presented in Figure 12.27. Wind turbines supplying alternating current receivers can be equipped with DC to AC inverters.

A wind turbine equipped with an AC generator can produce high-quality DC power after voltage regulation and rectification. Because the rotational speed of a wind turbine is determined by wind speed, the voltage and frequency of the generated output also varies. For this reason, such wind turbines must be provided with an intermediate DC circuit and a voltage regulator that works with a DC/AC inverter (Fig. 12.28). In such a system, electricity can be stored in electrochemical batteries or electric double-layer capacitors.

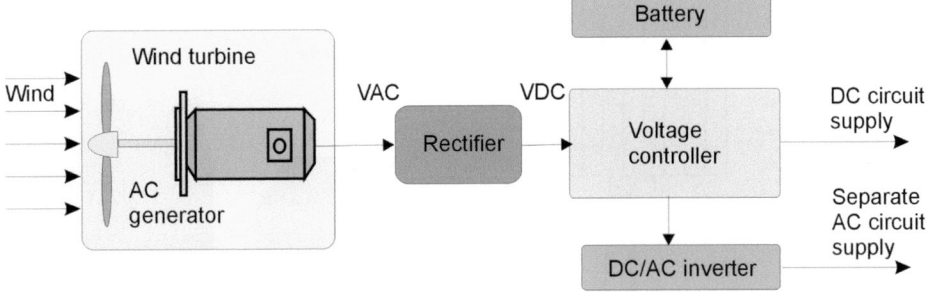

Figure 12.29.   Diagram of an autonomous wind turbine system with an AC generator.

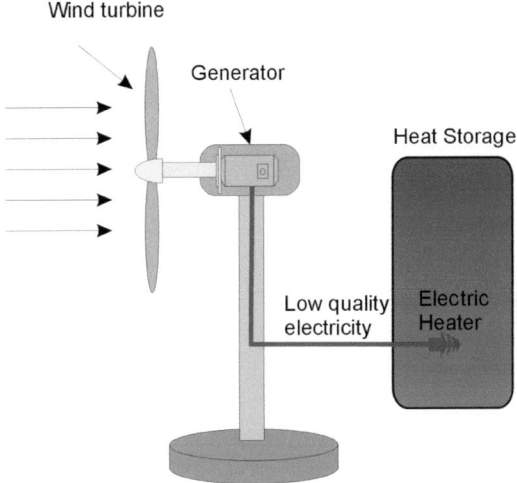

Figure 12.30.   A system which converts wind energy to heat (with the involvement of electricity).

12.8.2   *Solution 2 – Conversion of wind energy to heat*

Small-scale wind turbines can be used to convert wind energy to heat (with the involvement of electricity) (Fig. 12.30).

A wind turbine converts wind energy to electricity, and uses electric heaters to produce heat. Such systems can be used to heat process, household and industrial water. Other media may be applied, such as paraffin, to extract heat from their phase transmission. The advantage of such a solution is that low-quality energy generated under variable wind conditions can be used to produce heat.

12.8.3   *Solution 3 – Conversion of wind energy to electricity with excess energy storage in the form of hydrogen*

Fuel cell technology creates new options for energy storage. The first element in the process of converting wind energy is a wind turbine connected to a power generator. The electricity produced can be used directly to supply power receivers or it can be used to produce hydrogen through electrolysis. The hydrogen is stored in a reservoir, and it can be used in fuel cells to generate electricity alone or both electricity and heat (Fig. 12.31).

In small-scale wind turbines, any excess electricity is not used to create hydrogen stores because of the comparatively high costs involved. In the near future, however, it can be expected that

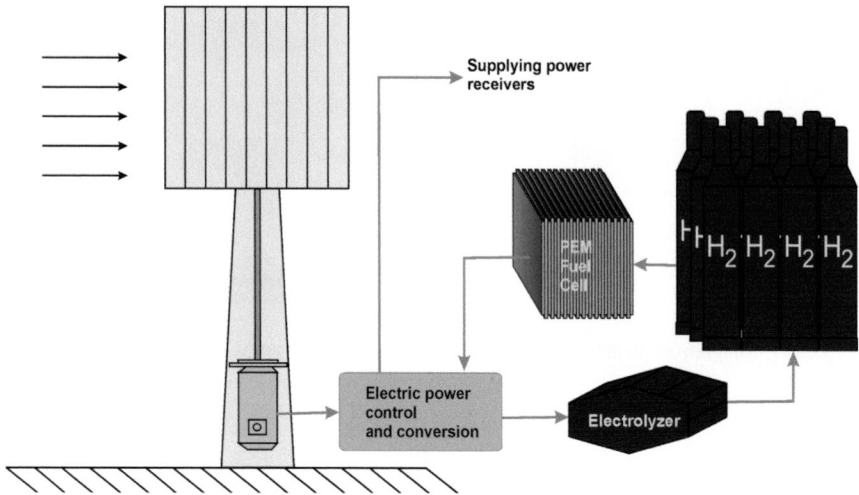

Figure 12.31.   A system that converts wind energy to electricity and stores excess electricity as hydrogen (Nalepa *et al.*, 2011).

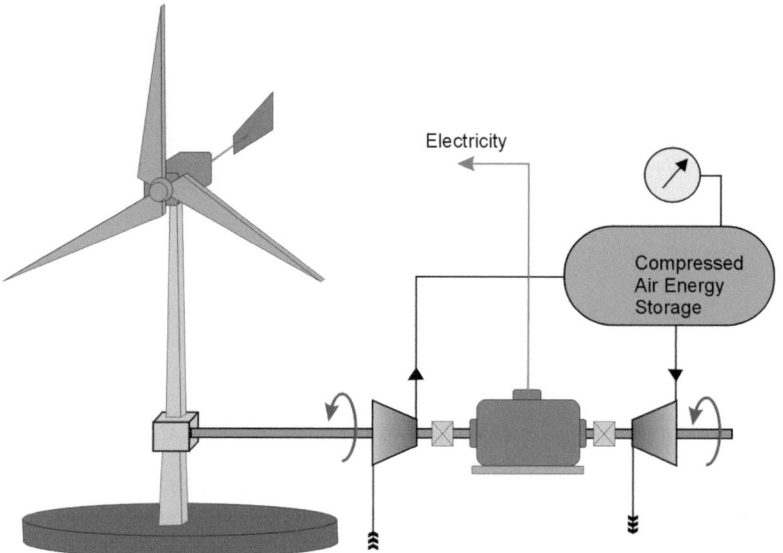

Figure 12.32.   A system that converts wind energy to electricity and stores excess energy in the form of compressed air (Nalepa *et al.*, 2014).

innovative solutions will emerge which support the use of hydrogen-stored energy for powering combustion engines, electric vehicles and fuel cells that can supply electricity to the power grid.

12.8.4   *Solution 4 – Conversion of wind energy to electricity with excess energy storage in the form of compressed air*

The system presented in Figure 12.32 stores excess energy in the form of compressed air. Using this solution, wind energy that is not converted to electricity can be used at a later date.

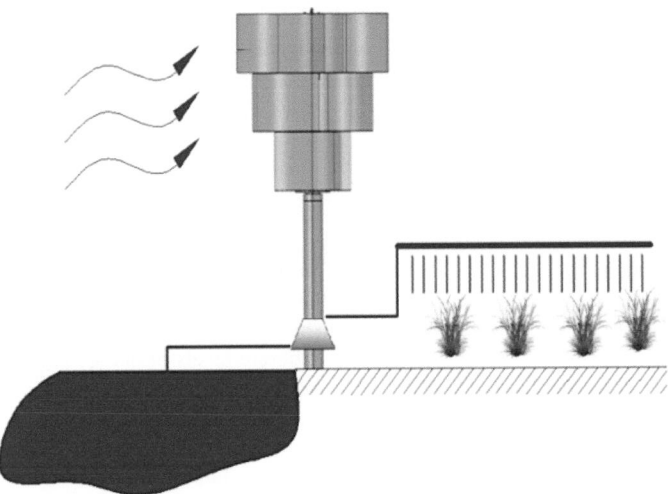

Figure 12.33.   A system that converts wind energy to mechanical energy can be applied to irrigate fields and pump water (Miąskowski *et al.*, 2014).

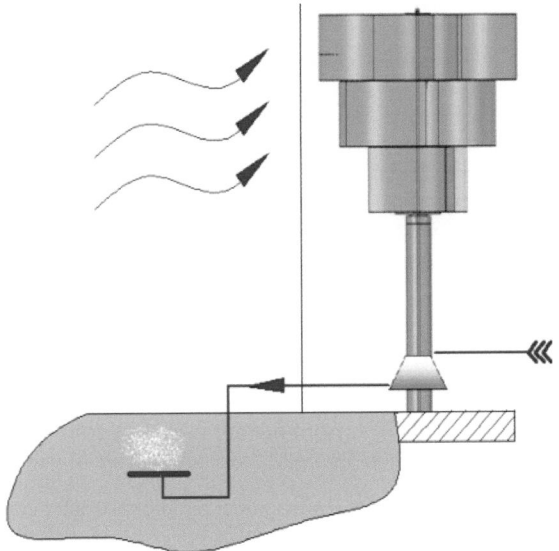

Figure 12.34.   A system that converts wind energy to mechanical energy can be applied to aerate lakes (Miąskowski *et al.*, 2014).

12.8.5   *Solutions 5 and 6 – Conversion of wind energy to mechanical energy*

Systems that convert wind energy to mechanical energy can be used to power water pumps. A wind turbine converts wind energy to mechanical energy, which is used to pump:

- Liquids (Fig. 12.33), for example, to irrigate water crops that are grown in open fields or under cover.
- Air (Fig. 12.34), for example, to aerate fishponds, recycle lakes and aerate settlement tanks in sewage treatment plants.

The solutions presented can be used to develop modular wind turbine systems, the individual components and parameters of which can be selected to accommodate the specific power needs and weather conditions of a given site.

12.9   CONCLUSIONS

The preparation of the material in this chapter led to the dynamic development of wind energy installations in many countries and a growth of interest in these types of solution by potential investors. This work was intended mainly for consumers, who wanted to know not only the theoretical foundations of wind energy acquisition but, most importantly, the technical capabilities of applied solutions dedicated to installations at different levels of energy production, including solutions provided for small wind turbines allowing the fulfillment of individual energetic needs at the level of a single household or farm.

We have not only discussed examples of existing use or the technical solutions that are possible, but also the storage capacity of generated energy, which is significant in the case of large fluctuations in wind energy supply.

In terms of household devices of relatively low power, vertical-axis wind turbines are more widely used. Their basic advantages are, among others, low production costs, simple construction, quiet operation and no need for alignment with wind direction.

Developments in wind turbine technology create opportunities to build more installations that transform wind energy into useful energy, mainly electricity. The concepts for system solutions outlined in the latter part of this chapter, acquiring energy from the wind in the form most useful to the consumer, are only a part of the potential solutions. The approach presented by the authors, in which the functional elements of the systems are separated, encourages the possibility of independent research in relation to each of these elements, and their combination to obtain the useful forms of energy desired. Division into functional elements also allows systems to be built that generate diversified forms of useful energy, for example, electrical and mechanical energy (e.g. pumping water for crop watering or air for water reservoir aeration), as well as storing surplus energy in different forms too.

REFERENCES

Abiven, C., Palma, J.L. & Brady, O. (2011) High-frequency field measurements and time-dependent computational modelling for wind turbine siting. *Journal of Wind Engineering and Industrial Aerodynamics*, 99, 123–129.

Adams, A.S. & Keith, D.W. (2013) Are global wind power resource estimates overstated? *Environmental Research Letters*, 8(1), 015021.

Adaramola, M.S. (2014) *Wind Turbine Technology: Principles and Design*. CRC Press, Boca Raton, FL.

Altan, B.D. & Atılgan, M. (2008) An experimental and numerical study on the improvement of the performance of Savonius wind rotor. *Energy Conversion and Management*, 49, 3425–3432.

Altan, B.D., Atılgan, M. & Özdamar, A. (2008) An experimental study on improvement of a Savonius rotor performance with curtaining. *Experimental Thermal and Fluid Science*, 32, 1673–1678.

Betz, A. (1966) *Introduction to the Theory of Flow Machines*. Translated by D.G. Randall. Pergamon Press, Oxford.

Bhutta, M.M.A., Hayat, N., Farooq, A.U. & Ali, Z. (2012) Vertical axis wind turbine – a review of various configurations and design techniques. *Renewable and Sustainable Energy Reviews*, 16, 1926–1939.

Carta, J.A., Ramírez, P. & Velázquez, S. (2009) A review of wind speed probability distributions used in wind energy analysis: case studies in the Canary Islands. *Renewable and Sustainable Energy Reviews*, 13(5), 933–955.

Cuerva, A. & Sanz-Andres, A. (2000) On sonic anemometer measurement theory. *Journal of Wind Engineering and Industrial Aerodynamics*, 88, 25–55.

Deglaire, P., Engblomb, S., Ågrena, O. & Bernho, H. (2009) Analytical solutions for a single blade in vertical axis turbine motion in two-dimensions. *European Journal of Mechanics B Fluids*, 28, 506–520.

Desert Power (n.d.) Introducing PacWind's revolutionary vertical axis wind turbines. Desert Power Inc., Palm Desert, CA. Available from: http://www.desertpowerinc.com/pacwind.htm [accessed March 2017].

Franković, B. & Vrsalović, I. (2001) New high profitable wind turbine. *Renewable Energy*, 24, 491–499.

Gipe, P. (2004) *Wind Power – Renewable Energy for Home, Farm, and Business*. Chelsea Green Publishing, White River Junction, VT.

Golfman, Y. (2012) *Hybrid Anisotropic Materials for Wind Power Turbine Blade*. CRC Press, Boca Raton, FL.

GUAL (n.d.) Info & Docs. GUAL Industrie, France. Available from: http://www.gual-statoeolien.com/English/cdromang.html [accessed March 2017].

Guidati, G., Ostertag, J. & Wagner, S. (2000) Prediction and reduction of wind turbine noise – an overview of research activities in Europe. *2000 ASME Wind Energy Symposium, 10–13 January 2000, Reno, NV*.

GWEC (2017) Global wind statistics 2016. Global Wind Energy Council. Available from: http://www.gwec.net/wp-content/uploads/vip/GWEC_PRstats2016_EN_WEB.pdf [accessed March 2017].

Hansen, M.O.L. (2015) *Aerodynamics of Wind Turbines*. Routledge, Abingdon, UK.

Hansen, M.O.L., Sørensen, J.N., Voutsinas, S., Sørensen, N. & Madsen, H.A. (2006) State of the art in wind turbine aerodynamics and aeroelasticity. *Progress in Aerospace Sciences*, 42(4), 285–330.

Harries, M. (2002) Disseminating wind pumps in rural Kenya – meeting rural water needs using locally manufactured wind pumps. *Energy Policy*, 30, 1087–1094.

Hau, E. & von Renouard, H. (2013) *Wind Turbines: Fundamentals, Technologies, Application, Economics*. Springer Science & Business Media, New York, NY.

Hill, D.R. (1991) Mechanical engineering in the medieval Near East. *Scientific American*, May 1991.

HomeEnergy (n.d.) Wind energy: Energy Ball V100; Energy Ball V200. Available from: https://www.homeenergy.nl/en-gb/about [accessed March 2017].

Inhabitat (n.d.) Helix wind turbine: small wind gets smart. Available from: http://inhabitat.com/helix-wind-turbine-small-wind-gets-smart/ [accessed March 2017].

Irabu, K. & Roy, J.N. (2007) Characteristics of wind power on Savonius rotor using a guide-box tunnel. *Experimental Thermal and Fluid Science*, 32, 580–586.

IRENA (2012) Wind power. *Renewable Energy Technologies Series*, Volume 1: *Power Sector*, Issue 5/5. Available from: https://www.irena.org/DocumentDownloads/Publications/RE_Technologies_Cost_Analysis-WIND_POWER.pdf [accessed March 2017].

Islam, M., Ting, D.S.K. & Fartaj, A. (2008) Aerodynamic models for Darrieus-type straight-bladed vertical axis wind turbines. *Renewable and Sustainable Energy Reviews*, 12, 1087–1109.

Kamoji, M.A., Kedare, S.B. & Prabhu, S.V. (2009) Performance tests on helical Savonius rotors. *Renewable Energy*, 34, 521–529.

Kishinamia, K., Taniguchi, H., Suzuki, J., Ibano, H., Kazunou, T. & Turuhami, M. (2005) Theoretical and experimental study on the aerodynamic characteristics of a horizontal axis wind turbine. *Energy*, 30, 2089–2100.

KOMEL (n.d.) Renewable energy sources. Available from: http://www.komel.katowice.pl/ENGLISH/produkcja3.html [accessed March 2017].

Kose, R., Ozgur, M.A., Erbas, O. & Tugcu, A. (2004) The analysis of wind data and wind energy potential in Kutahya, Turkey. *Renewable and Sustainable Energy Reviews*, 8, 277–288.

La Rotta, J. & Pinilla, A. (2007) Performance evaluation of a commercial positive displacement pump for wind-water pumping. *Renewable Energy*, 32, 1790–1804.

LAB-EL (n.d.) Laboratory Electronics, Reguly, Poland. Available from: http://www.label.pl/en/ix.dmet.html [accessed March 2017].

Lewandowski, W.M. (2007) Pro-ecological renewable energy sources [in Polish]. Wydawnictwa Naukowo-Techniczne, Warsaw, Poland.

Lucas, A. (2006) *Wind, Water, Work: Ancient and Medieval Milling Technology*. Brill Publishers, Leiden, The Netherlands.

Manwell, J.F., McGowan, J.G. & Rogers, A.L. (2002) *Wind Energy Explained – Theory, Design and Application*. John Wiley & Sons, Hoboken, NJ.

Matsushima, T., Takagi, S. & Muroyam, S. (2006) Characteristics of a highly efficient propeller type small wind turbine with a diffuser. *Renewable Energy*, 31, 1343–1354.

Mavroidis, I., Griffiths, R.F. & Hall, D.J. (2003) Field and wind tunnel investigations of plume dispersion around single surface obstacles. *Atmospheric Environment*, 37, 2903–2918.

Menet, J.-L. (2004) A double-step Savonius rotor for local production of electricity: A design study. *Renewable Energy*, 29, 1843–1862.

Miąskowski, W., Adamkowski, A., Nalepa, K., Pietkiewicz, P., Henke, A., Góralczyk, A. & Kaniecki, M. (2014) Microgeneration of energy from wind and water [in Polish]. Wydawnictwo Instytutu Maszyn Przepływowych PAN, Gdansk, Poland.

Morgan, E.C., Lackner, M., Vogel, R.M. & Baise, L.G. (2011) Probability distributions for offshore wind speeds. *Energy Conversion and Management*, 52(1), 15–26.

Nagai, B.M., Ameku, K. & Roy, J.N. (2009) Performance of a 3 kW wind turbine generator with variable pitch control system. *Applied Energy*, 86, 1774–1782.

Nałęcz, T. (2007) Fluid mechanics laboratory [in Polish]. Olsztyn, Poland.

Nalepa, K., Miąskowski, W., Pietkiewicz, P., Piechocki, J. & Bogacz, P. (2011) Small wind energy guide [in Polish]. WMAE, Olsztyn, Poland.

Nalepa, K., Miąskowski, W., Pietkiewicz, P., Neugebauer, M. & Wilamowska-Korsak, M. (2014) Storage, conditioning and conversion of energy from renewable sources [in Polish], Wydawnictwo Instytutu Maszyn Przepływowych PAN, Gdansk, Poland.

Ohyaa, Y., Karasudania, T., Sakurai, A., Abeb, K.-I. & Inoue, M. (2008) Development of a shrouded wind turbine with a flanged diffuser. *Journal of Wind Engineering and Industrial Aerodynamics*, 96, 524–539.

Palma, J.L., Castro, F.A., Ribeiro, L.F., Rodrigues, A.H. & Pinto, A.P. (2008) Linear and nonlinear models in wind resource assessment and wind turbine micro-siting in complex terrain. *Journal of Wind Engineering and Industrial Aerodynamics*, 96, 2308–2326.

Pawlak, Z. (n.d.) Poprzeczna dwustopniowa turbina przeplywowa [Transverse two-stage flow turbine]. Submitted to the Patent Office in 1985 – P251710. Available from: http://patenty.republika.pl/turbina/turbina.htm [accessed March 2017].

Quietrevolution (n.d.) qr5 – product information. Available from: http://www.quietrevolution.com [accessed March 2017].

Ropatec (n.d.) Ropatec wind turbines. Available from: http://www.ropatec.it [accessed March 2017].

Schaffarczyk, A.P. (2014) *Introduction to Wind Turbine Aerodynamics, 7, Green Energy and Technology.* Springer-Verlag, Heidelberg, Germany.

SWIND (n.d.) Male elektrownie wiatrowe [Small wind turbines]. Available from: http://www.swind.pl/index.php?link=male_elektrownie_wiatrowe [accessed March 2017].

TYPMAR (n.d.) Off grid 3 kW magnetic levitation wind turbine with lightning arrestor. Available from: http://www.cntimar.com/sale-7870086-off-grid-3kw-magnetic-levitation-wind-turbine-with-lightning-arrestor.html [accessed March 2017].

Valdès, L.-C. & Raniriharinosy, K. (2001) Low technical wind pumping of high efficiency. *Renewable Energy*, 24, 275–301.

Valdès, L.-C. & Ramamonjisoa, B. (2006) Optimised design and dimensioning of low-technology wind pumps. *Renewable Energy*, 31, 1391–1429.

van Bussel, G.J.W. & Mertens, S.M. (2005) Small wind turbines for the built environment. *EACWE4 – The Fourth European & African Conference on Wind Engineering; ITAM AS CR, 11–15 July 2005, Prague, Czech Republic.* Paper #210.

Wang, T. & Coton, F.N. (2001) A high resolution tower shadow model for downwind wind turbines. *Journal of Wind Engineering and Industrial Aerodynamics*, 89, 873–892.

Windpowermonthly (n.d.) The top 10 biggest wind turbines in the world. Available from: http://www.windpowermonthly.com/10-biggest-turbines [accessed March 2017].

Windside (n.d.) Windside vertical axis wind turbines. Available from: http://www.windside.com [accessed March 2017].

CHAPTER 13

Windmills for water pumping, irrigation and drainage

István Patay & Norbert Schrempf

13.1 INTRODUCTION

Wind energy is a clean and renewable energy source that has been used in agriculture for centuries. The origins of harnessing wind as energy began over 3000 years ago with the invention of the windmill. Windmills were the only mechanical energy sources that drove equipment and tools in many farms until the beginning of the 20th century. The main tasks of classical windmills were to produce grain, prepare feed and pump water at the farms. Windmills have functioned as electric power generators since the late 19th century, mainly for pumping water.

A renaissance in wind technologies has occurred over recent decades throughout the world. The basis of these new developments has been the use of modern materials, innovative and simple design, and high quality manufacture.

Nowadays, two main wind technologies are used in isolated and remote locations:

- Windmill technology for direct water pumping.
- Electric wind power technology for pumping and general use.

Direct wind pumping technology can be used successfully for water supply in farms, aquaculture, wild animal farming, irrigation and drainage. But the old type constructions have some disadvantages in comparison with electric wind pump technologies. For example, the suction conditions are limited in direct wind pumping equipment. Because of this, some types of water source cannot be used.

Electric wind power technologies are undergoing intensive development. The main reason for this is that the electric systems have flexibility, suitable control, and effective and low specific investigation and maintenance costs. Electric wind pumps are applicable for all uses of water supply, irrigation and drainage.

13.2 WIND TECHNOLOGIES FOR WATER PUMPING

The base of wind power technologies is the transformation of wind kinetic energy to mechanical driving energy by some kind of rotor or windmill. There are many types of rotor construction for pumping and electricity generation. The most effective type for a given application depends on the local wind conditions, the water source and the aim of the water pumping.

13.2.1 *Direct water pumping*

Direct water pumping means a general construction where the pump is driven directly by the wind rotor. This system gives the most simple energy transformation process: wind kinetic energy → mechanical energy → hydraulic energy of water. A typical arrangement and construction is shown in Figure 13.1.

The multi-bladed farm windmill was developed in the mid-1800s and was designed primarily for pumping water from wells. This type of wind pump is usually completed by some jack or

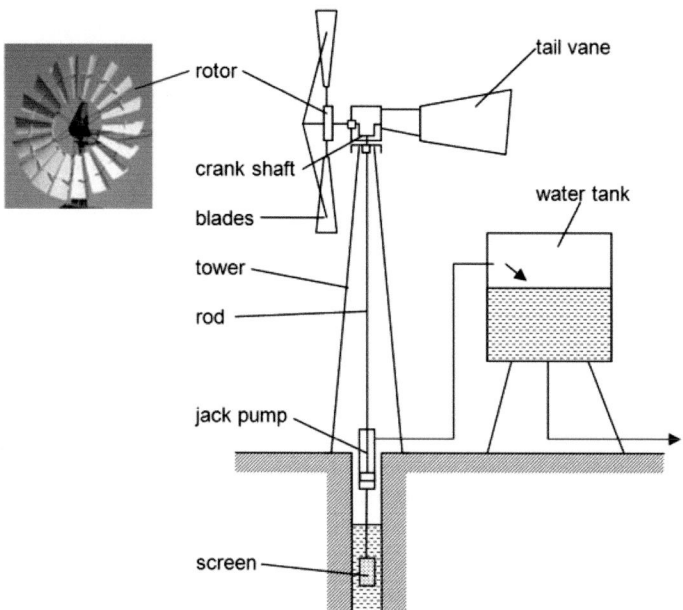

Figure 13.1.   Direct wind pumping system with multi-bladed rotor.

piston pump; the small-scale wind pumps often have a diaphragm pump. Both of types of pump need alternative movements, which are provided directly by the wind rotor.

Air flow through the rotor cross section causes a cascading effect, enhancing the drag and increasing rotor efficiency to 30%. Rotor diameter ranges from 2 to 6 m and units contain 16–24 curved-shaped metal blades. Some features of the multi-bladed rotors give advantages for pumping of water, including (Goswami, 1986):

- High starting torque.
- Simple design and construction.
- Simple control requirements.
- Durability.

The undesirable features are:

- Exerts high rotor drag loads on tower.
- Not readily adaptable to loads other than water pumping.

Another rotor type for the operation of water pumps is the Savonius rotor (or S-rotor, Fig. 13.2). This rotor was invented in the early 1920's and has received considerable attention because of its simple construction. The Savonius rotor also has a high starting torque, which is advantageous for pumping, but because of the low rotational speed and control problems, it is not well suited to electrical generation. Savonius rotors can be used for applications including not only pumping, but also other direct loads, for example, driving compressors or pond agitators. The wind compressors driven by Savonius can be applicable to airing the water at any aquacultural production.

13.2.2   *Characteristics of windmills*

The kinetic energy of the wind is determined by the speed of the air. The general equation for kinetic energy is:

$$E_{\mathrm{k}} = \frac{1}{2}mv^2 = \frac{1}{2}\rho V v^2 \quad [\mathrm{J}] \tag{13.1}$$

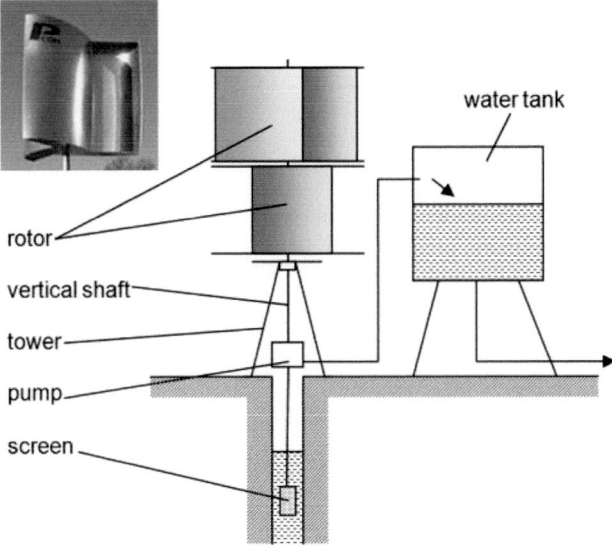

Figure 13.2.   Water pumping with a two-stage Savonius rotor.

where m is the moving mass [kg], v is the speed [m s^{-1}], ρ is the density [kg m^{-3}] and V is the volume [m^3]. If A is the cross section of the wind (perpendicular to the direction of the wind) and ρ_a is the density of the air, the kinetic energy stream (or the wind power) is:

$$P_w = \frac{1}{2}\rho_a A\, v^3 \quad [\text{W}] \tag{13.2}$$

If $A = 1$, Equation (13.2) gives the specific wind power:

$$P_{ws} = \frac{1}{2}\rho_a v^3 \quad [\text{W M}^{-2}] \tag{13.3}$$

Equation (13.3) is used as a basis to design windmills and their applications. From this formula, a strong dependency of the specific wind power on the wind speed is clear. While the wind speed changes over time, the wind relationship can be described with statistical methods. The wind speed distribution is the proportion of time, $t(v)$, for which the wind speed exceeds the value v. The weibull distribution formula is (Johansson *et al.*, 1993):

$$t(v) = \exp\left[\left(\frac{-v}{v_c}\right)^b\right] \tag{13.4}$$

The parameter v_c is the characteristic wind speed and b is the Weibull shape parameter. In many regions of the world, b is about 2; when $b = 2$, v_c typically exceeds the mean wind speed by about 10% (Johansson *et al.*, 1993). The wind speed relationship is usually known from meteorological data as an average wind speed (v_a). The local average wind speed can be given for any time period, but the annual average wind speed is most frequently used. Because wind is so variable, the mean energy density at a site is much higher than the energy density calculated with the average wind speed.

In practice, the specific wind power for $b = 2$ can be estimated using the formula:

$$P_{ws} = \frac{3}{\pi}\rho_a v_a^3 \quad [\text{W M}^{-2}] \tag{13.5}$$

Table 13.1.   Average values of Hellmann's exponent (Patay *et al.*, 2007).

Type of surface/roughness	α
Flat open field	0.12
Open area with some wind breaks	0.16
Flat woody area	0.28
Settlement with low buildings	0.35
Settlement with high buildings	0.50

The wind speed depends on the height of the ground surface. The connection between the height and wind speed (the vertical wind profile) depends on the roughness of the surface. If at any reference height (h_{ref}) the wind speed is known (v_{ref}), the wind speed at a different height, h, is:

$$v = v_{ref} \left(\frac{h}{h_{ref}} \right)^{\alpha} \quad [\text{m s}^{-1}] \tag{13.6}$$

where α is the exponent of height or Hellman's exponent. Equation (13.6) shows that the wind speed increases with height; the intensity of the increase is influenced by the roughness of the surface, as shown in Table 13.1. Because the exponent of height has a significant effect on energy production, the selection of the location for a wind pump is important.

The basic parameter of any wind rotor is the effective power. The ratio of the wind power and the effective power is the power coefficient, c_p:

$$c_p = \frac{P_e}{P_w} \tag{13.7}$$

The theoretical limit of c_p (c_{pmax}) is 0.593 (the Betz-limit); a higher value of the power coefficient is not possible. The effective power of a windmill can be calculated from Equation (13.7):

$$P_e = \frac{1}{2} c_p A \rho_a v^3 \quad [\text{W}] \tag{13.8}$$

The power coefficient is not a constant value; it depends on the load condition of the rotor (Manyonge *et al.*, 2012). The actual rotor speed is defined by the load; the power coefficient is a function of the ratio of the blade tip speed and the wind speed (Fig. 13.3). Both the multi-bladed and the Savonius rotors have relatively low c_p values, especially if the load is not optimal. This shows the importance of energetic agreement between the rotor and the pump.

The hydraulic power of wind pumps (P_H) can be calculated if the power coefficient of the rotor and the efficiency of the pumps (including the driving mechanisms efficiency), η_p, are known:

$$P_H = P_{WS} A c_P \eta_P = \frac{3}{\Pi} c_P \eta_P \rho_a A v_A^3 \quad [\text{W}] \tag{13.9}$$

Otherwise, the hydraulic power is:

$$P_H = Q g \rho_w H \quad [\text{W}] \tag{13.10}$$

where Q is the flow ratio [$\text{m}^3 \text{ s}^{-1}$], g is the acceleration due to gravity ($g = 9.81$ [m s^{-2}]), ρ_w is the density of water ($\rho_w = 10^3$ [kg m^{-3}]) and H is the total head of water [m]. Based on Equations (13.9) and (13.10) the theoretical Q-H equation of a wind pump is:

$$Q = \frac{3 c_p \eta_p \rho_a A v_a^3}{\pi g \rho_w H} \quad [\text{m}^3 \text{s}^{-1}] \tag{13.11}$$

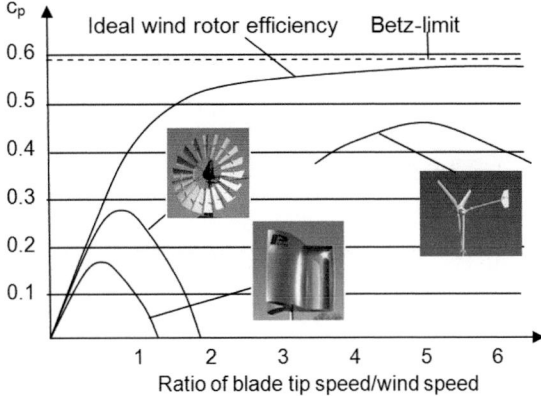

Figure 13.3.   The power coefficients of multi-bladed, Savonius and wing-type rotors.

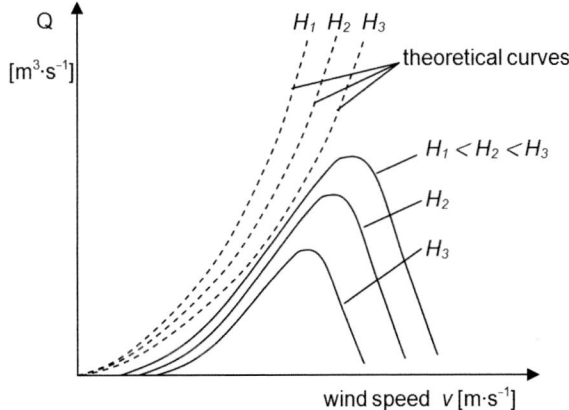

Figure 13.4.   Theoretical and real Q-H curves of wind pumps.

There are some parameters in Equation (13.11) that change during the operation of a wind pump, depending mainly on the wind speed. Because of this, the theoretical and the real Q-H curves show significant differences (Fig. 13.4).

Practically, H means the height of the water level in the tank. Because the height of the water tank is limited, the hydrostatic pressure determined by H is usually low. For this reason, the pressure demand of the irrigation system cannot be high. Therefore, the direct wind pumps are suitable for surface and small-scale micro-irrigation. The energy demands of irrigation technologies were discussed in Volume 2 of this book series (Chapter 12).

13.2.3   *Wind electric pumping*

Direct pumping, as used in classical windmills, has an important disadvantage: the pressure of the water supply is limited by the reservoir or water tank. Because of the low hydrostatic pressure, many applications are not possible with direct pumping techniques. In recent years, more and more manufacturers have made the switch from century old windmill technologies to wind power electric pumps. The most important difference between the windmill pump and the wind electric systems is that the water source and the power source (wind generator) can be at separate localities.

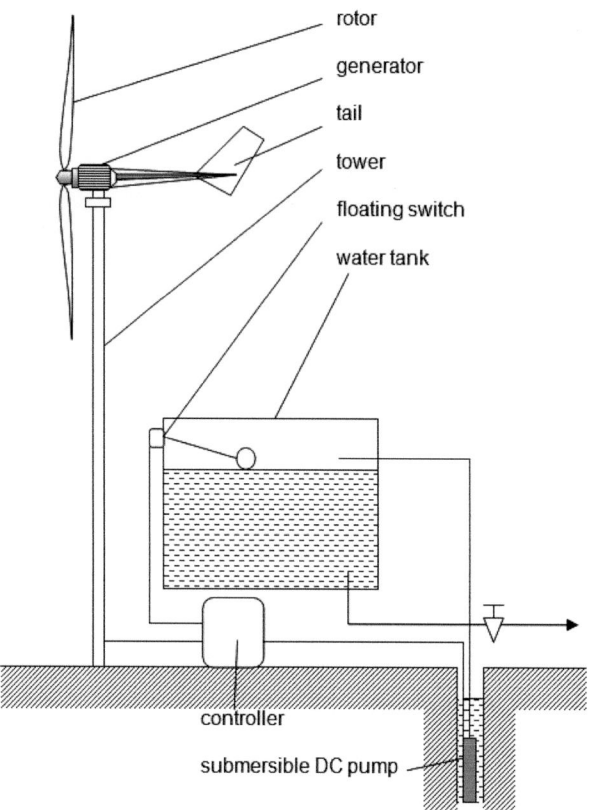

rotor

generator

tail

tower

floating switch

water tank

controller

submersible DC pump

Figure 13.5.   Direct connected wind electric pumping system.

In this way, the system efficiency can be higher, because of the favorable wind speed conditions for the wind generator at the place of installation.

There are two technologies for wind electric pumping:

- Direct pumping with a water tank.
- Indirect pump operation with a battery bank.

The direct pumping system (Fig. 13.5) incorporates a wind turbine and submersible pump that is directly connected via an interface control unit, which ensures optimum system performance and maximizes flow rates over all wind conditions, plus many other effective features. Because of the electric system, there are some advantages contrary to the mechanical wind pumps:

- Controller accepts input with wind charger.
- High wind speed electrical protection.
- Controller maximizes flow rates over all wind conditions.
- Pump status indicated by display.
- Pump or wall mount weather-proof controller enclosure.
- High efficiency proven windmill turbine.
- Built-in safety features.
- Reservoir float switch to increase pump life.
- Long life, durable and reliable components with minimum maintenance.

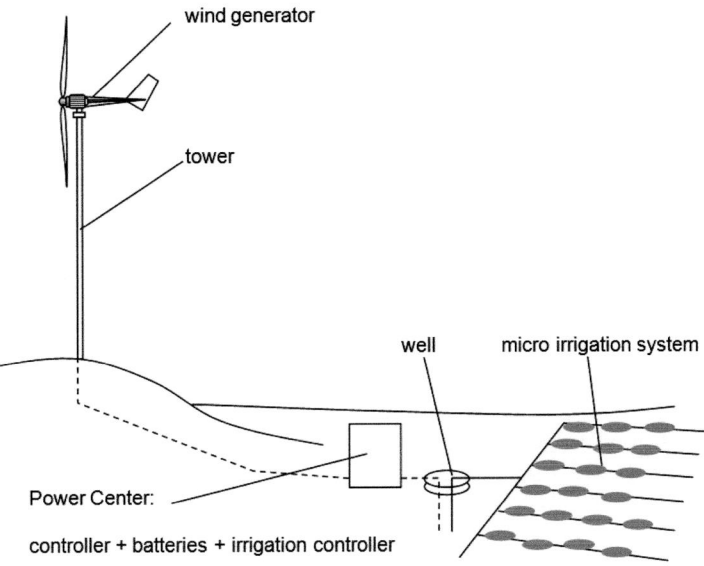

Figure 13.6. Wind electric micro-irrigation.

The high efficiency of the system is ensured by the wing-type rotor and the direct current (DC) pump. The power coefficient of the rotor is over 40% ($c_p = 0.4$ to 0.45) and the efficiency of the DC motor pump is over 90%. For this the system, efficiency can reach the favorable value of 35–40%.

The other solution for water pumping and irrigation is the wind pumping system with electric energy storage (Fig. 13.6). This system is similar to the solar photovoltaic (PV) pump described in Volume 2 of this book series (Chapter 12). The electric energy produced by the wind generator is stored in batteries. The process of charging is regulated by a controller, which ensures the optimal operation conditions of the DC generator at different wind speeds.

The produced electric energy can be calculated based on the nominal or peak power of the wind generator for a required pumping period. The actual effective electric power of the wind generator is:

$$P_e = \frac{1}{2} c_p \eta_g A \rho_a v^3 \quad [\text{W}] \tag{13.12}$$

where η_g is the efficiency of the generator. For a specific time period, the produced electric energy can be estimated using the practical value of capacity factor (c_f):

$$c_f = \frac{n_P}{n} \tag{13.13}$$

where n_P is the number of hours when the wind generator produces electric energy at the nominal power and n is the number of total hours in the time period. So c_f is a statistical parameter that can be determined from the production data. Average experimental c_f values for wind generators are included in Table 13.2. The capacity factor depends mainly on locality (wind speed conditions), but is also influenced by the rotor features.

The electric energy produced from a P_n [kW] power wind generator during a specific time period, characterized by the number of days (n_d) is:

$$E_e = 24 P_n c_f n_d \quad [\text{kWh}] \tag{13.14}$$

Table 13.2.   Capacity factor of wind generators (based on European Wind Energy Association (EWEA) reports and production data).

Type of locality	c_f
Seas and large water surfaces	0.29–0.35
Coastal land area	0.25–0.30
Continental area	0.15–0.25

The accuracy of the estimation of the energy production is dependent on the length of the time period: for a short time period the accuracy is low and increasing the time period will increase the accuracy.

The nominal power of the wind generator required depends on the energy demand of the pumping process. The hydraulic power of a pump is determined by the total pressure (p_t) or pressure head (H_t) and the flow rate (Q), as shown in Equation (13.10). If the volume of pumped water is V [m³] at H_t total head [m], the energy demand is:

$$E_p = \frac{1}{3.6 \times 10^6} V \rho_w \, g \, H_t \quad [\text{kWh}] \tag{13.15}$$

Because of the energy losses, the electric energy demand is:

$$E_{ed} = \frac{E_p}{\eta_s} = \frac{E_p}{\eta_{st}\eta_p} \quad [\text{kWh}] \tag{13.16}$$

where η_s is the system efficiency, η_{st} is the efficiency of the storage and η_p is the efficiency of the motor-pump unit. If the frequency of pumping is n_d [days], the nominal power of the wind generator, based on Equation (13.14), is:

$$P_n = \frac{E_{ed}}{24 \, c_f \, n_d} \quad [\text{kW}] \tag{13.17}$$

The nominal power of any wind generator is given by the manufacturer in W or kW.

13.2.4 *Solar-wind electric pump system*

Solar PV or wind electric pumping is usually suitable for irrigation in small-scale farming (<2 ha) at remote, off-grid conditions. By adding a solar PV array with a wind turbine and partitioning the large-scale irrigation system between a winter crop and a summer crop, the goal of a cost-competitive large-scale irrigation system powered by renewable energy may be attainable. Adding on-farm uses for the excess wind and solar energy during the winter months, to produce valuable products on the farm, enhances the prospects of a profitable system (Vick, 2010).

The integrated solar PV and wind electric power system uses solar and wind energy simultaneously, providing safety and increased pumping capacity. In this way, the energy production over a year can be kept relatively steady. The possibility of the energy supply for irrigation throughout the year provides advantages in crop production for many regions of the world.

The solar-wind electric system consists of a solar PV array and a wind generator (Fig. 13.7). The produced electric energy can be concentrated and collected in a common power center. The power center has multiple parts; usually, the main parts are: controller, rectifier, battery bank, dump load and a DC/alternating current (AC) inverter.

The wind generator produces AC, so a rectifier is needed between the generator and the battery bank to convert the AC to DC. The dump load controller, also known as a diversion load controller, diverts electricity from a battery bank once the charging source (wind turbine and PV) has fully charged the battery bank. An inverter is needed if the irrigation is based on AC or AC and DC pumping.

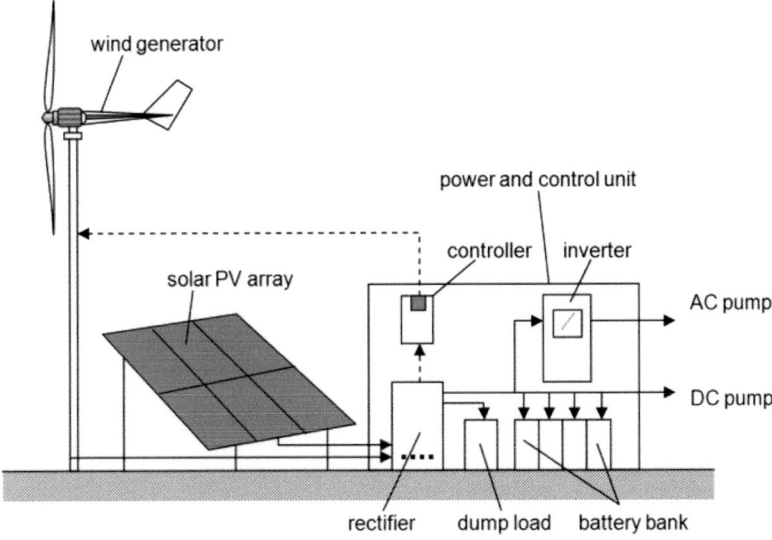

Figure 13.7.   Solar-wind electric power system for irrigation.

Development of the battery bank needs special consideration in every renewable energy system. There are three main types of batteries that are commonly used in renewable energy systems, each with their own advantages and disadvantages. Flooded or "wet" batteries are the most cost efficient and the most widely used batteries in PV applications. They require regular maintenance, need to be used in a vented location, and are extremely well suited for renewable energy applications. Sealed batteries come in two varieties: gel cells and absorbed glass mats (AGMs). The gel cell uses a silica additive in its electrolyte solution that causes it to stiffen or gel, eliminating some of the issues with venting and spillage. The AGM construction method suspends the electrolyte in close proximity with the plate's active material. These batteries are sealed, requiring virtually no maintenance. They are more suitable for remote applications, where regular maintenance is difficult, and enclosed locations, where venting is an issue. All three types of battery may be suitable for the solar-wind electric power system, if the selected batteries are deep cycle batteries (WholeSale Solar, 2014).

Solar-wind electric power systems for irrigation and drainage can be suggested for large-scale applications. With this technique, it is possible to operate nearly all irrigation equipment in arable lands (for example, even the large-scale equipment too) and in plantations.

13.3   APPLICATIONS

13.3.1   *Water supply of fish pond*

There is a fish pond with a $d_w = 2$ mm day^{-1} water deficit, because of the evaporation, in summer time. The evaporated water can be replenished from a well. Without connection to the electric grid, the pumping of water can be solved using a local, renewable energy source. The task is to select a windmill pump with the power required for the water supply. The surface of the pond, S, is 1 ha, the average wind speed during the pumping season at 10 m height, $v_{a(10)}$, is 4 m s^{-1}, and the Hellmann-exponent, α, is 0.2. The estimated power coefficient, c_p, based on Figure 13.3 is 0.2 (multi-bladed rotor) and the capacity factor, c_f, (Table 13.2) is 0.2. The total head, H_t, of needed is 10 m.

The daily water volume required is:

$$V_d = S\, d_w = 10^4 \times 2 = 20{,}000\ \text{dm}^3\ \text{day}^{-1} = 22\ \text{m}^3\ \text{day}^{-1} \tag{13.18}$$

The planned height of the windmill axis (the height of tower; H_w) is 20 m. The average wind speed at this height is:

$$v_{a(20)} = v_{a(10)} \left(\frac{H_w}{H_{\text{ref}}}\right)^{\alpha} = 4\left(\frac{20}{10}\right)^{0.2} = 4.6\ \text{m s}^{-1} \tag{13.19}$$

The specific wind power at the height of wind rotor axis is:

$$P_{ws} = \frac{3}{\pi} \cdot \rho_a \cdot v_{a(20)}^3 = \frac{3}{\pi} \times 1.23 \times 4.6^3 = 114\ \text{W m}^2 \tag{13.20}$$

The hydraulic power demand is:

$$P_h = \rho_w \cdot g \cdot H_t \cdot Q = 10^3 \times 9.81 \times 10 \times 2.3 \times 10^{-4} = 22.6\ \text{W} \tag{13.21}$$

where Q is the average flow rate, calculated from the daily water volume:

$$Q = \frac{V_d}{24 \times 3.6 \times 10^3} = \frac{20}{24 \times 3.6 \times 10^3} = 2.3 \times 10^4\ \text{m}^3\ \text{s}^{-1} \tag{13.22}$$

The minimal effective windmill power needed for the required pumping capacity is:

$$P_{n\min} = \frac{P_h}{c_p c_f \eta_p} = \frac{22.6}{0.2 \times 0.2 \times 0.8} = 565\ \text{W} \tag{13.23}$$

where $\eta_p = 0.8$, the pump efficiency. The required minimal rotor cross section can be calculated from the specific wind power and the effective windmill power:

$$A_{\min} \cong \frac{P_{n\min}}{P_{ws}} = \frac{565}{114} = 5\ \text{m}^2 \tag{13.24}$$

The minimal rotor diameter needed is:

$$D_{r\min} = \sqrt{\frac{4 \times A_{\min}}{\pi}} = \sqrt{\frac{4 \times 5}{\pi}} \cong 2.5\text{m} \tag{13.25}$$

13.3.2 *Water tank capacity*

A water tank capacity of $V = 50\ \text{m}^3$ for gravity irrigation of a plantation is given. A wind electric pumping system is needed to pump water from a well, as seen in Figure 13.5. The total pumping head, H_t, is 20 m and the average wind speed at the height of the planned wind generator, v_a, is 3.8 m s^{-1}. The average frequency of the irrigation process, n_d, is 5 days. To select the suitable wind electric pumping system, power and pump data are needed.

The average flow rate during the tank filling time is:

$$Q = \frac{V}{24 \times n_d} = \frac{50}{24 \times 5} = 0.42\ \text{m}^3\ \text{h}^{-1} \tag{13.26}$$

The hydraulic power demand is:

$$P_h = Q\, H_t\, \rho_w\, g = 0.42 \times (3.6 \times 10^{-3}) \times 20 \times 10^3 \times 9.81 = 23\ \text{W} \tag{13.27}$$

The electric energy demand of the pump is:

$$P_e = \frac{P_h}{\eta_p} = \frac{23}{0.8} = 29\ \text{W} \tag{13.28}$$

where the η_p at 0.8 indicates 80% pumping efficiency can be available.

If the three-bladed rotor type is selected, a power coefficient, c_p, of 0.35 and a capacity factor, c_f, of 0.18 can be estimated. The minimal rotor power required is:

$$P_{nmin} = \frac{P_e}{c_p \cdot c_f} = \frac{29}{0.35 \times 0.18} = 460\,\text{W} \tag{13.29}$$

REFERENCES

Goswami, D.Y. (ed.) (1986) *Alternative Energy in Agriculture*, Vol. II. CRC Press, Boca Raton, FL.

Johansson, T.B., Kelly, H., Reddy, A.K.N. & Williams, R.H. (eds) (1993) *Renewable Energy. Sources for Fuels and Electricity*. Island Press, Washington, DC.

Manyonge, A.W., Ochieng, R.M., Onyango, F.N. & Shichikha, J.M. (2012) Mathematical modelling of wind turbine in a wind energy conversion system: power coefficient analysis. *Applied Mathematical Sciences*, 6 (91), 4527–4536.

Omara, A.I., Sourell, H., Irps, H. & Sommer, C. (2004) Low-pressure irrigation system powered by wind energy. *Journal of Applied Irrigation Science*, 39(1), 83–91.

Patay, I., Tóth, L., Schrempf, N. & Fogarasi, L. (2007) Effect of the vertical wind profile upon the operation of wind power plant. *Hungarian Agricultural Engineering*, 20, 26–29.

Stetson, V.E. & Mecham, B.Q. (eds) (2012) *Irrigation*, 6th edn. Irrigation Association, Falls Church, VA.

Vick, B.D. Developing a hybrid solar/wind powered irrigation system for crops in the Great Plains. American Solar Energy Society. Available from: http://www.cprl.ars.usda.gov [accessed January 2014].

WholeSale Solar. Available from: http://www.wholesalesolar.com/batteries.html [accessed December 2016].

Yiasoumi, B. (2016) Selecting an irrigation pump, Agfact E5.8, 3rd edn. NSW Government. Available from: http://www.dpi.nsw.gov.au/agriculture/resources/water/irrigation/systems/pumps/selecting [accessed December 2016].

CHAPTER 14

Lakes as a heat source for heat pumps – a model study to determine the ecological impact of summer heat transfer

Renata Brzozowska, Maciej Neugebauer & Janusz Piechocki

14.1 INTRODUCTION

Lakes are geological forms that are formed due to several geological activities related to volcanism, tectonism, glaciation and erosion, and wind erosion. Lakes can also be formed during the evolution of river basins, sea coast changes and meteoritic impact. At the human life scale, they may seem a permanent form in the landscape, but at the geological scale lakes are an ephemeral form of water retention (Hutchinson, 1957; Lossow, 1996). The largest portion of the present lakes of the northern hemisphere is a remnant of the Pleistocene ice age (Choiński, 1995).

Natural water reservoirs are very valuable parts of the landscape. The water volume creates a special habitat for numerous plants and animals, which live around them in highly organized trophic nets. Especially in urban areas, people use lakes for many different purposes, for example, as a source of drinking water or for industrial use, for fishery and angling, and for recreational purposes (Brzozowska *et al.*, 2012). However, human impacts on lake ecosystems can be negative, mainly because of increasing external loading of nutrients into lake water, which significantly accelerates eutrophication processes. A particularly negative anthropogenic influence is connected to using lakes as raw wastewater receivers (Grochowska and Brzozowska, 2013; Klapper, 1991). This can have excessive eutrophication effects and can strongly change the whole lacustrine ecosystem, which is seen in, for example, mass algae production known as algal blooming, oxygen depletion, changes in biodiversity and a species-dominant structure.

A relatively new idea is the use of lakes by humans as a heat source for energetic purposes. This is an interesting idea; however, the indispensable condition is that it should have the minimum possible negative impact on the functioning of the whole lake ecosystem. The Water Framework Directive of the European Community (EC, 2000) maintains that water is one of the most valuable resources for mankind and every use of it should be conducted according to sustainable development principles. At the beginning of the 21st century, mankind was faced with two serious problems: (i) the limited availability of low-cost energy sources and (ii) environmental pollution. An alternative to solving these problems involves the search for low-cost and efficient energy sources that do not pollute the natural environment. From this perspective, one of the proposed solutions is geothermal systems with heat pumps, which draw heat from the ambient surroundings – referred to as the bottom heat source (e.g. ground, groundwater, surface water, etc.) – and transport it to the place of destination (e.g. for domestic heating purposes). In the summer, heat pump operation can be reversed and the system may be used for cooling (i.e. heat is drawn from the interior of the house and is stored in the bottom heat source).

It should be noted, however, that every heat exchange process has an impact on the environment, and intense heat uptake from a bottom source may have adverse consequences, such as the lowering of the ground-freezing boundary. The study described here involves a simulation to determine the environmental impact of the process of heat uptake and transfer when a lake is used as the heat reservoir.

One of the areas of human activity where heat is needed is agriculture. An additional problem with the availability of heat in agricultural production is its temporal and spatial specificity. Very often heat is needed in places far away from the big (industrial) heating services and agricultural heat demand during the year is very variable. The result is that the use of local heat sources in agricultural production may be more cost-effective than buying it from system manufacturers. In this sense, geothermal systems are very promising as a heat source for agriculture. Only electrical power needs to be added to control their actions, which is much easier to transmit over long distances or can be produced locally from generators.

The lower heat source can be the ground; however, in the case of agricultural production there are limitations in its appropriate use (even in the case of vertical systems). Another source of heat available for heat pump systems is surface water. This solution is particularly interesting because of the possibility of using portable heat exchangers for surface water. This provides easier portability of the system from place to place to meet needs and/or to alternate the location of the heat source, for example, from one water reservoir (lake, pond) to another in the event of an excessive reduction in temperature in one of the reservoirs.

14.2   HEAT PUMPS

14.2.1   *Theoretical basis of operation of the heat pump*

A heat pump is a device, the function of which is to extract heat from a source at a lower temperature (lower) and transfer it to a target at a higher temperature (upper). To make this achievable, it is necessary to provide the operating power for the heat pump, since the spontaneous flow of heat from a lower temperature body to a higher temperature body is not possible. The ratio of heat given back into the upper source to the energy needed to drive the heat pump is always greater than one. The efficiency of an ideal heat pump depends on the temperature difference of the two heat sources. The smaller it is, the more efficiently the pump is working (Zawadzki, 2003). Heat pumps can be divided into three types:

- Compressor heat pump
- Thermoelectric heat pump
- Absorption heat pump.

Heat pumps belonging to the first group are the most commonly used in space heating (Kavanaugh and Rafferty, 2014; Petit and Collins, 2011; Rubik, 2011).

14.2.2   *Construction of compressor heat pumps*

The main elements of the most popular heat pump compressor systems are:

- Evaporator
- Condenser
- Compressor
- Expansion valve.

The purpose of the evaporator is to receive heat from a low-temperature source and to transfer it to the medium. The essence of the proper operation of the evaporator is to provide such hydraulic conditions to the secondary evaporator (boiling refrigerant) as to get the nucleate boiling and enable the free flow of vapor.

Vaporizers are divided according to the following criteria (Rubik, 2011):

- The type of construction:
  - pipe
  - shell-and-tube
  - plate
  - spiral;

- The type of working fluid in the primary (medium heat):
  - o water and aqueous solutions
  - o air.

The condenser pressurizes the working fluid (the temperature of working fluid is rising) and it is transferred to the upper heat source. Condensers are divided according to the following criteria (Rubik, 2011):

- The type of construction:
  - o shell-and-tube
  - o coil-pipe
  - o minimum reserves
  - o plate
  - o counter
  - o spiral;
- The type of media to be heated in the upper heat source:
  - o air (heating in air)
  - o water (in heating systems or hot water).

Compressors are machines which are the means for driving the power supply, transferring heat from the source at a lower temperature to a higher temperature target. Compressor operation involves the suction of the working medium, which is in the gas phase, having been previously vaporized in the evaporator, compressing it and then raising its temperature.

The most commonly used compressors in heat pumps are positive displacement compressors and flow (turbo) compressors. Among the first of these groups, we can distinguish (Rubik, 2011):

- Reciprocating compressors
- Rotary compressors
- Scroll compressors
- Screw compressors.

The expansion valve is designed to expand and adjust the flow rate at which the working medium hits the condenser to the evaporator (Rubik, 2011). The following expansion valves are used most frequently:

- Electronic
- Thermostatic
- Automatic (pressostatic).

In addition to the elements of the heat pump compressor systems described above, there are many others, for example, thermostats, pressure switches, filters, oil separators, tanks of liquid refrigerant, safety valves, shut-off and feedback and microprocessor controllers (Rubik, 2011).

14.2.3   *The ideal thermodynamic cycle of the heat pump compressor*

The heat pump extracts heat from a low-temperature source and transmits it to a high-temperature target. This process occurs through the involvement of electricity supplied to the compressor from an external source.

Implementation of the heat transport is realized by the existence of three circuits:

- Thermodynamic cycle
- The upper heat source circuit
- Brine heat.

The closed-loop thermodynamic process generally occurs by changing the physical state of the thermodynamic factors of evaporation, compression and condensation by expansion (circuit Linde; see Fig. 14.1).

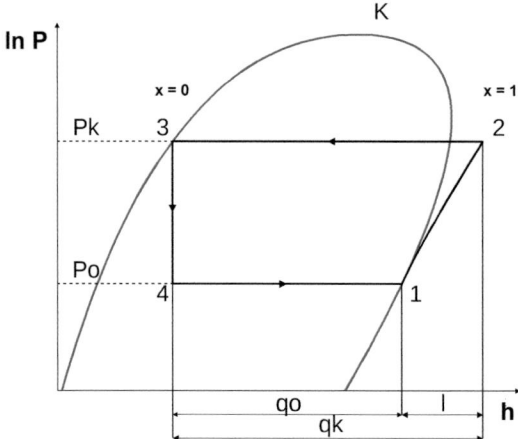

Figure 14.1.   Graph of the thermodynamic transformation cycle of the Linde system. 1–2: the process of compression; 2–3: liquefaction process; 3–4: expansion process; 4–1: pairing process; qk: unit thermal power given by the factor in the condenser; qo: due refrigeration unit; l: unit labor compression factor.

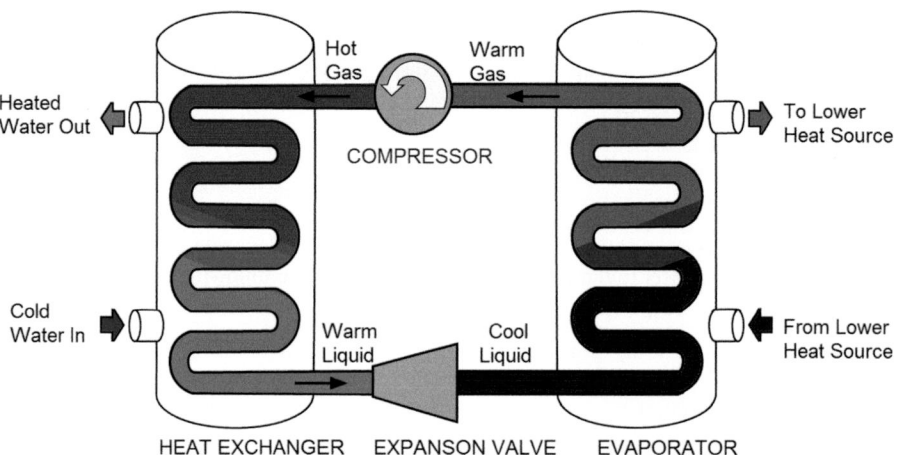

Figure 14.2.   Thermodynamic cycle in the compressor heat pump (Zawadzki, 2003).

The connecting element between the thermodynamic cycles of circulation of the heat source is the evaporator. The condenser connects an evaporator with the upper heat source.

The lower circuit is designed to transfer heat from the low-temperature heat source to the evaporator. As a result, the working fluid is vaporized in a power cycle circulating the liquid. Then a part of the working medium is sucked into and compressed in the compressor, thereby increasing its temperature. Hot steam under high pressure goes to the condenser where, due to the temperature difference, it is condensed and gives a substantial part of its heat to the upper heat source circuit (e.g. water heating). The thermodynamic medium, already in the liquid phase (even though it is still under high pressure), reaches the expansion valve. Here, it is decompressed, thereby decreasing its temperature rapidly. Then the cooled working fluid goes to the low-pressure part of the system, and again the evaporator absorbs heat from the heat source (Fig. 14.2). In this way, the cyclic process of transferring heat from a lower to a higher temperature level occurs (Zawadzki, 2003).

14.2.4   *Working media*

The working medium, also called the thermodynamic medium, circulates in the heat pump and mediates the transport of heat from the source at a lower temperature to that at a higher temperature. The working media in heat pump installations are refrigerants that should meet the additional requirements arising from the specific action of the particular heat pump and should (Rubik, 2011):

- Be chemically stable in the operating temperature range, and chemically inert relative to the heat pump's construction materials.
- Be non-flammable, non-explosive and non-toxic.
- Have the lowest possible condensing pressure.
- Have a high volumetric efficiency heating, which indirectly allows reduction of the size of the compressor.
- Not have a negative effect on the environment.
- Not show any chemical activity with respect to the construction materials or lubricating oil.

Working media currently used in heat pumps are (Maina and Huan, 2013; Rubik, 2011):

- *Chlorofluorocarbons* (CFCs): carbon compounds in which all hydrogen atoms in the molecule have been replaced by chlorine and fluorine atoms. They are stable compounds, but decompose the ozone layer and are dangerous for the environment.
- *Hydrochlorofluorocarbons* (HCFCs): carbon compounds where not all hydrogen atoms have been replaced by chlorine and fluorine atoms. Decompose in the lower layers of the atmosphere.
- *Hydrobromofluorocarbons* (HBFCs): compounds in which hydrogen atoms have been partly replaced by fluorine and bromine. Pose a significant threat to the environment.
- *Hydrofluorocarbons* (HFCs): organic compounds, in which some hydrogen atoms are replaced by fluorine atoms (no information about the degree of harm).
- *Fluorocarbons* (FCs): in these compounds, the hydrogen atoms have been replaced with fluorine. They are not harmful to the ozone layer.
- *Saturated hydrocarbons* (HCs).

14.2.5   *Lower heat source*

The source of low-temperature heat needed to evaporate the working media or fluid should have the following characteristics (Rubik, 2011):

- Easy availability and low cost of construction of the heat pump system.
- High thermal capacity.
- Lack of pollution-causing corrosion or deposits on parts of the installation.
- Ideally, able to maintain a high and constant temperature.

Basically, the low-temperature heat sources are divided into (Rubik, 2011):

- The waste heat:
  - air and gases
  - sewage
  - return water in heating systems;
- Renewable energy:
  - atmospheric air
  - ground
  - solar radiation
  - surface water
  - groundwater.

14.2.5.1   *Atmospheric air*

The easiest available source of low-temperature heat is atmospheric air, which is often used in heat pump installations. Because air has a low coefficient of heat transfer and high temperature fluctuation both daily and annually, it should be applied only to power evaporator heat pumps with low-to-medium power.

Since air has a low heat capacity, a large amount of air has to be forced through the evaporator, which involves the need for a fan. As a result, this increases the energy required to power the heat pump system as well as the level of noise.

In a situation where the ambient air temperature falls and reduces the efficiency of the heat pump, while the demand for the thermal power needed to heat the premises increases, the installation may not provide adequate comfort. To alleviate this, sometimes dual source systems are applied (bivalent or multi-source) (Rubik, 2011).

14.2.5.2   *Solar energy*

Very often, for domestic hot water and central heating uses, solar radiation is utilized as a low-temperature heat source. The heat needed to power the heat pump evaporator is transported by a carrier from a solar collector. The efficiency of such systems depends on several meteorological factors (Rubik, 2011).

14.2.5.3   *Waste heat*

Waste heat, which is derived from wastewater, exhaust air ventilation or return network water, is rarely used in Poland or anywhere else in the world as a low-temperature heat source. This solution, used in ventilation system or plumbing, is unsuitable for small houses. This heat source is most often used in industrial plants and large farms (Rubik, 2011). According to Forsen (2005), more than 90% of the heat pumps installed in the world are in private houses. Waste heat is regarded as a preferred heat source (Chua *et al.*, 2010; Hoffmann, 2015); its use with heat pumps allows the recovery of otherwise unused and wasted heat from production processes. However, it is used very rarely, which may be due to high installation costs arising from the lack of uniform technical solutions (Forsen, 2005).

14.2.5.4   *Ground*

The ground is used as a lower source for heat pumps with relatively low heating power. Heat absorption from the ground to power an evaporator is generally done through a secondary heat transfer medium, which flows through the so-called exchanger (collector) ground forced circulation pump. The carrier is typically water, aqueous salt solutions (brine) or aqueous glycol solutions.

Heat accumulates in the soil to a depth of approximately 10 m and at this depth the temperature is equal to the average annual air temperature. However, in Poland, average atmospheric conditions are approximately 7–8°C and placing horizontal ground heat exchangers at these depths to harvest such temperatures is not economically justified, so they are put at a depth of 1 to 2 m instead. At this level, the ground temperature varies sinusoidally in a year and is 11–17°C in the summer and 1–5°C in the winter.

The big advantage of the land as a low-temperature heat source is the ability to create an installation in conjunction with solar collectors.

On the basis of the way they are installed, ground collectors can be divided into (Fig. 14.3):

- Horizontal
- Spiral
- Vertical (borehole heat exchangers).

The selection of the type of collector for a heat pump system depends mainly on the prevailing local conditions, such as soil type, groundwater level, land area available, and availability of land area (Rubik, 2011).

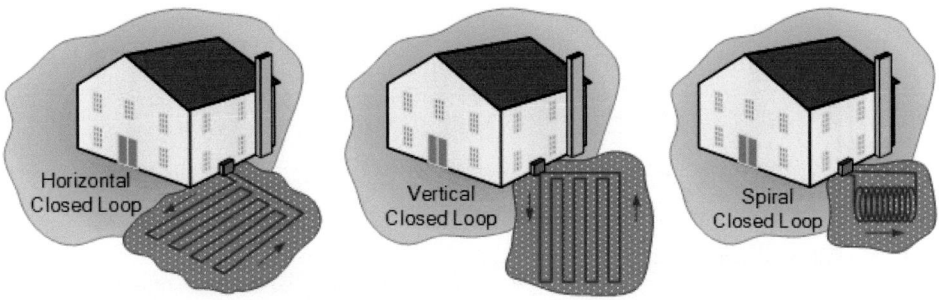

Figure 14.3.   Ground heat collectors (Lund, 2002).

Figure 14.4.   Preparing a loop on the reservoir shore (http://strandlund.com/wp-content/uploads/2014/01/pic04-full1.jpg).

14.2.5.5   *Water*

Groundwater is low-temperature heat source that is very easily accessible. Its main advantage is the relatively small temperature variation during the year. In Poland, the typical temperature of the groundwater is 5–12°C.

Groundwater can be fed directly into an evaporator, or an indirect heat exchanger can be used if there is a high degree of salinity. The disadvantage is a relatively high investment cost.

It is usually necessary to build two wells: one to collect water to supply the evaporator, and the second a discharge well (absorbent) to drain the water back into the ground. There are also solutions that use a single pit (Rubik, 2011).

Two types of heat pump system are designed to use surface water as a heat source: closed-loop systems and open-flow systems.

14.2.5.5.1   *Closed-loop system*

A closed system submerged in water can be less expensive and more effective than any other closed loop (geosourceone.com). It does not require as much work as vertical or horizontal earth loops, because there is no need to create trenches or holes to hide the pipe, and consequently it has a significant reduction in installation costs (earthenergysystems.com). Examples of execution of the installation are shown in Figures 14.4 to 14.6.

Figure 14.5. Whole loop towed to the center of the reservoir (http://en.wikipedia.org/wiki/Geothermal_
heat_pump#/media/File:Pond_Loop_Being_Sunk.jpg).

Figure 14.6. Reservoir after flooding/sinking of the loop and completion of works (http://www.
earthenergysystems.com/Silver_Rose_pond_-_filled.jpg).

The principal requirement for ponds or lakes is that the water reservoir must have a minimum depth of 3 m. During droughts, the water level should not fall below the minimum of 3 m of depth. The area of the reservoir must usually be at least 1.5 times greater than the floor area of the building (house) to be heated (Fig. 14.7) (www.geoconn.com/7.html).

Another limitation is the distance from the building to the water reservoir – the smaller the better. In practice, it is assumed that this distance should not be greater than 70 m.

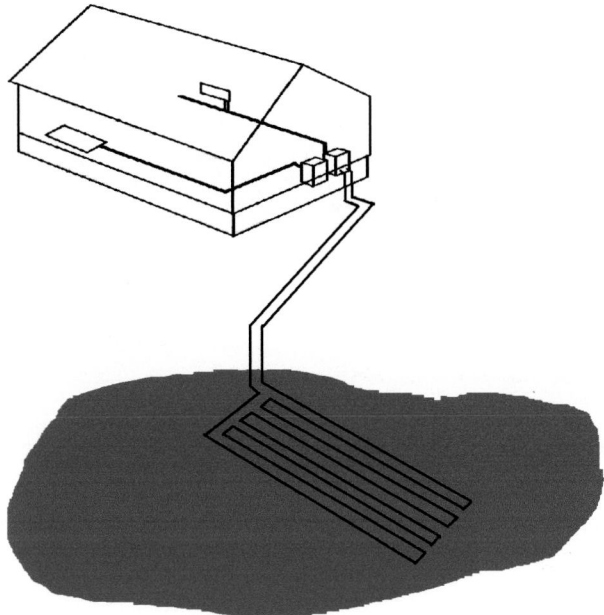

Figure 14.7.   Comparison of the surface area of the house to the water reservoir (Swedish Heat Pump Association (Svenska Varmepumpfbreningen, SVEP), www.geoconn.com/7.html).

It is difficult to place the loop in a river, because the water there is in constant motion. Rivers are also more vulnerable to flooding and drought, and changes in levels and flow rates. The loops are put in place using concrete anchors, which hold the pipe at the bottom. In addition, an exhaust must be directed oppositely to the flow of the river. The pipeline is protected by mesh fencing around the installation, which protects the coil from damage by material carried downstream.

Loops are immersed more efficiently than an open system, producing a more constant temperature. The principle of operation is that a closed loop is immersed such that fluid flowing in the tubes absorbs heat from the water, and delivers it to the heat pump. The fluid used in a closed loop must be 100% environmentally friendly and refrigerant R-410A is commonly used as a heat carrier. Pipelines are constructed mostly of PVC material with a diameter of $3/4''$. To serve their purpose, loops of pipelines must be an average of 100 to 150 m in length. Every loop must be at a distance of at least 0.1 m from each other, and they are kept in this position by means of nylon cables.

14.2.5.5.2   Open-flow systems

Another heat pump system that uses surface water as a heat source is the open-flow system. Under ideal conditions, an open loop may be the most economical system. It can be used for commercial applications or home use in detached houses. Surface water is sucked up by a pump and forced through the evaporator of the heat pump. Taken mass of flowing water is relatively big because of small decrease in water temperature, amounting to 4–5°C, and the unit amount of heat obtained is 4500–5900 Wh m^{-3}. Schemes of surface water intakes to power heat pump evaporators are shown in Figures 14.8 to 14.10. An open system can be used for any type of surface water. Using this system in a lake – if it contains a hypolimnion layer – can draw water from both the bottom and water discharge situated near the surface of the lake. Such a solution will stimulate the circulation of water in the lake and contribute to better oxygenation of the water from the hypolimnion layer.

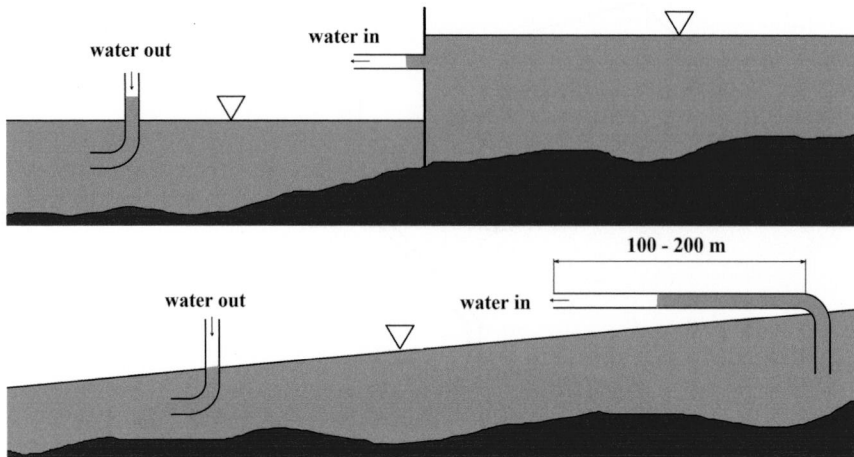

Figure 14.8.   A schematic diagram of a surface water intake showing the intake of river water without a pump (Rubik, 1999).

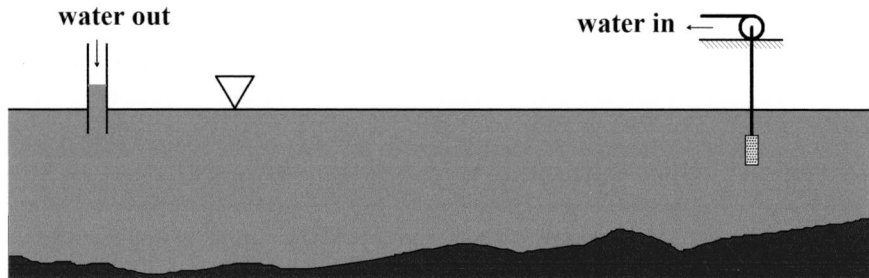

Figure 14.9.   A schematic diagram of a surface water intake involving the pumping of river water (Rubik, 1999).

At the same time, the open system is especially suitable for use with rivers (as a heat source). Because the temperature of the water in the river is shaped by heat exchange with the environment (air), these installations can be placed at many sites along the course of a river (Rubik, 2011).

The disadvantage of surface water as a heat source is power consumption during periods of low temperatures and with minimal flows. Due to the potential contamination of surface waters, the choice of the system should be properly tested for contaminants. As a rule, it is necessary to use indirect heat and appropriate filter systems, and this reduces the energy efficiency of the heat pump and increases investment costs (Rubik, 2011). It is also possible that the water will not be suitable for an open system if the compounds contained in it could lead to rapid corrosion and damage to the heat pump.

Figure 14.11 shows the energy change in the efficiency of heat pumps as a function of the temperature of the heat source for the four types of heat pump operating systems (lower source–upper circuit): air-to-water, air-to-air, water-to-air, and water-to-water. The figure shows not only the increase in the efficiency of a heat pump with an increasing heat-source temperature, but also provides a comparison of these four heat pump types. Water-to-water systems and water-to-air systems have an efficiency that is more than 6% higher than the other two systems.

Figure 14.10.   Intake of surface water and groundwater drain (www.geoconn.com).

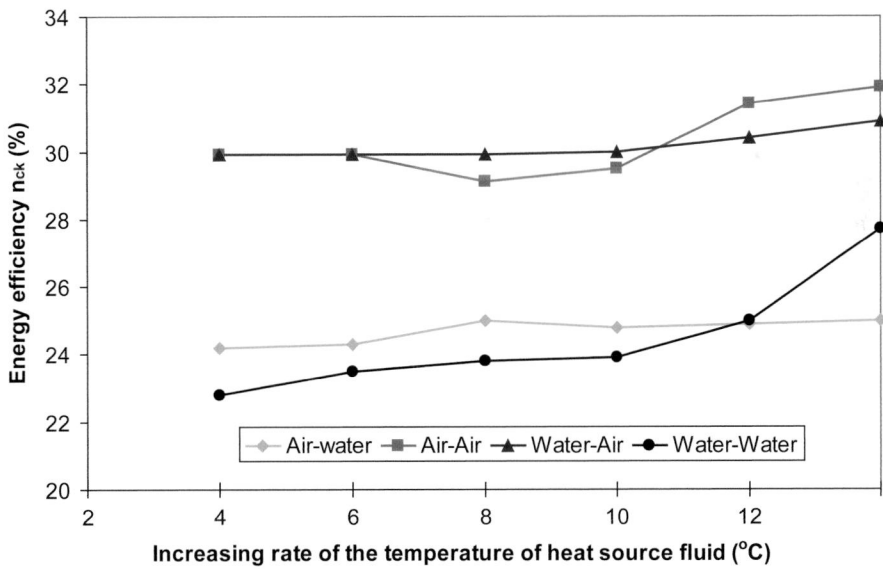

Figure 14.11.   Energy efficiency of heat pumps versus heat-source temperature (Çakır *et al.*, 2013).

14.2.6   *Examples of heat pump installations that use surface water as a heat source*

14.2.6.1   *Using heat pumps for agriculture*

At the outset of this chapter, we described how, regardless of the lower heat source selected, heat pumps can be used in agriculture and their functionality is the same. However, due to the varying costs of installation, the profitability of the installations may differ. Nevertheless, their application (as in other situations) depends on several factors (Kurpaska, 2011), which include:

- Situations in which heat needs to be transferred over long distances; it might prove more economical to generate heat at the place of demand, rather than to transfer it.
- The availability of a low-temperature heat source, to act as the lower source.

Both of these factors may be significant in agriculture. However, in addition to the conventional heat sources described in Section 14.1, there are other heat source possibilities present in agriculture:

- Heat from products being removed (e.g. milk that must be cooled immediately after milking).
- Animal feces.
- Ventilation air – both from livestock buildings and from, for example, greenhouses.
- Air from dryers (Kurpaska, 2011).

Heat recovered in this way can be used conventionally for heating residential areas for people, or for providing hot water for domestic use (Hepbasli and Kalinci, 2009), but also for:

- Heating and drying air before introduction to a dryer (Goh *et al.*, 2011).
- Heating ventilation air.
- Heating substrates (soil) for vegetable production.

Analyzing the literature, it can be concluded that heat pumps in agriculture are mostly used conventionally, for heating rooms. For example, Ozgener (2010) describes a system using wind turbines, solar systems and heat pumps operated in combination for heating needs. Such an application of heat pumps in agriculture does not differ from conventional uses of heat pumps for heating single-family houses or larger commercial buildings.

Another group of heat pump applications, related more closely to agriculture, is the use for drying (Colak and Hepbasli, 2009a, 2009b), or as an element supporting the drying process. Examples of applications can be found throughout the worldwide literature. For example, in the work of Patel and Kar (2012), solutions are presented that use heat pumps for supporting other drying methods:

- Heat pumps and solar energy use.
- Heat pumps and drying with infrared radiation.
- Heat pumps and drying with radio waves.
- Heat pumps and drying with microwaves.
- Heat pumps and drying in a fluidized bed.
- Drying with the use of chemical heat pumps.

14.2.6.2 *Examples of the use of heat pumps in agriculture using surface water as a heat source*
Surface water as a heat source for heat pump systems in agriculture, as in most applications of heat pumps, is rarely used (Forsen, 2005). The advantage of using heat pumps, particularly in agriculture, is that they can be used when remote from conventional heat sources. On the other hand, surface water, if present in the vicinity of the farm (or other place of heating demand), is one of the most favorable sources for agricultural applications. This is because they do not need investment in earthworks for their application. All ground-based heat exchangers are limited by the available farm size and production and also lowers the ground's average annual temperature, which has a negative effect on plant growth. These problems do not occur if surface water is used instead. According to Kukkonen (2000), by the 1990s, 20% of heat pumps in Finland used a lake as a heat source. Later installations have mainly used vertical wells (http://www.ehpa.org/market-data/2013/?eID=dam_frontend_push&docID=2224).

When using surface water (lakes, rivers, etc.) as the lower heat source, the heat pump systems can be either closed or open systems; however, in the latter case it is necessary to use special filters that will purify the water taken. On the other hand, closed systems need to be careful to avoid the

Figure 14.12.   Special heat exchanger for lakes (http://www.limnion.com/pdfs/HLLM_0912_1_print_rev. pdf).

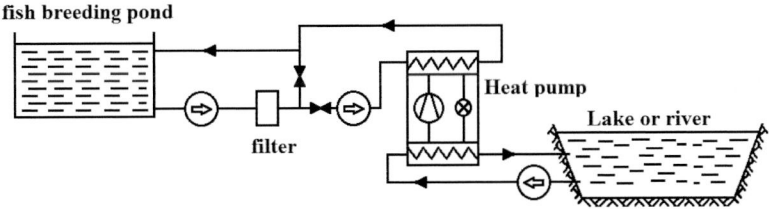

Figure 14.13.   The use of heat pumps for heating ponds for thermophilic fish (http://www.modr. mazowsze.pl/42-ekologia/ekologia-energetyka-odnawialna/1330-zastosowanie-pomp-ciepl a-w-rolnictwie).

heat exchanger immersing itself in bottom sediment because it will change the water circulation around the hose and thus heat transfer. The solution may be to use special heat exchangers (see, for example, Fig. 14.12), which can be combined with each other to intensify heat transfer. Their advantage is that they work at a certain height above the bottom of the lake.

One example of the use of surface water as a heat source in agriculture could be the heating of fishponds. Such ponds are usually located near to natural water reservoirs, which can also be used as a heat source (Fig. 14.13). Such a solution has been described by the Agricultural Advisory Centre in Mazovia, a province in Poland. The use of heat pumps allows the maintenance of a constant temperature of 22°C in the fishpond. This allows the growth of heat-loving fish regardless of the weather conditions or amount of sunlight.

A similar solution was applied in New Zealand, where ponds for freshwater shrimp are heated to a temperature of 24°C. The heat source is wastewater from the Wairakei power plant and from the nearby river Waikato (this water has a temperature of 10°C). The solution enables production of more than 16 tons of shrimp per year (Thain *et al.*, 2006). The use of heat pumps for heating ponds can also enable, in a hot period, cooling of the water in the ponds to maintain a constant temperature (Baird *et al.*, 1993).

These systems, though still rarely used, are increasingly being offered to customers because of their advantages, including reduction of CO_2 emissions into the atmosphere (http://www.heatpumpsuppliers.com/aquaculture-heat-pumps.html).

In many countries of Europe and throughout the world, heating greenhouses has become standard practice in the last 25 years. Due to the rising prices of traditional energy sources, customers are increasingly advised to heat greenhouses using heat pumps. Such an environmentally friendly solution is also recommended by the UN's Food and Agriculture Organization (FAO), in particular for developing countries (Nguyen *et al.*, 2015). But, in Europe too, it is advisable to move away from oil or gas heating (the most common) to renewable sources of energy, including heat pumps (Garcia *et al.*, 2012). Hunt *et al.* (2013) provide a description of a program that calculates the cost of heating greenhouses using different energy sources. A description of several such installation scenarios is given in Badgery-Parker (2013). These calculations show that when the heating system of greenhouses will be changed from liquified petroleum gas (LPG) to the heat pump system, the costs of such investment will be returned in less than two years, and the degree of cost-effectiveness is 116%. The results of these calculations and the examples show that the use of heat pumps for heating greenhouses has become an increasingly important alternative. Research (Mugnozza *et al.*, 2012) into the heating of greenhouses using heat pumps and photovoltaic systems shows that this solution saves 40% energy compared to traditional heating systems of greenhouses, and it does not emit CO_2.

According to the research of Chiasson (2005), heating greenhouses is more cost-effective using an open system (in this work the water source was groundwater). For a closed system, the cost was calculated at 0.53 US\$ m^{-3} and for the open system was 0.35 US\$ m^{-3}.

Heat pumps in agricultural production can also be used in gardening. Hyun *et al.* (2014) analyzed the amount of potential heat available for use in large-scale horticultural farms. They assumed that the all estimated amount of heat available in South Korea from the surface water is equal to 3140 TJ. These calculations show the potential of this source of heat, in an effort to encourage investors to use it.

Finally, it should be made clear once again that the use of surface water in different countries may be subject to different rules and regulations. In each case, the construction and use of heat removal systems (both open and closed) requires an analysis of the legislation in force in the respective country.

14.2.6.3   *Stockholm, Sweden*

One of the biggest seawater heat pump systems is installed and connected to the municipal district heating in Stockholm. Between 1984 and 1986, six Unitop® 50FY heat pumps were commissioned by Friotherm AG (Figs. 14.14 and 14.15). Unfortunately, there is no precise information about the profitability of this investment, but the typical payback time of such installations is less than five years. The heat pumps provide a heat output of 180 MW and contributed significantly to the growth of renewable energy in the energy balance of the city.

Originally, all the aggregates used refrigerant R22 as their working fluid. To prevent the leakage of refrigerant to the atmosphere all units are equipped with seal oil systems. Retrofit of the heat pumps to refrigerant R134a as a working medium started in 2003.

In this system, seawater is supplied at a rate of 2.5 m^3 s^{-1} to the falling film evaporator, where heat is extracted by lowering the water's temperature by 2°C. This allows about 22 MW_t of low-temperature heat to be obtained in each of the six installed pumps. To upgrade this heat to 80°C, each heat pump needs 8 MW_e of electrical power (this power is also transferred to the district heating system). In summer, the seawater is taken from the surface, and in the winter from a depth of 15 m where there is a constant temperature of 3°C. The specially designed evaporator allows the transfer of heat from a high rate of water flow to the working fluid with a very low temperature gradient. The working medium evaporates at a rate of 110 kg s^{-1} at a constant temperature of −3°C. It is then compressed and condenses at a temperature of 82°C, releasing heat from the heat

Figure 14.14.   Friotherm Unitop® 50FY heat pump (http://www.friotherm.com/webautor-data/41/vaertan_e008_uk.pdf).

Figure 14.15.   Friotherm Uniturbo® 50FY 2-stage compressor with integrated lubricating oil tank (http://www.friotherm.com/webautor-data/41/vaertan_e008_uk.pdf).

source along with the energy used to drive the compressor. In the condenser, the total energy of 30 MW is transferred from each of the six heat pumps to the district heating water, increasing its temperature from 57 to 80°C.

14.2.6.4   *Xinghai, China*
Another example is in China. In Xinghai, the heating and air conditioning system was upgraded using Friotherm's Unitop heat pumps (Friotherm AG, Frauenfeld, Switzerland). The system is connected to office buildings, shopping centers, recreation areas, and indoor swimming pools.

Since the beginning of 2007, three Unitop 33/28 heat pumps (Fig. 14.16) have provided heat in the winter and cooling in the summer for an area of almost 300,000 m². The heat pump plant is located in the basement of a commercial two-story building. Heat is extracted from treated sewage water from the adjacent sewage treatment plant. The heating capacity of each heat pump is 8300 kW, with an absorbed power of 2500 kW. During winter, the units work in heat pump mode and the compressors operate in series. In summer, the units are used as air-conditioning chillers with the compressors operating in parallel. The cooling capacity is 10,000 kW, with an absorbed electrical power of 1800 kW. Excess heat is discharged to the sewage water.

Figure 14.16.   Friotherm Unitop® 33/28 heat pump in Xinghai, China (http://www.friotherm.com/webautor-data/41/xinghai_e013_uk.pdf).

Figure 14.17.   Friotherm Unitop® 50 heat pump in Helsinki, Finland (http://www.friotherm.com/webautordata/41/katri_vala_e012_uk.pdf).

14.2.6.5   *Helsinki, Finland*

HELEN is one of the largest energy companies in Finland; it produces and sells electricity, district heating and cooling, as well as distributing and selling other forms of energy. The company has almost 400,000 customers and won an award for the most efficient production of energy in the world.

In 2005, HELEN installed five Friotherm Unitop® 50 heat pumps (Fig. 14.17) in a rock cavern excavated beneath the Katri Vala Park in the Sörnäinen district of Helsinki, with a total thermal power of 150 MW.

In summer, the heat pumps produce chilled water at 4°C. The total cooling capacity of the five heat pumps is 60,000 kW. The low-temperature heat is upgraded by the heat pumps to 88°C and used for the production of sanitary water with a heating capacity of 90,565 kW. Excess heat is discharged into the sea.

In winter, sewage water at 10°C is used as a heat source and is cooled to 4°C before being discharged into the sea. The low-temperature heat is boosted to 62°C with a Coefficient Of Performance (COP) value of 3.5. The total heating capacity of the heat pump plant in winter mode is 83,850 kW. In winter, the electrical power absorbed by each heat pump is 4770 kW, while in summer it is 6113 kW.

To achieve optimal operation of the heat pumps, finned-tube heat exchangers were used. Each heat pump is equipped with a separate control system from Siemens AG (Munich, Germany) and a touch screen, allowing ready control of activities and changing of settings. The operation of the entire system is shown in Figure 14.18.

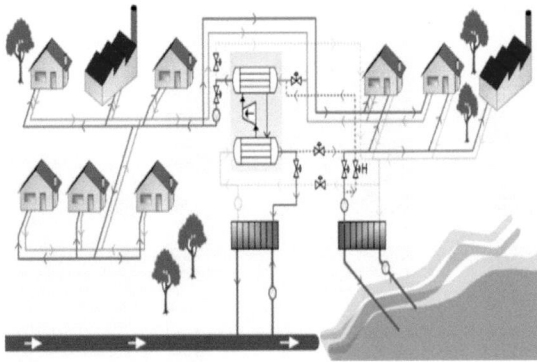

Figure 14.18.   Simplified diagram of the Helsinki heat pump system using sewage water as a heat source (http://www.friotherm.com/webautor-data/41/katri_vala_e012_uk.pdf).

Figure 14.19.   Friotherm Unitop® 33 chillers in the workshop.

14.2.6.6   *Paris, France*

In Paris, Friotherm AG installed eight Unitop® 33 chillers (Fig. 14.19); the cooling plant is built on five levels of a 30 m deep cylinder below the earth's surface (Fig. 14.20). The eight units produce 52,000 kW of cooling capacity. The evaporators are connected in series and chilled water is cooled in two stages, from 10 to 6°C and from 6 to 2°C, respectively. The cooling of the chillers is performed by river water from the Seine by means of intermediate heat exchangers fed by a total water volume of 11,200 m^3 h^{-1}. Five intermediate heat exchangers are used between clean condensed water and Seine water, each with 12.4 MW capacity. This cooling plant provides chilled water for the district cooling system of Paris.

14.2.6.7   *Zurich, Switzerland*

In Zurich, heat pumps are connected to the city's district heating system. The first heat pumps in district heating were installed in the 1940s and 1950s (Leijendeckers, 1985). One example is the heating system of the ETH university in Zurich. The first installation was commissioned in 1940.

After Friotherm AG renewed the plant in 1988, two new heat pumps were installed using the Limmat River, which flows through the city, as a heat source. The entire flow of the Limmat River is 8640 m^3 h^{-1}. When the Limmat water temperature is 3.5°C, the heat pumps can produce 10,000 kW of heat at 70°C with a COP value of 3.09.

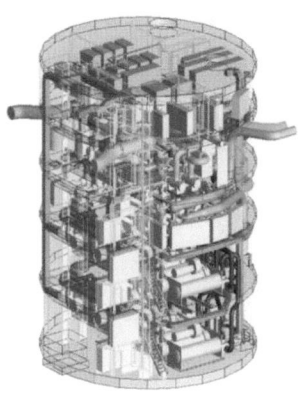

Figure 14.20.   Schematic and building site of Paris pump complex – 30 m deep, 21 m diameter – and view of location after construction.

Figure 14.21.   Zurich heat pump unit operating since 1988 (Zogg, 2008).

At a temperature of 15°C, heating capacity rises to 13,000 kW and the COP value to 3.39. Water is cooled at this time by 0.7°C and discharged back into the river. The diameter of the pipe through which the water is supplied to the heat pump is 1.1 m (Figs. 14.21 and 14.22) (Zogg, 2008).

14.2.6.8   *Gródek, Poland*

Another interesting example of the use of a lake as a heat source was described by Zimny and Fiszer (2003) and Zimny and Michalak (2008). They describe the heating system in the school in Gródek on the Dunajec River, a village situated near Lake Rożnowskie. The hybrid heating system (compressor heat pump + solar + gas boiler water as a reserve peak and emergency provision) consists of the following elements:

- Viessmann Vitocal 300 type WW 20 heat pump.
- Vitosol 100 Type S-2, 5 solar system.
- DeDitrich boiler, type DTG 250-9.
- Vitocell V100.
- 1500 L hot water tank.
- Emitting heat exchanger circuit in the lake water.
- Three heating circuits.
- Usable hot water circuit.

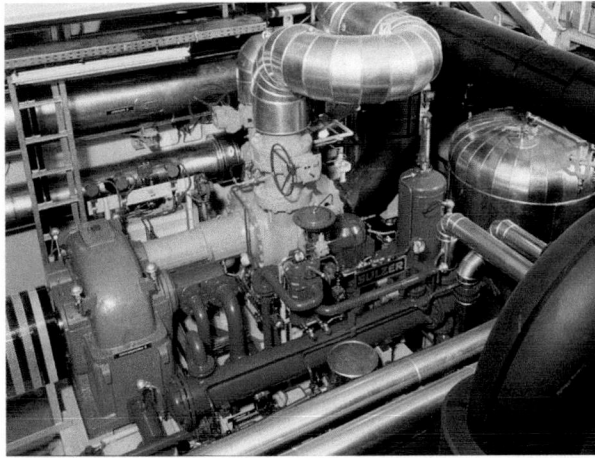

Figure 14.22.   Heat pump unit operating since 1988 (Zogg, 2008).

The basis of the heating system is a heat pump of 106.8 kW heating power. It is fed from two sources of heat. The first is water from nearby Lake Rożnowskie, the second is a ground collector. The control system automatically selects the optimum heat source so as to minimize the amount of electricity consumed by the heat pump. The next element of the system is controlled by a system of solar panels located on the roof of the building. The solar installation is filled with glycol, which allows operation throughout the year. When the sun collectors raise the temperature sufficiently, the circulating pump is automatically started which provides heating from hot water in the collector tray. Depending on the needs, the heat is directed to the heating system or hot water. Another source of heat is a gas boiler with a heating power of 92 kW. It is activated only when no heat comes from the two renewable heating devices. The boiler is maintained at all times, ready to work. The output of the boiler also has its own individual counter, which specifies the amount of heat taken out of it. The security system of the boiler is plugged into the control system. Every failure is recorded and transmitted to the host system.

Zimny and Fiszer (2003) and Zimny and Michalak (2008) also examine the costs of heating the school building with different energy sources in terms of their efficiency. The calculations show that the cheapest source of heat is the heat pump powered by electricity in night tariff. Unfortunately, the studies do not show the effect of the installation of the heat pumps on the ecosystem of the lake.

It should also be noted that in Poland the use of surface water as a heat source is very rare. This is due primarily to the fact that surface waters in Poland, most of which are state-owned, require special permits both in open and closed systems for the use of them as a source of heat. These issues are governed by the "Water Law", "Environmental Protection Law", "Geological and Mining Law" and the "Law on Spatial Planning and Management". In any country, before installing this type of system, one should check the legal requirements relating to obtaining appropriate permits.

14.3   LAKE CHARACTERISTICS AND ANTHROPOGENIC PRESSURE OF HEAT PUMPS ON LAKES

According to Berntsson (2002), the ecological and environmental aspects of the heat pump in general, and those on lakes, are both local and global.

Global: Influence on the *COP* (coefficient of performance) for the heat pump calculated as the amount of heat taken from the condenser divided by the electrical energy supplied to the compressor, which in turn affects the primary energy consumption. This can be compared to the

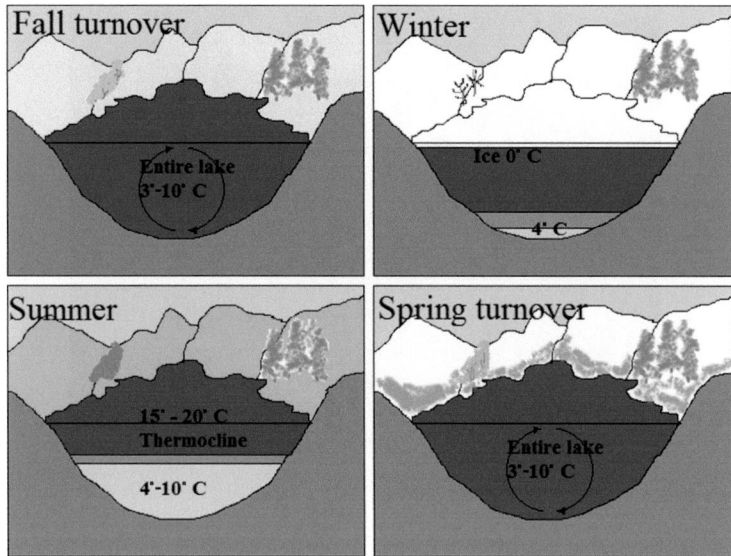

Figure 14.23.   Water mixing during lacustrine year seasons in temperate lakes.

consumption of other sources of alternative energy and thus translated in terms of the greenhouse effect.

Local: Of course, the *COP* also translates into a reduction of primary energy consumption compared to other heating systems, and to reduce emissions of NO_x, polycyclic aromatic hydrocarbons (PAH), SO_x, solid particles, etc. The local impact also includes environmental damage associated with some failures, leaks, etc.

14.3.1   *The physical properties of water and its influence on lakes' thermal regimes*

The physical properties of water are very favorable for energy accumulation, since its specific heat value is very high ($4189.9 \, \text{J} \, \text{kg}^{-1} \, \text{K}^{-1}$).

The density anomaly of water (maximum density at $3.98°C$) determines a lacustrine year cycle. Depending on geographical latitude, the thermal rhythm of lentic water is influenced by air temperature. The top (about 2 m thick) water layer is that part of lake water volume that is directly influenced by air temperature. Under natural conditions the heat transfer toward deeper parts of the water column is shaped by convection currents (as a direct effect of density differentiation). Wind, wave motion and water currents are also involved in the heat transfer in lake water (Choiński, 1995).

Thanks to the density anomaly of water, in the lakes of temperate latitudes there exists four lacustrine year seasons (Fig. 14.23). The heat energy transfer in those lakes occurs perpendicularly to a water mirror, causing vertical thermal stratification. The vertical thermal equalization in lake water (homoiothermal conditions) takes place during the spring and autumn seasons. In that time, lake water temperature amounts to $4°C$ and the mixing of the whole water volume is possible by wind. During the summer season the warmer water is present on the top layers and colder water, as it is more dense, is localized at the bottom zone (temperature about $4°C$) (Choiński, 1995).

On the other hand, in the summer, lakes form the so-called phenomenon of the thermocline, which is when the top layer of water is at a temperature much higher than the deeper water layers. This results in a lack of mixing of the two layers (Figs. 14.23 and 14.34). In this case, this also causes a decrease in oxygen content of the lower layer – often to zero. Applying heat to the lower layers of water in the summer (at the time of formation of the thermocline) can cause sufficient movement of the masses of water heated up to overcome, or even eliminate, the thermocline and improve the aerobic condition of the lake.

In the case of circulation periods (homoiothermy – spring and autumn) the use of open systems for heat pumps should not cause significant changes in water temperature. Water with a temperature of 4°C, which is withdrawn from the hypolimnion (near the bottom), will pass through a heat exchanger and, after cooling, it is discharged into the lake at surface area. This will generate a higher temperature gradient at the surface of the lake and more intense heat received from the air, as the average daily temperature is higher than 4°C in spring and autumn. Additionally, if the water aeration system is applied before discharging water into the lake it will improve the lake water oxic conditions. The open system in the stratified lakes during the summer season (in the case of using the heat pump system for cooling (air conditioning)) will withdraw water with a temperature of 4°C (due to the thermocline). This water will be heated during the operation of the heat pump and aerated by the additional aeration system and later discharged to the surface water layer of the lake. As a result, the epilimnion volume will increase.

For non-stratified lakes, air cooling system efficiency will be lower than for stratified lakes due to the fact that the whole water volume in those lakes has a temperature similar to ambient air. The use of aerating installations for open systems should significantly improve the oxic conditions in the lake water. The water circulation within epilimnion and hypolimnion zones will be improved during summer and winter seasons – similar to natural spring and autumn destratification. Collection of heat in winter, using both the open and closed systems, will reduce the lake water temperature, increasing the thickness of the ice cover. This phenomenon will result in longer retention of the ice cover. However, if an open system with simultaneous aeration is applied it should not have a negative impact on the biology of the lake. The use of closed systems in spring and autumn will not affect the natural environmental conditions (it will slightly reduce the average lake water temperature only). Then using the closed system in the summer will result in a small circulation within the hypolimnion zone without breaking the thermocline.

14.3.2　*Morphometric factors controlling lakes*

14.3.2.1　*Mean depth*

The mean depth is the ratio between the lake water volume and the lake surface:

$$z_{ave} = \frac{V}{A} \tag{14.1}$$

where:
z_{ave}: mean lake depth [m]
V: lake water volume [thousand m^3]
A: lake area [thousand m^2]

Mean depth is one of the most useful parameters in limnology. Lakes with a higher mean depth value (more than 5 m) are classified as stratified lakes – having water stratification during the summer season. Shallow lakes with a low (less than 5 m) mean depth do not have a full stratification period and their water circulates constantly during the year (Choiński, 1995; Hutchinson, 1957). Many works discuss this and models have been built about lake resistance to pollution based on that parameter (z_{ave}). One of the most popular sets of models are Vollenweider's models, which define accepted and critical N and P loads, and provide insight into the process rates of lake eutrophication (Vollenweider, 1968).

The choice of the heat pump type (open or closed system) should be considered according to lake morphometric parameters, such as mean depth (Z_{ave}) and depth index (D_i).

14.3.2.2　*Maximum effective length and width of lake*

The term of an effective length and width characterizes the maximum distances on the lake water mirror on which wind can act directly on the water. Usually there are maximum length and width values, reduced on the emergent macrophytes reach. Wind is the main factor influencing lake water mixing (Choiński, 1995).

14.3.2.3   Relative depth

The relative depth shows mutual relationships between vertical and horizontal lake basin dimensions. Three equations are used for the calculation of this parameter: Halbfass, Ivanov and Hutchinson equations (Choiński, 1995):

Halbfass equation:

$$Z_r = \frac{Z_{max}}{\sqrt{A}} \tag{14.2}$$

Ivanov equation:

$$Z_r = \frac{z_{ave}}{\sqrt{A}} \tag{14.3}$$

Hutchinson equation:

$$Z_r = 50 Z_{max} \sqrt{\frac{\pi}{A}} \tag{14.4}$$

where:
   Z_r: relative depth of lake
   Z_{max}: maximum depth of lake [m]
   z_{ave}: mean depth of lake [m]
   A: lake area [m^2]

These parameters provide significant information about lake water mixing conditions. The lakes with a high value of this parameter have harder and more difficult water mixing.

14.3.2.4   Depth index

The depth index gives information about a lake basin shape. It is a ratio between mean depth and maximum depth of lake (Choiński, 1995):

$$D_i = \frac{\overline{z_{ave}}}{Z_{max}} \tag{14.5}$$

where:
   D_i: depth index
   z_{ave}: mean depth of lake [m]
   Z_{max}: maximum depth of lake [m]

Depending on the depth index values, the possible shapes of lake basins are (Fig. 14.24):

- $D_i < 1/3$ – concave type of lake basin.
- $D_i = 1/3$ – cone type of lake basin.
- $D_i = 1/2$ – paraboloidal type of lake basin.
- $D_i = 2/3$ – hemispherical type of lake basin.

The lake basin shape informs us about the possible lake thermal stratification type and about the water mixing type. Lakes with a low value of that index are stratified and hardly mixed and lakes with high values are non-stratified and easily mixed.

In the case of lakes with D_i values less than ½, using the open systems should be much better, because of better mixing within the epilimnion and hypolimnion layers during stratification, and the improvement of oxic water conditions if the aeration system of water discharged into the lake is implemented.

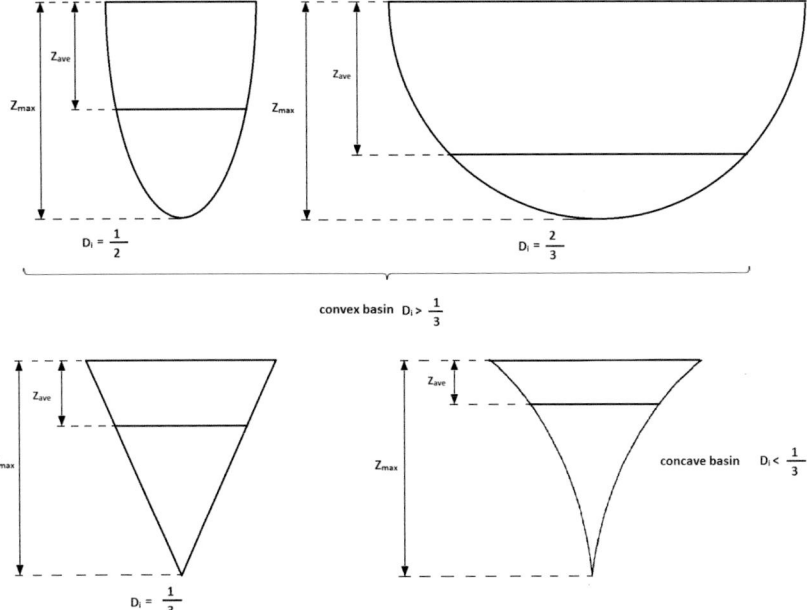

Figure 14.24.   Lake basin shape depending on the depth index value (Choiński, 1995).

14.3.3   *The temperature influence on in-lake processes*

According to the van't Hoff-Arrhenius equation, temperature influences the biological, chemical and physical processes (Zieliński and Krzemieniewski, 2005), because the reaction rate for those processes rises with increasing temperature:

$$k_{T_1} = k_{T_2} e^{\frac{E}{RT_1 T_2} T_1 - T_2} \tag{14.6}$$

where:
   k: rate constant of reaction
   T_1: temperature
   T_2: temperature
   R: universal gas constant
   E: activation energy

Temperature is an important factor in the biological nitrogen transformation processes (Brzozowska and Gawrońska, 2009; Cerco, 1989; Gelda *et al.*, 2000). With a temperature increase the ammonification (ammonia release from organic compounds) increases (Bowden, 1984; Boynton *et al.*, 1980; Cerco, 1989). In particular, nitrification bacteria are sensitive to its influence, and *Nitrosomonas* sp. bacteria (nitrifying bacteria of the Ist phase) have a larger temperature tolerance than *Nitrobacter* sp. bacteria (nitrifying bacteria of the IInd phase). However, denitrification can run in a wide temperature range. In a temperate climate the denitrification process rate increases linearly with temperature within the range 5–35°C (Tiren, 1977; Tomaszek and Czerwieniec, 1995).

Temperature also influences phosphate loading from the bottom sediment of lakes (Cerco, 1989; Suplee and Cottner, 2002; Zdanowski *et al.*, 2002), which can be an important phosphorus source in eutrophic lakes, with a huge amount of phosphorus stored in bottom deposits. Kajak (2001) states that over 90% of the phosphorus and nitrogen content in the whole lake ecosystem is included in lake bottom sediments. Although many authors present different opinions about temperature's role in an internal P load, Fisher *et al.* (1982) observed phosphate uptake by sediment

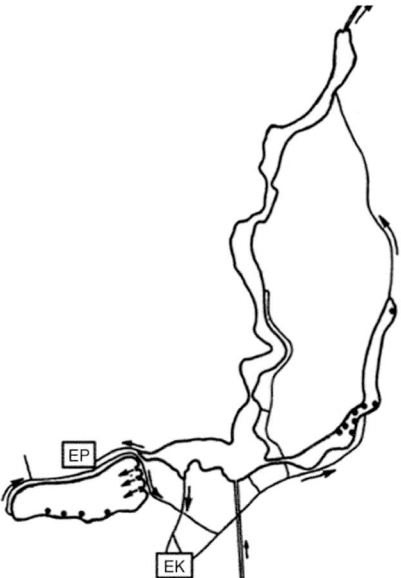

Figure 14.25.   Map of Konin Lakes System: EK – Konin Power Plant; EP – Pątnów Power Plant. The arrows
show the direction of water flow in channels connecting the lakes (IFI, 1958).

at low temperature (4°C) and its release at high temperature (20°C). Boynton *et al.* (1980) also noted that phosphate release increases with increasing temperature. However, Cerco (1989) did not observe a significant temperature influence on the phosphate loading from sediment. However, Liikanen *et al.* (2002) maintain that the temperature effect on phosphate release from sediment is significant at both low and high temperatures.

Undoubtedly, water temperature helps to determine the amount of dissolved gases in it, including oxygen, which is the most important one for the majority of living organisms. The presence of oxygen or its lack rules the processes taking place in the lake. In higher temperatures, physical oxygen solubility decreases and it can have an influence on the biota. The water organisms, such as, for example, Salmonidae fish, prefer cold and oxygenated water and can die at water temperatures exceeding about 20°C.

Oxygen is a limiting factor for the nitrification process; for the correct running of this process an oxygen concentration above $1 \, mg \, dm^{-3}$ in water is needed. However, oxygen is also an inhibiting factor for the denitrification process, which requires anaerobic or anoxic conditions (Rysgaard *et al.*, 1994). The oxygen conditions, more than temperature, decide about phosphorus internal loading, because of the fact that the oxygen concentration regulates the direction of redox processes in the lake (Marsden, 1989; Nürnberg, 1994; Nürnberg and Peters, 1984).

Human interference in lake ecosystems can modify the thermal properties of lakes. Up to now it took place mainly in two ways: the first is using lakes as cooling water receivers (Dunalska *et al.*, 2012; Kraszewski, 2004), the second is the thermal destratification during the lake restoration procedure called artificial mixing (Grochowska and Brzozowska, 2013; Höhener and Gächter, 1994; Klapper, 1991, 2003).

It is possible to study the influence of lake water heating on the physical, chemical and biological conditions of lakes in the Konińskie Lake Complex, located in the western part of Poland, in the Wielkopolskie Lake District (Dunalska *et al.*, 2012; Kraszewski, 2004).

The Konińskie Lakes Complex consists of five lakes: Ślesińskie Lake, Mikorzyńskie Lake, Licheńskie Lake, Pątnowskie Lake and Gosławskie Lake (Fig. 14.25). The lakes are connected by channels (Dunalska *et al.*, 2012; Kraszewski, 2004). The division of water from a power plant into particular lakes is regulated depending on needs. The Konin power plant takes water from

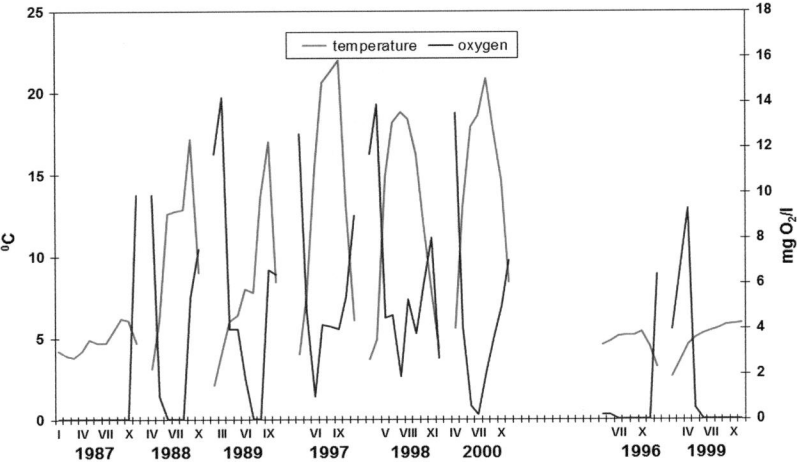

Figure 14.26.   Temperature in the near-bottom water layer of Długie Lake during the experimental years (with artificial mixing – 1987, 1988, 1989, 1997, 1998, 2000) and the control years (1996, 1999) (data from Grochowska, 2001).

Pątnowskie Lake and uses it for cooling condensers. The heated water flows into a preliminary cooling reservoir or directly to lakes, as follows: Licheńskie Lake (up to 50% of discharge), Pątnowskie Lake and Mikorzyńskie Lake. The Pątnów power plant takes water from the western part of Gosławskie Lake. The discharge of heated water is directed to the eastern part of the lake (up to 50% of water volume) and the rest flows into the Licheński channel (Dunalska *et al.*, 2012).

The results of multiannual observations of the Konińskie Lakes Complex show that the discharge of cooling water changes the thermal regime of lakes (disappearing of thermal stratification, and then the limitation of oxygen content in the near-bottom zone). It is one of the most important factors influencing the environmental conditions and causing transformations on the trophic levels of all of the lakes (Stawecki *et al.*, 2007).

The changed conditions favored the appearance and mass development of invasive species that originate from subtropical zones and which are not native to Polish lakes. For example, in the Konińskie Lakes Complex the mussel *Sinanodonta woodiana* (Lea) and macrophyte *Valisneria spiralis* found very good environmental conditions for mass development (Hutorowicz *et al.*, 2007; Kraszewski, 2004).

Lake restoration procedures are sometimes applied on degraded lakes, which are not able to rebuild good water quality after a limitation of anthropogenic pollution. Artificial mixing is the most popular restoration method throughout the world – artificial circulation has been implemented on more than 90% of restored lakes (Klapper, 1991, 2003). The Department of Water Protection Engineering, University of Warmia and Mazury in Olsztyn, carried out the restoration of Długie Lake in Olsztyn from 1987 to 2003. The lake was renovated using two methods: artificial mixing (1987–2000, with two control years without artificial circulation – 1996 and 1999) and the phosphorus inactivation method (2001–2003) (Grochowska and Brzozowska, 2013).

Artificial circulation is a method that changes the natural thermal stratification of lakes. In the summer season lake water is heated by solar energy and in the artificially mixed lake the whole water volume is very warm. The increasing water temperature causes the acceleration of organic matter mineralization processes. A very important thing is to guarantee good oxygen conditions in the lake water – the dissolved oxygen concentration in the water should exceed $2\,mg\,O_2\,L^{-1}$, which is the boundary value between oxic and anoxic conditions. The oxic conditions in the water are necessary for precipitation of phosphorus compounds with iron and manganese to sediment. As shown in Figure 14.26, that condition was mainly fulfilled during the second stage of artificial mixing (1997–2000). And in the control years the lake returned to natural thermal stratification, with hypolimnetic water temperature ca. 4°C.

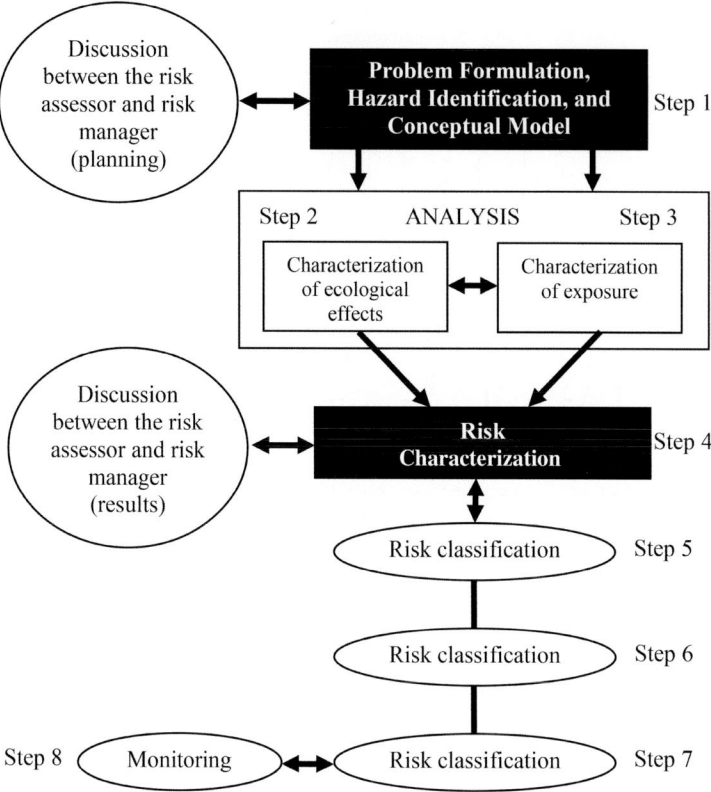

Figure 14.27.   Steps of risk assessment of the environmental impact (Ritter *et al.*, 2002).

The implementation of the artificial mixing with destratification of Długie Lake in Olsztyn caused the intensification of biological nitrogen transformation and the reduction in the total nitrogen contents in lake water to ca. 80–90%, which is mainly due to coupled nitrification-denitrification processes, the acceleration of organic matter mineralization processes, and the decrease of phosphorus content in the lake water also to ca. 80–90%, compared to the period before restoration.

A lake is rarely used as a heat source. This is due to various problems. One of them is the as-short-as-possible distance between the lake (the low-temperature heat source) and the heated house (the upper heat source). This requires that the position of the building to be heated is close to the lake (Berntsson, 2002). Another problem, which is difficult to estimate, is the impact on the ecosystem of the lake. Receiving part of the heat in the winter will result in a thicker layer of ice and a longer time with the surface of the lake covered with ice. Any such investment requires an analysis of the particular case. As shown above, based on the literature review, this is a difficult task, requiring the construction of a suitable model that takes into account the specific circumstances of a particular lake. The steps of this analysis are shown in Ritter *et al.* (2002) (Fig. 14.27).

When a larger quantity of heat is transferred (subject to the individual characteristics of the lake), the process has significant consequences on the lake, depending on its type:

- In pure, non-stratified lakes, the process has harmful consequences – an increase in water temperature of only a few degrees could alter the lake's biological conditions and disturb its ecological balance.

- In eutrophic or polluted stratified lakes, which are marked by temperature and oxygen stratification (lakes with anoxic hypolimnion), this process could have a positive effect. The bottom water layers are heated to the temperature of the surface strata, which facilitates mixing, e.g. due to wind, and increases oxygen content. It should be noted, however, that the biological condition of bottom layers that are artificially saturated with oxygen will change.
- Further research is also required to determine the quantity of heat that can be safely drawn from a lake in the winter (for domestic heating purposes, with the use of a heat pump) in lakes to which heat is not transferred in the summer and which are used as heat accumulators without causing major disturbances in the thermal balance (measured by the thickness of the lake's ice cover), i.e. without upsetting the lake's ecosystem.

14.3.4 *Lake modeling*

This section provides general examples related to the modeling of lakes as a whole. The process, as is shown below, is very complex and each case separately run. Even for these same lakes, various physical phenomena will be modeled by forming separate models. An example of a model for the situation of consumption and heat dissipation of a particular lake for the design of heating using heat pumps is described in Section 14.4. There are no commercial programs available that allow the modeling of a lake in its entirety. Such models are created on the basis of general programs, such as the programs for finite element method (FEM). From the point of view of the use of lakes as a source of heat, particularly important are models showing the temperature distribution of the waters of the lake and the temperature dependence of external factors. Therefore, since there is not much such work, older studies were also used, but the results of temperature modeling, due to their confirmation in reality, are useful for the design of heat extraction from lakes by a heat pump. Modeling of thermal processes occurring in the lake is a difficult task. It requires consideration of many factors, such as:

- Heat exchange between water and the bottom of the lake (including changes in temperature of land and water with depth).
- The heat exchange between the water and the air (over the entire surface of the lake) – subject to wind and its impact, e.g. on the surface of the water by the formation of the waves.
- Heating of the water from the sun (infrared radiation), including the light-scattering mechanism on particulates (turbidity lakes) and reflection from the bottom.
- The effect of wind on the mixing of water in the volume of the lake.
- The geometry of the lake and its geographic location.
- The phenomenon of the so-called thermocline and the associated summer stratification.
- From the biological point of view on lakes, which is also of great importance, the organic matter and nutrients and the amount of oxygen dissolved in water are very important for the models.

The work of Hondzo and Stefan (1996) shows the results of simulation calculations of models based on observations of the 25-year relationship between the individual sizes characterizing the lake (Figs. 14.28 and 14.29). These studies are particularly worth showing since the models were based on measurements made over a 25-year period. The model used in the above-mentioned work was based on mathematical differential equations. At the same time, it shows what may be the temperature of the water at the bottom (Fig. 14.28), which is important in the case of the modeling and design of heat removal from lakes. At the same time, it has been concluded that the temperature of the surface layer of the water is more or less constant for the test lakes of different morphologies, and only depends on the weather in southern and northern Minnesota. In this work (Hondzo and Stefan, 1996), the impact of the trophic status of lakes for the average temperature of the water in the entire volume of the lake was also studied. It has been found that for oligotrophic lakes, this temperature is higher. This gives an indication that these lakes are better sources of heat for systems with heat pumps.

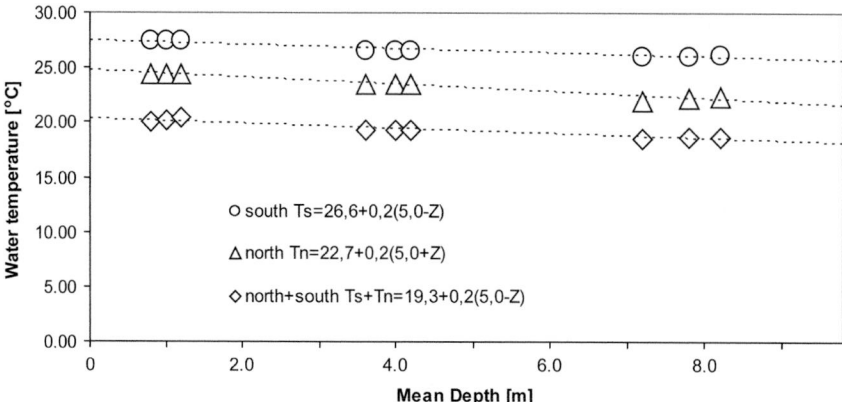

Figure 14.28.   Twenty-five year simulated maximum daily water temperature against mean lake depth (Hondzo and Stefan, 1996).

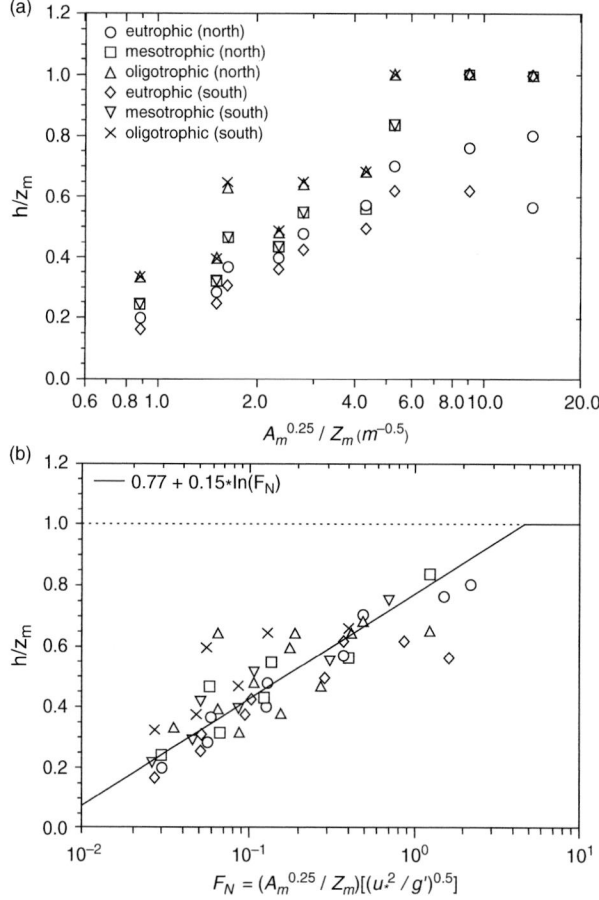

Figure 14.29.   Twenty-five year simulated maximum thermocline depth h normalized by maximum lake depth z_m plotted against: (a) lake geometry ratio; (b) wind-related densimetric Froude number (Hondzo and Stefan, 1996).

Shen (1998) modeled the breeze on the lake, depending on the time of day, type of vegetation and wind outside. This work used a two-dimensional, non-hydrostatic, compressible mesoscale model coupled with the SiB2 land-surface scheme. This is a kind of numerical model. One of the modeled phenomena was a change in the distribution of latent heat and the heat transfer from the lake water into the air. The results show that the amount of moisture being carried and receive heat by the breeze does not depend on the type of plants growing in the area of the lake. In the case of the use of lakes as a heat source, this gives the information that the lake will lose similar heat with the wind irrespective of the vegetation growing thereon. The work of Rimmer (2003) on the salinity of Lake Kinneret was also analyzed, and he built models depicting these changes. The model used was based on the fundamental equations of balance and the replacement of water and salt between the layers of water in the lake. It was a model of one-dimensional differential. This model has been calibrated for Lake Kinneret (Israel). The constructed models were used to predict water salinity within the next 10 years. These studies show that due to the large number of factors that affect the physical and chemical phenomena in the lake, one of the conclusions is that each model must be built and verified for individual lakes separately (Menshutkin *et al.*, 2013).

The work of Liu and Chen (2012) compared the results of modeling obtained with two different methods. One of them was the model circulation of 3-D mathematical modeling and for the second an implicit model was used – the artificial neural network (ANN). In any event, annual changes of temperature at a depth of 1, 2 and 3 m below the surface of the lake were modeled. A mathematical model was calibrated with the actual data and the ANN was taught on the prepared set of real data (the training set). Then the models obtained were verified, using data that had not been used to teach and calibrate the models. Figure 14.30 shows the results for the ANN model, and Figure 14.31 those for the analytical models. These results suggest that the greatest error is obtained from modeling the surface temperature (Figs. 14.30 and 14.31a); this is related to the effects of a large number of variables, which cannot be included in the model. The results of the modeling error 3 m below the surface is much lower (Fig. 14.30 and 14.31d). These studies were conducted on Yuan-Yang Lake in Taiwan. The results show in the case of modeling the temperature of water in lakes that the more simply created ANN model is as effective as the analytical 3-D model.

On the other hand, the work by León *et al.* (2007) shows the simulation results of temperature distribution in Great Slave Lake in Canada – examples are shown in Figure 14.32. A 3-D hydrodynamic model was used (ELCOM). In order to forecast weather phenomena, data from the Canadian Regional Climate Model (CRCM), which does not have a component that allows modeling lakes, obtained from modeling temperature, were compared with the measured data for a specific lake (Great Slave Lake in Canada). Model error relative to the real data was 15.5% in modeling the surface water temperature. The error of the modeling of the temperature distribution into the depth of the lake was even higher.

Similar simulations are described by Filoche (2007) for the Lac de Bourget, who used the measured air temperature and wind speed to model temperature distribution as a function on the depth of the lake using the finite element method. A computational grid was constructed using cubes sized 100×100 m in the horizontal direction and from 1 m at the top layers up to 30 m at the bottom of the lake in a vertical direction. The total number of "active" (wet) cells was approximately 66,000 (Filoche, 2007). The actual data, collected by a special set of thermistors, exhibits a strong movement inside the wave with the largest amplitude at a depth of 40 to 50 m. This movement is caused by strongly suppressed intense cooling water due to low air temperature and by the wind blowing over the lake. What is important in this is that numerical simulations are in very good agreement with the observations in terms of overall changes in the temperature of the lake. This model can be the basis for modeling thermal phenomena in lakes planned to be used as a heat source.

Lytras (2007) showed a simplified geometric model that allows modeling of the lake (Fig. 14.33), and changes in temperature and oxygen concentration as a function of depth in the particular seasons (Fig. 14.34). Shown at work, a simplified geometric model (adapted to the nature of the modeled lake) was used in Section 14.4 for modeling phenomena in Długie

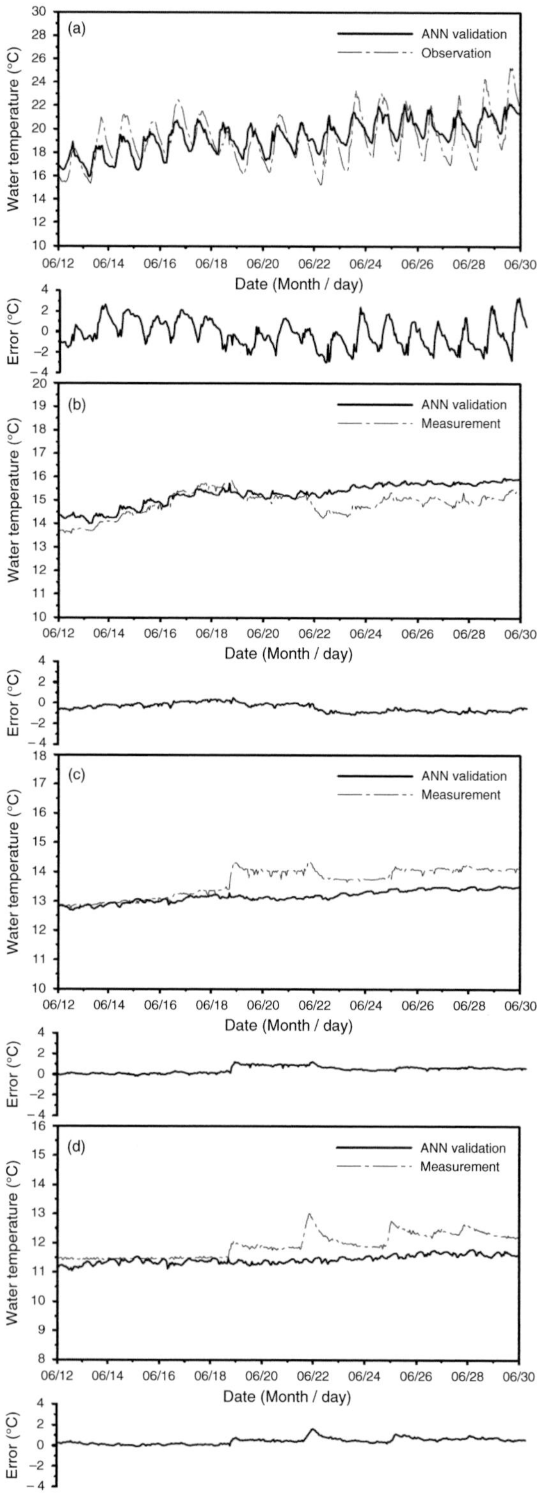

Figure 14.30.   Comparison of water temperature for ANN validation phase at the buoy station: (a) surface layer; (b) 1 m below water surface; (c) 2 m below water surface; (d) 3 m below water surface (Liu and Chen, 2012).

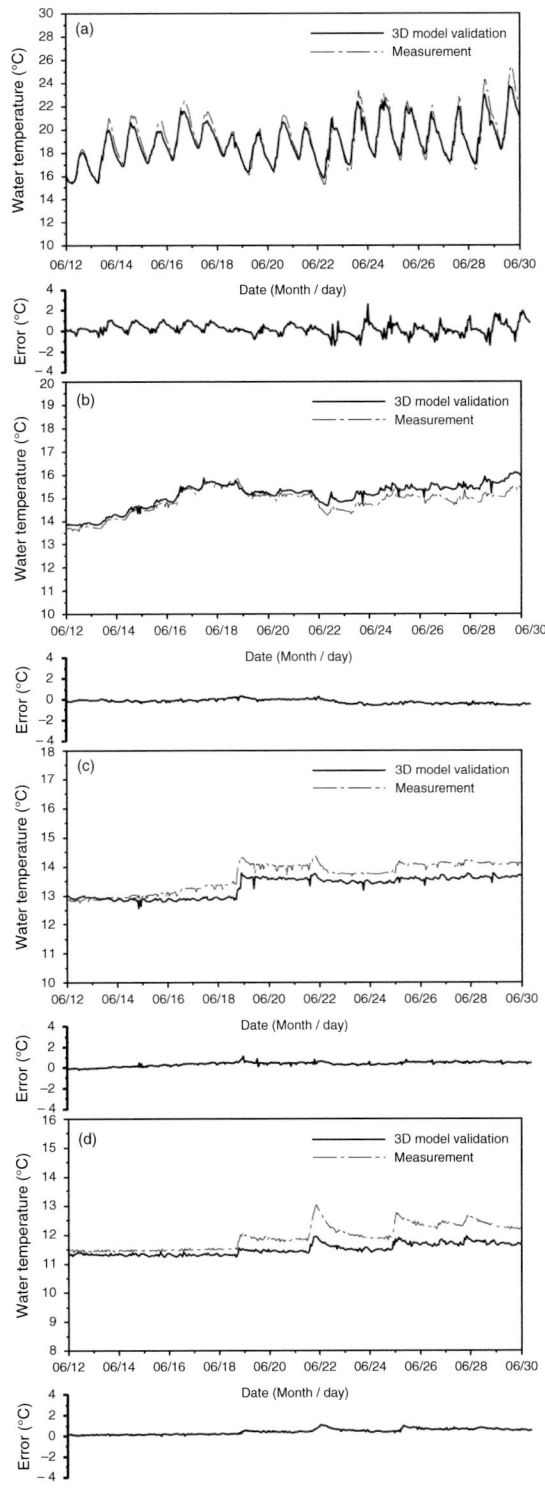

Figure 14.31. Comparison of predicted water temperature with three-dimensional circulation model and measured data at buoy station for 3-D model: (a) surface layer; (b) 1 m below water surface; (c) 2 m below water surface; (d) 3 m below water surface (Liu and Chen, 2012).

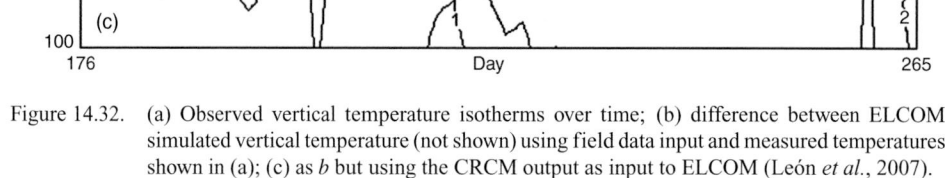

Figure 14.32.   (a) Observed vertical temperature isotherms over time; (b) difference between ELCOM simulated vertical temperature (not shown) using field data input and measured temperatures shown in (a); (c) as *b* but using the CRCM output as input to ELCOM (León *et al.*, 2007).

Lake. Simplifying the shape of the modeled lake does not adversely affect the correctness of the results of modeling. On the other hand, phenomena that occur at particular times of the year in the vertical section of the lake are also the basis for the assumptions and the construction of the model and its subsequent verification (see Section 14.3.2.4).

A mathematical model of the mass and heat circulation in large lakes was shown in the work of Kolodochka (2003). During the construction of the model, it is assumed that the components of dissolved and suspended matter in the lake water (both organic and inorganic) can be divided

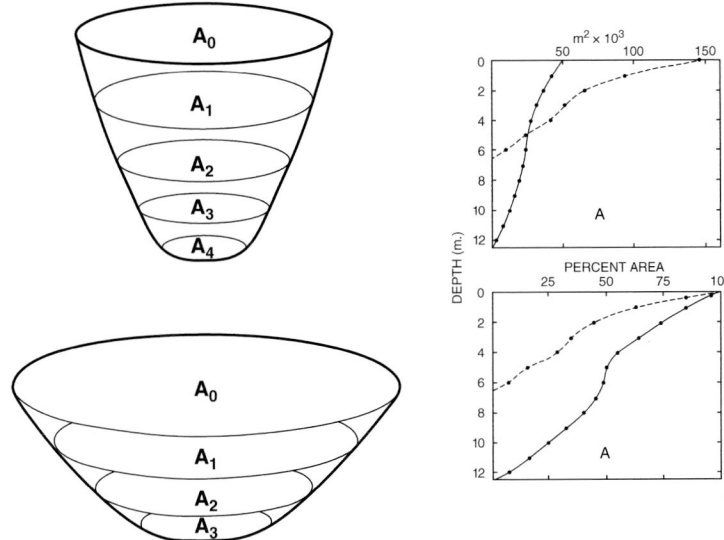

Figure 14.33.   Depth volume curves of oligotrophic Lawrence Lake and eutrophic Wintergreen Lake (Lytras, 2007).

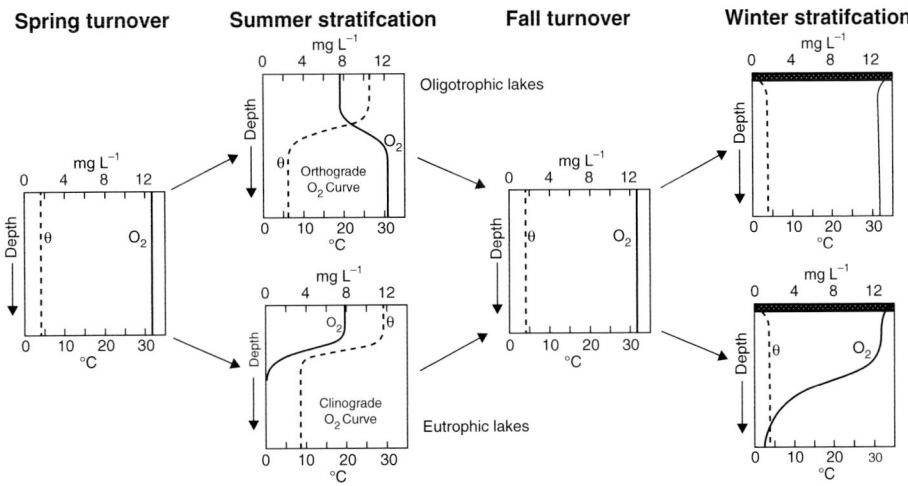

Figure 14.34.   Idealized vertical distributions of oxygen concentrations and temperature during the four main seasonal phases of a year in oligotrophic and a eutrophic dimictic lake (Lytras, 2007).

into three subclasses: chemical compounds, suspended sediments and mineral materials, and plankton. Each of these groups needs to be taken into account in the model, although of course, to some extent, one is better able to build simplified models actually reflecting the changes taking place in the waters of lakes (Kolodochka, 2003). As part of the calculations, the 2-D model was built based on the equations of mass and heat transport in the vertical axis. One advantage of this model is that it allows a closed space within the model to be extracted as homogeneous (e.g. with a similar temperature). These structures reflect well the internal movement of water masses, which improves the effectiveness of the model, while facilitating the modeling process itself. The model was verified on the basis of phenomena in Lake Onega (Russia).

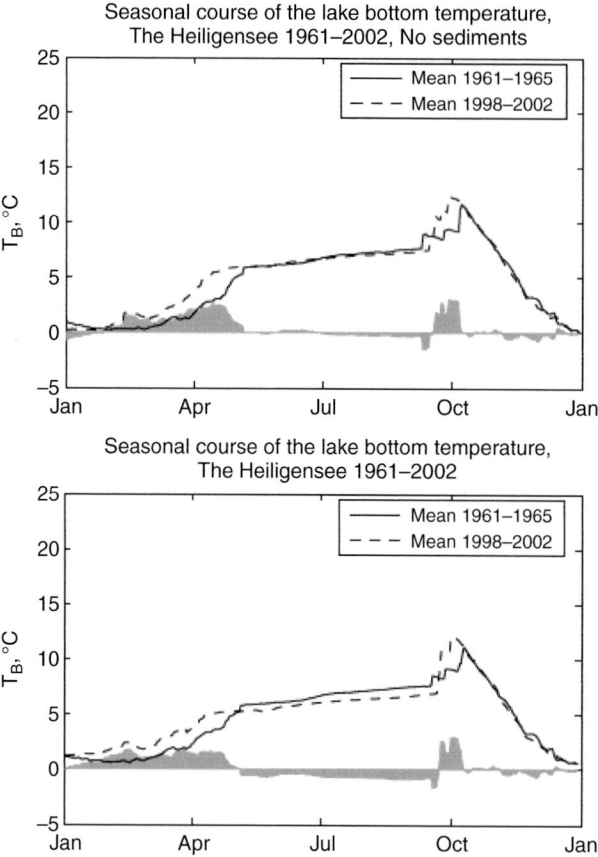

Figure 14.35.   Comparison of five-year means of modeled seasonal temperature course on the border of water and sediment for Lake Heiligensee. The shadow represents the difference between the two runs. The first figure shows the calculation without sediment heat storage (Golosov and Kirillin, 2010).

However, Golosov and Kirillin (2010) have shown interesting models illustrating the changes of lake bottom water temperature depending on the air temperature. These models, according to the authors, can be used to determine the impact of climate change on the thermal conditions prevailing in lakes. Here the analytical model of heat distribution in a layer of sediment was used, to which a one-dimensional model of changes in temperature and mixing (flake) was added. The whole model was then validated for the studied lakes with the real data gathered over several years. Various cases were modeled, for example the influence of the heat distribution through the layer of sediments (to be included or not in the calculation). Examples of distributions obtained from the model are shown in Figure 14.35. The first figure shows the results of the temperature distribution for the case without considering the impact of sediments and the second figure includes this consideration.

14.3.4.1   *Comparison*
An important finding of the literature is that for each lake to build a model separately, even if the same mathematical relationships are used, this model should be parameterized in each case for the premises. The whole literature review shows that heat exchange between the different parts of the lake surroundings is at the present moment not possible to model with one universal model that takes into account all the variables.

The heat exchange between the water and the bottom of the lake should also take into account the heat stored in the sediments (Golosov and Kirillin, 2010), but it is difficult to calculate without actual data for a particular lake, because it depends on the composition, thickness and density of the sediments. These parameters are dependent on the lake.

Heat exchange between the water and the air (wind) does not depend on plants growing on the edges of the lake, but depends on the direction, strength and frequency of winds (Shen, 1998). The geometry of the lake itself also affects how much heat the lake gathers in the summer, how quickly it will give in autumn and winter, and, which is important, how the winds will cause mixing of water layers during the autumn and spring turnover (Lytras, 2007). From the viewpoint of heat reception, the biological parameters are not significant – as they do not affect the potential of the lake as a heat accumulator. But, receiving heat from the lake (or its delivery in summer) will affect the biological processes occurring in the lake (Menshutkin *et al.*, 2014).

Only certain models of the lakes, reported in the literature, take into account the phenomenon of thermocline (Filoche, 2007; León *et al.*, 2007).

The whole picture shows that the modeling of lakes is difficult; many authors focus on only one aspect. On the other hand, due to legal requirements, the use of lakes as a source of heat in many countries is regulated with additional laws. Getting permission is often associated with the presentation of the environmental analysis of the planned investment. Modeling to estimate what might change in the lake by its use as a heat source is at present the only option. Good alternative models (neither numerical nor analytical) are ANNs (Liu and Chen, 2012).

14.4   CALCULATIONS

14.4.1   *Preliminary assumptions*

For the calculation, the Długie Lake, located in Olsztyn in the province of Warmia and Mazury, was selected.

Some initial assumptions were adopted for the calculation, as shown in Figure 14.36, which presents the bathymetric plan for Długie Lake that consists of three separate reservoirs. For calculation purposes, the middle deepest water reservoir was selected (Fig. 14.37), which has dimensions of 800 m length, 240 m width, and a maximum depth of 14.5 m. It is also assumed that, throughout the period of use, the energy stored in the lake in the winter is not supplied to it from an external source (e.g. no energy in the form of sunlight), which corresponds to a situation where water is frozen on the lake's surface. It was assumed that the lake surface in winter had a temperature of 0°C, and at the bottom it maintained a constant temperature of 4°C.

After selecting the reservoir, a cross-section of it was made along the width A–A and length B–B (Fig. 14.37). The cross-section is divided into layers. Assumed temperature was averaged for each of the layers, in increments of 0.5°C, together with the depth changing the dimensions and temperature of the lake. Mean values used for the calculation are shown in Figures 14.38 and 14.39. Paraboloidal shape was adopted on the basis of the calculated depth index values D_i (depth index $D_i = z_{ave}/Z_{max}$ (Choiński, 1995) = 0.69) for $D_1 \approx \frac{1}{2}$. The temperature distribution in the lake in winter was adopted on the basis of similar measurements for other lakes (in fact, it is also variable depending on the outdoor temperature, thickness and extent of ice cover and snow).

14.4.2   *Calculation of the total amount of energy stored in the lake*

We start with the following parameter values:

- Specific heat of water: $4187 \, J \, kg^{-1} \, K^{-1}$
- Water density: $1 \, kg \, dm^{-3}$
- Heat pump data used in the calculation (manufacturer): Heat pump "WWC 280x Compact"
- Heat output/COP^*: 27 kW/5.2

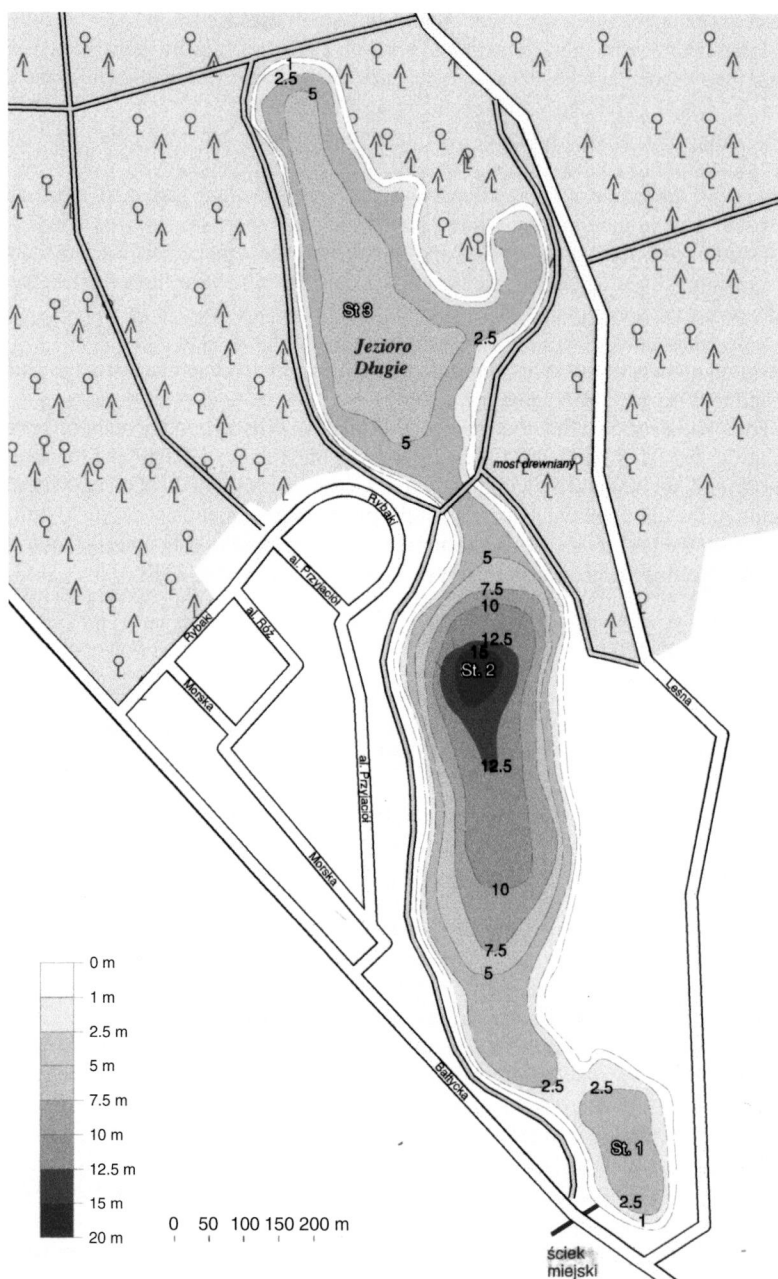

Figure 14.36.   The bathymetric plan for Długie Lake (IFI, 1958).

- Flow of the brine (min/nom/max): 5300/5300/9300 $dm^3 h^{-1}$
- Heating water flow rate (min/nom/max): 2300/4600/5800 $dm^3 h^{-1}$
- Refrigerant type/volume: R407C/4.4 dm^3
- Limit temperature of hot water: 60°C

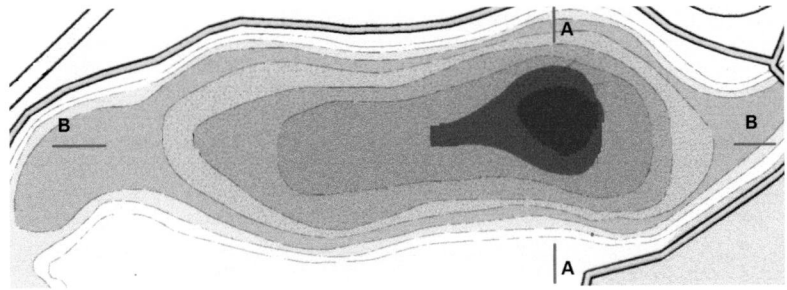

Figure 14.37.   The middle part of the lake with the selected cross-sections.

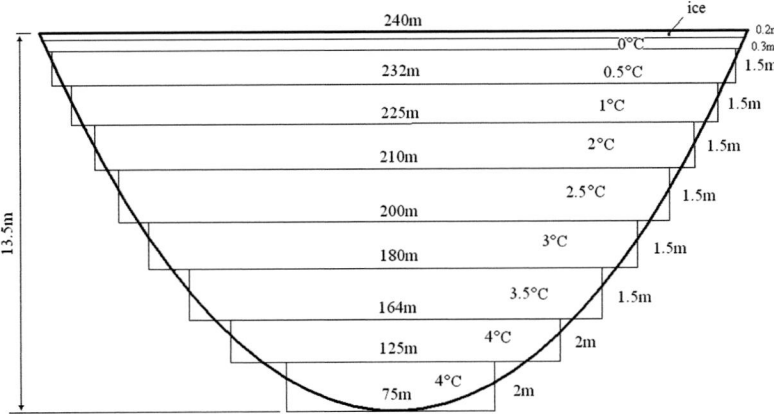

Figure 14.38.   Cross-section of Długie Lake at A-A (width).

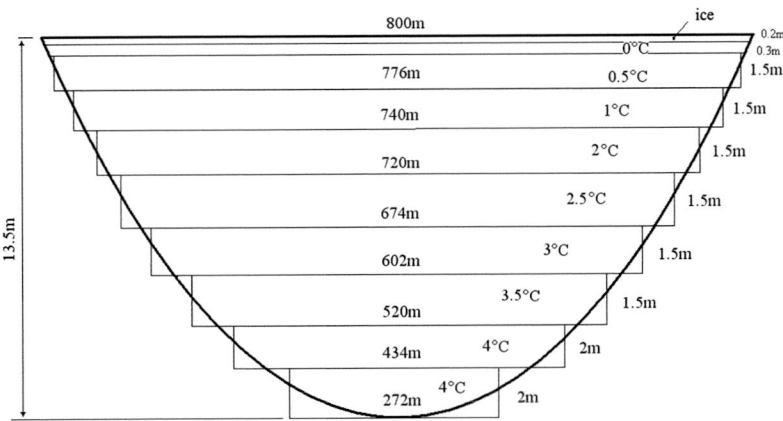

Figure 14.39.   Cross-section of Długie Lake at B-B (length).

- Limit temperature of abstracted water: $+7°C$ to $+25°C$ (after phone consultation with a technician the limit of abstracted water is $+3°C$)
- Free circuit pressure C.O. the difference 7K: -3300 dm^3/0.4 MPa
- Weight/dimensions (W D. H): 365 kg/750 × 650 × 1650 mm
- Voltage/high current (at the start): 400 V/12.5 A

Table 14.1.   Data for calculations.

Lp	Depth, h [m]	Thickness of layer, a [m]	Temperature difference, ΔT [K]	Lake width, b [m]	Lake length, c [m]
1	0.2	0.2	0	800	240
2	0.5	0.3	0	800	240
3	2	1.5	0.5	776	232
4	3.5	1.5	1	740	225
5	5	1.5	2	720	210
6	6.5	1.5	2.5	674	200
7	8	1.5	3	602	180
8	9.5	1.5	3.5	520	164
9	11.5	2	4	434	125
10	13.5	2	4	272	75

- Power consumption: 5.29 kW
- Compressor efficiency: 0.85
  (*Coefficient of performance)

Using the pre-set assumptions, all the data needed was collected for the calculation (Table 14.1). After applying the basic formulas and their transformations, it was calculated how much heat was in the selected part of the lake in the winter. In Table 14.1, the value of ΔT is the temperature difference between the temperature of water in the concrete layer, and at 0 degrees – the maximum possible temperature difference for each layer that can be used for the heat pump. Column data a is the thickness of each layer (Figs. 14.38 and 14.39). Column data b and c are the width and length, respectively, of each lake layer (Figs. 14.38 and 14.39). The volume of each layer of the lake was calculated (assuming an ellipsoidal shape for each layer):

$$L = \pi \tfrac{1}{2} b \tfrac{1}{2} c \quad [\text{m}^2] \tag{14.7}$$

$$V = L\, a \quad [\text{m}^3] \tag{14.8}$$

On the basis of the calculated volume of each layer, and held by its temperature, is calculated the maximum amount of heat that can be collected from each layer, assuming that it will be cooled to zero degrees. Stored heat was calculated from the formula:

$$Q = c_w\, m \Delta T \quad [J] \tag{14.9}$$

where:
 c_w: specific water heat
 m: weight of each layer
 ΔT: the above described temperature difference

The mass of water in each layer was calculated on the assumption that 1 dm^3 of water weighs 1 kg. The results are summarized in Table 14.2.

- *COP* calculation:

$$COP = \frac{E_S}{E_E} = \frac{27}{5.29} = 5.1 \text{ (the real value)} \tag{14.10}$$

- Carnot efficiency:

$$\eta_C = \frac{1}{COP} = \frac{1}{5.1} = 0.19 \tag{14.11}$$

Table 14.2. Calculations.

Lp	Volume, V [m³]	Mass, m [Mg]	Heat, Q [GJ]
1	30 159	30 159	0
2	45 239	45 239	0
3	212 095	212 095	444
4	196 153	196 153	821
5	178 128	178 128	1492
6	158 808	158 808	1662
7	127 659	127 659	1604
8	100 468	100 468	1472
9	85 216	85 216	1427
10	32 044	32 044	537
Sum	1 165 969	1 165 969	9459

- Heat power (heat collected from the bottom of the lake):

$$Q_O = Q_g \frac{\varphi - 1}{\varphi} = 27 \frac{5.2 - 1}{5.2} = 21.8 \ [\text{kW}] \qquad (14.12)$$

The time of discharging the heat is:

$$T = Q / P_{\text{pump}} \quad [\text{s}] \qquad (14.13)$$

where:

P_{pump}: power of heat pump

The total amount of energy (assuming a lack of heat exchange between the lake and the surroundings) one can pick up from the water of the lake is 9,458,978 MJ. The simplifying assumptions, i.e. the heating efficiency being equal to one, 25 heat pumps with a capacity of 27 kW, taking the same amount of heat by a time of about 199 days, would freeze over the lake to the bottom. Such a situation is obviously unacceptable from an ecological point of view.

14.4.3 *Receiving heat in winter – increasing the thickness of the ice layer*

The earlier calculations show us how much energy is stored in Długie Lake and for how much time we would be able to use it until there would be a total freeze of the reservoir.

Using the previously defined calculation method, according to the amount of energy we can get out of the lake in the winter, we can calculate taking the heat in such quantities that the ice sheet will increase from 20 cm (Figs. 14.38 and 14.39) to 50 cm, which should not interfere with the biological status of the lake (especially considering that the heat pump working in an open system could also aerate water discharged back into the lake). It is possible to calculate the amount of heat received, assuming a mass of water equal to the weight of water in layer 2 given in Table 14.2. Also, it is assumed that the heat source will have a constant temperature of 4°C all the time.

With these assumptions, calculations were carried out by two heat pumps with a capacity of 27 kW, which could work without interruption for a period of about 199 days (over 6 months). This gives the opportunity to warm up around 470 m² of living space throughout the heating season.

14.4.4 *The introduction of heat into the lake in the summer*

The study involved a simulation to establish whether the transfer of heat to the bottom of the lake in the summer would have an effect on lake stratification, i.e. whether the summer thermocline layer can be broken down in dimictic lakes. The thermocline layer (metalimnion) separates the water of the lake into two strata: the warm surface layer (epilimnion) containing oxygen and the cold over-bottom layer (hypolimnion), which, due to lake eutrophication or pollution, contains no or very little oxygen.

The overall heat balance of the lake can be written as:

$$\Delta Q = Q_r + Q_d + Q_o + Q_t + Q_p \tag{14.14}$$

where:

ΔQ: heat balance

Q_r: heat from the radiation

Q_d: heat exchange inflows and outflows associated surface waters

Q_o: heat from the rain

Q_t: heat from the air

Q_p: the heat balance caused by evaporation and condensation (Choiński, 1995)

The following assumptions were made to model the impact of heat transfer to the lake on the ecosystem:

- The elements and chemical compounds found in water do not affect the process.
- There is no heat exchange between the water and the bottom of the lake.
- Heat flux from solar radiation heats water subject to the turbidity coefficient.
- Heat flux from ambient air heats only a limited area of surface water.
- A heat pump with a capacity of 27 kW working in reverse cycle.

These simplifying assumptions assume the only source of heat for the lake water is sunlight. For the process of heat receiving from solar radiation by lake water, the water turbidity is a very important factor. The more turbid the water, the more effectively the heat from sunlight is transferred into the water.

A rectangular model grid was adopted with an elementary cell grid size of $25 \times 25 \times 5$ m, producing a total of 870 elements for the water body selected for modeling. Data regarding irradiation and average daily temperatures was collected from a meteorological station between April and August 2003 – a total of 153 days. The outside temperature was used to calculate the heat loss from the surface of the lake. Using this data, the simulation was started for spring turnover (initial temperature for all grid elements was set at 4°C, which corresponds to the spring period of water mixing in the lake). The increase in temperature, caused by solar radiation, depends on the turbidity. The more turbid the water is, the more it warms up, but on the other hand, the rays of the sun penetrate to a smaller depth. Heat losses through the surface of the lake depend on the temperature difference between the surface layer and the air temperature. Using these relationships and assumptions, a simulation was performed to calculate the changes in water temperatures during the summer. The obtained simulation results were compared with summer temperature distribution data from a field study of 25th August 2003 (Lossow *et al.*, 2005) (Fig. 14.40). The thermocline is shown in Figure 14.40. Based on the above, the turbidity coefficient at which simulation results approximate real survey data was determined. Figures 14.41 and 14.42 show simulation results for turbidity coefficients of 0.30 and 0.64 for 146 days of modeling (25th August 2003). As can be seen in Figures 14.41 and 14.42, a turbidity coefficient of 0.64 gives the best match between the results of experimental and empirical studies.

The next stage of the simulation involved the determination of the impact of heat transfer to the lake. It was assumed that throughout the entire cooling season, i.e. from mid-May to the end of August (110 days), a previously computed quantity of energy, i.e. 34×10^{12} J, can be transferred to the lake, producing around 31×10^{10} J day^{-1} on average. Modeling was performed only in respect to a turbidity coefficient of 0.64. Sample results for day 93 of modeling are shown in Figure 14.43. Modeling results for day 150, with heat transfer to the lake, are shown in Figure 14.44.

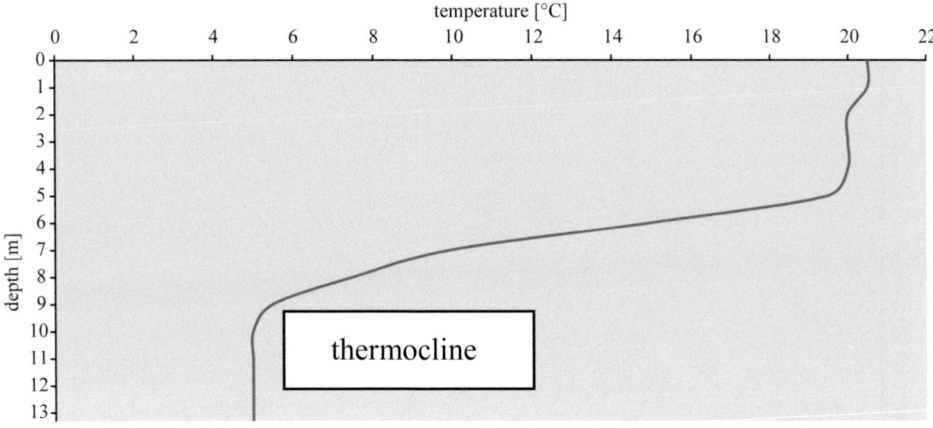

Figure 14.40. Real temperature distribution in Lake Długie – summer measurements (adapted from Lossow *et al.*, 2005).

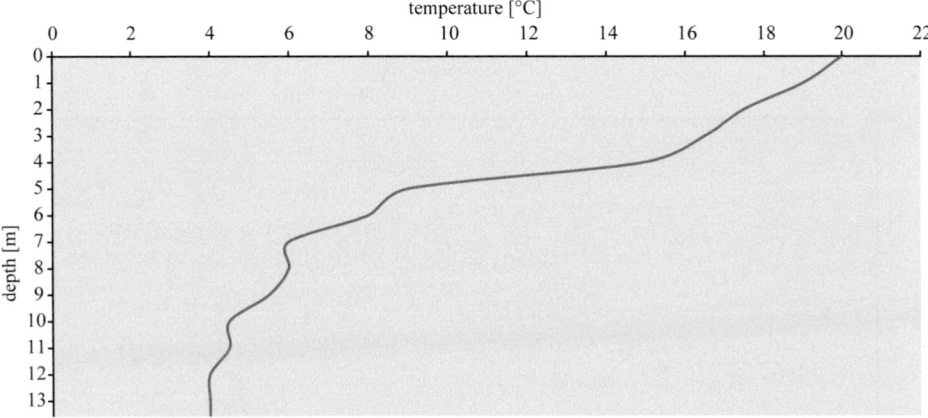

Figure 14.41. Results of temperature distribution simulation for day 146 of modeling; turbidity coefficient of 0.30.

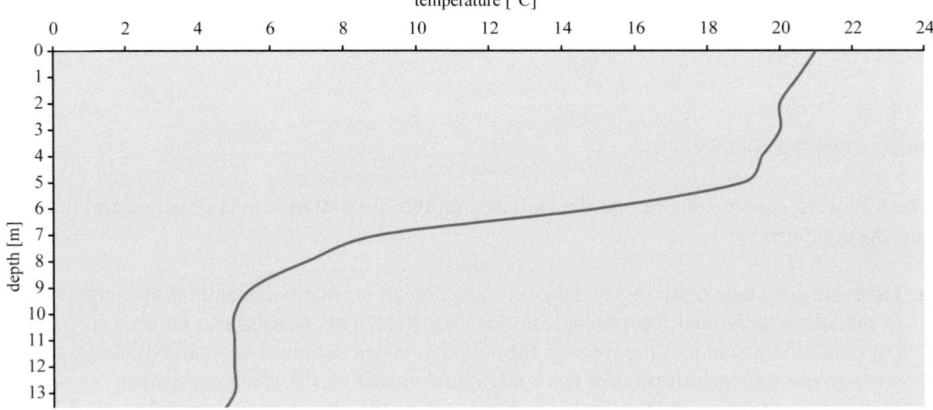

Figure 14.42. Results of temperature distribution simulation for day 146 of modeling; turbidity coefficient of 0.64.

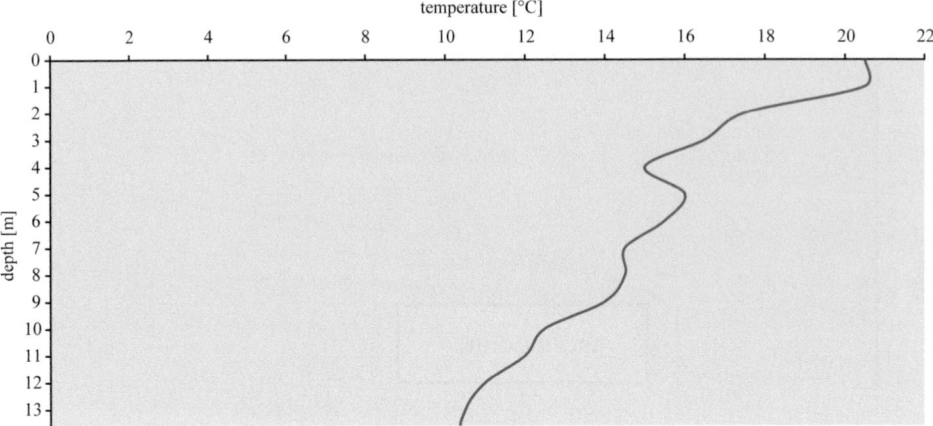

Figure 14.43.   Results of temperature distribution simulation for day 93 of modeling; turbidity coefficient of 0.64, with heat transfer to the bottom of the lake.

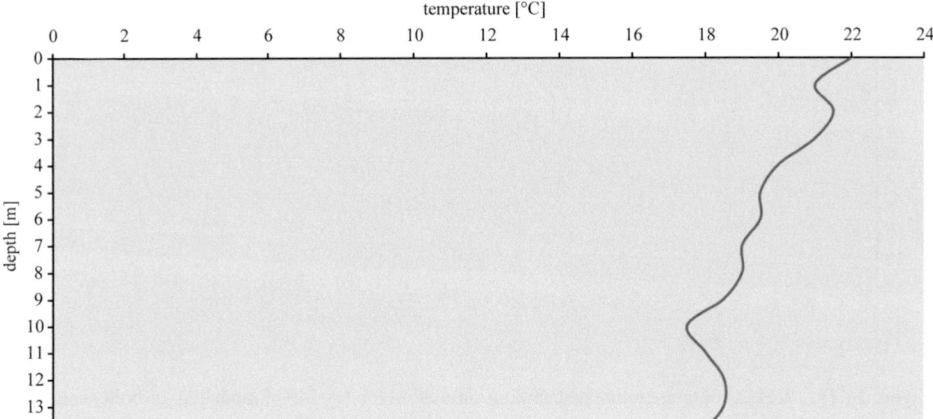

Figure 14.44.   Results of temperature distribution simulation for day 150 of modeling; turbidity coefficient of 0.64, with heat transfer.

14.5   CONCLUSIONS

The following conclusions can be drawn based on the simulation results obtained and the field surveys conducted:

- Lakes are good heat reservoirs and they can accumulate substantial quantities of energy, subject to the lake's individual morphological characteristics. This process can be used for cooling (air conditioning) indoor premises in the summer. When deployed in small buildings, such as onshore restaurants, this process has a negligible impact on the lake's ecosystem.
- In the case of lakes with D_i values less than ½, using the open systems should be much better, because of better mixing within the epilimnion and hypolimnion layers during stratification, and the improvement of water oxic conditions if the aeration system of water discharged into the lake is implemented.

- When larger quantities of heat are transferred (subject to the individual characteristics of the lake), the process has significant consequences for the lake, depending on its type:
  - In pure, not stratified lakes, the process has harmful consequences – an increase in water temperature of only several degrees could alter the lake's biological conditions and disturb its ecological balance.
  - In eutrophic or polluted stratified lakes, which are marked by temperature and oxygen stratification (lakes with anoxic hypolimnion), this process could have a positive effect. The bottom water layers are heated to the temperature of the surface strata, which facilitates mixing, e.g. due to wind, and increases oxygen content. It should be noted, however, that the biological condition of bottom layers that are artificially saturated with oxygen will change, and the nature of such changes requires further investigation.
- In the case of using lakes as a heat source for heat pumps, the ecological impact of such solutions on the lake is limited to changes in temperature distribution:
  - In the winter the thickness of the ice increases and reduces the volume of the bottom layer at a temperature of $4°C$. These changes are unfavorable ecologically, because a thick layer of ice will take longer to melt in the spring and increase the risk of killing fish and other living creatures that inhabit the lake. Therefore, any use of the lake as a heat source should be preceded by simulation studies to increase the thickness of the ice sheet – as has been carried out in Section 14.4.3. However, using the open system with simultaneous aeration should prevent fish-kill.
  - In the summer, the heat transfer to the lower layers of the water in the case of using a heat pump to work as a cooling system increased the bottom water temperature, which can have the following impact on the lake (all the issues described in Section 14.4.3):
    - For oligotrophic lakes – it can increase the temperature of water in the whole volume of lakes, and – this is the situation of environmentally very unfavorable – as described in Section 14.4 the increase of the water temperature in summer will entail a change in living conditions for all animal and plant in the lake (including the replacement of occurring species by another, with better tolerance to higher water temperature and a lower oxygen level).
    - For eutrophic lakes with occurring hypolimnion layer and emerging thermocline, it might be advantageous since bottom water (hypolimnion layer without oxygen – and therefore as dead) heating can cause artificial mixing of the water which will result in the oxygen conditions improvement in the whole volume of the lake.
- Further research is also required to determine the quantity of heat that can be safely drawn from a lake in the winter (for domestic heating purposes, with the use of a heat pump), in lakes to which heat is not transferred in the summer and which are used as heat accumulators, without causing major disturbances in the thermal balance (measured by the thickness of the lake's ice cover), i.e. without upsetting the lake's ecosystem.
- The algorithm for dealing with the use of the lake as a heat and cold source:
  - Check the conditions of the location (distance from the house to be heated to the water reservoir, its surface and depth) – Section 14.2.5.5.
  - The choice of the type of heat pump and the specific model – Sections 14.2.2 and 14.2.3.
  - The choice of work system – open or closed – Section 14.2.5.5.
  - For a closed system – calculation of the surface and the length of the heat source heat exchanger – Section 14.2.5.5.
  - The calculation of the power consumed by the heat pump with brine – Section 14.4.2.
  - The calculation of the increase in the thickness of the ice cover in the event of the selected heat pump working at full power for the heating season – Section 14.4.3.
  - The assessment of the impact of increased ice thickness on the ecological status of the lake.
  - Verification of the type of lake – eutrophic or oligotrophic – and thus the possibility of using the system designed to cool your home in summer.
  - Calculation of the effects of the heat pump as a cooling system in the summer on the rise of the lake water temperature – Section 14.4.4.

- o Assessment of the impact of increased water temperature on the ecological status of the lake.
- o If points g) and j) give a negative response (i.e. the ecological impact is negligible, the increase in the thickness of the ice cover for the entire lake is of the order of 1 cm or less, an increase in water temperature in the summer of the entire mass is of the order of 0.1°C) investment can proceed.
- It should be added that the cases of using the lake as a reservoir of heat and cold with a simultaneous attempt to improve the trophic conditions are not known at the moment to the authors of this chapter. Simulation studies on the model and tests at laboratory scale showed that the use of the lake as a source of coolness in the summer should reduce (or even eliminate) the phenomenon of the summer stratification (confirmed by the occurrence of thermocline).

NOMENCLATURE

C_W	water specific heat [$J\,kg^{-1}\,K^{-1}$]
ρ	water density [$g\,cm^{-3}$]
V	volume [m^3]
a	layer thickness [m]
b	length [m]
c	width [m]
h	depth [m]
Q	cumulated heat [J]
m	mass [kg]
T	temperature [K]
ΔT	temperature difference [K]
L	capacity [L]
t	time [s]
P	power [W]
COP	coefficient of performance
E_s	energy imparted in the condenser [kW]
E_e	input power [kW]
η_c	Carnot cycle efficiency [ad]
Q_0	thermal power [kW]
Q_g	heating capacity [kW]
$\varphi - COP$	specified by manufacturer

REFERENCES

Badgery-Parker, J. (2013) *Ground source heat as an option for greenhouses*. Powerplants Australia, Hallam, Australia. Available from: http://www.hydroponics.com.au/ground-source-heat-as-an-option-for- greenhouses [accessed July 2016].

Baird, C.D., Bucklin R.A., Watson C.A. & Chapman F.A. (1993) *Heat pump for heating and cooling water for aquacultural production*. Circular 1096, University of Florida, Florida Cooperative Extension Service/Institute of Food and Agricultural Sciences, Gainesville, FL. Available from: http://darc.cms.udel.edu/AquaPrimer/heatpump.pdf [accessed July 2016].

Berntsson, T. (2002) Heat sources – technology, economy and environment. *International Journal of Refrigeration*, 25, 428–438.

Bowden, W. (1984) A nitrogen – 15 isotope dilution study of ammonium production and consumption in a marsh sediment. *Limnology and Oceanography*, 29, 1004–1015.

Boynton, W., Kemp, W. & Osborne, C. (1980) Nutrient fluxes across the sediment – water interface in the turbid zone of coastal plain estuary. In: Kennedy, V. (ed.) *Estuarine Perspectives*. Academic Press, New York. pp. 93–109.

Brzozowska, R. & Gawrońska, H. (2009) The influence of a long-term artificial aeration on the nitrogen compounds exchange between bottom sediments and water in Lake Długie. *Oceanological and Hydrobiological Studies*, 38(1), 1–7.

Brzozowska, R., Grochowska, J. & Gawrońska, H. (2012) The role of bottom deposits in the long term durability of lake restoration effects – the case of degraded urban lake (Długie Lake, Poland). In: dos Santos Afonso, M. & Torrez Sanchez, R.M. (eds) *Ciencia y tecnologia ambiental: un efoque integrador [Environmental science and technology: an integrative approach]*. Asociación Argentina para el Progreso de las Ciencias, Buenos Aires, Argentina. pp. 114–119.

Çakır, U., Çomal, K., Çomaklı, Ö. & Karslı, S. (2013) An experimental exergetic comparison of four different heat pump systems working at same conditions: as air to air, air to water, water to water and water to air. *Energy*, 58, 210–219.

Cerco, C.F. (1989) Measured and modelled effects of temperature, dissolved oxygen and nutrient concentration on sediment-water nutrient exchange. *Hydrobiologia*, 174, 185–194.

Chiasson, A.P.E. (2005) *Greenhouse heating with geothermal heat pump systems*. Geo-Heat Center, Klamath Falls, OR. Available from: http://www.oit.edu/docs/default-source/geoheat-center-documents/publications/heat-pump/tp118.pdf?sfvrsn=2 [accessed July 2016].

Choiński, A. (1995) *Zarys Limnologii Fizycznej Polski [The Treatise on Physical Limnology of Poland]*. Wyd. Nauk. UAM, Poznań, Poland.

Chua, K.J., Chou, S.K. & Yang, W.M. (2010) Advances in heat pump systems: review. *Applied Energy*, 87, 3611–3624.

Colak, N. & Hepbasli, A. (2009a) A review of heat pump drying. 1. Systems, models and studies. *Energy Conversion and Management*, 50(9), 2180–2186.

Colak, N. & Hepbasli, A. (2009b) A review of heat-pump drying (HPD). 2. Applications and performance assessments. *Energy Conversion and Management*, 50, 2187–2199.

Dunalska, J.A., Górniak, D., Jaworska, B. & Gaiser, E.E. (2012) Effect of temperature on organic matter transformation in a different ambient nutrient availability. *Ecological Engineering*, 49, 27–34.

EC (2000) Water Framework Directive of the European Parliament and of the Council of 23 October 2000 establishing a framework for community action in the field of water policy. 2000/60/EC. Available from: http://eur-lex.europa.eu/legal-content/EN/TXT/?uri=CELEX:32000L0060 [accessed July 2016].

Filocha, M. (2007) *Modeling of the internal waves in Lac du Bourget (France)*. Non published material from the project founded by European Council (FP5 Contract No.: EVK1-CT1999-00004 – EUROLAKES project).

Fisher, T., Carlsson, P. & Barber, R. (1982) Sediment nutrient regeneration in three North Carolina estuaries. *Estuarine, Coastal and Shelf Science*, 14, 101–116.

Forsen, M. (2005) *Heat pumps technology and environmental impact, Part 1*. European Commission, report prepared by: Swedish Heat Pump Association, SVEP. Available from: http://ec.europa.eu/environment/ecolabel/about_ecolabel/reports/hp_tech_env_impact_aug2005.pdf [accessed July 2016].

Garcia, G.R., Verhoef, A., Vidale, P.L., Main, B.G.G. & Wu, Y. (2012) Interactions between the physical soil environment and a horizontal ground coupled heat pump, for a domestic site in the UK. *Renewable Energy*, 44, 141–153.

Gelda, R.K., Brooks, C.M., Effler, S.W. & Auer, M.T. (2000) Interannual variations in nitrification in an hypereutrophic urban lake: occurrences and implications. *Water Research*, 34(4), 1107–1118.

Goh, L.J., Othman, M.Y., Mat, S., Ruslan, H. & Sopian, K. (2011) Review of heat pump systems for drying application. *Renewable and Sustainable Energy Reviews*, 15(9), 4788–4796.

Golosov, S. & Kirillin G. (2010) A parameterized model of heat storage by lake sediments. *Environmental Modelling and Software*, 25, 793–801.

Grochowska, J. (2001) *Możliwości Odnowy silnie zdegradowanego Jeziora Metodą wieloletniego sztucznego Napowietrzania na przykładzie Jeziora Długiego w Olsztynie [The possibilities of highly degraded lake renovation by multiannual artificial aeration on the example of Długie Lake in Poland]*. PhD Thesis, University of Warmia and Mazury (UWM), Olsztyn, Poland.

Grochowska, J. & Brzozowska, R. (2013) The influence of different recultivation methods on the water buffer capacity in a degraded urban lake. *Knowledge and Management of Aquatic Ecosystems*, 410(1), 2–13.

Hepbasli, A. & Kalinci, Y. (2009) A review of heat pump water heating systems. *Renewable and Sustainable Energy Reviews*, 13(6–7), 1211–1229.

Hoffmann, K. (2015) *Optimising heat recovery from industrial processes with heat pumps*. European Heat Pump Association, EHPA. Available from: http://www.ehpa.org/about/news/article/optimising-heat-recovery-from-industrial-processes-with-heat-pumps [accessed July 2016].

Höhener, P. & Gächter, R. (1994) Nitrogen cycling across the sediment: water interface in an eutrophic, artificially oxygenated lake. *Aquatic Sciences*, 56(2), 115–132.

Hondzo, M. & Stefan, H.G. (1996) Long-term lake water quality predictors. *Water Research*, 30(12), 2835–2852.

Hunt, D., Badgery-Parker, J. & Dembowski, B. (2013) *Greenhouse energy use and assessment.* QDAFF, Brisbane, Australia. Available from: https://sites.google.com/site/greenhouseenergyefficiency/publications [accessed July 2016].

Hutchinson, G.E. (1957) *A Treatise on Limnology, Volume 1 Geography, Physics and Chemistry.* Wiley, New York. p. 1015.

Hutorowicz, A., Hutorowicz, J. & Brzozowska, R. (2007) Chemical composition of the top-layer bottom deposits from underneath assemblages of the subtropical species *Vallisneria spiralis* L. in the phytolittoral zone of two heated lakes. *Archives of Polish Fisheries*, 15(4), 457–464.

Hyun, I.T., Lee, J.H., Yoon, B.Y., Lee, K.H. & Nam, Y. (2014) The potential and utilization of unused energy sources for large-scale horticulture facility applications under Korean climatic conditions. *Energies*, 7(8), 4781–4801.

IFI (Inland Fisheries Institute) (1958) Bathymetric map and morphometric data of Długie Lake in Olsztyn. Unpublished internal material of IFI, Olsztyn, Poland.

Kajak, Z. (2001) *Hydrobiologia – Limnologia. Ekosystemy Wód Śródlądowych [Hydrobiology-limnology. Freshwater Ecosystems].* PWN Warszawa, Warsaw, Poland.

Kavanaugh, S. & Rafferty, K. (2014) Geothermal heating and cooling: design of ground-source heat pump systems. American Society of Heating, Refrigerating and Air-Conditioning Engineers (ASHARE).

Klapper, H. (1991) *Control of Eutrophication in Inland Waters.* Ellis Hornwood, New York.

Klapper, H. (2003) Technologies for lake restoration. *Journal of Limnology*, 62(Suppl 1), 73–90.

Kolodochka, A.A. (2003) Schemes of profile models of heat and mass transfer in large lakes. *Water Resources*, 30(1), 34–41; translated from *Vodnye Resursy*, 30:1, 40–48.

Kraszewski, A. (2004) *Struktura i funkcjonowanie populacji małży chińskich Anodonta woodiana (Lea 1834) w systemie podgrzanych jezior konińskich [Structure and functioning of the Chinese Musses Anodonta woodiana (Lea 1834) in the Konin heated Lake System].* PhD Thesis, Inland Fisheries Institute (IFI), Olsztyn, Poland.

Kukkonen, I.T. (2000) Geothermal energy in Finland. *Proceedings World Geothermal Congress, WGC 2000,* 28 May–10 June 2000, Beppu-Morioka, Japan. Available from: http://www.geothermal-energy.org/pdf/IGAstandard/WGC/2000/R0778.PDF [accessed July 2016].

Kurpaska, S. (2011) Utilisation of low-temperature heat in agricultural using heat pumps: current state and prospects. *Agricultural Engineering*, 7(132), 65–72.

Leijendeckers, P.H.H. (1985) Heat pump technology. In: Baldwin, A.R. (ed.) *Proceedings of World Conference on Emerging Technologies in the Fats and Oils Industry*, 3–8 November 1985, Cannes, France. American Oil Chemists' Society.

León, L.F., Lam, D.C.L., Schertzer, W.M., Swayne, D.A. & Imberger, J. (2007) Towards coupling a 3D hydrodynamic lake model with the Canadian regional climate model: simulation on Great Slave Lake. *Environmental Modelling and Software*, 22, 787–796.

Liikanen, A., Murtoniemi, T., Tanksanen, H., Väisänen, T. & Martikainen, P.J. (2002) Effects of temperature and oxygen availability on greenhouse gas and nutrient dynamics in sediment of a eutrophic mid-boreal lake. *Biogeochemistry*, 59, 269–286.

Liu, W.-C. & Chen, W.-B. (2012) Prediction of water temperature in a subtropical subalpine lake using an artificial neural network and three-dimensional circulation models. *Computers & Geosciences*, 45, 13–25.

Lossow, K. (1996) Znaczenie jezior w krajobrazie młodoglacjalnym Pojezierza Mazurskiego [The importance of lakes in the younger glacial landscape of Masurian Lake District]. *Zeszyty Problemowe Postępów Nauk Roln*, 431, 47–59.

Lossow, K., Gawrońska, H., Mientki, C., Łopata, M. & Wiśniewski, G. (2005) *Jeziora Olsztyna, Stan troficzny, Zagrożenia [Lakes of Olsztyn, trophic State, Threats].* EDYCJA, Olsztyn, Poland.

Lund, J.W. (2002) Direct utilization of geothermal resources. In: Chandrasekharam, D. & Bundschuh, J. (eds) *Geothermal Energy Resources for Developing Countries.* A.A. Balkema, Lisse, The Netherlands. pp. 129–147.

Lytras, E. (2007) Developing models for lake management. *Desalination*, 213, 129–134.

Maina, P. & Huan, Z. (2013) Effects of refrigerant charge in the output of a CO_2 heat pump. *African Journal of Science, Technology, Innovation and Development*, 5, 303–311.

Marsden M.W. (1989) Lake restoration by reducing external phosphorus loading: the influence of sediment phosphorus release. *Freshwater Biology*, 21, 139–162.

Menshutkin, V.V., Rukhovets, L.A. & Filatov, N.N. (2013) Ecosystem modeling of freshwater lakes (Review): 1. Hydrodynamics of lakes. *Water Resources*, 40(6), 606–620.

Menshutkin, V.V., Rukhovets, L.A. & Filatov, N.N. (2014) Ecosystem modeling of freshwater lakes Review): 2. Models of freshwater lake's ecosystem. *Water Resources*, 41(1), 32–45.

Mugnozza, G.S., Pascuzzi, S., Anifantis, A.S. & Verdiani G. (2012) Use of low-enthalpy geothermal resources for greenhouse heating: aAn experimental study. *Acta Scientiarum Polonorum Technica Agraria*, 11(1–2), 13–19.

Nguyen, M., Arason, S., Gissurarson, M. & Palsson, P.G. (2015) Uses of geothermal energy in food and agriculture. Opportunities for developing countries. Food and Agriculture Organization of the United Nations (FAO), Rome, Italy. Available from: http://www.fao.org/3/a-i4233e.pdf [accessed July 2016].

Nürnberg, G.K. (1994) Phosphorus release from anoxic sediments: hat we know and how we can deal with it. *Limnética*, 10(1), 1–4.

Nürnberg, G.K. & Peters, R.H. (1984) The importance of internal phosphorus loading to the eutrophication of lakes with anoxic hypolimnia. *Verhandlungen des Internationalen Verein Limnologie*, 22, 190–194.

Ozgener, O. (2010) Use of solar assisted geothermal heat pump and small wind turbine systems for heating agricultural and residential building. *Energy*, 35, 262–268.

Patel, K.K. & Kar, A. (2012) Heat pump assisted drying of agricultural produce: an overview. *Journal of Food Science and Technology*, 49(2), 142–160.

Petit, R.F. & Collins, T.L. (2011) *Heat Pumps: Operation, Installation, Service*. Esco Press, Mount Prospect, IL.

Rimmer, A. (2003) The mechanism of lake Kinneret salinization as a linear reservoir. *Journal of Hydrology*, 281, 173–186.

Ritter, L., Solomon, K., Sibley, P., Hall, K., Keen, P., Mattu, G. & Linton B. (2002) Sources, pathways, and relative risks of contaminants in surface water and groundwater: a perspective prepared for the Walkerton Inquiry. *Journal of Toxicology and Environmental Health* A, 65, 1–142.

Rubik, M. (1999) Pompy Ciepła [Heat Pumps]. Poradnik. Wyd. II. Ośrodek Informacji "Technika Instalacyjna w Budownictwie", Warsaw, Poland.

Rubik, M. (2011) Pompy ciepła w systemach geotermii niskotemperaturowej [Heat pumps in low-temperature geothermal systems]. MULTICO Oficyna Wydawnicza Publishing House, Warszawa, Poland.

Rysgaard, S., Risgaard-Petersen, N., Sloth, P., Jensen, K. & Nielsen, L.P. (1994) Oxygen regulation of nitrification and denitrification in sediments. *Limnology and Oceanography*, 39(7), 1643–1652.

Shen, J. (1998) Numerical modelling of the effects of vegetation and environmental conditions on the lake breeze. *Boundary-Layer Meteorology*, 87, 481–498.

Stawecki, K., Pyka, J.P. & Zdanowski B. (2007) The thermal and oxygen relationship and water dynamics of the surface water layer in the Konin heated lakes ecosystem. *Archives of Polish Fisheries*, 15(4), 247–258.

Suplee, M.W. & Cotner, J.B. (2002) An evaluation on the importance of sulphate reduction and temperature to P fluxes from aerobic-surfaced lacustrine sediments. *Biogeochemistry*, 61, 199–228.

Thain, I., Reyes, A.G. & Hunt, T. (2006) *A practical guide to exploiting low temperature geothermal resources*. GNS Science Report 2006/09, New Zealand.

Tiren, T. (1977) Denitrification in sediment – water systems of various types of lakes. In: Golterman, H.L. (ed.) *Interactions between Sediments and Fresh Water. Proceedings International Symposium, 6–10 September 1976, Amsterdam, The Netherlands*. Dr. W. Junk B.V., The Hague, The Netherlands. pp. 363–369.

Tomaszek, J.A. & Czerwieniec, E. (1995) Znaczenie procesu denitryfikacji dla bilansu azotu w ekosystemach wodnych [The importance of denitrification process for nitrogen balance in the water ecosystems]. In: Zalewski, M. (ed.) *Procesy biologiczne w Ochronie i Rekultywacji nizinnych Zbiorników Zaporowych [Biological Processes in the Protection and Restoration of lowland Dam Reservoirs]*. Biblioteka Monitoringu Środowiska, Łódź, Poland.

Vollenweider, R.A. (1968) Scientific fundamentals of the eutrophication of lakes and flowing waters, with particular reference of nitrogen and phosphorus as factors in eutrophication. OECD, Directorate for Science, Technology and Innovation, Paris, France DAS (CSI) 68:27 (1968). pp. 1–182.

Zawadzki, M. (2003) Kolektory słoneczne, Pompy ciepła – Na Tak [Solar collectors, heat pumps, yes]. Polska Ekologia, ISBN 83-918540-0-0.

Zdanowski, B., Dunalska, J. & Stawecki, K. (2002) Variability of nutrient contents in heated lakes of the Konin area. *Limnological Review*, 2, 457–464.

Zieliński, M. & Krzemieniewski, M. (2005) Wpływ temperatury wypełnienia reaktora jako czynnika stymulującego szybkość przemian azotu w reaktorze z błoną biologiczną [The influence of reactor filling temperature as the nitrogen transformation rate stimulating factor in the reactor with biological film]. *Biotechnologia*, 1(68), 207–214.

Zimny, J. & Fiszer, T. (2003) Nowoczesne systemy grzewcze z wykorzystaniem odnawialnych źródeł energii – na przykładzie gimnazjum centralnego w gródku nad Dunajcem [Modern heating systems using renewable energy sources – on example, central middle school in Gródek on the Dunajec]. Conference material *Konferencyjne IX Ogólnopolskie Forum Odnawialnych Źródeł Energii*, 21–23 May 2003, Zakopane – Kościelisko, Poland.

Zimny, J. & Michalak, P. (2008) The work of a heating system with renewable energy sources (RES) in school building. *Environmental Protection Engineering*, 34(1), 81–88.

Zogg, M. (2008) Geschichte der Wärmepumpe [History of the heat pump]. *Schweizer Beiträge und internationale Meilensteine.* Bundesamt für Energie, Bern, Switzerland. Available from: http://zogg-engineering.ch/publi/GeschichteWP.pdf [accessed July 2016].

Subject index

Sustainable Energy Developments

Series Editor: Jochen Bundschuh

ISSN: 2164-0645

Publisher: CRC Press/Balkema, Taylor & Francis Group

1. Global Cooling – Strategies for Climate Protection
   Hans-Josef Fell
   2012
   ISBN: 978-0-415-62077-2 (Hbk)
   ISBN: 978-0-415-62853-2 (Pb)

2. Renewable Energy Applications for Freshwater Production
   Editors: Jochen Bundschuh & Jan Hoinkis
   2012
   ISBN: 978-0-415-62089-5 (Hbk)

3. Biomass as Energy Source: Resources, Systems and Applications
   Editor: Erik Dahlquist
   2013
   ISBN: 978-0-415-62087-1 (Hbk)

4. Technologies for Converting Biomass to Useful Energy –
   Combustion, gasification, pyrolysis, torrefaction and fermentation
   Editor: Erik Dahlquist
   2013
   ISBN: 978-0-415-62088-8 (Hbk)

5. Green ICT & Energy – From smart to wise strategies
   Editors: Jaco Appelman, Anwar Osseyran & Martijn Warnier
   2013
   ISBN: 978-0-415-62096-3

6. Sustainable Energy Policies for Europe – Towards 100% Renewable Energy
   Rainer Hinrichs-Rahlwes
   2013
   ISBN: 978-0-415-62099-4 (Hbk)

7. Geothermal Systems and Energy Resources – Turkey and Greece
   Editors: Alper Baba, Jochen Bundschuh & D. Chandrasekaram
   2014
   ISBN: 978-1-138-00109-1 (Hbk)

8. Sustainable Energy Solutions in Agriculture
   Editors: Jochen Bundschuh & Guangnan Chen
   2014
   ISBN: 978-1-138-00118-3 (Hbk)